Oil Crop Genomics

Huseyin Tombuloglu • Turgay Unver
Guzin Tombuloglu • Khalid Rehman Hakeem
Editors

Oil Crop Genomics

Springer

Editors
Huseyin Tombuloglu
Department of Genetics
Institute for Research and Medical
Consultations (IRMC)
Imam Abdulrahman bin Faisal University
(formerly Dammam University)
Dammam, Saudi Arabia

Guzin Tombuloglu
Al Khalidiyah Ash Shamaliyah
Dammam, Saudi Arabia

Turgay Unver
Ficus Biotechnology
Yenimahalle, Ankara, Turkey

Khalid Rehman Hakeem
Department of Biological Sciences
Faculty of Science
King Abdulaziz University
Jeddah, Saudi Arabia

ISBN 978-3-030-70419-3   ISBN 978-3-030-70420-9 (eBook)
https://doi.org/10.1007/978-3-030-70420-9

This Springer imprint is published by the registered company Springer Nature Switzerland AG
The registered company address is: Gewerbestrasse 11, 6330 Cham, Switzerland

This book is dedicated to our beloved parents

Your Lord had decreed, that you worship none save Him, and (that you show) kindness to parents. If one of them or both attain old age with you, say not "Fie" unto them nor repulse them, but speak unto them a gracious word. And lower unto them the wing of submission through mercy and say: My Lord! Have mercy on them both, as they did care for me when I was young.

[Al-Qur'an 17:23-24]

Foreword

Plants are an important source of fats and oils which are essential for the human diet. Genomics of oil biosynthesis in plants has generated great interest in recent years, especially in high oil-bearing plants such as soybean, olive, and oil palm. Considering that, genome-sequencing projects of these plants have been undertaken in recent years with the help of advanced genomics tools like next-generation sequencing. As well, several genome-sequencing projects of oil crops are in progress and many others are on the way. In addition to genome information, advanced genomics approaches such as transcriptomics, genomics-assisted breeding, genome-wide association study (GWAS), genotyping by sequencing (GBS), and CRISPR are improving our understanding of the oil biosynthesis mechanism and breeding strategies for oil extraction and production. However, there is no edited book that covers the genomes and genomics of oil crops, yet.

The book *Oil Crops Genomics* edited by Tombuloglu et al. (Springer) is an interesting and comprehensive scientific work covering recent studies on the genetics and genomics of selected oil crops. The book includes three parts: Part I: Genomes of Oil-Bearing Crops; Part II: Oil Crops Genomics; and Part III: Oil Crop Biotechnology. Part I (Chaps. 1, 2, 3, 4, 5, 6, 7 and 8) covers the genome sequence of oil crops such as soybean, cotton, olive, melon, oil palm, argan tree, and linseed (flax). The second part (Chaps. 9, 10, 11, 12, 13 and 14) concentrates on the genomics of oil crops without genome-sequence information, such as coconut, safflower, mustard, poppy, and Jatropha. Part III (Chaps. 15, 16, 17, 18, 19 and 20) represents comprehensive information about the economics of oil crops and the most recent biotechnological methods.

This book combines up-to-date knowledge on genomics and genetics of oil crops. It is well prepared, organized, and represents a valuable source for graduate-level students, instructors, and researchers. Throughout this book, the latest genomics developments and discoveries as well as open problems and future challenges in oil crop genomics are highlighted for future studies. Also, this volume collects the most recent knowledge on oil crop genomics for researchers who study oil crop genome, genomics, biotechnology, pharmacology, medicine nutrition, food industry, or economy.

Ebtesam Abdullah
Al-Suhaimi, PhD
Institute for Research & Medical Consultations
Imam Abdulrahman bin Faisal University
Dammam, Saudi Arabia

Preface

Oil plants are indispensable food sources for human nutrition and health. Especially in recent years, genome analyzes of plants with high oil content such as soybean, olive and palm oil have been carried out intensively. With the determination of genome sequences, it is aimed to better understand the mechanism of oil biosynthesis in plants and to make plants with low oil yield more productive with biotechnological approaches.

The aforementioned developments in oil bearing crops and their novel and advanced studies have not been combined in a single book. While oil production alters depending on the season and years, it causes fluctuations in economical income. Increasing the oil production capacity of wild plants by using genomics-assisted breeding approaches is of great importance to combating this problem. Hence, this volume will be helpful for researchers who study oil crops, their breeding, and biotechnology.

The present book covers genome-sequenced oil crops as well as plants producing important oil metabolites. Throughout this book, the latest genomics developments and discoveries as well as open problems and future challenges in oil crop genomics and their economy are highlighted. We are hopeful that this book shall introduce readers to state-of-the-art developments and trends in oil crop genomic studies.

We are thankful to the contributors for readily accepting our invitation for not only sharing their knowledge and research, but for venerably integrating their expertise in dispersed information from diverse fields in composing the chapters and enduring editorial suggestions to finally produce this venture. We also thank the Springer-International team for their generous cooperation at every stage of the book production.

Dammam, Saudi Arabia — Huseyin Tombuloglu
Ankara, Turkey — Turgay Unver
Dammam, Saudi Arabia — Guzin Tombuloglu
Jeddah, Saudi Arabia — Khalid Rehman Hakeem

Contents

About the Editors

Huseyin Tombuloglu, PhD is an associate professor in the Institute of Research and Medical Consultation (IRMC) at Imam Abdulrahman bin Faisal University, Dammam, Saudi Arabia. He received his BSc degree in 2007 from Istanbul University, Department of Molecular Biology and Genetics, Turkey, and also he studied as an exchange student at the University of Groningen, the Netherlands. Dr. Tombuloglu obtained his MS degree (biology) in 2010 and PhD degree (biotechnology) in 2014. He became an assistant professor in 2014, and associate professor in 2018. He has more than 15 years of teaching and research experience in genetics, molecular biology, plant genomics, and biotechnology as well as bioinformatics. He is one of the members of the International Olive Genome Sequencing Consortium. Currently, he is in the research team member of the International Garlic Genome Sequencing Consortium (IGGS) (http://garlicgenome.org/). His current research is focused on genome sequencing of plants, data analysis, proteomics, and nanoparticle-plant interaction.

Turgay Unver, PhD received his PhD degree from Middle East Technical University (METU; Turkey) in 2008. He became an assistant professor at Cankiri Karatekin University in 2009, and obtained his associated professorship in 2011. His main research areas involve genome and transcriptome analyses, including microRNA. He has published >50 indexed articles with more than 4000 citations. In 2012, he was awarded the TWAS prize. He was selected as outstanding young scientist by the Turkish Academy of Science in 2013. Prof. Unver is currently running two biotechnology companies and bearing editorial duties for *Genomics* (Elsevier), *BMC Genomics*, and *Plos One*.

Guzin Tombuloglu, PhD received her MS degree (biology) in 2008 and PhD degree (biotechnology) in 2014. She has experience in transcriptome sequencing, plant abiotic stress tolerance, and molecular biology of plants. During her PhD, she studied transcriptomics identification of boron tolerance mechanism in barley. She has experienced several projects on abiotic stress, plant stress responses, boron toxicity, and transcriptomics. She has given several courses on teaching genetics, molecular biology, and biotechnology education for more than 15 years. She also worked as chairman of the Pathology Laboratory Techniques Programme and assistant manager at the Vocational School of Medical Sciences at university level.

Khalid Rehman Hakeem, PhD is a professor at King Abdulaziz University, Jeddah, Saudi Arabia. After completing his doctorate (botany; specialization in plant eco-physiology and molecular biology) from Jamia Hamdard, New Delhi, India, in 2011, he worked as assistant professor at the University of Kashmir, Srinagar, for a short period. Later, he joined Universiti Putra Malaysia, Selangor, Malaysia, and worked there as postdoctorate fellow in 2012 and fellow researcher (associate professor) from 2013 to 2016. Dr. Hakeem has more than 10 years of teaching and research experience in plant eco-physiology, biotechnology and molecular biology, medicinal plant research, plant-microbe-soil interactions, as well as in environmental studies. He is the recipient of several fellowships at both national and international levels. He has also served as a visiting scientist at Jinan University, Guangzhou, China. Currently, he is involved with a number of international research projects with different government organizations. So far, Dr. Hakeem has authored and edited more than 70 books with international publishers, including Springer Nature, Academic Press (Elsevier), and CRC Press. He also has to his credit more than 140 research publications in peer-reviewed international journals and 60 book chapters in edited volumes with international publishers. At present, Dr. Hakeem serves as an editorial board member and reviewer of several high-impact international scientific journals from Elsevier, Springer Nature, Taylor and Francis, Cambridge, and John Wiley Publishers. He is included in the advisory board of Cambridge Scholars Publishing, UK. Dr. Hakeem is also a fellow of Plantae group of the American Society of Plant Biologists; member of the World Academy of Sciences; member of the International Society for Development and Sustainability, Japan; and member of Asian Federation of Biotechnology, Korea. Dr. Hakeem has been listed in Marquis Who's Who in the World, between 2014 and 2020. Currently, Dr. Hakeem is engaged in studying the plant processes at eco-physiological as well as molecular levels.

Contributors

Naima Ait Aabd CRRA-Agadir, National Institute for Agricultural Research (INRA), Rabat, Morocco

Muhammad Aasim Faculty of Agricultural Sciences and Technologies, Sivas University of Science and Technology, Sivas, Turkey

Bilal Haider Abbasi Department of Biotechnology, Quaid-i-Azam University, Islamabad, Pakistan

Kamal Aberkani Polydisciplinary Faculty of Nador, University Mohammed Premier, Oujda, Morocco

Fawad Ali Department of Plant Sciences, Quaid-I-Azam University, Islamabad, Pakistan

Bayan Alotaibi Department of Genetics Research, Institute for Research and Medical Consultations (IRMC), Imam Abdulrahman Bin Faisal University, Dammam, Saudi Arabia

Noha Alqahantani Department of Genetics Research, Institute for Research and Medical Consultations (IRMC), Imam Abdulrahman Bin Faisal University, Dammam, Saudi Arabia

Sumayah Alsanie Biology Department, Imam Abdulrahman Bin Faisal University, Dammam, Saudi Arabia

Thamer AlShammari Department of Genetic Research, Institute for Research & Medical Consultations (IRMC), Imam Abdulrahman Bin Faisal University, Dammam, Saudi Arabia

Raghdah Alshumrani Department of Genetics Research, Institute for Research and Medical Consultations (IRMC), Imam Abdulrahman Bin Faisal University, Dammam, Saudi Arabia

Muhammad Ameen Institute of Soil and Environmental Sciences, University of Agriculture, Faisalabad, Pakistan

Touseef Amna Department of Biology & Biotechnology, Faculty of Science, Albaha University, Albaha, Saudi Arabia

Sumaira Anjum Department of Biotechnology, Kinnaird College for Women, Lahore, Pakistan

Wajiha Anum Department of Agronomy, Regional Agricultural Research Institute, Bahawalpur, Pakistan

Yusuf Arslan Department of Field Crops, Faculty of Agriculture and Natural Sciences, Bolu Abant Izzet Baysal University, Bolu, Turkey

Mehtap Aydin Genetics and Bioengineering Department, Yeditepe University, Istanbul, Turkey

Abdellah Idrissi Azami School of Medicine and Pharmacy, University Mohammed V, Rabat, Morocco

Oussama Badad Faculty of Sciences, University Mohammed V, Rabat, Morocco

Faheem Shehzad Baloch Faculty of Agricultural Sciences and Technologies, Sivas University of Science and Technology, Sivas, Turkey

Muruj Bamhrez Department of Genetics Research, Institute for Research and Medical Consultations (IRMC), Imam Abdulrahman Bin Faisal University, Dammam, Saudi Arabia

H. D. D. Bandupriya Department of Plant Sciences, Faculty of Science, University of Colombo, Colombo, Sri Lanka

Muzaffer Barut Department of Field Crops, Faculty of Agriculture, Çukurova University, Adana, Turkey

Joyce A. Cartagena Graduate School of Bioagricultural Sciences, Nagoya University, Nagoya, Japan

Peng Chee Cotton Molecular Breeding Laboratory, University of Georgia – Tifton Campus, Tifton, GA, USA

Haifeng Chen Key Laboratory of Biology and Genetic Improvement of Oil Crops, Ministry of Agriculture and Rural Affairs of PRC, Oil Crops Research Institute of Chinese Academy of Agriculture Sciences, Wuhan, China

Diny Dinarti Plant Molecular Biology Lab., Department of Agronomy and Horticulture, Faculty of Agriculture, Bogor Agricultural University, Bogor, West Java, Indonesia

Gabriel Dorado Dep. Bioquímica y Biología Molecular, Campus Rabanales C6-1-E17, Campus de Excelencia Internacional Agroalimentario (ceiA3), Universidad de Córdoba, Córdoba, Spain

Samantha Drouet Laboratoire de Biologie des Ligneux et des Grandes Cultures, INRAE USC1328, Orleans University, Orléans Cedex 2, France

Aissam El Finti Laboratory of Biotechnology and Valorization of Natural Resources (LBVRN), Faculty of Sciences, University Ibn Zohr, Agadir, Morocco

Azra Frkatovic Genos Glycoscience Research Laboratory, Zagreb, Croatia

Fatima Gaboun CRRA-Rabat, National Institute for Agricultural Research (INRA), Rabat, Morocco

Bedriye Nazli Genc Department of Medical Pathology, Faculty of Medicine, Istinye University, Istanbul, Turkey

Hassan Ghazal National Center for Scientific and Technological Research (CNRST), Rabat, Morocco

Christophe Hano Laboratoire de Biologie des Ligneux et des Grandes Cultures, INRAE USC1328, Orleans University, Orléans Cedex 2, France

Andi Nadia Nurul Latifa Hatta Plant Molecular Biology Lab., Department of Agronomy and Horticulture, Faculty of Agriculture, Bogor Agricultural University, Bogor, West Java, Indonesia

Rashid Iqbal Department of Agronomy, Islamia University of Bahawalpur, Bahawalpur, Pakistan

Asim Iqbal Department of Economics and Business Administration, Division of Arts and Social Sciences, University of Education, Lahore, Pakistan

Moazzam Jamil Department of Soil Science, Islamia University of Bahawalpur, Bahawalpur, Pakistan

Slimane Khayi CRRA-Rabat, National Institute for Agricultural Research (INRA), Rabat, Morocco

Mary Kinsel Department of Chemistry & Biochemistry, Southern Illinois University in Carbondale, Carbondale, IL, USA

Jenny Koebernick Department of Crop, Soil and Environmental Sciences, Auburn University, Auburn, AL, USA

Neha Kothari Fiber Quality Research, Cotton Incorporated, Cary, NC, USA

Cemal Kurt Department of Field Crops, Faculty of Agriculture, Çukurova University, Adana, Turkey

Kuaybe Yucebilgili Kurtoglu Faculty of Science, Department of Molecular Biology and Genetics, Istanbul Medeniyet University, Istanbul, Turkey

Eric Lainé Laboratoire de Biologie des Ligneux et des Grandes Cultures, INRAE USC1328, Orleans University, Orléans Cedex 2, France

Hanna Levchuk Flax Breeding Lab, Institute of Oilseed Crops of the National Academy of Agricultural Sciences of Ukraine, Ukraine

Rong Li Key Laboratory of Biology and Genetic Improvement of Oil Crops, Ministry of Agriculture and Rural Affairs of PRC, Oil Crops Research Institute of Chinese Academy of Agriculture Sciences, Wuhan, China

Edward Lubbers Cotton Molecular Breeding Laboratory, University of Georgia – Tifton Campus, Tifton, GA, USA

Amal Mahmoud Department of Biology, College of Science, Imam Abdulrahman Bin Faisal University, Dammam, Saudi Arabia

Yuliia Makhno Flax Breeding Lab, Institute of Oilseed Crops of the National Academy of Agricultural Sciences of Ukraine, Ukraine

Lucija Markulin Laboratoire de Biologie des Ligneux et des Grandes Cultures, INRAE USC1328, Orleans University, Orléans Cedex 2, France

Gian Powell B. Marquez College of Global Liberal Arts, Ritsumeikan University, Osaka, Japan

Ismail Maskromo Indonesian Palms Research Institute, Agency for Agricultural Research and Development, Manado, Indonesia

Cecilia McGregor Institute of Plant Breeding, Genetics & Genomics and Department of Horticulture, University of Georgia, Athens, GA, USA

Rachid Mentag CRRA-Rabat, National Institute for Agricultural Research (INRA), Rabat, Morocco

Geoffrey Meru Department of Horticulture, Tropical Research & Education Center, University of Florida, IFAS, Homestead, FL, USA

Jamila Mouhaddab Laboratory of Biotechnology and Valorization of Natural Resources (LBVRN), Faculty of Sciences, University Ibn Zohr, Agadir, Morocco

Abdelhamid El Mousadik Laboratory of Biotechnology and Valorization of Natural Resources (LBVRN), Faculty of Sciences, University Ibn Zohr, Agadir, Morocco

Fouad Msanda Laboratory of Biotechnology and Valorization of Natural Resources (LBVRN), Faculty of Sciences, University Ibn Zohr, Agadir, Morocco

Ghulam Murtaza Institute of Soil and Environmental Sciences, University of Agriculture, Faisalabad, Pakistan

Ghulam Mustafa Department of Economics and Business Administration, Division of Arts and Social Sciences, University of Education, Lahore, Pakistan

Muhammad Azhar Nadeem Faculty of Agricultural Sciences and Technologies, Sivas University of Science and Technology, Sivas, Turkey

Muhammad Naeem Department of Plant Breeding and Genetics, Faculty of Agriculture and Environmental Sciences, The Islamia University of Bahawalpur, Bahawalpur, Punjab, Pakistan

Insha Nahvi Department of Basic Sciences, Preparatory Year Deanship, King Faisal University, Hofuf, Saudi Arabia

Sadaf Nazir Department of Food Technology, Institute of Engineering & Technology, Bundelkhand University, Jhansi, Uttar Pradesh, India

Jinesh Patel Department of Crop, Soil and Environmental Sciences, Auburn University, Auburn, AL, USA

S. A. C. N. Perera Faculty of Agriculture, University of Peradeniya, Peradeniya, Sri Lanka

Stacy Pirro Iridian Genome, Bethesda, MD, USA

Qamar-uz-Zaman Department of Environmental Sciences, University of Lahore, Lahore, Pakistan

Annisaa Rahmawati Plant Molecular Biology Lab., Department of Agronomy and Horticulture, Faculty of Agriculture, Bogor Agricultural University, Bogor, West Java, Indonesia

Suriya Rehman Department of Epidemic Diseases Research, Institute for Research & Medical Consultations (IRMC), Imam Abdulrahman Bin Faisal University, Dammam, Saudi Arabia

Umair Riaz Soil and Water Testing Laboratory for Research, Agriculture Department, Government of Punjab, Bahawalpur, Pakistan

Hussein Sabit Department of Genetics Research, Institute for Research and Medical Consultations (IRMC), Imam Abdulrahman Bin Faisal University, Dammam, Saudi Arabia

Mohammad R. Sabzalian Department of Agronomy and Plant Breeding, College of Agriculture, Isfahan University of Technology (IUT), Isfahan, Iran

Tayyaba Samreen Institute of Soil and Environmental Sciences, University of Agriculture, Faisalabad, Pakistan

Samira Smajlovic Faculty of Science, Department of Biology, University of Zagreb, Zagreb, Croatia

Irfan Sohail Institute of Soil and Environmental Sciences, University of Agriculture, Faisalabad, Pakistan

Sudarsono Sudarsono Plant Molecular Biology Lab., Department of Agronomy and Horticulture, Faculty of Agriculture, Bogor Agricultural University, Bogor, West Java, Indonesia

Burcu Tarıkahya Hacıoğlu Department of Biology, Faculty of Science, Hacettepe University, Ankara, Turkey

Tatiana Tatusova National Center for Biotechnology Information, National Institutes of Health, Bethesda, MD, USA

Dilek Tokel Department of Economics, Faculty of Economics, Marmara University, Istanbul, Turkey

Huseyin Tombuloglu Department of Genetics Research, Institute for Research and Medical Consultations (IRMC), Imam Abdulrahman Bin Faisal University, Dammam, Saudi Arabia

Duangjai Tungmunnithum Department of Pharmaceutical Botany, Faculty of Pharmacy, Mahidol University, Bangkok, Thailand

Turgay Unver Ficus Biotechnology, Ankara, Turkey

Hugo Alfried Volkaert Center for Agricultural Biotechnology, Kasetsart University, Nakhon Pathom, Thailand

Abdurrahim Yılmaz Department of Field Crops, Faculty of Agriculture and Natural Sciences, Bolu Abant Izzet Baysal University, Bolu, Turkey

Hilal Yılmaz Department of Plant and Animal Production, Izmit Vocational School, Kocaeli University, Kocaeli, Turkey

Mehmet Zahit Yeken Department of Field Crops, Faculty of Agriculture and Natural Sciences, Bolu Abant Izzet Baysal University, Bolu, Turkey

Songli Yuan Key Laboratory of Biology and Genetic Improvement of Oil Crops, Ministry of Agriculture and Rural Affairs of PRC, Oil Crops Research Institute of Chinese Academy of Agriculture Sciences, Wuhan, China

Abdelaziz Zahidi Laboratory of Biotechnology and Valorization of Natural Resources (LBVRN), Faculty of Sciences, University Ibn Zohr, Agadir, Morocco

Insha Zahoor Department of Neurology, Henry Ford Hospital, Detroit, MI, USA

Houcine Zaid Faculty of Sciences, University Mohammed V, Rabat, Morocco

Sara Zare Department of Agronomy and Plant Breeding, College of Agriculture, Isfahan University of Technology (IUT), Isfahan, Iran

Xinan Zhou Key Laboratory of Biology and Genetic Improvement of Oil Crops, Ministry of Agriculture and Rural Affairs of PRC, Oil Crops Research Institute of Chinese Academy of Agriculture Sciences, Wuhan, China

Part I
Genomes of Oil-Bearing Crops

Chapter 1
Soybean Genome

Sumayah Alsanie

Contents

1.1 Introduction

Glycine max (L.) Merr. (soybean) is a subtropical plant native to southeastern Asia. It has been a part of the human diet in Asian countries for a very long period. European countries and the USA were introduced to this dietary staple comparatively very recently, the 1700s and 1800s, respectively (Joy et al. 1998). It is the widely grown crop in the world and the fourth most important harvested and produced crop (Schmutz et al. 2010).

According to the recent statistics, soybean constitutes 59% of oilseed production globally in 2019 (Fig. 1.1), and the USA and Brazil are the top producers of the soybean (Soystats 2020).

The soybean is native to East Asia. The widely cultivated soya bean *Glycine max* belongs to subgenus *Soja* of the genus *Glycine* (which has at least 25 perennial species). The other wild soybean, *Glycine soja* Sieb and Zucc, also belongs to the subgenus *Soja* of the genus *Glycine. Glycine soja* and *Glycine max* both are annual plants. The widely cultivated, modern soybean can no longer be traced to the

S. Alsanie (✉)
Biology Department, Imam Abdulrahman Bin Faisal University, Dammam, Saudi Arabia
e-mail: sualsanea@iau.edu.sa

H. Tombuloglu et al. (eds.), *Oil Crop Genomics*,
https://doi.org/10.1007/978-3-030-70420-9_1

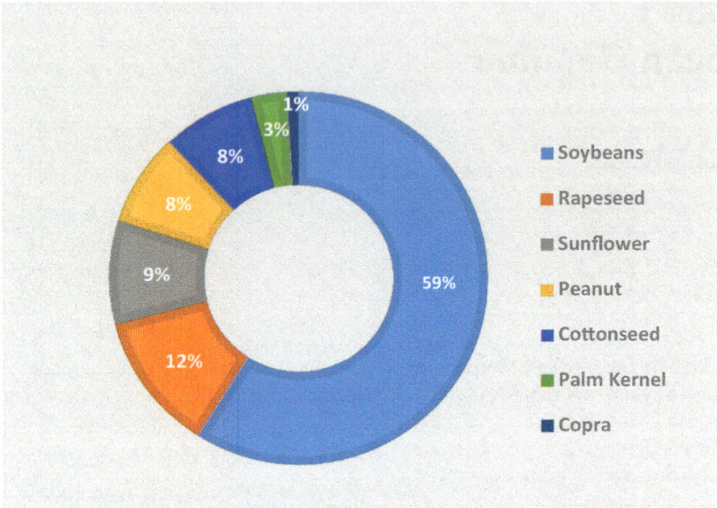

Fig. 1.1 Percentage of world oilseed production (2019). (Data from soystats2020.com, United States Department of Agriculture (USDA), foreign agricultural service (FAS))

wild-growing species (Shekhar et al. 2016). When those two species are crossed together, they produce fertile hybrids, which concludes that they have similar genomes (Singh and Hymowitz 1988). While the 26 wild perennial species originate in Australia, they differ morphologically, cytologically, and genomically from species that belong to subgenus *Soja* (Chung and Singh 2008).

Soybean is an economically important crop as one of the primary sources of oil and protein. Soybean is solvent-extracted for oil, and the remaining part is "toasted." The soy meal thus produced contains 50% protein, which is then used as animal feed (e.g., chicken, hog, turkey). Soybean is also an important part of the several processed foods. It gained importance during World War II as an equivalent alternative for protein foods and as a source of edible oil (Shurtleff and Aoyagi 2009).

The soybean crop requires a climate with hot summers (temperatures range 20–30 °C). The temperature outside of this range results in stunted growth. Soybean growth is optimum in moist soils with rich organic content but also grows in other types of soil. Soybean roots have nodules for nitrogen-fixing bacterium like other legumes that perform nitrogen fixation. The crop is favored for rotational cropping for its nitrogen fixation.

The soybean crop is affected by the day length – longer day results in taller plants, more nodes, and delayed flowering. Conversely, shorter days result in early flowering. It requires 100–130 days to grow, and there is no vegetative growth during the yield formation.

The soybean yield is highly dependent on water availability, fertilization, seed variety, and row spacing. Improved varieties can yield as high as 2.5 and 3.5 ton/ha, while under rain-fed conditions a soybean yield is around 1.5–2.5 ton/ha. Although irrigation doesn't affect the oil and protein content of the seed significantly, a slight increase in protein and a small decrease in oil content may occur (Food and Agriculture Organization of the United Nations (FAO) 2020).

Soybean has high (38–45%) protein and high (approximately 20%) oil content. After the oil extraction, the material remaining is called soybean meal, or soy meal, which contains 50% soy protein. Soybean constitutes major part of the livestock feed worldwide. About 97% of livestock feed is produced from soybean, which serves as protein source (Kansas Soybean Commission, https://kansassoybeans.org/about-the-checkoff/animal-ag/). Some of it is also used in some dog foods (Menniti et al. 2014). In the European Union, 60% of the protein fed to livestock comes from the soy meal (Heuzé et al. 2020).

Protein content in the soy flour is 49% and the fiber content is 18 gm. Soy flour is gluten-free. Also, protein and other nutrient content in the soy flour are higher compared to wheat flour (Nutrition data, https://nutritiondata.self.com/facts).

Apart from the livestock feed, soybean products have found a way in the human diet and have gained an important part. Commonly used products include soybean oil, soy sauce, soy milk, tofu, soy flour, tempeh, textured vegetable protein, and soy lecithin. In Japan, edamame is a minimally processed dish of soybean. Soy flour is used in sauces, baked food, and frying (to reduce oil absorption). The soy flour imparts tenderness, moistness, and texture to baked food; hence it is used in baked foods. Soybean has been part of food traditionally in the form of soy milk and tofu.

Soybeans can be processed into various forms to substitute for a variety of food items having a comparable texture and appearance. Some of the examples of dairy products that similar soybean products can substitute are milk, margarine, ice cream, yogurt, and cheese. It can also be an alternative to meat as in veggie burgers. It can also be a cost-effective replacement for meat and poultry products.

However, the digestible calcium content in soy milk is not significant. Hence, many calcium-enriched products of soy milk are also manufactured. Though the protein quality in the soy products is roughly equivalent to the natural products, the products can be fortified with vitamin and mineral to enrich nutrient content to the equivalent level. The soy-based meat substitute has been in use to replace the ground beef cost-effectively and without compromising on the nutrition (Van Wyk 2005; Rizzo and Baroni 2018; North Carolina soybean production association (ncsoy.org)). Soy nut butter is another product that can substitute for peanut butter (Shurtleff and Aoyagi 2012).

Soy-based infant formula is another usage of soybean. It is advantageous for infants who need breastfeeding or wanted a vegan diet. It is a good substitute for infants allergic to pasteurized cow milk proteins. Different powdered, ready-to-feed forms are available in the market (American Academy of Pediatrics 1998; Rizzo and Baroni 2018).

Considering the various uses of the soybean as food, demand for soybean is increasingly high (Soystats 2020). However, viral and pests infections can reduce production substantially.

On a global scale, approximately 191 million hectares of genetically engineered crops were planted in 26 countries in 2017 (International Service for the Acquisition of Agri-biotech Applications (ISAAA) 2018).

Soybeans have been genetically modified in recent few years for better yield and other advantages. Several products are made from the genetically modified soybeans. The genetic modification started in 1996 when Monsanto Company introduced a genetically modified variety by substituting *Agrobacterium* sp. (strain CP4) gene EPSP (5-enolpyruvyl-shikimic acid-3-phosphate) synthase (Padgette et al. 1995; Duke and Cerdeira 2010). It was a herbicide-tolerant variety. Also, studies were conducted to develop *CpMMV* (Cowpea Mild Mottle Virus)-resistant variety.

In Brazil, the adaption of genetically modified soya bean increased from about 34% of soya bean acres in 2004 to 96% in 2019 (Soystats 2020). Though the cultivation of genetically modified soybean has increased exponentially, it has affected the export of soya bean to the European Union as a considerable number of suppliers and consumers are reluctant to use genetically modified products like food or animal feed. Hence they require extensive certification before export.

An extensive genetic diversity is observed in the soybean due to its predisposition to adapt to a variety of environmental conditions (Li et al. 2010; Zhou et al. 2015). Breeders choose the best varieties to improve soybean cultivation and ultimately the soybean yield. Therefore, in order to advance the marker-assisted breeding programs of soybean, knowledge of fundamental genetics of agronomical traits is essential (Wang et al. 2009). In the future, plant breeding in combination with recent advances in genomics will drive the expectations for soybean improvement (Palmer and Hymowitz 2016).

1.2 Soybean Taxonomy and Morphology

The taxonomy of cultivated soybean is presented below:

Order	Fabales
Family	Fabaceae (Leguminosae)
Subfamily	Papilionoideae (Faboideae)
Tribe	Phaseoleae
Subtribe	Glycininae
Genus	*Glycine* Willd
Subgenus	*Soja* (Moench) F. J. Herm.
Species	*G. max*
Botanical name	*Glycine max* (L.) Merr
Synonyms	*G. gracilis, G. Soja*
Common name	Soybean, soya bean
(Joy et al. 1998; Ratnaparkhe et al. 2011)	

The Fabaceae or Leguminosae family is a large family of flowering plants which includes around 20,000 species (Lewis et al. 2005). The three subfamilies of Fabaceae are Caesalpinieae, Mimosoideae, and Papilionoideae. The subfamily

Papilionoideae includes the species grown as crops such as soybean, pea, common bean, mung bean, cowpea, etc. (Lewis et al. 2005; Bruneau et al. 2008).

Genus *Glycine* Willd has approximately 25 perennial species (Greek word *glykys* = sweet). The term *Glycine* was coined by Linnaeus (Hymowitz and Newell 1981). The taxonomy tree was first arranged by Bentham (1864, 1865) (Table 1.1) who divided the genus *Glycine* into three sections: *Johnia*, *Soja*, and *Leptocyamus*. The cultivated soybean species was included in section *Soja*.

Table 1.1 Development of *Glycine* taxonomy according to different researchers

Bentham (1864, 1865)	Hermann (1962)	Verdcourt (1966, 1970)	Newell And Hymowttz (1980)	Current
Section *Johnia* *G. javanica* –	Subgenus *Glycine* *G. javanica* *G. petitiana*	Subgenus *Bracteata* *G. wightii* = *Neonotonia*	–	–
Section *Soja*	Subgenus *Soja*	Subgenus *Soja*	Subgenus *Soja* (Moench) F. J. Herm.	Subgenus *Soja* (Moench) F. J. Herm.
G. soja (cultivated)	*G. ussuriensis*	*G. soja* (previously *G. ussuriensis*)	*G. soja* Sieb. and Zucc.	*G. soja* Sieb. and Zucc.
G. hedysaroides = *Ophrestia*	–	–	–	–
G. pentaphylla = *Ophrestia*	–	–	–	–
G. lyalli = *Ophrestia*	–	–	–	–
–	*G. max* (cultivated)	*G. max*	*G. max* (L.) Merr.	*G. max* (L.) Merr.
Section *Leptocyamus*	Subgenus *Leptocyamus*	Subgenus *Glycine*	Subgenus Glycine	Subgenus *Glycine*
G. falcata	*G. falcata*	*G. falcate*	*G. falcata* Benth.	*G. falcata* Benth.
G. clandestina	*G. clandestine*	*G. clandestina*	*G. clandestina* Wendl.	*G. clandestina* Wendl.
G. clandestina var. *sericea*	*G. clandestina* var. *sericea*	*G. clandestina* var. *sericea*	*G. clandestina* var. *sericea* Benth.	–
G. latrobeana	*G. latrobeana*	*G. latrobeana*	*G. latrobeana* (Meissn.) Benth.	*G. latrobeana* (Meissn.) Benth.
G. tabacina	*G. tabacina*	*G. tabacina*	*G. tabacina* (Labill.) Benth.	*G. tabacina* (Labill.) Benth

(continued)

Table 1.1 (continued)

Bentham (1864, 1865)	Hermann (1962)	Verdcourt (1966, 1970)	Newell And Hymowttz (1980)	Current
G. tabacina var. *latifolia*	–	–	*G. latifolia* (Benth.) Newell and Hymowitz Hymowitz	*G. latifolia* (Benth.)
G. tabacina var. *uncinata*	–	–	–	–
G. sericea	*G. canescens*	*G. canescens*	*G. canescens* F. J. Herin	*G. canescens* F. J. Herin
G. tomentosa	*G. tomentella*	*G. tomentella*	*G. tomentella* Hayata	*G. tomentella* Hayata
–	–	–	–	*G. argyrea* by Tindale (1984) + 18 species (see Table 1.8)

Classified genus *Glycine* into three subgenera:

Glycine – This included the species from Africa and southeastern Asia
Soja – This included the soybean and its annual wild progenitor
Leptocyamus – Composed of nine perennial Australian species (Hermann 1962)

In 1966, Verdcourt readjusted the nomenclature and renamed *Glycine* to *Bracteata*, also subgenus *Leptocyamus* to *Glycine* (Verdcourt 1966; Hymowitz and Newell 1981). But in 1977, Lackey removed *Bracteata* as the characteristics of *G. wightii* were different from *Glycine* and hence transferred to a new genus (Lackey 1977a, b, c). Shortly after that, Newell and Hymowttz divide the genus into two subgenera (*Soja* and *Glycine*) (Newell and Hymowttz 1980). Later, 19 species were added to subgenus *Glycine* (Tables 1.1 and 1.9), while, subgenus *Soja* remains the same (Sherman-Broyles et al. 2014). More information is available at the websites of US Department of Agriculture and the Plant List:

The plant list: http://www.theplantlist.org/tpl1.1/search?q=glycine
US Department of Agriculture (USDA), Natural Resources Conservation Service: https://plants.usda.gov/java/nameSearch

1.2.1 Soybean Morphology

The widely cultivated soybean, *G. max*, is a derivative of either *G. ussuriensis* (*soja*) or some Asiatic ancestor. It differs from other wild species in terms of inflorescence, seed pods, and stem structure. Unlike other species inflorescence has axillary clusters or greatly reduced racemes without a bract at the base, and seed pods are broad, often curved, and contain two to four seeds. Additionally, *G. max* has a stout primary stem and sparse branches and does not twine or climb like other species (Hermann 1962). The characteristics of soybean are highly influenced by climatic conditions, soil quality, and genetics (Ratnaparkhe et al. 2011).

1.2.1.1 The Plant

Soybean is an annual plant (Ratnaparkhe et al. 2011), a bushy herb that can grow up to 2 m (Joy et al. 1998).

1.2.1.2 Roots

Soybean has a characteristic taproot system (see Fig. 1.2a). Initially, the root grows as a taproot which later has secondary, tertiary, and higher-order roots. The root nodules found in the soybean roots form a symbiotic relation with *Bradyrhizobium japonicum* (a nitrogen-fixing bacterium) and fix nitrogen in the soil (Carlson and Lersten 1987; Miladinović and Đorđević 2011), and this feature causes the legumes to have an increase in protein content (Singh and Chung 2016).

Fig. 1.2 The soybean plant. (**a**) Figure shows the shoot and root systems of soybean plant; photo by Adam Parker. (**b**) The color and shape differentiate between soybean seeds. (Photo by J. Miladinović (Miladinović and Đorđević 2011))

1.2.1.3 Stem

The stem is erect, stout, green, and covered with thick, fine brown, or gray hairs. It is mostly branched with the first node carrying the cotyledons (Fig. 1.2a) and the second node holding unifoliolate leaves. Hereinafter nodes have alternated trifoliolate leaves (Miladinović and Đorđević 2011).

1.2.1.4 Leaves

Soybean has three types of leaves – cotyledons, unifoliolate leaflets, and trifoliolates (see Fig. 1.2a). The cotyledons leaves are parts of the embryo, green or yellow, and round in shape. Unifoliolate leaves are the first pair of oval-shaped leaves. The trifoliolate leaves have three to four oval or spear-shaped leaflets per leaf. It is usually light to dark green and alternately arranged. They fall before the seeds are mature (Hermann 1962; Joy et al. 1998; Miladinović and Đorđević 2011).

1.2.1.5 Flowers

Flowers of soybean are very small (4–6 mm) that grow as a cluster of three to five flowers in the axil of a branch. The color varies from white to purple. Flowers are self-fertilizing (Miladinović and Đorđević 2011). It is a typical papilionaceous flower, with characteristic corolla composed of a standard, two wings, and two keel petals.

1.2.1.6 Fruit

The size and shape of the seed pods differ a lot with the varieties. Typically, the pods (legumes) are oblong, slightly bent (Frank and Fehr 1981) at the end, hairy, and yellowish-brown in color (Hermann 1962) and grow in clusters of 3–5 (Joy et al. 1998). Each legume contains usually one to three seeds, except that plant with *na* allele has pods with four seeds (Ratnaparkhe et al. 2011).

1.2.1.7 Seeds

Soybean seeds have different shapes (ovoid, spherical, or irregular rhomboid) and colors (Fig. 1.2b). Generally, seeds are 6–11 × 5–8 mm in diameter consisting of an embryo and a seed coat (Hermann 1962; Joy et al. 1998). The embryo is the main source of oil and protein (Wang et al. 2019a). The embryo consists of two large cotyledons, plumules, leaf primordia, epicotyl, hypocotyl, and radical.

Table 1.2 Examples of some phenotypes and their corresponding genes

Trait	Locus	Allele type	Phenotype	References
Nodulation	*Rj2*	*rj2(Rfg1)*	Causes limitation to some strains of *S. Fredii*, but does not limit *B. japonicum*	Xie et al. (2019)
Flower color	*W1*	*w1*	White flower	
Seed coat color	*I*	i^i	Colorless	
Seed coat color	*G*	*g*	Green seed color changes after maturation	
Related to soybean flowering time	*Gm11_ 10950924*	*Glyma.11G142900*	Transcription factor MYB59 related	Li et al. (2019)
Flowering time	*E1*	*E1/E1*	Late maturity	Owen (1927)
Stem growth habit	*Dt1*	*dt1*	Condition the determinate habit	Liu et al. (2010)

The seed coat protects the embryo. It is smooth/wrinkle in texture, glossy/matte, and water-resistant. It can be green, brown, black, or yellow. There is a scar marking on the seed coat known as hilum which is the point of attachment of funicle. The hilum is either linear or oval-shaped and can be black, brown, gray, yellow, green, or the same color as the seed coat. Hilum has a small pore at the end; micropyle, through which embryo absorbs water, exchanges gases and also serves as an outlet for radical to come out (Dzikowski 1936; Joy et al. 1998; Miladinović and Đorđević 2011).

Zhang et al. (2018) found a relationship between the shiny seed coat and the yield of oil. The seed coat of several legumes is covered with powdery bloom to protect it from predators. The seed coat bloom in wild soybean (*G. soja*) is controlled by Bloom1 (*B1*). Furthermore, Zhang et al. added that during the domestication of soybean, a nucleotide mutation in the coding region of *B1* leads to a shiny seed coat and increased the seed oil content (Table 1.2).

1.3 Soybean Cytogenetics

The soybean (*G. max*) genome is the product of a diploid ancestor ($n = 11$), which went through aneuploid loss ($n = 10$), followed by polyploidization ($2n = 20$) and diploidization ($n = 20$). It is a partially diploidized tetraploid (Lackey 1980; Singh and Hymowitz 1988). There is possibility of genome duplications or hybridization (Shoemaker et al. 1996, 2002; Blanc and Wolfe 2004; Tian et al. 2004). It is difficult to count the mitotic and meiotic chromosomes due to their smaller size. By using a chromosome image analyzing system (CHIAS), Yanagisawa et al. (1991) were able to separate 40 soybean chromosomes into 5 groups (A, B, C, D, E). Group A consists of a pair of nucleolus organizer chromosomes, group B includes two submetacentric chromosomes with a gap at the center of the long-arm contraction, and 10, 14, and 12 chromosomes make up the C, D, and E groups, respectively.

The monocular chromosome number of *Glycine max* is 20 (6 pairs metacentric and 14 pairs submetacentric), and the lengths vary from 1.99 to 1.26 μm in metaphase (Clarindo et al. 2007). Cytogenetic analysis reveals that large part of the genome is heterochromatin (>35%) with short arms of six pairs completely heterochromatic DNA (Adams and Burdon 2012).

Zou et al. (2003) assigned 11 molecular linkage groups to soybean chromosomes by using primary trisomics ($2x + 1 = 41$) and simple sequence repeats markers. Proposed chromosome numbers to linkage group assignments can be seen in Table 1.3. Linkage group length is in centimorgan (cM) based on the Soybean Consensus Map 3.0 (Choi et al. 2007), produced by Perry Cregan's group at the USDA-ARS, Soybean Genomics and Improvement Lab. (SoyBas.org).

Table 1.3 Proposed linkage groups of soybean chromosomes and the gene's number of each chromosome

Chromosome number	Linkage group[a]	Length[a] (cm)	Status[a]	Number of genes[b]
1	D1a	98.41	Previously assigned	2749
2	D1b	140.63	Genetic length	3278
3	N	99.51	Previously assigned	2837
4	C1	112.32	Previously assigned	2782
5	A1	86.75	Previously assigned	2667
6	C2	136.51	Genetic length	3360
7	M	135.15	Genetic length	2895
8	A2	146.67	Previously assigned	3745
9	K	99.60	Previously assigned	2987
10	O	132.89	Genetic length	3236
11	B1	124.24	Genetic length	2739
12	H	120.50	Genetic length	2510
13	F	120.03	Previously assigned	4078
14	B2	108.18	Genetic length	2489
15	E	99.88	Genetic length	2916
16	J	92.27	Genetic length	2295
17	D2	119.19	Previously assigned	2758
18	G	105.00	Previously assigned	3272
19	L	101.14	Previously assigned	2859
20	I	112.77	Previously assigned	2632

[a]From SoyBase, https://soybase.org/LG2Xsome.php
[b]According to Glycine_max_v2.1 from NCBI, https://www.ncbi.nlm.nih.gov/genome/?term=Glycine+max

1.4 Soybean Genome

1.4.1 Soybean Genome Assembly

Crop plants often undergo polyploidization and contain relatively higher repeat elements (Adams and Wendel 2005; Bennetzen and Wang 2014). Plant genome sequencing has gained momentum with innovative technologies. Newer long-read technologies complement the short-read technology. Genome sequencing is also supported by the scaffold data derived from mate-pair libraries that cover difficult-to-assemble repetitive sequences. Though it makes contigs assembly possible, the assemblies need to be anchored to a genome map. So the genomes are aligned with the related species (Brozynska et al. 2016).

The cultivated soybean's (*Glycine max*) genome size in 1.1–1.15 Gb ($n = 20$) (Arumuganathan and Earle 1991). A plant genome is comparatively difficult to assemble than animal genome due to genome size, polyploidy, and highly repetitive genomic regions (Imelfort and Edwards 2009; Valliyodan et al. 2017).

Researchers attempted several genome assembly protocols (Table 1.4), by making use of genetic map as a tool. One of the first linkage maps, which was created by Mudge et al. (2004), as a way to estimate the soybean genes, they used the Poisson distribution of bacterial artificial chromosomes (BACs) identified with restriction fragment length polymorphism (RFLP) probes.

The nature and the number of different polymorphic markers used to develop a genetic linkage map decide its utility. The recent sequencing technology helps to obtain several single nucleotide polymorphisms (SNPs) throughout the genome that are potential markers for high-density genetic maps. SNP markers are now the most

Table 1.4 Examples of genome assembly protocols

Genome assembly protocol	References
Poisson distribution of BACs identified with RFLP probes, from a pool of more than 100,000 BAC clones	Mudge et al. (2004)
Whole-genome shotgun approach and integrated it with physical and high-density genetic maps to create a chromosome-scale draft sequence assembly	Schmutz et al. (2010)
Constructed two high-density linkage maps based on a high number of single nucleotide polymorphism loci with recombinant inbred lines (RILs); the marker locations on the linkage map were compared with that of the chromosome-scale assembly of the soybean genome, Glyma1.01	Song et al. (2016)
Sequenced Chinese soybean genome for "Zhonghuang 13" by a combination of different techniques, including single-molecule real-time (SMRT) sequencing, chromosome conformation capture sequencing (Hi-C), optical mapping, and next-generation sequencing (HiSeq). The assembled genome (Gmax_ZH13) was compared with Glycine_max_v2.0	Shen et al. (2018)
Chinese soybean genome for "Zhonghuang 13" was sequenced by using a combination of various methods which include CANU (v1.7.1) for assembling PacBio subreads to PacBio contigs, HERA to generate longer contigs, Juicer and 3D-DNA and LACHESIS to secure the hybrid scaffolds into chromosomes with Hi-C reads	Shen et al. (2019)

recent molecular genetic map of soybean which includes a large number of SNPs from genic and nongenic regions (Song and Cregan 2017).

A soybean genetic map was constructed using restriction fragment length polymorphism (RFLP) markers (Keim et al. 1992; Lark et al. 1993; Shoemaker and Specht 1995). However, the development process was slower due to the polymorphism and the bulky genome. In 1999, Cregan et al. employed simple sequence repeat (SSR) markers to integrate the soybean linkage map into 20 homologous linkage groups. Further, the Cregan et al. map was integrated with the other five maps using JoinMap to develop a composite map using several markers (Song et al. 2004). Further Choi et al. (2007) analyzed gene distribution and SNPs in the soybean genome. Hwang et al. (2009) developed a high-density integrated map of 1810 SSR or sequence-tagged site (STS) markers.

Researchers have demonstrated a limited synteny block between mung bean, cowpea, common bean, and soybean (Boutin et al. 1995; Lee et al. 2001). Additionally, some researchers also reported macrosynteny among legumes like the common bean, soybean, and *Medicago Sativa*. Though some chromosome rearrangements have shortened synteny blocks, chromosomes of *M. truncatula* helped align the chromosomes of papilionoid species. However, *M. truncatula* and soybean had very few synteny blocks (Choi et al. 2004).

In 2006, the DOE-JGI Community Sequencing Program (CSP) commenced the soybean genome program using large-scale shotgun sequencing tech. Approximately 13 million attempted Sanger shotgun reads were produced and deposited in the National Center for Biotechnology Information (NCBI) (Phytozome, https://phytozome.jgi.doe.gov/pz/portal.html#!info?alias=Org_Gmax).

In 2010, the first whole-genome sequence of soybean (variety: *Glycine max* L. Merr.) – Glyma1.01 – was assembled (see Tables 1.4 and 1.5) based on the linkage maps constructed (Schmutz et al. 2010). It captured approximately 975 Mb of sequence with 236 unanchored scaffolds of length 10–100 kb, 51 unanchored scaffolds of length 100 kb, and genotyped 60–240 recombinant inbred lines (RILs) (Song et al. 2004; Hyten et al. 2010). This may have left large gaps, incorrect marker order, and low resolution in the linkage maps; hence improvisation was needed (Song et al. 2016).

The new assembly techniques and construction of high linkage maps helped to improve gene annotations and reduce unmapped scaffolds. Apart from the genomes available at the Phytozome website (Goodstein et al. 2012), several genome assemblies of *Glycine max* (L.) Merr. are available at the NCBI (Table 1.5). In addition, 4 genome assemblies were identified for *Glycine max* cultivars Enrei, Lee, Zhonghuang 13, and EMBRAPA BRS 537 which were added to NCBI database. The release dates of genome assemblies are summarized in Table 1.5.

Generally, the step for genome sequence construction starts with generation of contigs and formation of scaffolds containing contigs and gaps and finishes with the creation of physical map (Table 1.4) and pseudomolecules corresponding to each of chromosomes (International Cassava Genetic Map Consortium 2015; Lee et al. 2020). The integration of chromosome conformation information and BioNano optical mapping, generated either in vitro (Chicago) technologies (Jarvis et al.

Table 1.5 Genome assemblies and their release date (2008–2020)

Genome assembly name	*G. max* cultivar	Date of release	Submitter
Glyma0	Williams 82	January 2008	US DOE Joint Genome Institute (JGI-PGF) In Phytozome, https://phytozome-next.jgi.doe.gov/info/Gmax_Wm82_a4_v1
Glyma1.01		December 2008, complete 2010	
Wm82.a2.v1		January 2014	
Wm82.a4.v1		2015	
V1.0	Williams 82	2010	US DOE Joint Genome Institute (JGI-PGF) In NCBI, https://www.ncbi.nlm.nih.gov/genome/?term=txid3847[orgn]
V1.0 and V1.1		2012	
Glycine_-max_v2.0		2015	
Glycine_max_v2.1		2018	
Glycine max_Enrei_2.0	Enrei	2015	National Institute of Agrobiological Sciences
Glyma.Lee.gnm1	Lee	2015/2018	*Glycine max* cv Lee and *Glycine soja* PI 483463 sequencing consortium
Gmax_ZH13	Zhonghuang 13	2018	Institute of Genetics and Developmental Biology, Chinese Academy of Sciences
Gmax_ZH13_v2.0		2020	
EMB_GmBRS537_1.0	EMBRAPA BRS 537	2020	Brazilian Agricultural Research Corporation

Sources: Phytozome, NCBI

2017) or in vivo (Hi-C) (Teh et al. 2017; Mascher et al. 2017), was shown to be able to anchor the scaffolds into chromosomal or subchromosomal levels (Shi et al. 2019).

Recently, an updated soybean genome sequencing (Glycine_max_v2.1) was deposited in NCBI in 2018 containing 22 assembled chromosomes (20 ch. + 2 non-nuclear ch.) and unplaced scaffolds (see the number of genes of each linkage group in Table 1.3 and the feature statistics of this release in Table 1.6). Later in that same year, the sequencing of Chinese soybean genome for Zhonghuang 13 (Gmax_ZH13) was published. To compare between the two sequences, researchers were able to conclude through detailed analysis that Glycine_max_v2.0 had 12,761 gaps and 1169 unplaced scaffolds, whereas Gmax_ZH13 had 815 gaps in the chromosomes and 549 unplaced scaffolds. Furthermore, the most discrete inversion exists at chromosome 11 from 27.78 to 30.00 Mb in Gmax_ZH13, which assembled as 22.23–24.6 Mb in Glycine_max_v2.0 in an opposite direction. In addition, the largest translocation was 2.22 Mb locating from 13.32 to 15.56 Mb on chromosome 5 in Gmax_ZH13, which was anchored to Glycine_max_v2.0 from 18.33 to 20.52 Mb on chromosome 5. In another area for comparison, it was found that some genetic variations were associated with phenotypic changes in both *G. max* Williams 82 and *G. max* Zhonghuang 13 (Shen et al. 2018). For example, the soybean flower color has been reported to be controlled by the F3'5'H (SoyZH13_13G057600) gene (Zabala and Vodkin 2007), and this gene is responsible for the white flower color in Williams 82 and the purple color in Zhonghuang 13 (Shen et al. 2018). Finally, the

Table 1.6 The feature statistics of recent soybean genome assembly releases for three cultivars (Williams 82, Zhonghuang 13, and Embrapa BRS 537) of *G. max*, compared to the first release (V1) based on the data collected from NCBI website

Assembly Features	V1	Glycine_max_ v2.1	Gmax_ZH13_ v2.0	EMB_ GmBRS537_1.0
Total sequence length	973,492,571	979,046,046	1,011,378,317	1,116,181,090
Total ungapped length	955,203,020	955,932,237	1,008,981,415	1,115,677,742
Gaps between scaffolds	0	387	0	0
Number of scaffolds	1168	1579	45	153
Scaffold N50	47,781,076	4,417,540	51,952,677	56,308,508
Scaffold L50	10	66	10	10
Number of contigs	16,307	17,187	628	5186
Contig N50	189,443	182,849	17,978,897	1,396,888
Contig L50	1492	1548	21	208
Total number of chromosomes and plasmids	21	22	22	20
Number of component sequences (WGS or clone)	16,307	17,187	45	153

genome sequencing (Gmax_ZH13) was updated to a newer version named Gmax_ZH13_v2.0 (see Tables 1.5 and 1.6). In the improved version, remarkable changes were noticed which include an increase in the contig N50 size of Gmax_ZH13_v2.0 by 6.5 times and a decrease in the gap number as well as gap length, and it was found that the mapping ratio of WGS HiSeq reads and RNA isoform sequencing reads reached to 99.89% and 99.81%, confirming the high completeness of Gmax_ZH13_v2.0 (Shen et al. 2019) (Table 1.6 shows the feature statistics of recent soybean genome assembly releases from NCBI website).

1.4.2 The Sequence of Soybean's Chloroplast and Mitochondria Genome

The genome of chloroplast in most land plants is highly preserved (Palmer 1991; Raubeson and Jansen 2005). Generally, chloroplasts have a one copy of circular chromosome that has a quadripartite framework which includes two copies of an inverted repeat that comes between the large and small single-copy regions (Wakasugi et al. 1994). The chloroplast genome of *Glycine* contains 111 unique genes that account for 152,218 base pairs (bp) in length, which include a couple of identically sequenced inverted repeats of 25,574 bp separated by a large single-copy region of 83,175 bp and a small single-copy region of 17,895 bp (Saski et al. 2005).

The length of mitochondrial genome (mtDNA) is highly variable in seed plants (Gray et al. 1999; Lang et al. 1999), extending from 208 kb (white mustard) (Palmer

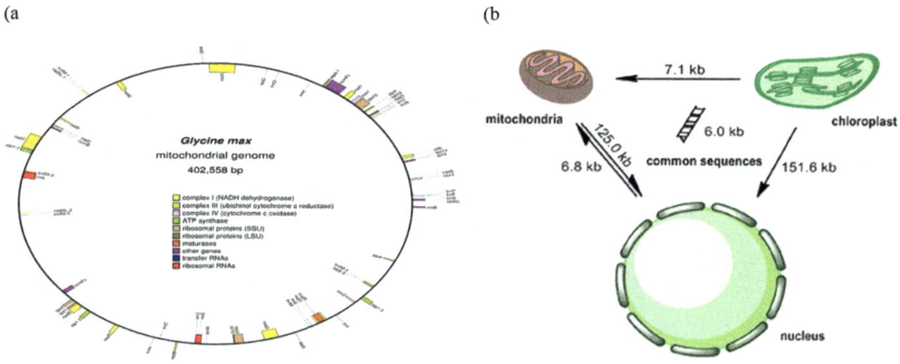

Fig. 1.3 (**a**) The circular map of the *Glycine max* mitochondrial genome. Features on the clockwise- and anticlockwise-transcribed strands are drawn on the inside and outside of the circle, respectively. The figure was drawn using OGDraw v1.2 (Lohse et al. 2007). doi: 10.1371/journal.pone.0056502.g001. (**b**) DNA transfers between the nuclear, chloroplast, and mitochondrial genomes in soybean. doi: https://doi.org/10.1371/journal.pone.0056502.g008. (Resource of the photos: Chang et al. 2013)

and Herbo 1987) to 11.3 Mb (*Silene conica*) (Sloan et al. 2012). Since the genomes possess high numbers of repeats, such as tandem, short, and large repeats (Kubo and Newton 2008; Alverson et al. 2010, 2011). As a result, the mitochondrial genome is longer in plants compared to animals (generally, 16 KB) (Boore 1999). Inheritable changes of the mitochondrial genome are a result of short repeats which are responsible for the permanent recombination of the mitochondrial genome (Andre et al. 1992; Newton et al. 2004).

The soybean mitochondrial genome includes 110 genes, which are divided into 58 genes with known function and 52 genes with unknown function (see Fig. 1.3). The genome contains sequences of numerous distinguishable sources, including 6.8 kb and 7.1 kb DNA fragments that have been transferred from the nuclear and chloroplast genomes, respectively, and some horizontal DNA transfers. In addition, the DNA transfers are not merely restricted to the mitochondria, but fairly large DNA segments can also be transferred to the nucleus, including transportation of 125.0 kb and 151.6 kb identified from the soybean mitochondrial and chloroplast genomes, respectively (Chang et al. 2013).

1.4.3 Website Links

There are many useful website sources that discuss soybean genome. Listed in Table 1.7 are some of the most viewed links.

Table 1.7 The most viewed links of soybean genome

Website names	Links	References
SoyBase	https://soybase.org/	Cannon et al. (2012)
SoyNet	http://www.inetbio.org/soynet	Kim et al. (2017)
Soybean-VCF2Genomes	http://pgl.gnu.ac.kr/soy_vcf2genome/	Ha et al. (2019)
Soybean Knowledge Base (SoyKB)	http://soykb.org/	Joshi et al. (2012)
Ensembl Plants	http://plants.ensembl.org/index.html	Bolser et al. (2016)
National Center for Biotechnology Information (NCBI)	https://www.ncbi.nlm.nih.gov/	Wheeler et al. (2007)
Phytozome	https://phytozome.jgi.doe.gov/pz/portal.html	Goodstein et al. (2012)
JGI Genome Portal	https://genome.jgi.doe.gov/portal/	Grigoriev et al. (2012) and Nordberg et al. (2014)

1.5 Soybean Genetic Resources (Germplasm)

In 1971, Harlan and deWet developed the gene pool concept which has played an essential role in producing high-yielding cultivars profiteering from germplasm resources by either conventional method or transformation methodology. Based on the success rate of hybridization among species, the three proposed gene pools are primary (GP-1), secondary (GP-2), and tertiary (GP-3) gene pool.

The primary source of variation in domesticated crops is the crop wild relatives (CWRs) (Brozynska et al. 2016). According to Harlan and deWet (1971), GP-1 is defined as the ability of biological species to easily cross within the gene pool to produce an active and fertile first generation (F1). Soybean (GP-1) would include cultivars, landraces, and *G. soja* genotypes. GP-2 includes species that can cross with GP-1 giving F_1 hybrids that have some fertility. Following this definition GP-2 is not currently found in soya bean. GP-3 is the extreme limit of genetic resources suggesting that gene transfer between GP-1 and GP-3 is almost impossible or calls for radical techniques that may result in sterile or lethal hybrids. In *Glycine*, GP-3 accounts for the 26 wild perennial species (Table 1.8). In order to exploit the subgenus *Glycine* germplasm in soybean breeding, methods were developed for yielding F1, amphidiploid, BC1, BC2, BC3, and fertile soybean plants that resulted from the crossing between soybean and the subgenus *Glycine* species (Singh 2010).

Perennial subgenus *Glycine* species (Table 1.8), which are predominantly found in Australia, are classified as either diploid ($2n = 38, 40$) or alloploid ($2n = 78, 80$) depending on the chromosomal number. *Glycine* species were further categorized according to their capacity in producing fertile hybrids as well as the degree to which meiotic chromosomes are paired (Ratnaparkhe et al. 2011). Following this concept, seven genome groups were designated (Table 1.8). Within a group, the

Table 1.8 The genus *Glycine* (chromosome number, genome symbol, and destruction) (Ratnaparkhe et al. 2011)

Species	Molecular group	2n	Growth form/ genome group	Distribution	Treatment
Subgenus *Soja* (Moench) F.J. Hermann (2 sp.)			Annual		
G. soja Sieb. and Zucc.		40	G	China, Korea, Taiwan Japan, Russia	
G. max (L.) Merr		40	G_1	Cultigen; worldwide	
Subgenus *Glycine* (26 sp.)			Perennial		
G. albicans Tindale and Craven		40	I	Australia	Tindale and Craven (1988)
G. aphyonota B.		40	I_3		Pfeil and Craven (2002)
G. arenaria Tindale		40	H		Tindale (1986b)
G. argyrea Tindale		40	A_2		Tindale (1984)
G. canescens F.J. Hermann		40	A		
G. clandestina Wendl		40	A_1		
G. curvata Tindale		40	C_1		Tindale (1986a)
G. cyrotoloba Tindale		40	C		Tindale (1984)
G. falcata Benth		40	F		
G. gracei B.E. Pfeil and Craven		40	?		Pfeil et al. (2006)
G. hirticaulis Tindale and Craven		40	H_1, (??)		Tindale and Craven (1988)
		80			
G. lactovirens Tindale and Craven		40	I_1		Tindale and Craven (1988)
G. latifolia (Benth.) Newell and Hymowitz		40	B_1		Newell and Hymowitz (1980)
G. latrobeana (Meissn.) Benth		40	A_3		
G. microphylla (Benth.) Tindale		40	B		Tindale (1986a, b)
G. montis-douglas B.E. Pfeil and Craven		40	?		Pfeil et al. (2006)
G. peratosa B.E. Pfeil and Tindale		40	A_5		Pfeil et al. (2001)
G. pescadrensis Hayata		40	AB_1	Australia, Taiwan, Japan	Pfeil et al. (2006)

(continued)

Table 1.8 (continued)

Species	Molecular group	2n	Growth form/ genome group	Distribution	Treatment
G. pindanica Tindale and Craven		40	H_2	Australia	Tindale and Craven (1993)
G. pullenii B. Pfeil, Tindale, and Craven		40	H_3		Pfeil and Craven (2002)
G. rubiginosa Tindale and B.E. Pfeil		40	A_4		Pfeil et al. (2006)
G. stenophita B. Pfeil and Tindale		40	B_3	Australia, (Japan?)	Doyle et al. (2000)
G. syndetika B.E. Pfeil	D4	40	A6	Australia	Pfeil et al. (2006)
G. dolichocarpa Tateishi and Ohashi		40	D_1A	Taiwan	Tateishi and Ohashi (1992)
G. tabacina (Labill.) Benth		40	B_2	Australia	
		80	BB_1; BB_2; B_1B_2	Australia, West Central and South Pacific Islands	
G. tomentella Hayata	D1,D2	38	E	Australia	
	D3	40	D		
	D5B	40	H_2		
	D5A	40	D_2		
	T1	78	D_3E		
	T5	78	AE		
	T6	78	E H_2		
	T2	80	D A_6	Australia, Taiwan	
	T3	80	D D_2	Australia, Papua New Guinea, Timor	
	T4	80	D H_2	Australia, Philippines, Taiwan	

species are reproductively compatible but are reproductively isolated from other genome groups (Singh and Hymowitz 1985). These groups were identified by capital letters A–G (Hymowitz et al. 1998). Moreover, based on isozyme similarities and crossing study in diploid and tetraploid, *G. tomentella* accessions have been separated into different groups [(D1–D6), (T1–T6)] (Doyle and Brown 1985; Doyle et al. 1986). Later on D6 became *G. arenaria* (Tindale 1986b), D4 was recognized as *G. syndetika* (Pfeil et al. 2006), and T2 as *G. dolichocarpa* (Tateishi and Ohashi 1992).

The CWRs evolve under natural selection adapting to the environmental changes. The CWRs provide dynamic resources for genetics; hence its conservation is essential (Brozynska et al. 2016).

Since the cultivation of soybean has started, more than 45,000 accessions have been developed. Of this only o.o2% (~80) of accessions have been employed in the cultivar development in North America (Carter et al. 2004). The use of very few accessions in the development of cultivars has led to a diversity-reduced bottleneck (Hyten et al. 2006). Due to the genetic bottleneck, wild and cultivated soybeans exhibit several lineage-specific genes and genes with copy number variation (CNV) or large-effect mutations (Lam et al. 2010; Li et al. 2013, 2014). Also, genetic diversity is observed across the seed from different geographic locations (Zhou et al. 2015). Furthermore, a study compared the Chinese and American soybean to demonstrate a diverse genetic basis (Liu et al. 2017). Thus the genetic variations are associated with agronomic traits (Li et al. 2014; Zhou et al. 2015; Wang and Tian 2015; Wei and Cao 2016). Considering these facts, a reference genome will not be representative of the genetic diversity of soybean.

A line of observations from Wang et al. (2019b) study suggests that the wide genomic introgression obtained from interspecific hybridization and subsequent backcrossing during or after soybean domestication is determined by the relative intensities of the two distinct selection pressures (natural and artificial selection) (for more clarification see Figs. 1.4 and 1.5).

Today, soybean cultivars are the result of sexual hybridization followed by inbreeding and selection. This process is considered to be the foundation for soybean cultivar development (Palmer and Hymowitz 2016). In order to recognize the link between the domesticated crops and CWRs, molecular analysis can be used (Jobin- Décor et al. 1997). For instance, Beche et al. (2020) were able to identify the usefulness of alleles related to the agronomic traits from wild soybean (*G. soja*) germplasm in enhancing soybean breeding programs, through using nested association mapping (NAM) method.

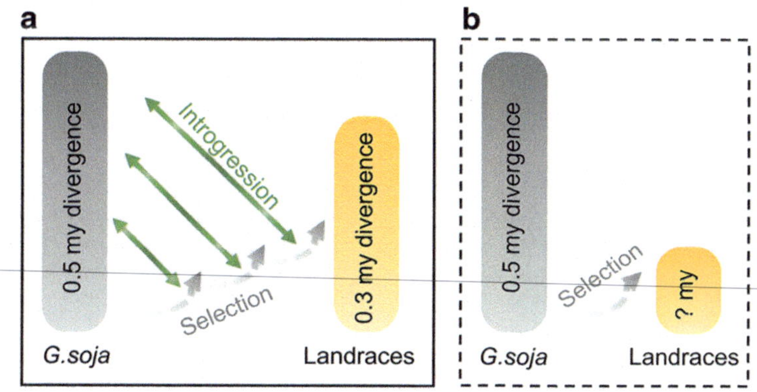

Fig. 1.4 Models of the domestication procedure of soybean. (**a**) Series of hybridization including diverse *G. soja* accessions opposed genetic bottleneck during domestication and participated in the landraces diversification and adaptation to different environments; green arrows indicate rounds of hybridization involving diverse *G. soja* accessions. The segmented arrows represent the frequent selection against introgression. (**b**) A hypothetical design of the domestication procedure of soybean without interspecific introgression which would result in a stronger domesticated bottleneck (Wang et al. 2019b)

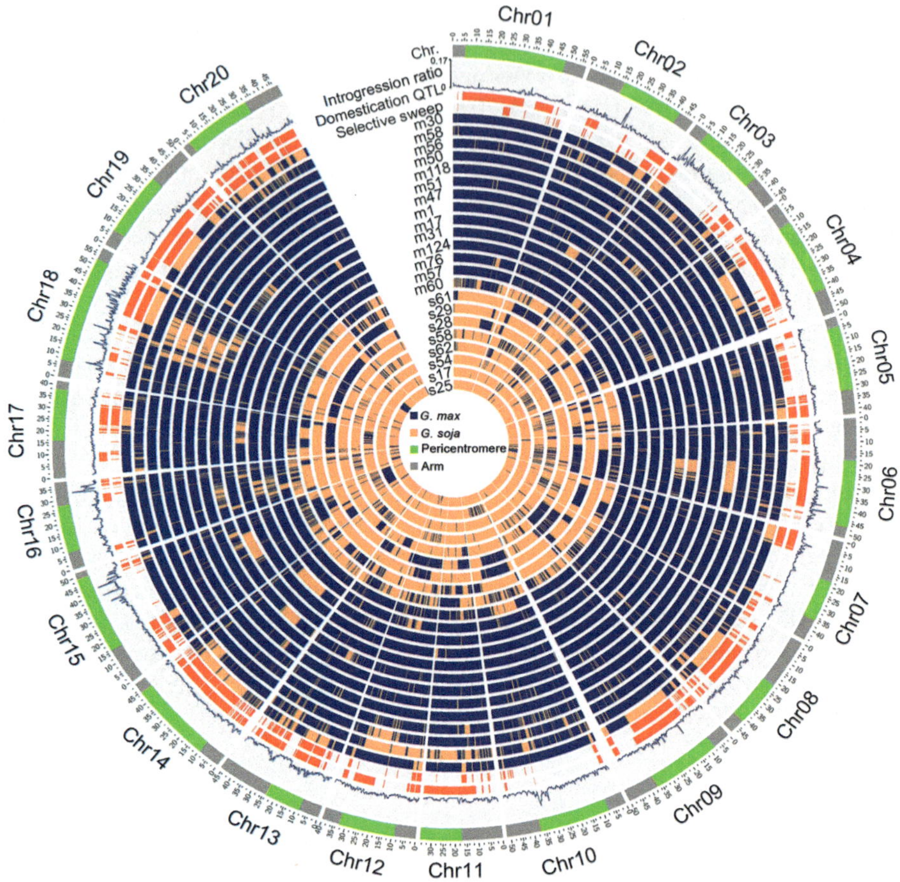

Fig. 1.5 Genome-wide distribution of interspecific introgression and genomic characteristics. Tracks from outer to inner circles indicate chromosomes including (**a**) chromosome arms (gray color) and pericentromeric regions (green color), (**b**) chromosomal distribution of introgression rates in the whole population, (**c**) chromosomal distribution of domestication-related QTL as indicated by red bars in corresponding ring, (**d**) chromosomal distribution of selective sweeps as indicated by red bars in corresponding track, (**e**) chromosomal distribution of interspecific introgression in each of the 12 *G. max* and 10 *G. soja* accessions, as indicated by the diagram, whose genomes each contain more than 5% introgressed fragments. In these 22 tracks, the *G. max* segments were shown in blue and the *G. soja* segments were shown in orange (Wang et al. 2019b)

In agriculture, the development of genetic molecular tools showed an important role in the documentation of genomic backgrounds of their plant yield by making the process easier and more precise than before.

Ha et al. (2019) made use of previously published soybean 222 validation germplasm consisting of *Glycine max*, *Glycine soja*, and breeding lines including F2 and RIL with some redundancy to construct reference genotype matrix. They were able to build "Soybean-VCF2Genomes" which is a web application, where users can upload their single-sample variant call format (VCF) file for analysis (Fig. 1.6).

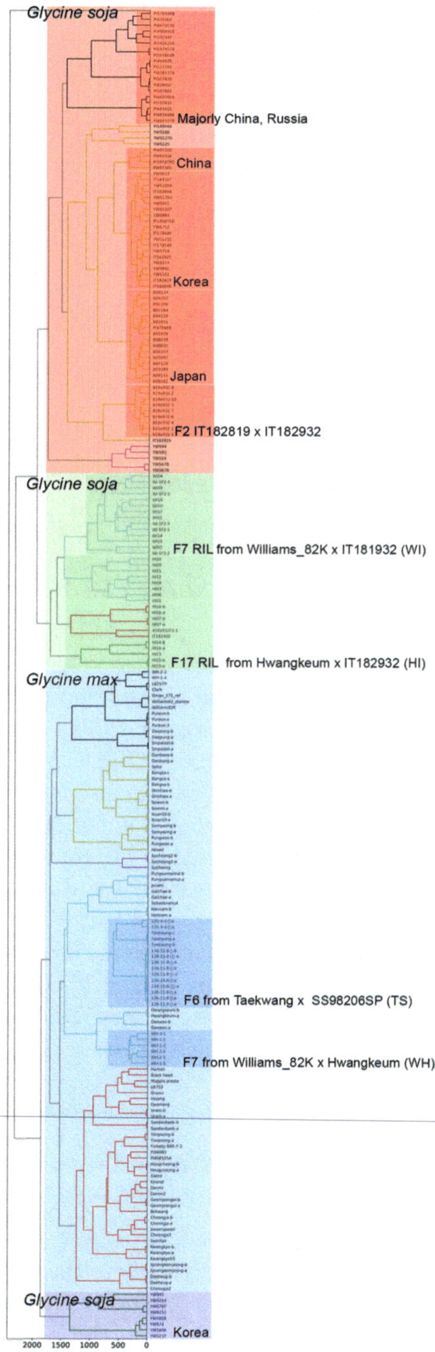

Fig. 1.6 Soybean phylogenetic tree structured by genotype matrix. (According to Ha et al. 2019)

Table 1.9 A list of some germplasm resources for soybeans

Institution	Links
USDA Soybean Germplasm Collection US National Plant Germplasm System	https://www.gbif.org/grscicoll/collection/6e5b27ae-183f-47c1-8a60-7dda5fe05b11 https://www.ars-grin.gov/
World Vegetable Center, Taiwan	https://avrdc.org/seed/improved-lines/grain-soybean/
Institute Crop Germplasm Research (ICGR), Beijing	http://www.cgris.net/icgr/icgr_english.html
N.I.Vavilov Institute of Plant Genetic Resources (VIR), Russia	http://vir.nw.ru/test/vir.nw/index.php?option=com_content&view=article&id=84&Itemid=511&lang=en
Department of Genetic Resources I, National Institute of Agrobiological Resources, Japan	https://www.genesys-pgr.org/wiews/JPN003
ICAR-Indian Institute of Soybean Research	https://iisrindore.icar.gov.in/
Embrapa Genetic Resources and Biotechnology, Brazil	https://www.embrapa.br/en/recursos-geneticos-e-biotecnologia

1.5.1 A List of Some Germplasm Resources for Soybeans

Plant genetic resources stored in large germplasm collections located outside of the plant's natural habitat are an important source of unprecedented, potentially useful genetic variations that can be utilized in plant breeding programs (Wang et al. 2017).

The USDA Soybean Germplasm Collection preserves 21,810 accessions of the genus *Glycine*, with 21 species. The majority of these are the cultivated soybean, *Glycine max*, and the remainder are 1179 wild soybean, *Glycine soja*, and 1005 accessions of 19 perennial *Glycine* species. The collection of soybean germplasm found in China represents 14% of the world's accessions (http://map.seedmap.org/solutions/conservation/seed-banks/chinas-institute-of-cropgermplasm-resources/).

Many new soybean varieties have been bred in European countries. Genebanks of Europe comprise more than 20,000 accessions of genus *Glycine*. The largest collection numbering more than 7000 accessions is preserved at the N.I.Vavilov Institute of Plant Genetic Resources in Russia (Table 1.9).

1.6 Economic Importance of Soybean

Soybean is an economically important crop as a source of dietary protein and oil. It has been part of the diet in Southeast Asia for a long period. It spread to the western world in the nineteenth century (Salgado and Donado-Pestana 2011). Brazil and the USA are the top exporters of the soybean (see Fig. 1.7).

Soybean has been a very important part of human nutrition for a long period. It contains oil, protein, dietary fiber, carbohydrates, and fat (Fig. 1.8). Though the

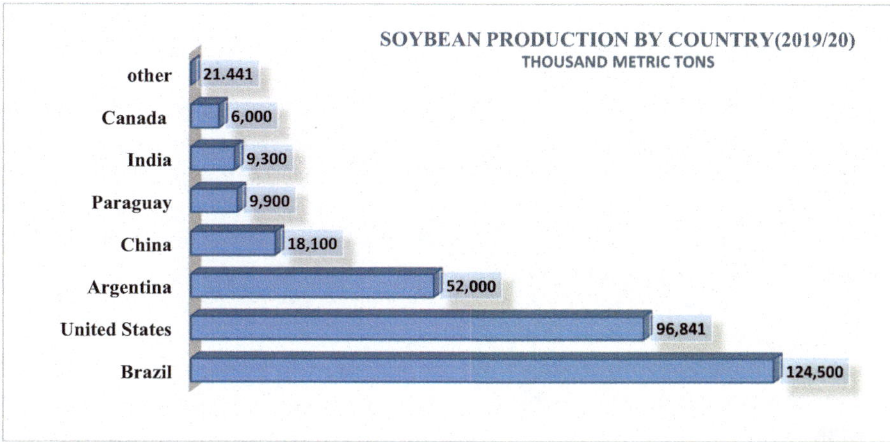

Fig. 1.7 Soybean production. (Data from Foreign Agricultural Service/USDA. https://apps.fas.usda.gov/psdonline/circulars/oilseeds.pd)

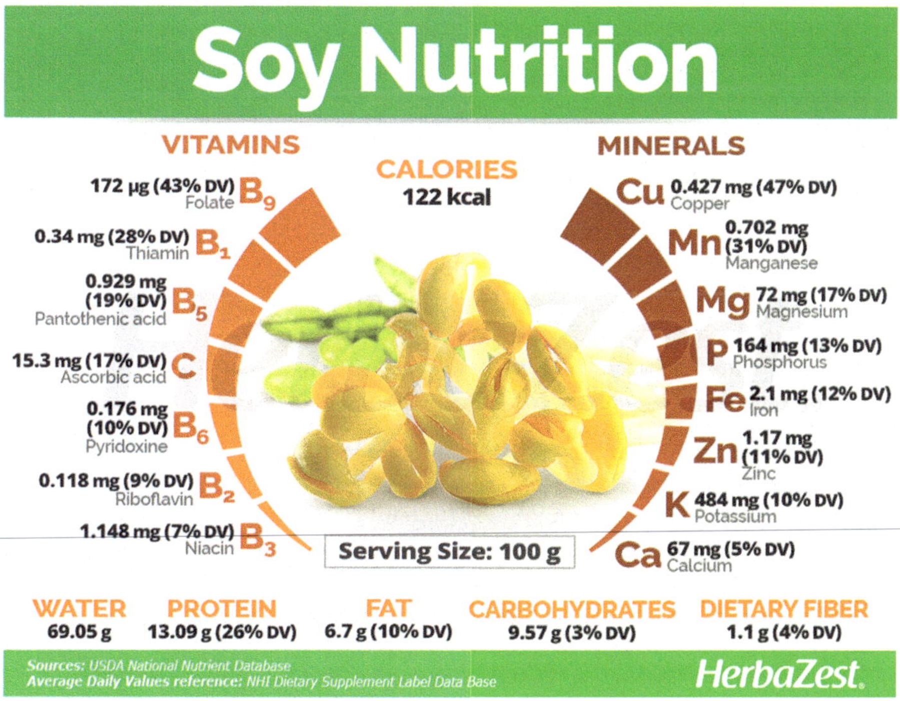

Fig. 1.8 Soy nutrition. (From: HerbaZest. https://www.herbazest.com/herbs/soy)

content percentage varies depending on the variety, environmental conditions, fertilizers, irrigation, etc., protein and oil content is typically 40% and 20%, respectively (Clemente and Cahoon 2009).

Soy protein can be added to some food products as a way to add extra nutritional value, like in infant formulas and meat-like products, etc. It has been estimated that the consumption of soybean oil are the following: 55% as cooking and salad oil, 24% as baking and frying oil, 11% as a substance for biodiesel production, 7% for other food and industrial uses, and lastly 4% as a component in margarine production (Wisner 2014; Medic et al. 2014).

Breeders have always relied on the protein content for a selection of varieties for breeding, but a recent increase in the oil demand for various reasons has changed the criterion to include the oil content and the quality. The public awareness toward dietary fat has led to increased efforts to increase the oxidative stability of the soybean oil to avoid trans fat due to hydrogenation. This also enhances V-3 fatty acid content in the oil (Clemente and Cahoon 2009; Lee et al. 2012). It has been observed that the oil and protein content of the soybean are inversely related. The carbon flux during the embryogenesis is shifted to one of these; it is always affected by genetics and environmental factors (Schwender et al. 2003). Interestingly, the location of seeds on the plant can also impact carbon flux during embryogenesis (Bennett et al. 2003). The seed from pods positioned at the top of the plant is observed to have a higher protein and lower oil content compared to the seed from pods from the bottom of the plant (Escalante and Wilcox 1993).

Soybean oil is widely used because of its excellent nutritional qualities, wide-use functionality, availability, and low cost. Soy oil is also an important ingredient of paints, lubricants, plastics, and biofuels (Johnson and Smith 2011). There are several processes to extract soybean oil, resulting in different products.

There are three main processes to extract soybean oil:

1. Extracting oil from soybean flakes by the solvent
2. Mechanical extraction by a screw press (expeller)
3. Combines extruding and expelling of soybean flakes and uses solvent for oil extraction

A 60-pound bushel of soybean will yield about 11 pounds of crude oil and 47 pounds of soybean meal (Johnson and Smith 2011). Soybean meal is used as an important protein source in animal feed for poultry, cattle, and other farm animals. It forms the two-third part of the animal feed worldwide (Oil World 2015), as well as 69.3% from global protein meal consumption in 2019 (Soystats 2020). High-protein types are obtained from dehulled seeds and contain 47–49% protein and 3% crude fiber (as-fed basis).

Other type of soybean meal is hulls which contain <47% protein and >6% crude fiber (Cromwell 2012). The amino acid composition of the soy meal varies among samples due to variety, the geographical location of soybean production, and the processing methods (Parsons et al. 1991, 2000; de Coca-Sinova et al. 2008; Baker et al. 2011).

Soybean seed oil is controlled genetically by quantitative trait loci (QTLs) along with various environmental factors (Burton 1987; Diers et al. 1992). Through extensive investigations (Diers et al. 1992; Spencer et al. 2004; Shibata et al. 2008; Mao et al. 2013; Pathan et al. 2013; Wang et al. 2015), researchers discovered that there are more than 322 oil QTLs and 228 fatty acid QTLs across all 20 chromosomes, which can be accessed in the SoyBase database (https://soybase.org/). However, recently a new study conducted by Yao et al. (2020) has led to the discovery of a novel locus for palmitic acid on chromosome 10, which is QTL, *qPA10_1* (5.94–9.98 Mb).

About 85% of the produced soybean is utilized to produce meal and vegetable oil. Though the large part of the produced meal is used in animal feed, 2% of the meal is processed into edible soy flours and proteins. Asian countries utilize approximately 6% of soybeans directly as human food. The soy oil produced is mostly used for human consumption (Oilseed and Grain News), and some foods are packed in soybean oil like pack of tuna (ncsoy.org).

Soybeans are also a rich source of phytochemicals (e.g., phytoestrogens) that are beneficial for human and animal health. Phytoestrogens are nonsteroidal compounds found in plants, demonstrating similar biologic activity – estrogenic and/or antiestrogenic activity (Salgado and Donado-Pestana 2011). The soy isoflavones (estrogens) act as potent antioxidants able to reduce the oxidation of LDL cholesterol and to induce vascular reactivity. According to some studies, soy isoflavones improve endothelial function and arterial relaxation (Seok et al. 2008; Salgado and Donado-Pestana 2011).

Soybean is enriched with vitamin K1, folate, molybdenum, copper, phosphorus, manganese, thiamin, phytic acid, and saponins. It also plays a role in the prevention of menopause, breast cancer, osteoporosis (Singh and Chung 2016), and prostate cancer by decreasing the risk of having the disease (Applegate et al. 2018).

Genetically modified soya bean is developed to give the plant resistance against pests and diseases, tolerance to herbicides, and an increase in the nutritional value of the plant (Rosculete et al. 2018). Approximately, 50% of 50,000 hectares in 18 biotech mega-countries were adopted by biotech soybean in 2018 (ISAAA 2018).

Soybean seed breeding and genetic engineering enhanced the nutritional and functional properties of soybean oil which makes it more beneficial in the different fields of its usage (Medic et al. 2014). One of the soybean oil applications is biodiesel production, which makes up about 50% of biodiesel feedstock. In 2016, $11.42 billion in US economic impact was contributed by the biodiesel industry (United Soybean Board).

In conclusion, soybean is an economically important seed with increasing demand worldwide. The genome sequencing efforts by researchers have led to several sequences, linkage maps, and ultimately newly developed varieties of the high-yielding soybean to fulfill the requirements.

References

Adams RL, Burdon RH (2012) Molecular biology of DNA methylation. Springer, Cham

Adams KL, Wendel JF (2005) Polyploidy and genome evolution in plants. Curr Opin Plant Biol 8(2):135–141

Alverson AJ, Wei X, Rice DW, Stern DB, Barry K, Palmer JD (2010) Insights into the evolution of mitochondrial genome size from complete sequences of *Citrullus lanatus* and *Cucurbita pepo* (Cucurbitaceae). Mol Biol Evol 27(6):1436–1448

Alverson AJ, Zhuo S, Rice DW, Sloan DB, Palmer JD (2011) The mitochondrial genome of the legume *Vigna radiata* and the analysis of recombination across short mitochondrial repeats. PLoS One 6(1):e16404

American Academy of Pediatrics. Committee on Nutrition (1998) Soy protein-based formulas: recommendations for use in infant feeding. Pediatrics 101(1 Pt 1):148

Andre C, Levy A, Walbot V (1992) Small repeated sequences and the structure of plant mitochondrial genomes. Trends Genet 8(4):128–132

Applegate CC, Rowles JL, Ranard KM, Jeon S, Erdman JW (2018) Soy consumption and the risk of prostate cancer: an updated systematic review and meta-analysis. Nutrients 10(1):40

Arumuganathan K, Earle ED (1991) Estimation of nuclear DNA content of plants by flow cytometry. Plant Mol Biol Report 9(3):229–241

Baker KM, Utterback PL, Parsons CM, Stein HH (2011) Nutritional value of soybean meal produced from conventional, high-protein, or low-oligosaccharide varieties of soybeans and fed to broiler chicks. Poult Sci 90(2):390–395

Beche E, Gillman JD, Song Q, Nelson R, Beissinger T, Decker J, Shannon G, Scaboo AM (2020) Nested association mapping of important agronomic traits in three interspecific soybean populations. Theor Appl Genet 133(3):1039–1054

Bennett JO, Krishnan AH, Wiebold WJ, Krishnan HB (2003) Positional effect on protein and oil content and composition of soybeans. J Agric Food Chem 51(23):6882–6886

Bennetzen JL, Wang H (2014) The contributions of transposable elements to the structure, function, and evolution of plant genomes. Annu Rev Plant Biol 65:505–530

Bentham G (1864) Flora Australiensis, vol 2. L. Reeve, London

Bentham G (1865) On the genera *Sweetia*, *Sprengel*, and *Glycine*, Linn., simultaneously published under the name of Leptolobium. Bot J Linn Soc 8(32):259–267

Blanc G, Wolfe KH (2004) Widespread paleopolyploidy in model plant species inferred from age distributions of duplicate genes. Plant Cell 16(7):1667–1678

Bolser D, Staines DM, Pritchard E, Kersey P (2016) Ensembl plants: integrating tools for visualizing, mining, and analyzing plant genomics data. In: Plant bioinformatics. Humana Press, New York, pp 115–140

Boore JL (1999) Animal mitochondrial genomes. Nucleic Acids Res 27(8):1767–1780

Boutin SR, Young ND, Olson T, Yu ZH, Vallejos CE, Shoemaker RC (1995) Genome conservation among three legume genera detected with DNA markers. Genome 38(5):928–937

Brozynska M, Furtado A, Henry RJ (2016) Genomics of crop wild relatives: expanding the gene pool for crop improvement. Plant Biotechnol J 14(4):1070–1085

Bruneau A, Mercure M, Lewis GP, Herendeen PS (2008) Phylogenetic patterns and diversification in the caesalpinioid legumes. Botany 86(7):697–718

Burton JW (1987) Quantitative genetics: results relevant to soybean breeding. In: Wilcox JR (ed) Soybeans: improvement, production, and uses, Agronomy (USA). ASA, CSSA, and SSSA, Madison

Cannon SB, Crow JA, Grant D (2012) SoyBase and the legume information system: accessing information about the soybean and other legume genomes. In: Designing soybeans for 21st century markets. AOCS Press, Urbana, pp 53–66

Carlson JB, Lersten NR (1987) Reproductive morphology. In: Wilcox JR (ed) Soybeans: improvement, production and uses. American Society of Agronomy, Madison, pp 95–134

Carter TE Jr, Nelson RL, Sneller CH, Cui Z (2004) Genetic diversity in soybean. In: Soybeans: improvement, production, and uses, vol 16. ASA, CSSA, and SSSA, Madison, pp 303–416
Chang S, Wang Y, Lu J, Gai J, Li J, Chu P, Guan R, Zhao T (2013) The mitochondrial genome of soybean reveals complex genome structures and gene evolution at intercellular and phylogenetic levels. PLoS One 8(2):e56502
Choi HK, Mun JH, Kim DJ, Zhu H, Baek JM, Mudge J, Roe B, Ellis N, Doyle J, Kiss GB, Young ND (2004) Estimating genome conservation between crop and model legume species. Proc Natl Acad Sci 101(43):15289–15294
Choi IY, Hyten DL, Matukumalli LK, Song Q, Chaky JM, Quigley CV, Chase K, Lark KG, Reiter RS, Yoon MS, Hwang EY (2007) A soybean transcript map: gene distribution, haplotype and single-nucleotide polymorphism analysis. Genetics 176(1):685–696
Chung G, Singh RJ (2008) Broadening the genetic base of soybean: a multidisciplinary approach. Crit Rev Plant Sci 27(5):295–341
Clarindo WR, De Carvalho CR, Alves BM (2007) Mitotic evidence for the tetraploid nature of *Glycine max* provided by high quality karyograms. Plant Syst Evol 265(1–2):101–107
Clemente TE, Cahoon EB (2009) Soybean oil: genetic approaches for modification of functionality and total content. Plant Physiol 151(3):1030–1040
Cregan PB, Jarvik T, Bush AL, Shoemaker RC, Lark KG, Kahler AL, Kaya N, VanToai TT, Lohnes DG, Chung J, Specht JE (1999) An integrated genetic linkage map of the soybean genome. Crop Sci 39(5):1464–1490
Cromwell GL (2012) Soybean meal: an exceptional protein source. Soybean Meal InfoCenter [Internet], Ankeny
de Coca-Sinova A, Valencia DG, Jiménez-Moreno E, Lázaro R, Mateos GG (2008) Apparent ileal digestibility of energy, nitrogen, and amino acids of soybean meals of different origin in broilers. Poult Sci 87(12):2613–2623
Diers BW, Keim P, Fehr WR, Shoemaker RC (1992) RFLP analysis of soybean seed protein and oil content. Theor Appl Genet 83(5):608–612
Doyle MJ, Brown AH (1985) Numerical analysis of isozyme variation in Glycine tomentella. Biochem Syst Ecol 13(4):413–419
Doyle MJ, Grant JE, Brown AH (1986) Reproductive isolation between isozyme groups of *Glycine tomentella* (Leguminosae), and spontaneous doubling in their hybrids. Aust J Bot 34(5):523–535
Doyle JJ, Doyle JL, Brown AH, Pfeil BE (2000) Confirmation of shared and divergent genomes in the Glycine tabacina polyploid complex (Leguminosae) using histone H3-D sequences. Syst Bot 1:437–448
Duke SO, Cerdeira AL (2010) Transgenic crops for herbicide resistance. In: Transgenic crop plants. Springer, Berlin/Heidelberg, pp 133–166
Dzikowski B (1936) Study of the soya bean *Glycine hispida* (Moench) Maxim. Part I. Morphology. Pamietnik Panstwowego Instytutu Naukowego Gospodarstwa Wiejskiego w Pulawach 16:69–100
Escalante EE, Wilcox JR (1993) Variation in seed protein among nodes of determinate and indeterminate soybean near-isolines. Crop Sci 33(6):1166–1168
FAO.org [Internet]. Food and Agriculture Organization of the United Nations; c2020 [cited 2020 May 20]. Available from: http://www.fao.org/land-water/databases-and-software/crop-information/soybean/en/
Frank SJ, Fehr WR (1981) Associations among pod dimensions and seed weight in soybeans 1. Crop Sci 21(4):547–550
Goodstein DM, Shu S, Howson R, Neupane R, Hayes RD, Fazo J, Mitros T, Dirks W, Hellsten U, Putnam N, Rokhsar DS (2012) Phytozome: a comparative platform for green plant genomics. Nucleic Acids Res 40(D1):D1178–D1186
Gray MW, Burger G, Lang BF (1999) Mitochondrial evolution. Science 283(5407):1476–1481

Grigoriev IV, Nordberg H, Shabalov I, Aerts A, Cantor M, Goodstein D, Kuo A, Minovitsky S, Nikitin R, Ohm RA, Otillar R (2012) The genome portal of the department of energy joint genome institute. Nucleic Acids Res 40(D1):D26–D32

Ha J, Jeon HH, Woo DU, Lee Y, Park H, Lee J, Kang YJ (2019) Soybean-VCF2Genomes: a database to identify the closest accession in soybean germplasm collection. BMC Bioinf 20(13):1–7

Harlan JR, de Wet JM (1971) Toward a rational classification of cultivated plants. Taxon 20(4):509–517

Hermann FJ (1962) A revision of the genus Glycine and its immediate allies. US Department of Agriculture

Heuzé V, Tran G, Noziere P, Lessire M, Lebas F (2020) Linseed meal. Feedipedia, a programme by INRA, CIRAD, AFZ and FAO. https://feedipedia.org/node/674. Last updated on March 4, 2020

Hwang TY, Sayama TA, Takahashi MA, Takada Y, Nakamoto YU, Funatsuki HI, Hisano H, Sasamoto S, Sato S, Tabata S, Kono I (2009) High-density integrated linkage map based on SSR markers in soybean. DNA Res 16(4):213–225

Hymowitz T, Newell CA (1981) Taxonomy of the genus *Glycine*, domestication and uses of soybeans. Econ Bot 35(3):272–288

Hymowitz T, Singh RJ, Kollipara KP (1998) The genomes of the Glycine. Plant Breed Rev 16:289–318

Hyten DL, Song Q, Zhu Y, Choi IY, Nelson RL, Costa JM, Specht JE, Shoemaker RC, Cregan PB (2006) Impacts of genetic bottlenecks on soybean genome diversity. Proc Natl Acad Sci 103(45):16666–16671

Hyten DL, Choi IY, Song Q, Specht JE, Carter TE, Shoemaker RC, Hwang EY, Matukumalli LK, Cregan PB (2010) A high density integrated genetic linkage map of soybean and the development of a 1536 universal soy linkage panel for quantitative trait locus mapping. Crop Sci 50(3):960–968

Imelfort M, Edwards D (2009) De novo sequencing of plant genomes using second-generation technologies. Brief Bioinform 10(6):609–618

International Cassava Genetic Map Consortium (2015) High-resolution linkage map and chromosome-scale genome assembly for cassava (Manihot esculenta Crantz) from 10 populations. G3 Genes Genomes Genet 5(1):133–144

ISAAA.org [Internet]. International Service for the Acquisition of Agri-biotech Applications, ISAAA Brief 54-2018: Executive Summary; c2020 [cited 2020 Jun 22]. Available from: https://www.isaaa.org/resources/publications/briefs/54/executivesummary/default.asp

Jarvis DE, Ho YS, Lightfoot DJ, Schmöckel SM, Li B, Borm TJ, Ohyanagi H, Mineta K, Michell CT, Saber N, Kharbatia NM (2017) The genome of Chenopodium quinoa. Nature 542(7641):307

Jobin-Décor MP, Graham GC, Henry RJ, Drew RA (1997) RAPD and isozyme analysis of genetic relationships between *Carica papaya* and wild relatives. Genet Resour Crop Evol 44(5):471–477

Johnson L, Smith K (2011) Soybean processing. Fact sheet. The Soybean Meal Information Center 19:2011

Joshi T, Patil K, Fitzpatrick MR, Franklin LD, Yao Q, Cook JR, Wang Z, Libault M, Brechenmacher L, Valliyodan B, Wu X (2012) Soybean Knowledge Base (SoyKB): a web resource for soybean translational genomics. BMC Genomics 13(S1):S15. BioMed Central

Joy PP, Thomas J, Mathew S, Skaria BP (1998) Medicinal plants. Kerala Agricultural University. Aromatic and Medicinal Plants Research Station, pp 4–6

Kansas Soybean Commission. https://kansassoybeans.org/about-the-checkoff/animal-ag/

Keim P, Beavis W, Schupp J, Freestone R (1992) Evaluation of soybean RFLP marker diversity in adapted germplasm. Theor Appl Genet 85(2–3):205–212

Kim E, Hwang S, Lee I (2017) SoyNet: a database of co-functional networks for soybean *Glycine max*. Nucleic Acids Res 45(D1):D1082–D1089

Kubo T, Newton KJ (2008) Angiosperm mitochondrial genomes and mutations. Mitochondrion 8(1):5–14

Lackey JA (1977a) A synopsis of Phaseoleae (Leguminosae, Papilionoideae). PhD dissertation, Iowa State University, Ames

Lackey JA (1977b) A revised classification of the tribe Phaseoleae (Leguminosae: Papilionoideae), and its relation to canavanine distribution. Bot J Linn Soc 74(2):163–178

Lackey JA (1977c) *Neonotonia*, a new generic name to include *Glycine wightii* (Arnott) Verdcourt (Leguminosae, Papilionoideae). Phytologia 37(3):209–212

Lackey JA (1980) Chromosome numbers in the Phaseoleae (Fabaceae: Faboideae) and their relation to taxonomy. Am J Bot 67(4):595–602

Lam HM, Xu X, Liu X, Chen W, Yang G, Wong FL, Li MW, He W, Qin N, Wang B, Li J (2010) Resequencing of 31 wild and cultivated soybean genomes identifies patterns of genetic diversity and selection. Nat Genet 42(12):1053–1059

Lang BF, Gray MW, Burger G (1999) Mitochondrial genome evolution and the origin of eukaryotes. Annu Rev Genet 33(1):351–397

Lark KG, Weisemann JM, Matthews BF, Palmer R, Chase K, Macalma T (1993) A genetic map of soybean (*Glycine max* L.) using an intraspecific cross of two cultivars: 'Minosy' and 'Noir 1'. Theor Appl Genet 86(8):901–906

Lee JM, Grant D, Vallejos CE, Shoemaker RC (2001) Genome organization in dicots. II. Arabidopsis as a 'bridging species' to resolve genome evolution events among legumes. Theor Appl Genet 103(5):765–773

Lee JD, Bilyeu KD, Pantalone VR, Gillen AM, So YS, Shannon JG (2012) Environmental stability of oleic acid concentration in seed oil for soybean lines with FAD2-1A and FAD2-1B mutant genes. Crop Sci 52(3):1290–1297

Lee K, Kim MS, Lee JS, Bae DN, Jeong N, Yang K, Lee JD, Park JH, Moon JK, Jeong SC (2020) Chromosomal features revealed by comparison of genetic maps of *Glycine max* and *Glycine soja*. Genomics 112(2):1481–1489

Lewis G, Schrire B, Mackinder B, Lock M (2005) Legumes of the world. Royal Botanic Gardens, Kew, London

Li YH, Li W, Zhang C, Yang L, Chang RZ, Gaut BS, Qiu LJ (2010) Genetic diversity in domesticated soybean (*Glycine max*) and its wild progenitor (*Glycine soja*) for simple sequence repeat and single-nucleotide polymorphism loci. New Phytol 188(1):242–253

Li YH, Zhao SC, Ma JX, Li D, Yan L, Li J, Qi XT, Guo XS, Zhang L, He WM, Chang RZ (2013) Molecular footprints of domestication and improvement in soybean revealed by whole genome re-sequencing. BMC Genomics 14(1):1–2

Li YH, Zhou G, Ma J, Jiang W, Jin LG, Zhang Z, Guo Y, Zhang J, Sui Y, Zheng L, Zhang SS (2014) De novo assembly of soybean wild relatives for pan-genome analysis of diversity and agronomic traits. Nat Biotechnol 32(10):1045–1052

Li M, Liu Y, Tao Y, Xu C, Li X, Zhang X, Han Y, Yang X, Sun J, Li W, Li D (2019) Identification of genetic loci and candidate genes related to soybean flowering through genome wide association study. BMC Genomics 20(1):987

Liu B, Watanabe S, Uchiyama T, Kong F, Kanazawa A, Xia Z, Nagamatsu A, Arai M, Yamada T, Kitamura K, Masuta C (2010) The soybean stem growth habit gene Dt1 is an ortholog of Arabidopsis TERMINAL FLOWER1. Plant Physiol 153(1):198–210

Liu Z, Li H, Wen Z, Fan X, Li Y, Guan R, Guo Y, Wang S, Wang D, Qiu L (2017) Comparison of genetic diversity between Chinese and American soybean (Glycine max (L.)) accessions revealed by high-density SNPs. Front Plant Sci 8:2014

Lohse M, Drechsel O, Bock R (2007) OrganellarGenomeDRAW (OGDRAW): a tool for the easy generation of high-quality custom graphical maps of plastid and mitochondrial genomes. Curr Genet 52(5–6):267–274

Mao T, Jiang Z, Han Y, Teng W, Zhao X, Li W (2013) Identification of quantitative trait loci underlying seed protein and oil contents of soybean across multi-genetic backgrounds and environments. Plant Breed 132(6):630–641

Mascher M, Gundlach H, Himmelbach A, Beier S, Twardziok SO, Wicker T, Radchuk V, Dockter C, Hedley PE, Russell J, Bayer M (2017) A chromosome conformation capture ordered sequence of the barley genome. Nature 544(7651):427–433

Medic J, Atkinson C, Hurburgh CR (2014) Current knowledge in soybean composition. J Am Oil Chem Soc 91(3):363–384

Menniti MF, Davenport GM, Shoveller AK, Cant JP, Osborne VR (2014) Effect of graded inclusion of dietary soybean meal on nutrient digestibility, health, and metabolic indices of adult dogs. J Anim Sci 92(5):2094–2104

Miladinović J, Đorđević V (2011) Soybean morphology and stages of development. In: Miladinović J, Hrustić M, Vidić M (eds) Soybean. Institute of Field and Vegetable Crops, Novi Sad and Sojaprotein, Bečej, Grafika, Novi Sad, pp 45–71

Mudge J, Huihuang Y, Denny RL, Howe DK, Danesh D, Marek LF, Retzel E, Shoemaker RC, Young ND (2004) Soybean bacterial artificial chromosome contigs anchored with RFLPs: insights into genome duplication and gene clustering. Genome 47(2):361–372

ncsoy.org [Internet]. North Carolina Soybean Production Association. Uses of Soybeans; c2019 [cited 2020 May 20]. Available from: https://ncsoy.org/media-resources/uses-of-soybeans/

Newell CA, Hymowitz T (1980) A taxonomic revision in the genus *Glycine* subgenus *Glycine* (Leguminosae). Brittonia 32(1):63–69

Newton KJ, Gabay-Laughnan S, De Paepe R (2004) Mitochondrial mutations in plants. In: Plant mitochondria: from genome to function. Springer, Dordrecht, pp 121–141

Nordberg H, Cantor M, Dusheyko S, Hua S, Poliakov A, Shabalov I, Smirnova T, Grigoriev IV, Dubchak I (2014) The genome portal of the Department of Energy Joint Genome Institute: 2014 updates. Nucleic Acids Res 42(D1):D26–D31

Nutrition data [Internet]. Know what you eat. c2018 [cited 2020 Jun 20]. Available from: https://nutritiondata.self.com/facts/legumes-and-legume-products/4384/2

Oil World [Internet]. Hamburg: ISTA Mielke GmbH; Oil World Annual [cited 2020 May 30]. Available from: https://www.oilworld.biz/

Oilseed and grain news [Internet]. Organic and Non-GMO Forum; c2017 [cited 2020 Jun 1]. Available from: https://www.oilseedandgrain.com/soy-facts

Owen FV (1927) Inheritance studies in soybeans. II. Glabrousness, color of pubescence, time of maturity, and linkage relations. Genetics 12(6):519

Padgette SR, Kolacz KH, Delannay X, Re DB, LaVallee BJ, Tinius CN, Rhodes WK, Otero YI, Barry GF, Eichholtz DA, Peschke VM (1995) Development, identification, and characterization of a glyphosate-tolerant soybean line. Crop Sci 35(5):1451–1461

Palmer JD (1991) Plastid chromosomes: structure and evolution. Mol Biol Plast 7:5–3

Palmer JD, Herbo LA (1987) Unicircular structure of the *Brassica hirta* mitochondrial genome. Curr Genet 11(6–7):565–570

Palmer RG, Hymowitz T (2016) Soybean: germplasm, breeding, and genetics. In: Reference module in food science. Elsevier, Amsterdam, pp 333–342

Parsons CM, Hashimoto K, Wedekind KJ, Baker DH (1991) Soybean protein solubility in potassium hydroxide: an in vitro test of in vivo protein quality. J Anim Sci 69(7):2918–2924

Parsons CM, Zhang Y, Araba M (2000) Nutritional evaluation of soybean meals varying in oligosaccharide content. Poult Sci 79(8):1127–1131

Pathan SM, Vuong T, Clark K, Lee JD, Shannon JG, Roberts CA, Ellersieck MR, Burton JW, Cregan PB, Hyten DL, Nguyen HT (2013) Genetic mapping and confirmation of quantitative trait loci for seed protein and oil contents and seed weight in soybean. Crop Sci 53(3):765–774

Pfeil BE, Craven LA (2002) New taxa in *Glycine* (Fabaceae: Phaseolae) from north-western Australia. Aust Syst Bot 15(4):565–573

Pfeil BE, Tindale MD, Craven LA (2001) A review of the *Glycine clandestina* species complex (Fabaceae: Phaseolae) reveals two new species. Aust Syst Bot 14(6):891–900

Pfeil BE, Craven LA, Brown AH, Murray BG, Doyle JJ (2006) Three new species of northern Australian *Glycine* (Fabaceae, Phaseolae), *G. gracei, G. montis-douglas* and *G. syndetika*. Aust Syst Bot 19(3):245–258

Phytozome [Internet]. The plant comparative genomics portal of the Department of Energy's Joint Genome Institute. c1997-2017 [cited 2020 May 5]. Available from: https://phytozome.jgi.doe.gov/pz/portal.html#!info?alias=Org_Gmax

Ratnaparkhe MB, Singh RJ, Doyle JJ (2011) Glycine, pp 83–116. In: Kole C (ed) Wild crop relatives: genomic and breeding resources. Springer, Berlin/Heidelberg, pp 83–116

Raubeson LA, Jansen RK (2005) Chloroplast genomes of plants. In: Henry R (ed) Plant diversity and evolution: genotypic variation in higher plants. CABI Publishing, Wallingford, pp 45–68

Rizzo G, Baroni L (2018) Soy, soy foods and their role in vegetarian diets. Nutrients 10(1):43

Rosculete E, Bonciu E, Rosculete CA, Teleanu E (2018) Detection and quantification of genetically modified soybean in some food and feed products. A case study on products available on Romanian market. Sustainability 10(5):1325

Salgado JM, Donado-Pestana CM (2011) Soy as a functional food. In: El-Shemy HA (ed) Soybean and nutrition. InTech, Croatia, pp 21–44

Saski C, Lee SB, Daniell H, Wood TC, Tomkins J, Kim HG, Jansen RK (2005) Complete chloroplast genome sequence of *Glycine max* and comparative analyses with other legume genomes. Plant Mol Biol 59(2):309–322

Schmutz J, Cannon SB, Schlueter J, Ma J, Mitros T, Nelson W, Hyten DL, Song Q, Thelen JJ, Cheng J, Xu D (2010) Genome sequence of the palaeopolyploid soybean. Nature 463(7278):178–183

Schwender J, Ohlrogge JB, Shachar-Hill Y (2003) A flux model of glycolysis and the oxidative pentosephosphate pathway in developing *Brassica napus* embryos. J Biol Chem 278(32):29442–29453

Seok YM, Baek I, Kim YH, Jeong YS, Lee IJ, Shin DH, Hwang YH, Kim IK (2008) Isoflavone attenuates vascular contraction through inhibition of the RhoA/Rho-kinase signaling pathway. J Pharmacol Exp Ther 326(3):991–998

Shekhar HU, Howlader ZH, Kabir Y (eds) (2016) Exploring the nutrition and health benefits of functional foods. IGI Global, Hershey

Shen Y, Liu J, Geng H, Zhang J, Liu Y, Zhang H, Xing S, Du J, Ma S, Tian Z (2018) De novo assembly of a Chinese soybean genome. Sci China Life Sci 61(8):871–884

Shen Y, Du H, Liu Y, Ni L, Wang Z, Liang C, Tian Z (2019) Update soybean Zhonghuang 13 genome to a golden reference. Sci China Life Sci 62:1257–1260

Sherman-Broyles S, Bombarely A, Powell AF, Doyle JL, Egan AN, Coate JE, Doyle JJ (2014) The wild side of a major crop: Soybean's perennial cousins from Down Under. Am J Bot 101(10):1651–1665

Shi J, Ma X, Zhang J, Zhou Y, Liu M, Huang L, Sun S, Zhang X, Gao X, Zhan W, Li P (2019) Chromosome conformation capture resolved near complete genome assembly of broomcorn millet. Nat Commun 10(1):1–9

Shibata M, Takayama K, Ujiie A, Yamada T, Abe J, Kitamura K (2008) Genetic relationship between lipid content and linolenic acid concentration in soybean seeds. Breed Sci 58(4):361–366

Shoemaker RC, Specht JE (1995) Integration of the soybean molecular and classical genetic linkage groups. Crop Sci 35(2):436–446

Shoemaker RC, Polzin K, Labate J, Specht J, Brummer EC, Olson T, Young N, Concibido V, Wilcox J, Tamulonis JP, Kochert G (1996) Genome duplication in soybean (*Glycine* subgenus *soja*). Genetics 144(1):329–338

Shoemaker R, Keim P, Vodkin L, Retzel E, Clifton SW, Waterston R, Smoller D, Coryell V, Khanna A, Erpelding J, Gai X (2002) A compilation of soybean ESTs: generation and analysis. Genome 45(2):329–338

Shurtleff W, Aoyagi A (2009) History of soybeans and soyfoods in South America (1884–2009): extensively annotated bibliography and sourcebook. Soyinfo Center

Shurtleff W, Aoyagi A (2012) History of soynuts, soynut butter, Japanese-style roasted soybeans (Irimame) and Setsubun (with Mamemaki) (1068–2012): extensively annotated bibliography and sourcebook. Soyinfo Center

Singh RJ, inventor; University of Illinois, assignee (2010) Methods for producing fertile crosses between wild and domestic soybean species. United States patent US 7,842,850

Singh RJ, Chung GH (2016) Landmark research for pulses improvement. Indian J Genet 76(4):399–409

Singh RJ, Hymowitz T (1985) The genomic relationships among six wild perennial species of the genus *Glycine* subgenus *Glycine* Willd. Theor Appl Genet 71(2):221–230

Singh RJ, Hymowitz T (1988) The genomic relationship between *Glycine max* (L.) Merr. and *G. soja* Sieb. and Zucc. as revealed by pachytene chromosome analysis. Theor Appl Genet 76(5):705–711

Sloan DB, Alverson AJ, Chuckalovcak JP, Wu M, McCauley DE, Palmer JD, Taylor DR (2012) Rapid evolution of enormous, multichromosomal genomes in flowering plant mitochondria with exceptionally high mutation rates. PLoS Biol 10(1):e1001241

Song Q, Cregan PB (2017) Classical and molecular genetic mapping. In: The soybean genome. Springer, Cham, pp 41–56

Song QJ, Marek LF, Shoemaker RC, Lark KG, Concibido VC, Delannay X, Specht JE, Cregan PB (2004) A new integrated genetic linkage map of the soybean. Theor Appl Genet 109(1):122–128

Song Q, Jenkins J, Jia G, Hyten DL, Pantalone V, Jackson SA, Schmutz J, Cregan PB (2016) Construction of high resolution genetic linkage maps to improve the soybean genome sequence assembly Glyma1. 01. BMC Genomics 17(1):33

Soystats.org [Internet]. The American Soybean Association; c2020 [cited 2020 Jun 1]. Available from: http://soystats.com/international-world-soybean-production/

Spencer MM, Landau-Ellis D, Meyer EJ, Pantalone VR (2004) Molecular markers associated with linolenic acid content in soybean. J Am Oil Chem Soc 81(6):559–562

Tateishi Y, Ohashi H (1992) Taxonomic studies on Glycine of Taiwan. J Jpn Bot 67:127–147

Teh BT, Lim K, Yong CH, Ng CC, Rao SR, Rajasegaran V, Lim WK, Ong CK, Chan K, Cheng VK, Soh PS (2017) The draft genome of tropical fruit durian (Durio zibethinus). Nat Genet 49(11):1633–1641

Tian AG, Wang J, Cui P, Han YJ, Xu H, Cong LJ, Huang XG, Wang XL, Jiao YZ, Wang BJ, Wang YJ (2004) Characterization of soybean genomic features by analysis of its expressed sequence tags. Theor Appl Genet 108(5):903–913

Tindale MD (1984) Two new eastern Australian species of *Glycine* Willd. (Fabaceae). Brunonia 7(1):207–213

Tindale MD (1986a) A new north Queensland species of *Glycine* Willd. (Fabaceae). Brunonia 9(1):99–103

Tindale MD (1986b) Taxonomic notes on three Australian and Norfolk Island species of *Glycine* Willd. (Fabaceae: Phaseolae) including the choice of a Neotype for *G. clandestina* Wendl. Brunonia 9(2):179–191

Tindale MD, Craven LA (1988) Three new species of *Glycine* (Fabaceae: Phaseolae) from north-western Australia, with notes on amphicarpy in the genus. Aust Syst Bot 1(4):399–410

Tindale MD, Craven LA (1993) *Glycine pindanica* (Fabaceae, Phaseolae), a new species from west Kimberley, Western Australia. Aust Syst Bot 6(4):371–376

United Soybean Board (USB) [Internet] [cited 2020 Jun 15]. Available from: https://www.unitedsoybean.org/media-center/issue-briefs/biodiesel/

USDA F. Oilseeds: world markets and trade (cited 2020 May 27). Available from: https://apps.fas.usda.gov/psdonline/circulars/oilseeds.pdf

Valliyodan B, Lee SH, Nguyen HT (2017) Sequencing, assembly, and annotation of the soybean genome. In: The soybean genome. Springer, Cham, pp 73–82

Van Wyk B-E (2005) Food plants of the world: an illustrated guide. Timber Press, Portland

Verdcourt B (1966) A proposal concerning *Glycine* L. Taxon 1:34–36

Verdcourt B (1970) Studies in the Leguminosae-Papilionoïdeae for the 'Flora of Tropical East Africa': III. Kew Bulletin, pp 379–447

Wakasugi T, Tsudzuki J, Ito S, Nakashima K, Tsudzuki T, Sugiura M (1994) Loss of all ndh genes as determined by sequencing the entire chloroplast genome of the black pine Pinus thunbergii. Proc Natl Acad Sci 91(21):9794–9798

Wang Z, Tian Z (2015) Genomics progress will facilitate molecular breeding in soybean. Sci China Life Sci 58(8):813–815
Wang X, Wang Y, Tian J, Lim BL, Yan X, Liao H (2009) Overexpressing AtPAP15 enhances phosphorus efficiency in soybean. Plant Physiol 151(1):233–240
Wang J, Chen P, Wang D, Shannon G, Shi A, Zeng A, Orazaly M (2015) Identification of quantitative trait loci for oil content in soybean seed. Crop Sci 55(1):23–34
Wang C, Hu S, Gardner C, Lübberstedt T (2017) Emerging avenues for utilization of exotic germplasm. Trends Plant Sci 22(7):624–637
Wang N, Yuan M, Chen H, Li Z, Zhang M (2019a) Effects of drought stress and rewatering on growth and physiological characteristics of invasive Aegilops tauschii seedlings. Acta Pratacul Sin 28(1):70–78
Wang X, Chen L, Ma J (2019b) Genomic introgression through interspecific hybridization counteracts genetic bottleneck during soybean domestication. Genome Biol 20(1):1–5
Wei L, Cao X (2016) The effect of transposable elements on phenotypic variation: insights from plants to humans. Sci China Life Sci 59(1):24–37
Wheeler DL, Barrett T, Benson DA, Bryant SH, Canese K, Chetvernin V, Church DM, DiCuccio M, Edgar R, Federhen S, Feolo M (2007) Database resources of the national center for biotechnology information. Nucleic Acids Res 36(Suppl_1):D13–D21
Wisner R (2014) Soybean oil and biodiesel usage projection balance sheet. Iowa State University Extension
Xie M, Chung CY, Li MW, Wong FL, Wang X, Liu A, Wang Z, Leung AK, Wong TH, Tong SW, Xiao Z (2019) A reference-grade wild soybean genome. Nat Commun 10(1):1–2
Yanagisawa T, Tano S, Fukui K, Harada K (1991) Marker chromosomes commonly observed in the genus *Glycine*. Theor Appl Genet 81(5):606–612
Yao Y, You Q, Duan G, Ren J, Chu S, Zhao J, Li X, Zhou X, Jiao Y (2020) Quantitative trait loci analysis of seed oil content and composition of wild and cultivated soybean. BMC Plant Biol 20(1):51
Zabala G, Vodkin LO (2007) A rearrangement resulting in small tandem repeats in the F3′ 5′ H gene of white flower genotypes is associated with the soybean W1 locus. Crop Sci 47:S-113
Zhang D, Sun L, Li S, Wang W, Ding Y, Swarm SA, Li L, Wang X, Tang X, Zhang Z, Tian Z (2018) Elevation of soybean seed oil content through selection for seed coat shininess. Nat Plant 4(1):30–35
Zhou Z, Jiang Y, Wang Z, Gou Z, Lyu J, Li W, Yu Y, Shu L, Zhao Y, Ma Y, Fang C (2015) Resequencing 302 wild and cultivated accessions identifies genes related to domestication and improvement in soybean. Nat Biotechnol 33(4):408–414
Zou JJ, Singh RJ, Lee J, Xu SJ, Cregan PB, Hymowitz T (2003) Assignment of molecular linkage groups to soybean chromosomes by primary trisomics. Theor Appl Genet 107(4):745–750

Chapter 2
Overview and Application of Soybean Genomics Study

Rong Li, Haifeng Chen, Songli Yuan, and Xinan Zhou

Contents

Abbreviations

2-DEG	two-dimensional gel electrophoresis
CNV	copy number variations
DMRs	differentially methylated regions
GEBV	genomic estimated breeding value
GS	Genomic selection
GWAS	Genome-wide association study
KTI	Kunitz trypsin inhibitor
MAS	marker-assisted selection
MCFS	moisture content of fresh seeds
MS/MS	tandem mass spectrometry

R. Li · H. Chen · S. Yuan (✉) · X. Zhou (✉)
Key Laboratory of Biology and Genetic Improvement of Oil Crops, Ministry of Agriculture and Rural Affairs of PRC, Oil Crops Research Institute of Chinese Academy of Agriculture Sciences, Wuhan, China

H. Tombuloglu et al. (eds.), *Oil Crop Genomics*,
https://doi.org/10.1007/978-3-030-70420-9_2

MSH1	MutS HOMOLOG 1
PAV	presence–absence variations
PFW	pod fresh weight
PTMs	post-translational modifications
SCN	Soybean cyst nematode
TALEN	transcription activator–like effector-based nucleases
TMTs	tandem mass tags
WGBS	whole-genome bisulfite sequencing
WGS	whole-genome sequencing
ZFN	zinc finger nuclease

2.1 Introduction

The cultivated soybean originated from annual wild soybean (*Glycine soja*) in China and Eastern Asia over 3000–~5000 years ago (Hymowitz 1970; Jeong et al. 2019). Soybean is valued mainly for the high protein and oil component, which can be used for food and livestock feed. Moreover, soybean oil also can be served as industry materials such as biofuel, lubricants, and plastics. Previous reports suggested that global soybean production needs to be doubled by 2050 to satisfy the increasing food consumption and industry demand (Tilman et al. 2011; Foley et al. 2011; Ray et al. 2013). Consequently, improving yield and quality have been the primary objective in soybean research and breeding programs.

Nowadays, China, other Asian countries, Europe, South America, and the United States are the main countries for soybean cultivation. The soybean yield of major countries in Asia (China, Korea, Vietnam, and Japan) have grown gradually since 1961, but reaching a plateau in recent years (Li et al. 2020; Ainsworth et al. 2012). The statistical data from the Food and Agriculture Organization of the United Nations revealed that the USA surpassed greatly those countries in soybean yield, meaning a huge space for improvement (Li et al. 2020). Together with the increasing demand for soybean production, innovative breeding technology in soybean is becoming more and more urgent.

Soybean growth and yields are affected by adverse environment such as drought, flood, salt, and temperature (Li et al. 2017; Boyer 1982; Maruyama et al. 2017). Recently, owing to the improvement of the next-generation sequencing technologies, cultivated soybean and Wild soybean whole-genome sequencing programs have been done (Schmutz et al. 2010; Lam et al. 2010; Qi et al. 2014; Xie et al. 2019), which providing abundant genetic resources for soybean breeding research. Besides, other omics studies (transcriptomics, proteomics, and epigenomics) have advanced traits studies related to yield and stress adaptation (Wang and Tian 2015; Li et al. 2017). These traits would promote the breeding of soybean with adaptability, quality, and productivity in the future.

In this chapter, we described the genomic sequencing programs and the achievements of other omics progresses in soybean in recent years. We also discuss some analysis methods and breeding strategies, including genome editing, genome-wide association study, and genomic selection, which have contributed greatly to genomic functional analysis and breeding in soybean.

2.2 Available Soybean Genomic Information

2.2.1 Cultivated Soybean Genome

Since the first soybean reference genome was sequenced and released in 2010 (Schmutz et al. 2010), the remarkable progress in soybean genomic studies over the past 10 years have generated a large quantity of data (Lam et al. 2010; Kim et al. 2010; Chung et al. 2014; Li et al. 2014; Qiu et al. 2014; Qi et al. 2014; Zhou et al. 2015; Shimomura et al. 2015; Shen et al. 2018a). The soybean Williams 82, which is a cultivated soybean cultivar from North America, was sequenced and assembled (Schmutz et al. 2010). Soybean (*Glycine max*) is paleopolyploid and the Williams 82 has a 1.1 Gb genome size. The number of predicted genes is 46,430, of which 78% occur in chromosome ends. The genome data of Williams 82 are available in Phytozome database (https://phytozome.jgi.doe.gov/pz/portal.html). The disclosure of Williams 82 genomic information was the milestone of soybean genomics, and it laid the cornerstone for the following functional genomics (Chan et al. 2012).

Previous studies have suggested that cultivated soybean from different geographic areas tend to show genetic diversity (Zhou et al. 2015), and one single reference genome can't represent all the genetic diversity of a species (Scherer et al. 2007; Golicz et al. 2016; Zhou et al. 2017; Hurgobin et al. 2018), so Williams 82 isn't enough to represent genetic information of different soybean varieties from different regions. Subsequent resequencing studies in China and Japan complemented soybean genomic studies (Lam et al. 2010; Shimomura et al. 2015; Shen et al. 2018a). The Japanese cultivated soybean Enrei, which is widely used in Japanese soybean research, was resequenced and assembled, and the sequencing data are available in DAIZUbase (http://daizu.dna.affrc.go.jp/enrei/). About 928 Mbs total bases and 60,838 gene models were obtained in Enrei genome, and the phylogenetic analysis exposed the ancestral relationships between Enrei and Williams 82. Besides, a high-quality reference genome of Zhonghuang 13, which has widespread planting areas in China, was released (Shen et al. 2018a). The Zhonghuang 13 genome has 1.025 Gb genome size, 52,051 protein coding genes and 36,429 transposable elements. Genome comparison analysis between Zhonghuang 13 and William 82 was made to discover genetic variations in both varieties.

2.2.2 *Wild Soybean Genome*

The genetic resources of wild soybean (*Glycine soja*) in China are valuable for soybean breeding and improvement. A total of 17 wild soybeans and 14 cultivated soybeans were resequenced in 2010, and comparative analysis of these wild and cultivated soybeans showed that wild soybeans have higher genetic diversities (Lam et al. 2010). A Korea wild soybean IT182932 was resequenced and assembled, a 3.76% difference between IT182932 and Williams 82 genomic sequence was found by mapping the resequencing reads to the cultivated Williams 82 reference genome. W05, a wild soybean from China, was performed whole-genome de novo sequencing (Qi et al. 2014) and state-of-the-art sequencing (Xie et al. 2019). In 2014, 10 semi-wild soybeans with a wild line were sequenced and analyzed, which exposed the hybridization origin of semi-wild and the high heterozygous rate (Qiu et al. 2014).

The characteristics of high allelic diversity lead to the wild soybean as important germplasm resource, and this wild soybean genomics can provide useful information for identifying functional genes. For example, comparing W05 genome with Williams 82 and Zhonghuang 13 genome, QTLs associated with yield components were identified. And these comparative analyses also revealed that soybean domestication may be involved with the copy number variations of Kunitz trypsin inhibitor (*KTI*) genes (Xie et al. 2019). Besides, a novel salt-tolerance gene GmCHX1 was identified from the W05 genome (Qi et al. 2014).

2.2.3 *Pan-Genome of Soybean*

Pan-genome study is a relatively new biological technology, and pan-genome construction and analysis rose gradually in recent years (Hirsch et al. 2014; Zhao et al. 2018; Tao et al. 2019; Garrison et al. 2018). Pan-genome analysis can uncover a great number of genetic variations such as presence–absence variations (PAV) and copy number variations (CNV) associating with agronomic traits (Hufford et al. 2012; Golicz et al. 2016; Li et al. 2014). In soybean, the first pan-genome was constructed in 2014 for seven *G. soja* soybeans, which originated from China, Japan, Korea, and Russia and were sequenced by HiSeq2000 and assembled using SOAPdenovo. This pan-genome analysis identified lineage-specific genes and copy number variations or large-effect mutations, which may be associated with agronomic traits such as biotic resistance, seed composition, flowering and maturity time, organ size, and final biomass (Li et al. 2014). Besides, a high-quality graph-based soybean pan-genome was constructed by assembling 26 representative wild and cultivated soybean accessions (Liu et al. 2020), providing a very important resource and platform for soybean functional genomics studies. By mapping against and comparing with genome Gmax_ZH13 (Shen et al. 2019) and Wm82, numerous sequence variations were identified. These studies will facilitate functional genomics and breeding in soybean.

2.3 Functional Genomics

The sequence information of the soybean reference genome was published in 2010, which open a new era of functional genomics studies in soybean. Functional genomics builds bridges between genome and traits/phenotypes. In recent years, with the application of omics technology in soybean functional genomics, some important genes related to significant agronomic traits were identified and studied, and the findings from these reports are probably to have application value in soybean breeding (Xia et al. 2013; Liu et al. 2015; Tran and Mochida 2010; Wang and Tian 2015).

2.3.1 Transcriptomics

The soybean whole-genome sequencing programs have helped facilitating in soybean transcriptome studies (Libault et al. 2010; Severin et al. 2010; Garg and Jain 2013), and transcriptome studies can disclose the gene expression level in various tissues and different developmental stages, which will help characterizing specific genes and defining their function in soybean. Therefore, transcriptomics studies can provide significant information for genome annotation.

Microarrays have been used to study transcriptomics. Nowadays, the whole transcriptome sequencing (RNA-Seq), which is a convenient and rapid method, has become a popular technology to study transcriptome and predict putative gene function (Garg and Jain 2013). The first two RNA-Seq studies in soybean were announced in 2010 (Libault et al. 2010; Severin et al. 2010), followed by a lot of RNA-Seq studies in soybean across multiple tissues and different developmental stages (Yim et al. 2015; Yuan et al. 2016, 2017; Zeng et al. 2018). The related transcriptome information of soybean was available in Soybase (https://soybase.org/soyseq/). In a recent systematic analysis of 1298 RNA-Seq samples, a comprehensive soybean transcriptome atlas was constructed and a web interface (http://venanciogroup.uenf.br/resources/) was made available for the community (Machado et al. 2020). Numerous tissue-specific genes were generated by comparing the global expression patterns of different tissues, especially for nodules-specific and endosperm-specific genes, which have contributed close to half the tissue-specific genes. Take nodule-specific genes as an example, 2 leghemoglobin and 10 nodulin genes were identified in Machado's study (Machado et al. 2020). Besides, other transcriptomics studies have provided candidate genes related to nodule development and/or nodule senescence (Yuan et al. 2017; Zhang et al. 2020a).

2.3.2 Proteomics

The proteomics study is more challenging and complex than transcriptomics, but proteomics research can provide more worthy functional analysis for soybean. Two-dimensional gel electrophoresis (2-DEG), tandem mass spectrometry (MS/MS), multidimensional liquid chromatography, label-free (quantifying peptide ion intensities or spectral counting), and tandem mass tags (TMTs) have become conventional techniques in soybean proteomics (Min et al. 2019; Komatsu et al. 2015; Hossain and Komatsu 2014).

The proteomics study in soybean functional analysis mainly focused on the comparative analyses of protein abundance in soybean under stress environment (Wang et al. 2020), the changes of protein expression profiling in various tissues and different developmental stages (Wang and Komatsu 2018), and the subcellular proteomics of soybean in response to stresses and in developmental stages (Wang and Komatsu 2016; Komatsu and Hashiguchi 2018). A review of proteomics in legume symbiosis revealed that symbiosis can improve stress-tolerance and disease-resistance (Larrainzar and Wienkoop 2017). A comparative proteomic analysis between wild soybean and a soybean mutant *SS2-2* suggested that protein-mediated defense-related suppression responses were existed in the roots cell inoculated with rhizobium (Lim et al. 2010). Besides, a review summarized the post-translational modifications (PTMs, especially for phosphorylation, glycosylation, and ubiquitination) of soybean proteins under flood stress (Hashiguchi and Komatsu 2016), suggesting that PTMs can be used to study stress response.

2.3.3 Epigenomics

The epigenetic variations have been found to regulate various biological processes in plant (Turck and Coupland 2014; Schmid et al. 2018; Latzel et al. 2012), and it can be inherited (Eichten et al. 2014), so the application of epigenetic variation provides potential for crop improvement (Springer 2013), and creates a novel breeding perspective with less genetic erosion (Tirnaz and Batley 2019). In soybean, an epigenetic breeding system through restraining MutS HOMOLOG 1 (*MSH1*) gene was conducted, and the results showed that the induced epigenetic variation can increase yield and stability (Raju et al. 2018), suggesting that epigenetics and transgenerational inheritance of nongenetic variation within soybean genomes maybe a new option for soybean breeding.

The epigenetic variation is mainly related to DNA methylation and histone modifications (Eichten et al. 2014; Seymour and Becker 2017). Small RNAs, which can regulate DNA methylation and histone modifications, are a research objective in epigenetics (Guleria et al. 2011; Li et al. 2017). Previous studies have shown that MicroRNAs were involved in regulating stress response (Guleria et al. 2011). In soybean, a genome-wide analysis of miRNAs suggested that miRNAs regulated the low temperatures stress response in mature nodules and symbiotic nitrogen fixation, helping to breed soybean variety with high nitrogen fixation ability under low temperatures (Zhang et al. 2014). Previous studies demonstrated that transcription factors (such as MYB proteins) related to salt-stress were modulated by DNA methylation and enriched with histone modifications (Song et al. 2012; Zhang et al. 2020b). The transcriptome study in soybean has identified some adaptation-related miRNAs and advanced miRNAs research (Li et al. 2017; Sun et al. 2016). Several miRNAs were found to play important roles in auxin–miRNA regulatory network by a global analysis of deep sequencing data, and a salt-responsive miRNA, miR399, can modulate soybean root developmental adaptation (Sun et al. 2016).

The epigenetic study also helps exploring the relationship between DNA methylation variation and genetic variation during soybean domestication and crop improvement. Benefiting from next-generation sequencing, the whole-genome bisulfite sequencing (WGBS) was widely applied in plant DNA methylation studies (Li et al. 2018). Forty-five soybean accessions (wild soybean, landraces, and cultivars) were sequenced by WGBS, and many differentially methylated regions (DMRs) were identified by methylomic analysis (Shen et al. 2018b). DMRs and selective genetic regions exhibited huge different genetic diversity.

2.4 Methods for Molecular Breeding and Functional Analysis

Genome editing, genome-wide association analysis, and genomic selection have become indispensable methods for soybean molecular breeding and functional genomics analysis. These approaches have developed rapidly in recent years, and lots of genetic information and candidate genes associated with agronomic traits of soybean have been exploited and analyzed.

2.4.1 Genome Editing

Genome editing is a powerful tool for functional genomic research and breeding in plant, the zinc finger nuclease (ZFN), transcription activator-like effector-based nucleases (TALEN), and clustered regularly interspaced short palindromic repeats/CRISPR-associated 9 (CRISPR/Cas9) were often used for targeting genome modification (Zhang et al. 2010; Curtin et al. 2011; Chen and Gao 2014; Li et al. 2017; Samanta et al. 2016). Specially, CRISPR/Cas9 is one of the most widely used and rapidly developing technology in recent years (Samanta et al. 2016), and this system is more efficient and easier to use than others (Chen and Gao 2014).

CRISPR/Cas9, which has been successfully applied in many plants, can aid in crop breeding and crop improvement under stress conditions by knocking genes with specific function (Chilcoat et al. 2017; Abdelrahman et al. 2018). In soybean, the first genome-editing plant by CRISPR/Cas9 technology was transformed by *A. rhizogenes*-mediated hairy root transformation (Cai et al. 2015). Later, the whole modified soybean plants using CRISPR/Cas9 system were made (Li et al. 2015). Then, several efforts have been used to optimize the CRISPR/Cas9 technology in soybean, which make it more acceptable (Liu et al. 2019; Bao et al. 2020). A lot of functional analyses in soybean have made by CRISPR/Cas9 technology in recent years (Du et al. 2016; Bao et al. 2019; Cai et al. 2018a). For example, four GmSPL9 genes were shown to be involved in regulating soybean architecture by using CRISPR/Cas9 technology (Bao et al. 2019), and four GmLHY genes were found in regulating plant height by using CRISPR/Cas9-mediated targeted mutagenesis (Cheng et al. 2019). With the progress of CRISPR/Cas9 technology and application, a few mutants have been obtained (Bao et al. 2019; Cheng et al. 2019; Cai et al. 2018b, 2020; Li et al. 2019a). The application of CRISPR/Cas9 technology in soybean has made huge progress in soybean, the first commercialized gene-editing plant product came out in recent years (Li et al. 2020).

CRISPR/Cas9 technology takes the way of knocking instead of introducing foreign genes. Taking food safety into consideration, genome editing with CRISPR/Cas9 technology has become a more widely accepted strategy for functional genomics and breeding in soybean. However, off-target effect is a limitation of CRISPR/Cas9 system. On the one hand off-target predictions were considered when designing sgRNAs by bioinformatic tools (Brazelton et al. 2015); On the other hand, off-target mutations were detected by identifying off-target sites (Jacobs et al. 2015). Several methods were used to reduce the limitation (Sun et al. 2015). The targeted mutagenesis using CRISPR/Cas9 technology was transformed into soybean hairy roots, which would help genes functional analysis in soybean roots and nodules.

2.4.2 Genome-Wide Association Analysis

With the advent and progress of the next-generation sequencing technologies, genome-wide association study (GWAS) has become a vital method for dissecting the QTLs related to important agronomic traits and exploring the candidate genes in plant (Singh et al. 2019; Li et al. 2019b; Patishtan et al. 2018). The aim of the GWAS is to connect the genotypic variations to difference in phenotypes/traits. The whole-genome sequencing (WGS) programs in soybean can produce numerous genetic markers, which will provide rich resources for GWAS analysis (Fang et al. 2017; Contreras-Soto et al. 2017). Moreover, several GWAS analyses in soybean have identified many SNPs and genes associated with agronomic traits such as seed unsaturated FA content (Zhao et al. 2019), seed protein and oil content (Li et al. 2019c), seed weight (Ikram et al. 2020; Guo et al. 2020), pest and disease resistance (Hanson et al. 2018), flowering time (Kim et al. 2020; Zhang et al. 2015), and sudden death syndrome (Swaminathan et al. 2019).

The main objective of soybean breeding is to improve yields. Four yield-related traits of 133 soybean landraces were conducted using the GWAS analysis, and five candidate genes were found to be associated with 100-pod fresh weight (PFW) and moisture content of fresh seeds (MCFS) (Li et al. 2019d). This study provided useful information for breeding in high-yield soybean. Moreover, as a functional genomic approach, GWAS was also applied to mine stress–linked genes in several food crop (Singh et al. 2019). In soybean, four traits and one locus for salt tolerance were identified by genome-wide association mapping, the associated candidate genes may improve salt tolerance (Do et al. 2019). Identifying a large number of SNPs by GWAS, the results revealed that the *GmSFT* (*Glyma.13 g248000*) gene maybe modulate seed-flooding tolerance in soybean (Yu et al. 2019). These studies have contributed greatly to soybean functional genomics, which promote soybean breeding in adverse regions in the future.

2.4.3 Genomic Selection

Genomic selection (GS) is a new breeding approach by using whole-genome prediction models, which exceed marker-assisted selection (MAS) (Desta and Ortiz 2014). The genomic estimated breeding value (GEBV) is used to estimate the genetic value of selection candidates in GS, and the genotypic and phenotypic data of the train population were predicted using statistical models (Nakaya and Isobe 2012). GS studies on several plants have successfully analyzed the agronomic traits such as grain yield and seed weight (Lozada and Carter 2020; Nielsen et al. 2016; Lozada et al. 2019).

In soybean, a breeding program about yield and seed composition traits has made using GS in 2019 (Stewart-Brown et al. 2019). Soybean cyst nematode (SCN) reduces greatly soybean yield, and the SNP markers related to SCN-tolerance were recently identified by using GWAS and GS (Ravelombola et al. 2020). These studies exhibited big potential of GS in soybean breeding programs.

2.5 Conclusion and Perspectives

The genomic sequencing technology has sped up the soybean genomics study in recent years and numerous genetic data and information were released. These abundant resources have advanced in mining genomic loci or genes related to yields and stress response. Other omics studies complement the genomics annotation and functional analysis, the transcriptome studies play vital roles in mining the tissue-specific genes and analyzing the differentially expressed genes during different developmental stages. The proteomics and epigenetics studies in soybean can also enrich functional analysis of traits related to yield and adaptation. Based on the genetic diversity in soybean, more genomics sequencing programs and omics study are still needed. The genome editing, GWAS, and GS can turn the genetic markers or candidate genes into functional analysis and breeding in soybean. Therefore, the renovation and study of these approaches play important roles in promoting soybean breeding.

Acknowledgments This work was supported by funds from the National Natural Science Foundation of China (grant 31701346) and the basic scientific research service fee special of the Central Scientific Research Institute (grant no. 1610172018001).

Conflict of Interest The authors declare that the research was conducted in the absence of any commercial or financial relationships that could be construed as a potential conflict of interest.

Author Contributions R.L., S.Y., and X.Z. designed this work. R.L. and S.Y. wrote the manuscript. H.C. contributed substantially to the completion of this work.

References

Abdelrahman M, Al-Sadi AM, Pour-Aboughadareh A, Burritt DJ, Tran LSP (2018) Genome editing using CRISPR/Cas9-targeted mutagenesis: an opportunity for yield improvements of crop plants grown under environmental stresses. Plant Physiol Biochem 131:31–36

Ainsworth EA, Yendrek CR, Skoneczka JA, Long SP (2012) Accelerating yield potential in soybean: potential targets for biotechnological improvement. Plant Cell Environ 35(1):38–52

Bao AL, Chen HF, Chen LM, Chen SL, Hao QN, Guo W et al (2019) CRISPR/Cas9-mediated targeted mutagenesis of GmSPL9 genes alters plant architecture in soybean. BMC Plant Biol 19(1):131

Bao AL, Tran LSP, Cao D (2020) CRISPR/Cas9-based gene editing in soybean. Methods Mol Biol 2107:349–364

Boyer JS (1982) Plant productivity and environment. Science 218(4571):443–448

Brazelton VA, Zarecor S, Wright DA, Wang Y, Liu J, Chen K et al (2015) A quick guide to CRISPR sgRNA design tools. GM Crops Food 6(4):266–276

Cai YP, Chen L, Liu XJ, Sun S, Wu CX, Jiang BJ et al (2015) CRISPR/Cas9-mediated genome editing in soybean hairy roots. PLoS One 10(8):e0136064

Cai YP, Chen L, Sun S, Wu CX, Yao WW, Jiang BJ et al (2018a) CRISPR/Cas9-mediated deletion of large genomic fragments in soybean. Int J Mol Sci 19(12):3835

Cai YP, Chen L, Liu XJ, Guo C, Sun S, Wu CX et al (2018b) CRISPR/Cas9-mediated targeted mutagenesis of GmFT2a delays flowering time in soya bean. Plant Biotechnol J 16(1):176–185

Cai YP, Wang LW, Chen L, Wu TT, Liu LP, Sun S et al (2020) Mutagenesis of GmFT2a and GmFT5a mediated by CRISPR/Cas9 contributes for expanding the regional adaptability of soybean. Plant Biotechnol J 18(1):298–309

Chan C, Qi XP, Li MW, Wong FL, Lam HM (2012) Recent developments of genomic research in soybean. J Genet Genomics 39(7):317–324

Chen K, Gao C (2014) Targeted genome modification technologies and their application in crop improvements. Plant Cell Rep 33(4):578–583

Cheng Q, Dong LD, Su TS, Li TY, Gan ZR, Nan HY et al (2019) CRISPR/Cas9-mediated targeted mutagenesis of GmLHY genes alters plant height and internode length in soybean. BMC Plant Biol 19(1):562

Chilcoat D, Liu ZB, Sander J (2017) Use of CRISPR/Cas9 for crop improvement in maize and soybean. Prog Mol Biol Transl Sci 149:27–46

Chung WH, Jeong N, Kim J, Lee WK, Lee YG, Lee SH et al (2014) Population structure and domestication revealed by high-depth resequencing of Korean cultivated and wild soybean genomes. DNA Res 21(2):153–167

Contreras-Soto RI, Mora F, de Oliveira MAR, Higashi W, Scapim CA, Schuster I (2017) A genome-wide association study for agronomic traits in soybean using SNP markers and SNP-based haplotype analysis. PLoS One 12(2):e0171105

Curtin SJ, Zhang F, Sander JD, Haun WJ, Starker C, Baltes NJ et al (2011) Targeted mutagenesis of duplicated genes in soybean with zinc-finger nucleases. Plant Physiol 156:466–473

Desta ZA, Ortiz R (2014) Genomic selection: genome-wide prediction in plant improvement. Trends Plant Sci 19(9):592–601

Do TD, Vuong TD, Dunn D, Clubb M, Valliyodan B, Patil G et al (2019) Identification of new loci for salt tolerance in soybean by high-resolution genome-wide association mapping. BMC Genomics 20(1):318

Du HY, Zeng XR, Zhao M, Cui XP, Wang Q, Yang H et al (2016) Efficient targeted mutagenesis in soybean by TALENs and CRISPR/Cas9. J Biotechnol 217:90–97

Eichten SR, Schmitz RJ, Springer NM (2014) Epigenetics: beyond chromatin modification and complex genetic regulation. Plant Physiol 165(3):933–947

Fang C, Ma YM, Wu SW, Liu Z, Wang Z, Yang R et al (2017) Genome-wide association studies dissect the genetic networks underlying agronomical traits in soybean. Genome Biol 18:161

Foley JA, Ramankutty N, Brauman KA, Cassidy ES, Gerber JS, Johnston M et al (2011) Solutions for a cultivated planet. Nature 478(7369):337–342

Garg R, Jain M (2013) Transcriptome analyses in legumes: a resource for functional genomics. Plant Genome 6:1–9

Garrison E, Siren J, Novak AM, Hickey G, Eizenga JM, Dawson ET et al (2018) Variation graph toolkit improves read mapping by representing genetic variation in the reference. Nat Biotechnol 36(9):875–879

Golicz AA, Batley J, Edwards D (2016) Towards plant pangenomics. Plant Biotechnol J 14(4):1099–1105

Guleria P, Mahajan M, Bhardwaj J, Yadav SK (2011) Plant small RNAs: biogenesis, mode of action and their roles in abiotic stresses. Genomics Proteomics Bioinformatics 9(6):183–199

Guo DL, Jiang HX, Yan WL, Yang LJ, Ye JL, Wang Y et al (2020) Resequencing 200 flax cultivated accessions identifies candidate genes related to seed size and weight and reveals signatures of artificial selection. Front Plant Sci 10:1682

Hanson AA, Lorenz AJ, Hesler LS, Bhusal SJ, Michel AP, Jiang GL et al (2018) Genome-wide association mapping of host-plant resistance to soybean aphid. Plant Genome 11(3)

Hashiguchi A, Komatsu S (2016) Impact of post-translational modifications of crop proteins under abiotic stress. Proteome 4(4):42

Hirsch CN, Foerster JM, Johnson JM, Sekhon RS, Muttoni G, Vaillancourt B et al (2014) Insights into the maize pan-genome and pan-transcriptome. Plant Cell 26(1):121–135

Hossain Z, Komatsu S (2014) Soybean proteomics. Methods Mol Biol 1072:315–331

Hufford MB, Xu X, Heerwaarden JV, Pyhäjärvi T, Chia JM, Cartwright RA et al (2012) Comparative population genomic of maize domestication and improvement. Nat Genet 44(7):808–811

Hurgobin B, Golicz AA, Bayer PE, Chan CK, Tirnaz S, Dolatabadian A et al (2018) Homoeologous exchange is a major cause of gene presence/absence variation in the amphidiploid Brassica napus. Plant Biotechnol J 16(7):1265–1274

Hymowitz T (1970) On the domestication of the soybean. Econ Bot 24:408–421

Ikram M, Han X, Zuo JF, Song J, Han CY, Zhang YW et al (2020) Identification of QTNs and their candidate genes for 100-seed weight in soybean (Glycine Max L.) using multi-locus genome-wide association studies. Genes (Basel) 11(7):E714

Jacobs TB, LaFayette PR, Schmitz RJ, Parrott WA (2015) Target genome modifications in soybean with CRISPR/Cas9. BMC Biotechnol 15:16

Jeong SC, Moon JK, Park SK, Kim MS, Lee K, Lee SR et al (2019) Genetic diversity patterns and domestication origin of soybean. Theor Appl Genet 132(4):1179–1193

Kim MY, Lee S, Van K, Kim TH, Jeong SC, Choi IY et al (2010) Whole-genome sequencing and intensive analysis of the undomesticated soybean (*Glycine soja Sieb. And Zucc.*) genome. PNAS 107(51):22032–22037

Kim KH, Kim JY, Lim WJ, Jeong S, Lee HY, Cho Y et al (2020) Genome-wide association and epistatic interactions of flowering time in soybean cultivar. PLoS One 15(1):e0228114

Komatsu S, Hashiguchi A (2018) Subcellular proteomics: application to elucidation of flooding-response mechanisms in soybean. Proteome 6(1):13

Komatsu S, Tougou M, Nanjo Y (2015) Proteomic techniques and management of flooding tolerance in soybean. J Proteome Res 14(9):3768–3778

Lam HM, Xu X, Liu X, Chen WB, Yang GH, Wong FL et al (2010) Resequencing of 31 wild and cultivated soybean genomes identifies patterns of genetic diversity and selection. Nat Genet 42(12):1053–1059

Larrainzar E, Wienkoop S (2017) A proteomic view on the role of legume symbiotic interactions. Front Plant Sci 8:1267

Latzel V, Zhang Y, Moritz KK, Fischer M, Bossdorf O (2012) Epigenetic variation in plant responses to defence hormones. Ann Bot 110(7):1423–1428

Li YH, Zhou GY, Ma JX, Jiang WK, Jin LG, Zhang ZH et al (2014) De novo assembly of soybean wild relatives for pan-genome analysis of diversity and agronomic traits. Nat Biotechnol 32(10):1045–1052

Li ZS, Liu ZB, Xing AQ, Moon BP, Koellhoffer JP, Huang LX et al (2015) Cas9-guide RNA directed genome editing in soybean. Plant Physiol 169(2):960–970

Li MW, Gao Y, Li KP, Fan K, Muñoz NB, Yung WS et al (2017) Using genomic information to improve soybean adaptability to climate change. J Exp Bot 68(8):1823–1834

Li Q, Hermanson PJ, Springer NM (2018) Detection of DNA methylation by whole-genome bisulfite sequencing. Methods Mol Biol 1676:185–196

Li CL, Nguyen V, Liu J, Fu WQ, Chen C, Yu KF et al (2019a) Mutagenesis of seed storage protein genes in Soybean using CRISPR/Cas9. BMC Res Notes 12(1):176

Li L, Mao XG, Wang JY, Chang XP, Reynolds M, Jing RL (2019b) Genetic dissection of drought and heat-responsive agronomic traits in wheat. Plant Cell Environ 42(2):2540–2553

Li SG, Xu HF, Yang JY, Zhao TJ (2019c) Dissecting the genetic architecture of seed protein and oil content in soybean from the Yangtze and Huaihe River valleys using multi-locus genome-wide association studies. Int J Mol Sci 20(12):3041

Li XN, Zhang XL, Zhu LM, Bu YP, Wang XF, Zhang X et al (2019d) Genome-wide association study of four yield-related traits at the R6 stage in soybean. BMC Genet 20:39

Li MW, Wang ZL, Jiang BJ, Kaga A, Wong FL, Zhang GH et al (2020) Impacts of genomic research on soybean improvement in East Asia. Theor Appl Genet 133(5):1655–1678

Libault M, Farmer A, Joshi T, Takahashi K, Langley RJ, Franklin LD et al (2010) An integrated transcriptome atlas of the crop model Glycine max, and its use in comparative analyses in plants. Plant J 63(1):86–99

Lim CW, Park JY, Lee SH, Hwang CH (2010) Comparative proteomic analysis of soybean nodulation using a supernodulation mutant, SS2-2. Biosci Biotechnol Biochem 74(12):2396–2404

Liu JZ, Graham MA, Pedley KF, Whitham SA (2015) Gaining insight into soybean defense responses using functional genomics approaches. Brief Funct Genomics 14(4):283–290

Liu JQ, Gunapati S, Mihelich NT, Stec AO, Michno JM, Stupar RM (2019) Genome editing in soybean with CRISPR/Cas9. Methods Mol Biol 1917:217–234

Liu YC, Du HL, Li PC, Shen YT, Peng H, Liu SL et al (2020) Pan-genome of wild and cultivated soybeans. Cell 182(1):162–176

Lozada DN, Carter AH (2020) Genomic selection in winter wheat breeding using a recommender approach. Genes (Basel) 11(7):E779

Lozada DN, Mason RE, Sarinelli JM, Brown-Guedira G (2019) Accuracy of genomic selection for grain yield and agronomic traits in soft red winter wheat. BMC Genet 20(1):82

Machado FB, Moharana KC, Almeida-Silva F, Gazara RK, Pedrosa-Silva F, Coelho FS et al (2020) Systematic analysis of 1,298 RNA-Seq samples and construction of a comprehensive soybean (Glycine max) expression atlas. Plant J. Online ahead of print

Maruyama K, Ogata T, Kanamori N, Yoshiwara K, Goto S, Yamamoto Y et al (2017) Design of an optimal promoter involved in the heat-induced transcriptional pathway in Arabidopsis, soybean, rice and maize. Plant J 89(4):671–680

Min CW, Gupta R, Agrawal GK, Rakwal R, Kim ST (2019) Concepts and strategies of soybean seed proteomics using the shotgun proteomics approach. Expert Rev Proteomics 16(9):795–804

Nakaya A, Isobe S (2012) Will genomic selection be a practical method for plant breeding? Ann Bot 110:1303–1316

Nielsen NH, Jahoor A, Jensen JD, Orabi J, Cericola F, Edriss V et al (2016) Genomic prediction of seed quality traits using advanced barley breeding lines. PLoS One 11(10):e0164494

Patishtan J, Hartley TN, de Carvalho RF, Maathuis FJM (2018) Genome-wide association studies to identify rice salt-tolerance markers. Plant Cell Environ 41(5):970–982

Qi XP, Li MW, Xie M, Liu X, Ni M, Shao GH et al (2014) Identification of a novel salt tolerance gene in wild soybean by whole-genome sequencing. Nat Commun 5:4340

Qiu J, Wang Y, Wu S, Wang YY, Ye CY, Bai XF et al (2014) Genome re-sequencing of semi-wild soybean reveals a complex soja population structure and deep introgression. PLoS One 9(9):e108479

Raju SKK, Shao MR, Sanchez R, Xu YZ, Sandhu A, Graef G et al (2018) An epigenetic breeding system in soybean for increased yield and stability. Plant Biotechnol J 16(11):1836–1847

Ravelombola WS, Qin J, Shi AN, Nice L, Bao Y, Lorenz A et al (2020) Genome-wide association study and genomic selection for tolerance of soybean biomass to soybean cyst nematode infestation. PLoS One 15(7):e0235089

Ray DK, Mueller ND, West PC, Foley JA (2013) Yield trends are insufficient to double global crop production by 2050. PLoS One 8(6):e66428

Samanta MK, Dey A, Gayen S (2016) CRISPR/Cas9: an advanced tool for editing plant genomes. Transgenic Res 25(5):561–573

Scherer SW, Lee C, Birney E, Altshuler DM, Eichler EE, Carter NP et al (2007) Challenges and standards in integrating surveys of structural variation. Nat Genet 39(7 Suppl):S7–S15

Schmid MW, Heichinger C, Schmid DC, Guthörl D, Gagliardini V, Bruggmann R et al (2018) Contribution of epigenetic variation to adaptation in Arabidopsis. Nat Commun 9(1):4446

Schmutz J, Cannon SB, Schlueter J, Ma JX, Mitros T, Nelson W et al (2010) Genome sequence of the palaeopolyploid soybean. Nature 463(7278):178–183

Severin AJ, Woody JL, Bolon YT, Joseph B, Diers BW, Farmer AD et al (2010) RNA-Seq Atlas of *Glycine max*: a guide to the soybean transcriptome. BMC Plant Biol 10:160

Seymour DK, Becker C (2017) The causes and consequences of DNA methylome variation in plants. Curr Opin Plant Biol 36:56–63

Shen YT, Liu J, Geng HY, Zhang JX, Liu YC, Zhang HK et al (2018a) De novo assembly of a Chinese soybean genome. Sci China Life Sci 61(8):871–884

Shen YT, Zhang JX, Liu YC, Liu SL, Liu Z, Duan ZB et al (2018b) DNA methylation footprints during soybean domestication and improvement. Genome Biol 19:128

Shen Y, Du H, Liu Y, Ni L, Wang Z, Liang C et al (2019) Update soybean Zhonghuang 13 genome to a golden reference. Sci China Life Sci 62(9):1257–1260

Shimomura M, Kanamori H, Komatsu S, Namiki N, Mukai Y, Kurita K et al (2015) The *Glycine max* cv. Enrei genome for improvement of Japanese soybean cultivars. Int J Genomics 358127:2015

Singh B, Salaria N, Thakur K, Kukreja S, Gautam S, Goutam U (2019) Functional genomic approaches to improve crop plant heat stress tolerance. F1000Reseach 8:1721

Song YG, Ji DD, Li S, Wang P, Li Q, Xiang FN (2012) The dynamic changes of DNA methylation and histone modifications of salt responsive transcription factor genes in soybean. PLoS One 7(7):e41274

Springer NM (2013) Epigenetics and crop improvement. Trends Genet 29(4):241–247

Stewart-Brown BB, Song Q, Vaughn JN, Li Z (2019) Genomic selection for yield and seed composition traits within an applied soybean breeding program. G3 (Bethesda) 9(7):2253–2265
Sun XJ, Hu Z, Chen R, Jiang QY, Song GH, Zhang H et al (2015) Targeted mutagenesis in soybean using the CRISPR-Cas9 system. Sci Rep 5:10342
Sun ZX, Wang YN, Mou FP, Tian YP, Chen L, Zhang SL et al (2016) Genome-wide small RNA analysis of soybean reveals auxin-responsive microRNAs that are differentially expressed in response to salt stress in root apex. Front Plant Sci 6:1273
Swaminathan S, Das A, Assefa T, Knight JM, Ferreira A, Silva D et al (2019) Genome wide association study identifies novel single nucleotide polymorphic loci and candidate genes involved in soybean sudden death syndrome resistance. PLoS One 14(2):e0212071
Tao YF, Zhao XR, Mace E, Henry R, Jordan D (2019) Exploring and exploiting pan-genomics for crop improvement. Mol Plant 12(2):156–169
Tilman D, Balzer C, Hill J, Befort BL (2011) Global food demand and the sustainable intensification of agriculture. PNAS 108(50):20260–20264
Tirnaz S, Batley J (2019) Epigenetic: potentials and challenges in crop breeding. Mol Plant 12(10):1309–1311
Tran LS, Mochida K (2010) Functional genomics of soybean for improvement of productivity in adverse conditions. Funct Integr Genomics 10(4):447–462
Turck F, Coupland G (2014) Natural variation in epigenetic gene regulation and its effect on plant development traits. Evolution 68(3):620–631
Wang X, Komatsu S (2016) Plant subcellular proteomics: application for exploring optimal cell function in soybean. J Proteome 143:45–56
Wang X, Komatsu S (2018) Proteomic approaches to uncover the flooding and drought stress response mechanisms in soybean. J Proteome 172:201–215
Wang Z, Tian ZX (2015) Genomics progress will facilitate molecular breeding in soybean. Sci China Life Sci 58(8):813–815
Wang XF, Zhao ZS, Guo N, Wang HT, Zhao JM, Xing H (2020) Comparative proteomics analysis reveals that lignin biosynthesis contributes to brassinosteroid-mediated response to *phytophthora sojae* in soybeans. J Agric Food Chem 68(19):5496–5506
Xia ZJ, Zhai H, Lü SX, Wu HY, Zhang YP (2013) Recent achievement in gene cloning and functional genomics in soybean. Sci World J 2013:281367
Xie M, Chung CYL, Li MW, Wong FL, Wang X, Liu AL et al (2019) A reference-grade wild soybean genome. Nat Commun 10(1):1216
Yim AK-Y, Wong JW-H, Ku Y-S, Qin H, Chan T-F, Lam H-M (2015) Using RNA-Seq data to evaluate reference genes suitable for gene expression studies in soybean. PLoS One 10(9):e0136343
Yu ZP, Chang FG, Lv WH, Sharmin RA, Wang ZL, Kong JJ et al (2019) Identification of QTN and candidate gene for seed-flooding tolerance in soybean [*Glycine max* (L.) Merr.] using genome-wide association study (GWAS). Genes (Basel) 10(12):957
Yuan SL, Rong L, Chen SL, Chen HF, Zhang CJ, Chen LM et al (2016) RNA-Seq analysis of differential gene expression responding to different rhizobium strains in soybean (Glycine max) roots. Front Plant Sci 7:721
Yuan SL, Rong L, Chen HF, Zhang CJ, Chen LM, Hao QN et al (2017) RNA-Seq analysis of nodule development at five different developmental stages of soybean (*Glycine max*) inoculated with *Bradyrhizobium Japonicum* strain 113-2. Sci Rep 7:42248
Zeng HQ, Zhang XJ, Zhang X, Pi EX, Xiao L, Zhu YY (2018) Early transcriptomic response to phosphate deprivation in soybean leaves as revealed by RNA-sequencing. Int J Mol Sci 19(7):2145
Zhang F, Maeder ML, Unger-Wallace E, Hoshaw JP, Reyon D, Christian M et al (2010) High frequency targeted mutagenesis in Arabidopsis thaliana using zinc finger nucleases. PNAS 107(26):12028–12033
Zhang SL, Wang YN, Li KX, Zou YM, Chen L, Li X (2014) Identification of cold-responsive miRNAs and their target genes in nitrogen-fixing nodules of soybean. Int J Mol Sci 15:13596–13614

Zhang JP, Song QJ, Cregan PB, Nelson RL, Wang XZ, Wu JX et al (2015) Genome-wide association study for flowering time, maturity dates and plant height in early maturing soybean (Glycine Max) germplasm. BMC Genomics 16(1):217

Zhang GY, Ahmad MZ, Chen BB, Manan S, Zhang YL, Jin HN et al (2020a) Lipidomic and transcriptomic profiling of developing nodules reveals the essential roles of active glycolysis and fatty acid and membrane lipid biosynthesis in soybean nodulation. Plant J. Online ahead of print

Zhang WX, Wang N, Yang JT, Guo H, Liu ZH, Zheng XJ et al (2020b) The salt-induced transcription factor GmMYB84 confers salinity tolerance in soybean. Plant Sci 291:110326

Zhao Q, Feng Q, Lu H, Li Y, Wang A, Tian QL et al (2018) Pan-genome analysis highlights the extent of genomic variation in cultivated and wild rice. Nat Genet 50(2):278–284

Zhao X, Jiang HP, Feng L, Qu YF, Teng WL, Qiu LJ et al (2019) Genome-wide association and transcriptional studies reveal novel genes for unsaturated fatty acid synthesis in a panel of soybean accessions. BMC Genomics 20(1):68

Zhou ZK, Jiang Y, Wang Z, Gou ZH, Lyu J, Li WY et al (2015) Resequencing 302 wild and cultivated accessions identifies genes related to domestication and improvement in soybean. Nat Biotechnol 33(4):408–414

Zhou P, Silverstein KA, Ramaraj T, Guhlin J, Denny R, Liu J et al (2017) Exploring structural variation and gene family architecture with do novo assemblies of 15 Medicago genomes. BMC Genomics 18(1):261

Chapter 3
Genetics and Genomics of Cottonseed Oil

Jinesh Patel, Edward Lubbers, Neha Kothari, Jenny Koebernick, and Peng Chee

Contents

3.1 Introduction

The cotton plant belongs to the genus *Gossypium* L. (Malvaceae), which includes about 50 diploid (2n = 2x = 26) and allopolyploid (2n = 4x = 52) species. Four *Gossypium* spp. were independently domesticated for crop production (Wendel and Cronn 2003; Wendel and Grover 2015). They include two allotetraploid species *G. hirsutum* L. and *G. barbadense* L. that are endemic to the Americas and two diploid species *G. arboreum* L. and *G. herbaceum* L. that are endemic to Africa and Asia. Phylogenetic studies have established that the allotetraploid tribe arose from a single hybridization involving two diploid species which resemble the progenitors of *G. herbaceum* or *G. arboreum* (A genome) and *G. raimondii* (D genome) (Wendel and Grover 2015). While the two diploid species are still being cultivated, the two

J. Patel · J. Koebernick
Department of Crop, Soil and Environmental Sciences, Auburn University, Auburn, AL, USA
e-mail: jdp00789@auburn.edu; jenny.koebernick@auburn.edu

E. Lubbers · P. Chee (✉)
Cotton Molecular Breeding Laboratory, University of Georgia – Tifton Campus, Tifton, GA, USA
e-mail: elubbers@uga.edu; pwchee@uga.edu

N. Kothari
Fiber Quality Research, Cotton Incorporated, Cary, NC, USA
e-mail: NKothari@cottoninc.com

H. Tombuloglu et al. (eds.), *Oil Crop Genomics*,
https://doi.org/10.1007/978-3-030-70420-9_3

allotetraploid species occupied an overwhelming majority of the world's production. Specifically, *G. hirsutum*, which is commonly referred to as "Upland" cotton, is responsible for over 90% of the total world's production and will be the focus in this chapter.

As a crop, the cotton plant is the most important source of natural fiber, supplying about 40% of the world fiber market (ICAC 2015), making cotton the single most important natural fiber in the textile industries. In recent years, over 36 million hectares of cotton are planted in over 100 countries, producing about 106 million bales (480 lb or 218 kg), which has an aggregate value of US$30 billion/year. More than 350 million people are engaged in jobs related to the production and processing of cotton. Thus, cotton is a vital agricultural commodity in the global economy. The top cotton-producing countries include China, the United States, India, Pakistan, and Brazil, with China being the biggest consumer of raw fiber.

Seed cotton, which is harvested as raw cotton, is separated at ginning into cottonseed and lint fiber, which is the epidermal seed hair. Because 85–90% of the cotton crop's value resides in the lint fiber, cotton farming has emphasized maximizing the production of fiber. The value of the cottonseed is considered as a separate commodity because of its characteristic uses. Cottonseed is processed into protein meal (45%), hulls (26%), oil (16%), and linters (9%) with the remaining lost in processing (Cherry and Leffler 1984). Of these components, the oil is the most significant product followed by the meal. The hulls and linters together have much less value than any of the other components by themselves.

Cottonseed oil is now garnering a small but important share of the vegetable oil market and has the potential to increase its share. It has a mild taste, is high in desirable unsaturated fats (e.g., linoleic acid), and is stable during cooking/frying/storing compared to many other vegetable cooking oils. Globally, cottonseed oil is the seventh-largest vegetable oil source in the world at 44.3 million metric tons of production/year or about 5.3% of the global annual vegetable oil production (Cherry and Leffler 1984). In 2015, India (1487 kilotons), China (1305 kilotons), and the United States (286 kilotons) were the top three producers of cottonseed oil (USDA-FAS 2015).

3.2 Genetic Improvement of Oil Content

Historically, cotton has been bred for its lint yield. In many parts of the world, breeding goals have also extended into improving fiber quality for earning premiums and enhancing the marketability. While some breeding programs in India and China also have emphasized improving the oil production from cottonseed along with fiber yield, the United States considers cottonseed as a by-product. However, there have been recent interests in the research community to improve cottonseed oil content as well as the fatty acid profile. A major challenge in breeding for oil is the phenotyping process. While the NMR testing method only takes approximately 4–5 min per sample for estimating oil content, it requires the use of delinted seeds, and acid delinting is a hazardous and laborious process. Therefore, approximately 110 cottonseed entries can be tested on the machine in a single day. Breeding programs generally have a large number of entries that are evaluated every season, making genomics an ideal approach to tackle improvement in quantitative and qualitative aspects of cottonseeds.

The first challenge in cottonseed oil improvement is the levels of genetic variability that is available within the cotton germplasm. Various cotton germplasms stored within the US National Cotton Germplasm Collection have been surveyed by Kohel (1978) and Hinze et al. (2015). Kohel (1978) used a wide-line nuclear magnetic resonance (NMR) using a Newport Quality Analyzer, while Hinze (2015) employed a newer, nondestructive, time-domain ^{1}H (TD)-NMR methodology developed by Horn et al. (2011) to measure oil and protein simultaneously. Both studies concluded that sufficient genetic variability for cottonseed oil exists in *Gossypium* spp. that can be utilized for making genetic gains. Hinze et al. (2015) evaluated 2256 accessions encompassing 9 genomes within 33 species and found that most of the variability for oil within the tested accessions resides in *G. hirsutum* and *G. barbadense*. Kothari et al. (2016) showed that, among the allotetraploid species, the oil content ranged from 13% to 27% and the protein values ranged from 16% to 36% (Fig. 3.1). Other studies that included Agarwal et al. (2002) and Khan et al. (2015), found significant variability for oil content in the cotton germplasm collections of India and Pakistan, respectively. These results showed that genetic improvement of

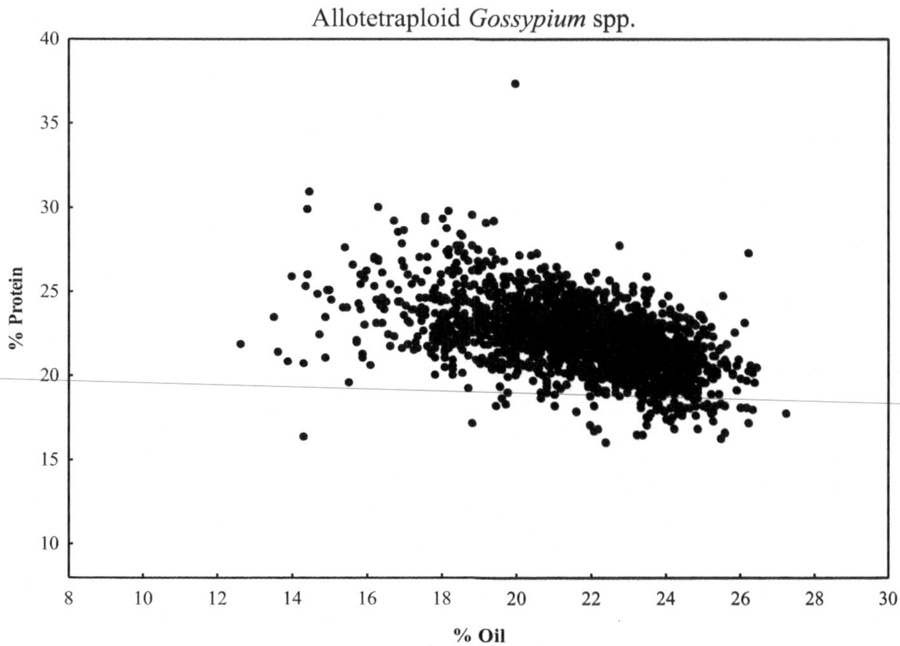

Fig. 3.1 Oil and protein content of 1980 allotetraploid cotton accessions from the nuclear magnetic resonance technique. (From Kothari et al. 2017)

cottonseed oil could be achieved through phenotypic breeding by transferring the trait of interests from unimproved germplasm into elite cultivars.

Once germplasm variability for any desired trait is verified, the next question is whether recombining the different genes/alleles available in the germplasm will provide improvement of the desired trait, in this case, change in quantity or quality of oil in the cottonseed. Determining the parental breeding value by measuring its combining ability gives the average improvement from the parent over a series of crosses (general combining ability) as well as the specific improvement from specific parents (specific combining ability). Using a diallel mating system with six cultivars, Singh et al. (2010) described significant general and specific combining abilities for improving cottonseed oil content. Kothari et al. (2016) used a line by tester analysis to determine combining abilities and found that cottonseed oil content was moderately heritable ($H^2 = 0.52$). Most of the phenotypic variation was associated with general combining ability, but there were examples of significant combining ability in several crosses. The correlation analyses that they performed also indicated that oil content could be increased along with improving seed index, boll weight, upper half-mean length, strength, uniformity, and elongation. Hinze et al. (2015) and Campbell et al. (2016) also showed a significant correlation between oil and seed index for the *G. hirsutum* accessions they studied. Finally, a recent study from a diallel mating design showed a positive combining ability for improving percent oil and fiber traits (Kothari et al. 2017). Therefore, these studies collectively showed that, at least in the US cotton germplasm, sufficient genetic variation exists in cottonseed oil for potential of genetic improvement in cotton breeding programs.

The second and perhaps more difficult challenge to overcome in cottonseed oil improvement is the ability to drive selection for oil traits without sacrificing other important traits such as lint yield and fiber quality. Obviously in cotton breeding, increased lint production is the primary effort, but Song and Zhang (2007), Yu et al. (2012), Zeng et al. (2015), Hinze et al. (2015), Liu et al. (2015a), and Campbell et al. (2016) all showed a negative correlation between seed oil and seed protein. Campbell et al. (2016) further showed a negative correlation with yield component traits such as seed cotton yield, lint %, seeds/boll, bolls/square meter, and fiber density and fiber quality traits such as fiber length, strength, maturity, and fineness. Campbell et al. (2014) and Campbell and Myers (2015) showed that lint percentage, an important yield component, emphasizes the difficulty expected in breeding for lint and seed oil traits simultaneously. Further, data from numerous studies herein as well as from the National Cotton Variety Trials show that the variation in protein content and oil content is negatively correlated across all germplasm. However, Campbell et al. (2016) also pointed out that two cultivars that they used as checks (FiberMax FM 960BR and Deltapine DP 444BR) did have high oil and high protein levels with a commercial level lint yield and quality. Therefore, it is possible to develop cotton lines with improved traits that have a negative correlation. There is now considerable interest in developing genetic and genomic tools aiming at assisting cotton breeders in transferring desirable cottonseed oil traits while maintaining other valuable traits during cultivar development.

3.3 Genetic Mapping and Quantitative Trait Loci

In improving the production of cottonseed oil, marker-assisted breeding is more effective than a basic phenotypic selection that is labor-intensive, costly, time-consuming, and sometimes futile. Further, the use of marker-assisted selection may assist in breaking the linkages between genes for oil traits and lint yield. Song and Zhang (2007) first described the location of QTLs for cottonseed nutritional quality traits by using 918 SSR and EST-SSR markers in an interspecific cross between *G. hirsutum* by *G. barbadense* (Song and Zhang 2007). They identified 11 significant QTLs for kernel percent, kernel oil content, kernel protein, and 7 amino acids. Only one QTL for kernel oil content was detected; the locus *qOP-D8–1* was mapped between two SSR markers BNL3860_190 and NAU1369_400, and it explained 29.5% of the phenotypic variation. Wu et al. (2009) used chromosome substitution lines to find that chromosomes 4 and 18 from the *G. barbadense* introgression into *G. hirsutum* were associated with increased oil percentage. Chromosomes 2, 6, 7, and 16 and chromosome arms 5sh, 14sh, 22sh, and 22lo were associated with decreased oil percentage. An et al. (2010) developed a pair of F_2 populations from two crosses, a fuzzless, lintless line (*N1*, *n2*, postulated *n3* for fuzzless) by a fuzzless, linted line and by FM 966 (Bayer CropScience, Lubbock, TX, USA), a commercial cotton cultivar. Within the FM 966 cross, there were five markers significantly associated with embryo oil %: STV164-173, CIR393-200, BNL3261-203, BNL119-224, and BNL3400-173. Unfortunately, the map was not developed adequately to show actual QTLs, and BNL3400-173 was not allocated to a location on the cotton genome. STV164-173, CIR393-200, BNL3261-203, and BNL119-224 were on chromosome 9sh, 7Lo, 12, and 20, respectively. These putative QTLs explained 32.7% of the phenotypic variation for the oil percentage. The sequence of the STV164-173 marker is similar to a gene *Pho1* that codes for α-1,4-glucan phosphorylase and shows a significant association with the oil percentage at a 0.05 significance level. *Pho1* is related to starch metabolism and gives reason to add this information in developing the biochemical pathways of oil production. Quampah et al. (2012) used a *G. hirsutum* intraspecific mapping population to identify four QTLs affecting seed oil. Two QTLs were on chromosome 18, one on chromosome 22, and the other on linkage group 11 (LGA03, chromosome 11). Quampah et al. (2012) also pointed out that care needs to be made when interpreting the genetic analyses because of the interaction of the maternal and embryonic tissue in seed production. Yu et al. (2012) used an advanced backcross inbred line (BIL) population from *G. barbadense* by *G. hirsutum* with 146 lines to identify 17 QTLs on 12 chromosomes (1, 3, 5, 11, and 12 and 15, 16, 19, 20, 21, 24, and 25) for seed oil content. The three major QTLs for oil (*qOil2-c12-1*, *qOil2-c21-1*, and *qOil4-c21-1*) explained a total of 22–26% of the phenotypic variation (Yu et al. 2012). They also noticed that all of the oil QTL alleles that had positive additive effects came from the *G. hirsutum* parent except for one QTL from the *G. barbadense* parent on chromosome 1, which could explain why negative transgressive segregation was prevalent in the population. In addition, a region on chromosome 25 had an additive effect to reduce total protein (*qGhPR-c25*) and an additive effect to increase linoleic acid (*qGhLA-c25*) (Yuan et al. 2018), but this QTL on chromosome 25 did not show any co-localization (Yu et al. 2012). Generally, the co-localization of genes for multiple traits with opposite effects for selection makes it difficult to enhance them simultaneously. This result generally supported the observations by Wu et al. (2009) in which the *G. hirsutum* donor parent species was the positive donor for oil in the chromosome substitution population. In this experiment, the seed oil content was significant but negatively correlated with the seed protein content; apparently, of the eight QTLs for both protein and oil contents detected, all loci were co-localized with opposite additive effects. Shang et al. (2016) developed a composite interval genetic map using two related RILs and their corresponding backcross populations in Upland cotton. Twenty-four novel QTLs for oil content were detected of which nine QTLs were found in more than one environment and are considered stable. For the two most stable oil content QTLs in the experiment, genetic analysis indicates that between 6.42% and 14.01% of phenotypic variation can be explained by *qOC-Chr23-1* and between 7.01% and 13.4% can be explained by *qOC-Chr23-2*. They concluded that these two loci are excellent targets for a program to enhance oil content in cottonseeds.

After oil content, the manufacturing or dietary value of cottonseed oil is mostly defined by its fatty acid profile. However, beyond that, there is a relationship between these quantity and quality measures that needs to be acknowledged. Both Liu et al. (2015a) and Yuan et al. (2018) noted that for the negative correlation of seed protein content with seed oil content to be properly understood, any underlying correlations between the seed protein content and the yield components of seed oil content (and vice versa) must be ascertained. For example, the negative correlation of protein content with oil content could be due to the production of oleic acid, not to any of the other fatty acids. One must be careful in interpreting correlation data when a trait is a composite value from multiple components (i.e., the yield is made from the yield components).

Recent advancements in understanding the biochemical pathways and the cellular/molecular mechanisms of seed oil biogenesis have assisted in the identification and cloning of genes that likely could play a role in the quality and quantity of cottonseed oil. Modification of the profiles of the fatty acids found in cottonseed could change both the quality and quantity of the oil. Liu et al. (2015a) developed a RIL genetic map with 1675 SSR and nine morphological markers that identified QTLs for crude oil percentage (15 QTLs ranging between 2.0% and 39.8% of the PV) and percentages of the four principal fatty acids: linolenic acid (8 QTLs ranging between 2.2% and 8.0% of the PV), oleic acid (10 QTLs ranging between 2.0% and 15.4% of the PV), palmitic acid (13 QTLs ranging between 4.2% and 13.3% of the PV), and stearic acid (12 QTLs ranging between 4.4% and 22.7% of the PV) as well as for crude protein and lint percentage. The authors noted that there is a negative correlation of linoleic acid content with the content of the other fatty acids, thus leading to the understanding that the linoleic pathway is more competitive than the pathways of the other fatty acids that were studied. This is another detail that will affect the understanding of the relationship between the quality and the quantity of the oil content.

The QTLs reported for oil content, protein content, oil quality, gossypol content, and amino acid composition are listed in Table 3.1. Much of the information relevant to QTL results can also be found in the two cotton databases, i.e., CottonGen (Yu et al. 2013) and Cotton QTLdb (Said et al. 2015a). The location of QTLs on the genome for traits of interest can be downloaded from such databases (Said et al. 2015a), and a consensus map can be developed. Here we summarized the QTL mapping results from various studies available in Cotton QTLdb (Table 3.1) and developed a consensus map using the BioMercator V3 software for seed weight (qSW, SW, SWpQtl), oil content (qOil, QTLoil, oil2, qCO, qOC), protein content (qPro, qCP, qPC, QTLprotein), gossypol content (qGos, qGC), and fatty acid compositions (linolenic acid content, qLA; oleic acid content, qOA; palmitic acid content, qPA; stearic acid content, qSA), depicting multiple regions in the genome that have co-localized QTLs for two or more traits (Fig. 3.2). This information can be further

Table 3.1 QTLs for oil content, protein content, oil quality, gossypol content, and amino acid composition mapped in *Gossypium* spp.

Publication	Parents	Traits	# QTLs	PVE %
Song and Zhang (2007)	TM-1 × Hai7124	Kernel %	1	46.28
		Kernel oil %	1	29.35
		Protein content	1	22.25
		Amino acid comp.	8	10.9–31.1
AN et al. (2010)	MD17 × FiberMax 966	Oil content	5	4.7–10.3
	Or MD17 × 181	Protein content	2	6.2–15.4
		Amino acid comp.	56	
Alfred et al. (2012)	HS46 × MARCABUCAG8US-1-88	Oil content	4	–
Yu et al. (2012)	SG 747 × Giza 75	Oil content	17	5.5–25.6
		Protein content	22	5.6–35
		Gossypol content	3	5.79–5.9
Liu et al. (2013)	HS46 × MARKCBUCAG8US-1-88	Amino acid comp.	35	2–35.3
Liu et al. (2015a)	Yumian 1 × T586	Crude protein	13	5.2–48.1
		Crude oil	15	3.9–42.8
		Linoleic content	8	5.4–15.1
		Oleic content	10	5.7–16.6
		Palmitic content	13	4.2–14.8
		Stearic acid content	12	4.4–22.7
Shang et al. (2016)	GX1135 × GX100–2	Oil content	14	5.3–16.3
	GX1135 × VGX100–2	Oil content	10	6.4–13.4

used in cotton breeding programs to identify molecular markers that target multiple traits.

A consensus mapping or a meta-analysis of QTLs is a way to arrange loci from multiple, published maps into one map and to identify hotspots for a specific trait or QTL cluster associated with diverse traits. In cotton, several meta-QTL analysis studies focusing on interspecific (*G. hirsutum* by *G. barbadense*) and intraspecific (*G. hirsutum* by *G. hirsutum*) crossing populations in cotton have been reported (Rong et al. 2007; Lacape et al. 2010; Said et al. 2013, 2015b; Zhang et al. 2015a; Abdelraheem et al. 2017). While most studies have focused on fiber quality, Said et al. (2013) and Said et al. (2015b) have included traits related to seed nutrient composition. Five QTL clusters for seed and fiber quality were found on chromosomes 5, 6, 16, 19, and 20. Two clusters were found for oil and protein content and fiber quality (chromosomes 5 and 7), one was found for amino acid content and fiber traits (chromosome 5), and one was found for amino acid and fruiting traits (chromosome 22). The most interesting QTL cluster was on chromosome 5 because it contained QTLs for almost all traits for seed composition (seed quality, amino acid composition, oil and protein content) as well as fiber traits (Said et al. 2015b). Such genomic regions need to be further validated for simultaneously improving oil and fiber quality.

3.4 Genome-Wide Association Study

Genome sequencing has become the leading edge of genetics. The first genome sequence of a flowering plant, *Arabidopsis thaliana* (L.) Heynh, was published almost two decades ago (Initiative 2000). Since then, there has been a rapid increase in the number of plant genomes sequenced each year. The first draft genome sequence from *Gossypium* species was from *G. raimondii* Ulbrich (cotton D-genome) due to its small genome size and being a putative progenitor of the allotetraploid species, including the cultivated species *G. hirsutum* (Paterson et al. 2012). Subsequently, researchers have also sequenced the cultivated cotton species, i.e., *Gossypium arboreum* (Li et al. 2014), *G. hirsutum* (Li et al. 2015; Zhang et al. 2015b), and *G. barbadense* (Liu et al. 2015c). These genome sequences are now a valuable resource for developing new genetic markers, acquiring high-density genetic linkage maps, fine-mapping of QTLs, identifying candidate genes, conducting GWAS (genome-wide association studies), and studying the effects of domestication on genome composition.

Linkage disequilibrium-based association mapping is a powerful tool for dissecting the genetic basis underlying complex traits. There has been only a handful of GWAS conducted to discover genomic regions related to oil quantity and quality. Badigannavar and Myers (2015) reported a total of 5 and 6 genomic regions associated with seed protein and oil content, respectively, with an association panel of 75 Upland cotton germplasm lines screened by 64 primers for amplified fragment length polymorphism (AFLP) markers (Badigannavar and Myers 2015). In another

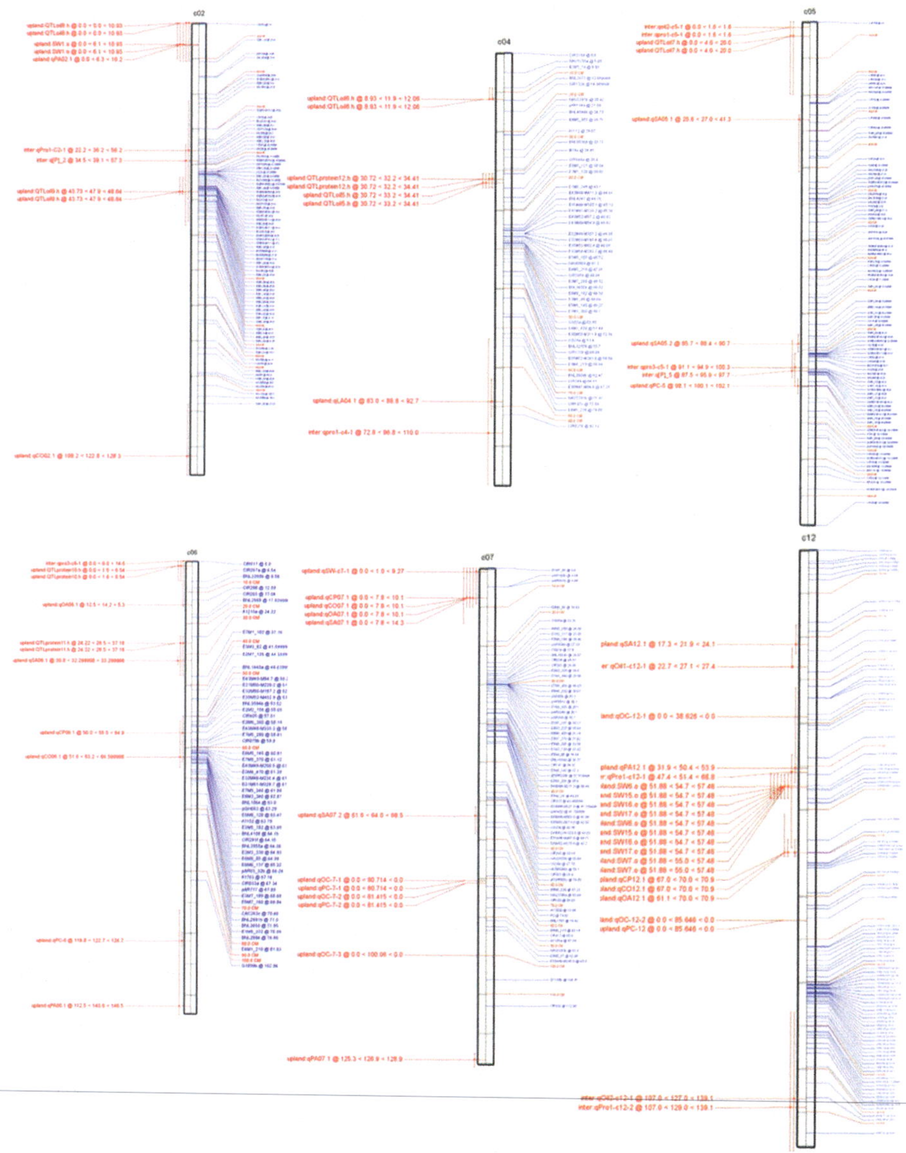

Fig. 3.2 Consensus map of the cotton genome depicting the QTL positions of oil traits in the At (Chromosome 1–13) and Dt subgenome (Chromosome 14–26). Traits mapped in the consensus map include seed weight (qSW, SW, SWpQtl), oil content (qOil, QTLoil, oil2, qCO, qOC), protein content (qPro, qCP, qPC, QTL protein), gossypol content (qGos, qGC), and fatty acid compositions (linolenic acid content, qLA; oleic acid content, qOA; palmitic acid content, qPA; stearic acid content, qSA)

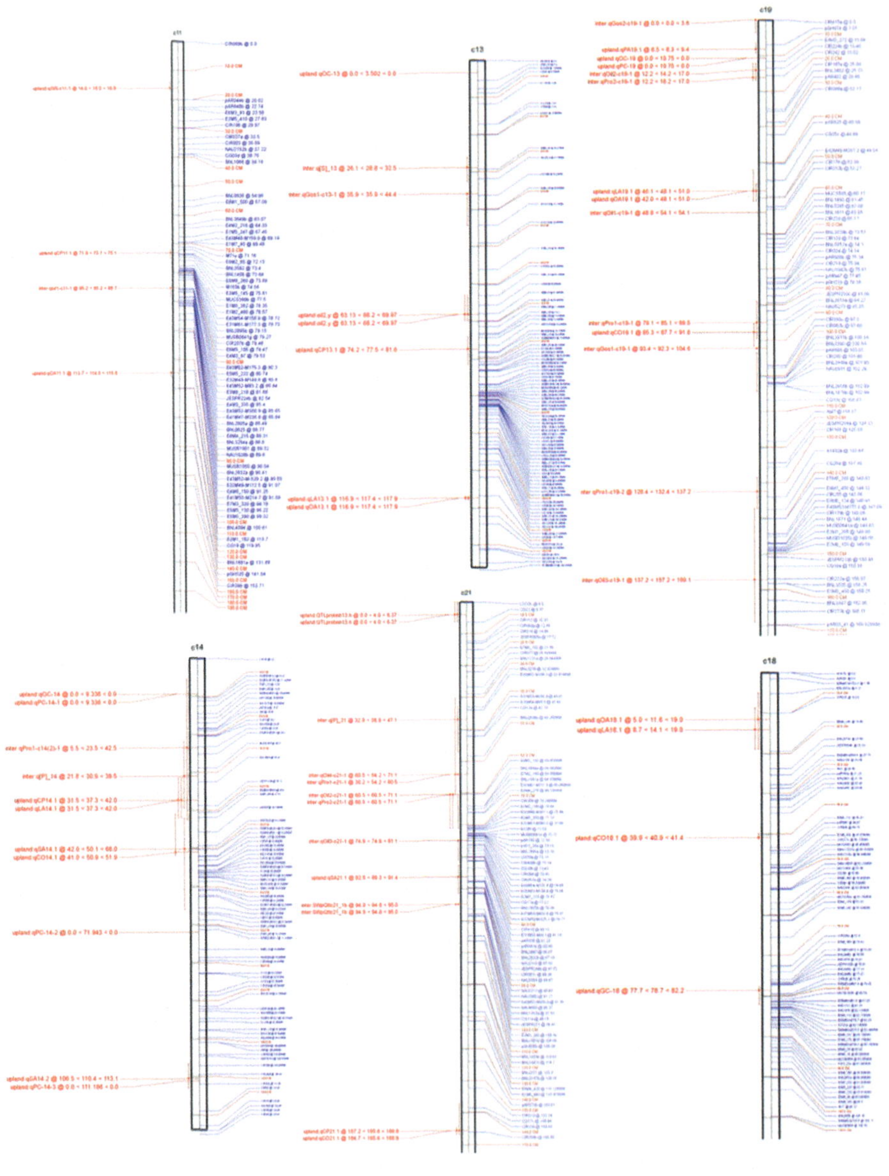

Fig. 3.2 (continued)

study, Hinze et al. (2015) utilized a nondestructive nuclear magnetic resonance technique to measure variation in seed protein and oil content in 30 delinted seeds of 2256 accessions of 33 different species of cotton (5 tetraploids and 28 diploids) applied to measure oil and protein content. Combining the phenotypic information with genotypic data generated using a core set of 105 SSR markers, they reported 2 genomic regions near TMB0043 (chromosome 21) and TMB1356 (chromosome

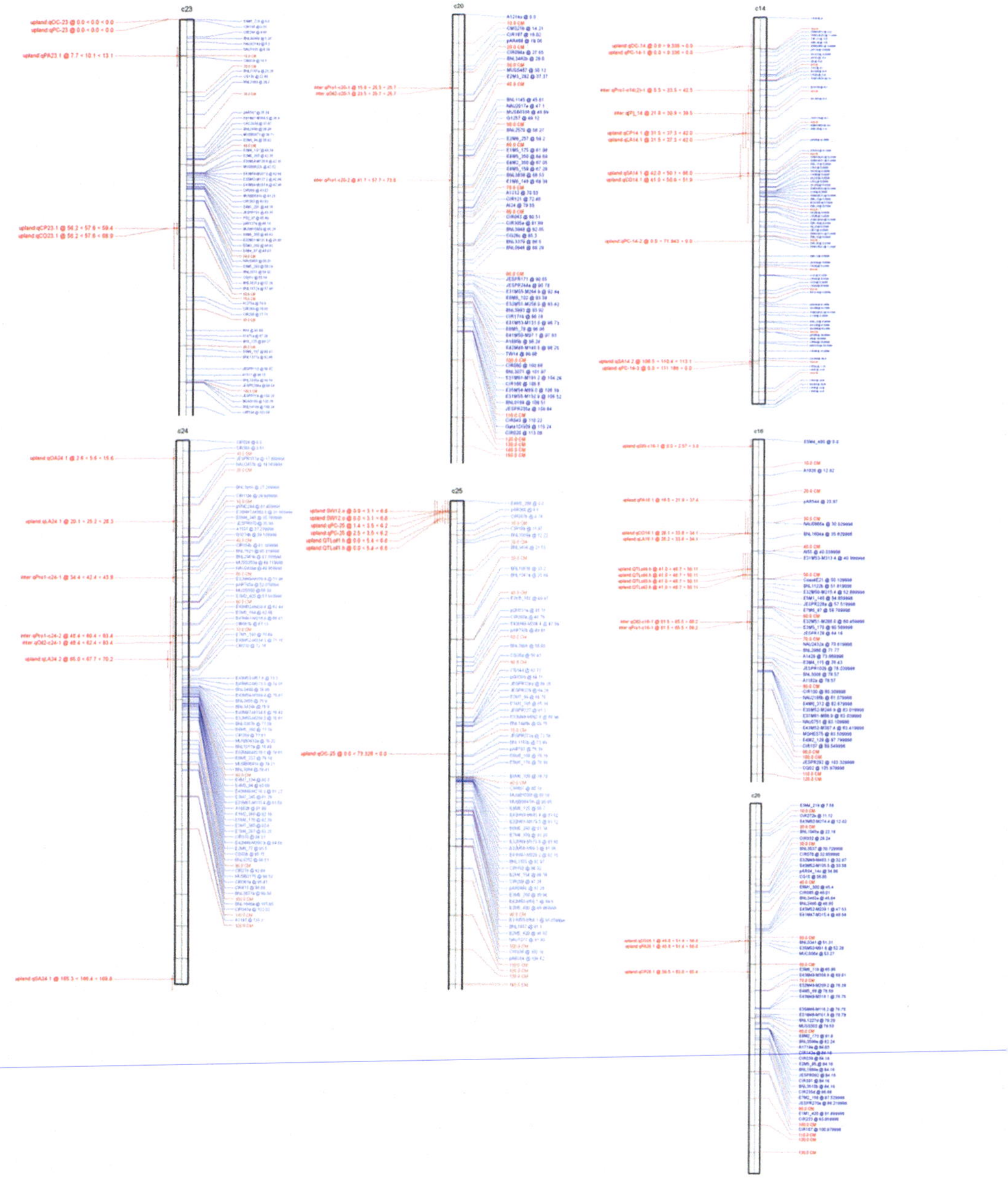

Fig. 3.2 (continued)

10) that were significantly associated with oil content, and 2 additional regions near BNL3090 (chromosome 15) and BNL3441 (chromosome 3) that were associated with protein content (Hinze et al. 2015). In a similar research (Liu et al. 2015b), 180 cotton accessions were genotyped using 228 SSRs and phenotyped for seed oil and protein content using the NIR (near-infrared reflectance) spectrum method in multiple environments. The researchers reported 12 to 15 genomic regions were associated with protein and oil content.

While prior GWAS utilizing genetic markers such as SSRs have often suffered from low genome coverage, more recent studies that employ resequencing or the cotton SNP chips have essentially overcome the issue of genome coverage. Du et al. (2018b) developed a cotton diversity panel consisting of 316 accessions and evaluated them for cottonseed traits such as seed protein, oil content, and fatty acids (palmitic, linoleic, oleic, myristic, and stearic). About 390 K SNPs were screened in the panel to decipher the complicated genetic architecture of these traits. A total of 16, 21, and 87 SNPs were found to be strongly associated with seed protein content, oil content, and fatty acids, respectively. Based on this genomic position of the associations, they concluded that fatty acids, except for stearic acid, were controlled by common genomic regions, possibly by genes that have pleiotropic effects (Du et al. 2018b). Yuan et al. (2018), focusing on understanding the genetic control of cottonseed nutrient traits and using ~78 K SNPs and a cotton association panel that consisted of 196 accessions, identified 28 QTLs associated with seed oil content, protein content, and fatty acid composition with 4 of these also discovered in other studies. Interestingly, only 6 of the 28 QTLs had higher LD decay, implying a narrow history of cotton breeding for cottonseed trait composition. They also reported that the protein enzyme, β-ketoacyl-ACP synthase (KAS), plays an essential role in the metabolic pathways of fatty acid biosynthesis in determining the content of palmitic acid (C16:0) and palmitoleic acid (C16:1) in the seed. Finally, in another study involving resequencing of 243 accessions of the diploid A genome cotton *G. arboreum*, Du et al. (2018a) identified a SNP variation in the GaKASIII gene that distinguishes conspicuous low and high oil content accessions. With high numbers of molecular markers in a wide range of cotton germplasm, this GWAS was able to narrow down the genomic regions associated with oil content and its composition to the extent that the researchers were able to scrutinize the region to identify candidate genes involved in the biosynthetic pathways.

3.5 Transcriptome Analysis and Candidate Genes

RNAseq (or RNA sequencing), which utilizes next-generation sequencing to profile the entire transcriptome, has become an important tool to decipher the gene expression and gene network of any biosynthesis process. With the availability of the current cotton genomes, it has become feasible to identify genes that are differentially expressed in response to temporal and spatial changes in environmental conditions. Several RNAseq studies have been conducted to identify genes and gene networks in oil biosynthesis (Jiao et al. 2013; Hovav et al. 2015; Hu et al. 2016; Zhao et al. 2018c). Hovav et al. (2015) used RNAseq to study four seed developmental time-points in *G. hirsutum* and determined that 16–20% of total genes expressed at different time points showed homeolog specific expression with a higher number of expressed genes from the At subgenome. They also reported that the gene ACL (ATP-citrate lyase) and *DGAT3* (Diacylglycerol Acyltransferase 3) may play a crucial role in oil biosynthesis. Zhao et al. (2018c) performed transcriptome sequencing on seed embryo at different development stages and identified important transcription factors like *WRI1* and *NF-YB6* that govern lipid metabolism. Two isoforms of the *GhFAD2-1* gene (Gh_A13G1850 and Gh_D13G2238) were highly expressed in cottonseed compared to other plant tissues suggesting a significant role of this gene in converting oleic acid (C18:1) to linoleic acid (C18:2). Further, comparative gene expression analysis between rapeseed (*Brassica napus*), oleaster (*Olea europaea*), and cotton showed that a much higher expression of the *FAD2* gene and a lower expression of the *FAD3* explained a higher concentration of linoleic acid and a lower concentration of linolenic acid in cottonseed oil (Zhao et al. 2018c). Other studies have also identified isoforms of the *FAD2* gene such as *GhFAD2-3* and *GhFAD2-4* to be associated with linoleic accumulation (Pirtle et al. 2001; Zhang et al. 2009). Manipulation of the expression of isoforms of *GhFAD2* gene similar to changes done in other oil-producing crops like soybean and peanut is likely to be a viable strategy to improve the quality of cottonseed oil. This approach may eventually increase the value of the cotton crop without compromising seed protein content, fiber yield, or quality (Patel et al. 2004; Pham et al. 2010; Haun et al. 2014). Alternatively, cotton breeders could introgress the naturally occurring mutants in *G. barbadense* accessions with the mutated *GbFAD2-1D* gene to obtain germplasm with a high oil and high oleic seed phenotype (Shockey et al. 2017; Sturtevant et al. 2017).

Genome-wide characterization and expression studies of gene families involved in oil biosynthesis have been conducted. A higher expression of *GhWRI1* in high-oil content cultivars as compared to low-oil content cultivars was noted as well as an overexpression of the gene in *Arabidopsis thaliana* that showed higher oil content and seed weight (Zhao et al. 2018b). A genome-wide identification of biotin carboxyl carrier protein (BCCP) genes and their expression in different tissues of the cotton plant found four class II *GhBCCP* genes that are highly expressed in developing ovules making them a potential target for improving cotton oil production (Cui et al. 2017b). In plants, cyclopropane synthases are involved in the catalytic reaction of cyclopropanation of unsaturated lipids to produce cyclopropane fatty acids (CPA), an important product for industrial applications (Bao et al. 2002; Yu et al. 2011). CPA found in cottonseed is malvalic acid (Johnson et al. 1967). Three genes (*GhCPS1*, *GhCPS2*, and *GhCPS3*) were identified in the cotton genome database that can encode cyclopropane synthases protein. Based on expression and gene transformation experiments, *GhCPS1* and *GhCPS2* seemed more relevant in the production of cyclopropane fatty acids (Yu et al. 2011). Lysophosphatidic acid acyltransferase (LPAAT) is one of the three acyltransferases required in the Kennedy pathway for diacylglycerol (DAG) biosynthesis, a precursor of triacylglycerol

(TAG) (Weiss et al. 1960; Bates 2012). Scanning cotton genomes identified 13 LPAAT genes in Upland cotton, five *GhLPAATs* that were present on Dt subgenome and eight that were on the At subgenome. Further sequence variation and gene expression analysis concluded that rather than using traditional breeding, genetic modification to overexpress single genes like *At-Gh13LPAAT5* will be effective in improving the production of total TAG and oil content (Wang et al. 2017). Stearoyl-acyl carrier protein desaturase (SAD) significantly affects the ratio of saturated and unsaturated fatty acid and thus the composition of fatty acid (Ohlrogge and Jaworski 1997; Zhang et al. 2015c). A genome-wide analysis found 18 *SAD* genes in *G. hirsutum* with both subgenomes (At and Dt) having 9 SAD genes each. *GhSAD2* and GhSAD4 showed a significantly different pattern of gene expression in high- and low-oil containing cultivars with further analysis affirming that the GhSAD4 of the At subgenome has a major role in fatty acid composition *GhSAD4* showed a significantly different pattern of gene expression in high- and low-oil containing cultivars with further analysis affirming that the GhSAD4 of the At subgenome has a major role in fatty acid composition *(*Shang et al. 2017*). In conclusion, gene expression studies either through* expression analyses of the cottonseed transcriptome or by genome-wide analysis of gene family associated with oil biosynthesis have identified specific genes and genomic regions that can be targeted to improve oil content and quality.

Classical genetic studies have shown that the increase of oil content has a negative impact on lint yield components (Turner et al. 1976; Wu et al. 2009; An et al. 2010; and Liu et al. 2015a). The extensive use of carbohydrates and energy by both processes might be a possible explanation for such a correlation. However, the discovery of QTLs such as *qOC-Chr5* that contribute to both increased oil content and improved fiber yield (Shang et al. 2016) and the detection of QTL clusters for oil and fiber traits in nearby vicinities on multiple chromosomes by meta-analysis studies (Said et al. 2013; Said et al. 2015b) suggest that simultaneous improvements for both traits could be achievable through breeding.

3.6 Genetic Transformation

Because fiber quantity and quality will always be prioritized over cottonseed oil in cotton production, improving oil content and quality would have to be a balancing act. Genetic engineering offers a means to manipulate cottonseed oil traits without disrupting the delicate genetic balance on fiber yield and quality that has been created through decades of modern plant breeding. Several studies have shown the potential of improving oil content and quality by manipulating the expression of genes in fatty acid biosynthesis pathways using genetic engineering techniques.

Acetyl-CoA carboxylase (ACCase) exists in homomeric and heteromeric forms and plays a vital role in initiating fatty acid biosynthesis by catalyzing acetyl-CoA to produce malonyl-CoA (Konishi and Sasaki 1994; Konishi et al. 1996; Sasaki and Nagano 2004). In plants, the acetyl-CoA heteromeric form is made up of three nuclear-coded subunits, namely, biotin carboxyl carrier protein (BCCP), biotin carboxylase (BC), and α-carboxyltransferase (α-CT), while the fourth subunit, β-carboxyltransferase (β-CT), is a plastid-coded subunit encoded by the chloroplast (Reverdatto et al. 1999; Ke et al. 2000). Expression of the four genes in ovules is positively correlated to oil accumulation and in transgenic plants, overexpressing *GhBCCP*, *GhBC*, and *GhCTβ* showed an increase in oil content in cottonseeds ranging from 9.2% to 21.9% (Cui et al. 2017a). Phosphoenolpyruvate carboxylase (PEPCase), a carboxy-lyases family enzyme, is found in plants in which oxaloacetic acid is needed for the tricarboxylic acid (TCA) cycle and affects the oil and protein content. RNAi silencing of the PEPC isoform *GhPEPC1* increased the oil content by 16.7% without affecting other plant characteristics (Xu et al. 2016). In a similar study, silencing another PEPC isoform *GhPEPC2* through the RNAi technique has improved the oil content by 7.3% while reducing the protein content by 5.7% in the seed kernel, thus proving that manipulating such genes can help to regulate oil and protein content (Zhao et al. 2018a). These discoveries suggest that silencing both genes at once by targeting a conserved sequence of the two isoforms might enhance the oil content without influencing the fiber traits. Transcription factors like *WRINKLED* (WRI1) (Pouvreau et al. 2011; Zhao et al. 2018b, c), *ABSCISIC ACID INSENSITIVE* (ABI) (Suzuki et al. 2007; Wang et al. 2007a), Dof-type transcription factor (Dof) (Wang et al. 2007b; Su et al. 2017), *LEAFY COTYLEDON* (LEC). (Mendoza et al. 2005; Mu et al. 2008), FUSCA3 (FUS3) (Elahi et al. 2015), and NUCLEAR FACTOR Y (NF-Y) (Yeap et al. 2017; Zhao et al. 2018c) have been identified in different plant species involved in oil biosynthesis. Manipulating the expression of transcription factors might be an interesting approach to change the oil content. Overexpression of the cotton transcription factor *GhDof1* improved the oil content, reduced the protein content, and also provided resilience to abiotic stress like salinity and cold (Su et al. 2017). Thus, manipulating the expression of one transcription factor, that has pleiotropic effects can serve to improve multiple traits.

Significant improvement for stearic and oleic concentration in cottonseed oil was achieved by downregulating the expression of two key fatty acid desaturase genes, ghSAD-1 encoding stearoyl-acyl carrier protein Δ9-desaturase and ghFAD2-1 encoding oleoyl-phosphatidylcholine ω6-desaturase in seed using hairpin RNA-mediated gene silencing (Liu et al. 2000, 2002). Likewise, an RNAi technique was deployed to silence *GhFAD2-1* and *GhFATB* at the same time; the resulting transgenic cotton lines had a higher amount of oleic acid content (almost 1.5 times) and a reduced amount of palmitic acid and linoleic acid in cottonseed oil (Liu et al. 2017a). Another study using RNAi techniques was successfully used to downregulate β-ketoacyl-ACP synthase II (KASII) which increased the concentration of the C16 fatty acid content and double the percentage of palmitic acid accumulation in cottonseed oil (Liu et al. 2017b).

3.7 Gossypol

Gossypol is a toxin present in glands of different tissues of cotton plants including seed. It is a phenolic compound thought to serve as a self-defense mechanism of cotton against some pests and diseases. Therefore, even though cottonseed oil has many desirable characteristics, gossypol in cottonseed and oil is a major hindrance to its use in food products (Bell and Stipanovic 1977; Hedin et al. 1992; Jenkins and Wilson 1996). Average cottonseed contains 0.52–1.01% gossypol (Calhoun et al. 2004), sufficient amount to be harmful to some animals when ingested (Blom et al. 2001; Eisele 1986), and this has limited the use of cottonseed products to adult ruminants in rationed quantities (Berardi and Goldblatt 1980; Kim et al. 1996; Santos et al. 2003). Therefore, any discussion of cottonseed for human or animal consumptions would not be complete without addressing gossypol.

Reducing gossypol content in the seed could increase for cottonseed in ruminant rations and might expand its use to other animals. There are ways to reduce seed gossypol that include mechanical/chemical processes (Damaty and Hudson 1975; Gardner Jr et al. 1976; Mayorga et al. 1975), but these treatments add expense and reduce the nutritional value (Lusas and Jividen 1987). Another strategy was used by McMichael (1959, 1960) to develop glandless genetic cotton germplasm. Unfortunately, the program to develop cultivars from this germplasm was not successful because these cultivars were subsequently more susceptible to pests (Hess 1977; Lusas and Jividen 1987). However, related strategies other than the complete elimination of the glands have shown potential (Romano and Scheffler 2008; Sunilkumar et al. 2006; Vroh Bi et al. 1999). Another strategy utilized these two facts: (1) gossypol exists in two enantiomers (+) and (−) with the (−) enantiomer more toxic (more biologically active and was eliminated more slowly), and (2) *Gossypium* species produce both forms in varying proportions. Although the goal is to decrease the overall toxicity of seed gossypol, (+) cultivars still need to provide some active protection. Puckhaber et al. (2002) working with the plant pathogen *Rhizoctonia solani* and Stipanovic et al. (2006) with corn earworm (*Helicoverpa zea*) larvae have noted that the (+) and (−) enantiomers are equally effective protective agents in these cases.

Glandless varieties that have failed as they are more susceptible to insect attacks suggest the need for tissue-specific elimination of gossypol (Jenkins et al. 1966). A δ-cadinene synthase gene encodes the enzyme δ-cadinene synthase that catalyzes the reaction to produce the precursor of gossypol, (+)-δ-cadinene. A RNAi technique was used to disrupt the expression of the δ-cadinene synthase gene by using a seed-specific promotor and produce ultra-low gossypol cottonseed but maintaining the normal levels of gossypol in the rest of the plant. Such transgenic cotton lines were found to be stable in controlled and field environments for multigenerations (Sunilkumar et al. 2006; Rathore et al. 2012; Palle et al. 2013). The availability of ultra-low gossypol cotton varieties will no doubt elevate the value of cottonseed as a source of high-quality vegetable oil.

3.8 Summary

In summary, multiple genomic regions have been identified using molecular markers in either bi-parental populations or association panels that can be targeted to improve oil content and its composition. Cautious use of natural and artificial mutants that have altered cotton oil in cotton breeding programs can simultaneously improve fiber and oil traits (Shockey et al. 2017; Sturtevant et al. 2017; Auld and Bechere 2018). Global transcriptome analysis and genome-wide association study of genes related to the oil biosynthesis pathways have specifically identified genes that can be targeted to improve the value of cotton as an oil crop through plant breeding and/or genetic engineering techniques. Target-specific genome modification techniques like CRISPR/CAS9 could also be deployed to overcome the linkage drag barrier for the improvement of cotton oil without sacrificing lint yield and quality.

References

Abdelraheem A, Liu F, Song M, Zhang JF (2017) A meta-analysis of quantitative trait loci for abiotic and biotic stress resistance in tetraploid cotton. Mol Gen Genomics 292:1221–1235

Agarwal DK, Sing P, Mayee CD, Nita K (2002) Genetic improvement of cottonseed oil. Central Institute for Cotton Research Technical Bulletin 21:3–10

An C, Jenkins JN, Wu J, Guo Y, McCarty JC (2010) Use of fiber and fuzz mutants to detect QTL for yield components, seed, and fiber traits of upland cotton. Euphytica 172:21–34

Auld DL, Bechere E (2018) Use of the naked-tufted mutant in Upland cotton to improve fiber quality, increase seed oil content, increase ginning efficiency, and reduce the cost of delinting. Euphytica 167:333–339

Badigannavar A, Myers GO (2015) Genetic diversity, population structure, and marker-trait associations for seed quality traits in cotton (*Gossypium hirsutum*). J Genet 94:87–94

Bao X, Katz S, Pollard M, Ohlrogge J (2002) Carbocyclic fatty acids in plants: biochemical and molecular genetic characterization of cyclopropane fatty acid synthesis of *Sterculia foetida*. PNAS 99:7172–7177

Bates PD (2012) The significance of different diacylgycerol synthesis pathways on plant oil composition and bioengineering. Front Plant Sci 3:147

Bell AA, Stipanovic RD (1977) The chemical composition, biological activity, and genetics of pigment glands in cotton. In: Proceeding of the Beltwide cotton production research conference, Jan. 10–12, Atlanta, pp 244–258

Berardi LC, Goldblatt LA (1980) Gossypol. In: Liener IE (ed) Toxic constituents of plant foodstuffs, 2nd edn. Academic, New York, pp 183–237

Blom JH, Lee KJ, Rinchard J, Dabrowski K, Ottobre J (2001) Reproductive efficiency and maternal-offspring transfer of gossypol in rainbow trout (*Oncorhynchus mykiss*) fed diets containing cottonseed meal. J Anim Sci 79(6):1533–1539

Calhoun MC, Wan PJ, Kuhlmann SW, Baldwin BC Jr (2004) Variation in the nutrient and gossypol content of whole and processed cottonseed. *Proceedings of the 2004 Mid-South Ruminant Nutrition Conference*

Campbell BT, Myers GO (2015) Quantitative genetics. In: Fang D, Percy R (eds) Cotton, Agron. Monogr. 57, 2nd edn. ASA, CSSA, and SSSA, Madison

Campbell BT, Greene JK, Wu J, Jones DC (2014) Assessing the breeding potential of day-neutral converted racestock germplasm in the Pee Dee cotton germplasm enhancement program. Euphytica 195:453–465

Campbell BT, Chapman KD, Sturtevant D, Kennedy C, Horn P, Chee PW, Lubbers E, Meredith WR Jr, Johnson J, Fraser D, Jones DC (2016) Genetic analysis of cottonseed protein and oil in a diverse cotton germplasm. Crop Sci 56:2457–2464

Cherry JP, Leffler HR (1984) Seeds. In: Kohel RJ, Lewis CF (eds) Cotton. ASA, CSSA, and SSSA, Madison

Cui Y, Liu Z, Zhao Y, Wang Y, Huang Y et al (2017a) Overexpression of heteromeric GhACCase subunits enhanced oil accumulation in Upland cotton. Plant Mol Biol Report 35:287–297

Cui Y, Zhao Y, Wang Y, Liu Z, Ijaz B et al (2017b) Genome-wide identification and expression analysis of the biotin carboxyl carrier subunits of heteromeric acetyl-CoA carboxylase in *Gossypium*. Front Plant Sci 8:624

Damaty SM, Hudson BJ (1975) Preparation of low-gossypol cottonseed flour. J Sci Food Agr 26:109–115

Du X, Huang G, He S, Yang Z, Sun G et al (2018a) Resequencing of 243 diploid cotton accessions based on an updated a genome identifies the genetic basis of key agronomic traits. Nat Genet 50:796–802

Du X, Liu S, Sun J, Zhang G, Jia Y et al (2018b) Dissection of complicate genetic architecture and breeding perspective of cottonseed traits by genome-wide association study. BMC Genomics 19:451

Eisele GR (1986) A perspective on gossypol ingestion in swine. Vet Hum Toxicol 28:118–122

Elahi N, Duncan RW, Stasolla C (2015) Decreased seed oil production in FUSCA3 *Brassica napus* mutant plants. Plant Phys Biochem 96:222–230

Gardner HK Jr, Hron RJ Sr, Vix HLE, Ridlehuber JM (1976) Process for producing an edible cottonseed protein concentrate. United States Patent, US3972861A

Haun W, Coffman A, Clasen BM, Demorest ZL, Lowy A et al (2014) Improved soybean oil quality by targeted mutagenesis of the fatty acid desaturase 2 gene family. Plant Biotechnol J 12:934–940

Hess DC (1977) Genetic improvement of gossypol-free cotton varieties. Cereal Food World 22:98–103

Hinze LL, Horn PJ, Kothari N, Dever JK, Frelichowski J et al (2015) Nondestructive measurements of cottonseed nutritional trait diversity in the US National Cotton Germplasm Collection. Crop Sci 55:770–782

Horn PJ, Neogi P, Tombokan X, Ghosh S, Campbell BT, Chapman KD (2011) Simultaneous quantification of oil and protein in cottonseed by low-field time-domain nuclear magnetic resonance. J Am Oil Chem Soc 88:1521–1529

Hovav R, Faigenboim-Doron A, Kadmon N, Hu G, Zhang X et al (2015) A transcriptome profile for developing seed of polyploid cotton. Plant Genome 8:1–15

Hu G, Hovav R, Grover CE, Faigenboim-Doron A, Kadmon N et al (2016) Evolutionary conservation and divergence of gene coexpression networks in *Gossypium* (cotton) seeds. Genome Biol Evol 8:3765–3783

Jenkins JN, Wilson FD (1996) Host plant resistance. In: King EG, Phillips JR, Coleman RJ (eds) Cotton insects and mites: characterization and management. The Cotton Foundation, Memphis, pp 563–597

Jenkins JN, Maxwell FG, Lafever HN (1966) The comparative preference of insects for glanded and glandless cottons. J Econ Entomol 59:352–356

Jiao X, Zhao X, Zhou XR, Green AG, Fan Y et al (2013) Comparative transcriptomic analysis of developing cotton cotyledons and embryo axis. PLoS One 8:e71756

Johnson AR, Pearson JA, Shenstone FS, Fogerty AC, Giovanelli J (1967) The biosynthesis of cyclopropane and cyclopropene -fatty acids in plant tissues. Lipids 2:308–315

Ke J, Wen TN, Nikolau BJ, Wurtele ES (2000) Coordinate regulation of the nuclear and plastidic genes coding for the subunits of the heteromeric acetyl-coenzyme A carboxylase. Plant Physiol 122:1057–1072

Khan FZ, Rehman SU, Abid MA, Malik W, Hanif CM, Bilal M, Qanmber G, Latif A, Ashraf J, Farhan U (2015) Exploitation of germplasm for plant yield improvement in cotton (*Gossypium hirsutum L.*). J Green Physiol Genet Genom 1:1–10

Kim HL, Calhoun MC, Stipanovic RD (1996) Accumulation of gossypol enantiomers in ovine tissues. Comp Biochem Phys B 113:417–420

Kohel RJ (1978) Survey of *Gossypium hirsutum* L. germplasm collections for seed-oil percentage and seed characteristics. U.S. Dep. Agric. ARS-S 187

Konishi T, Sasaki Y (1994) Compartmentalization of two forms of acetyl-CoA carboxylase in plants and the origin of their tolerance toward herbicides. PNAS 91:3598–3601

Konishi T, Shinohara K, Yamada K, Sasaki Y (1996) Acetyl-CoA carboxylase in higher plants: most plants other than *Gramineae* have both the prokaryotic and the eukaryotic forms of this enzyme. Plant Cell Physiol 37:117–122

Kothari N, Campbell BT, Dever JK, Hinze LL (2016) Combining ability and performance of cotton germplasm with diverse seed oil content. Crop Sci 56:19–29

Kothari NK, Dever J, Hinze L (2017) Combining abilities for seed oil and protein content in cotton. In: Proceeding of the Beltwide cotton conference, Dallas

Lacape JM, Llewellyn D, Jacobs J, Arioli T, Becker D et al (2010) Meta-analysis of cotton fiber quality QTLs across diverse environments in a *Gossypium hirsutum* x *G. barbadense* RIL population. BMC Plant Biol 10:132

Li F, Fan G, Wang K, Sun F, Yuan Y et al (2014) Genome sequence of the cultivated cotton *Gossypium arboreum*. Nat Genet 46:567–572

Li F, Fan G, Lu C, Xiao G, Zou C et al (2015) Genome sequence of cultivated Upland cotton (*Gossypium hirsutum* TM-1) provides insights into genome evolution. Nat Biotechnol 33:524–530

Liu Q, Singh S, Green A (2000) Genetic modification of cottonseed oil using inverted-repeat gene-silencing techniques. Biochem Soc Trans 28:927–929

Liu Q, Singh SP, Green AG (2002) High-stearic and high-oleic cottonseed oils produced by hairpin RNA-mediated post-transcriptional gene silencing. Plant Physiol 129:1732–1743

Liu D, Liu F, Shan X, Zhang J, Tang S et al (2015a) Construction of a high-density genetic map and lint percentage and cottonseed nutrient trait QTL identification in upland cotton (*Gossypium hirsutum* L.). Mol Gen Genomics 290:1683–1700

Liu G, Mei H, Wang S, Li X, Zhu X et al (2015b) Association mapping of seed oil and protein contents in Upland cotton. Euphytica 205:637–645

Liu X, Zhao B, Zheng HJ, Hu Y, Lu G et al (2015c) *Gossypium barbadense* genome sequence provides insight into the evolution of extra-long staple fiber and specialized metabolites. Sci Rep 5:14139

Liu F, Zhao YP, Zhu HG, Zhu QH, Sun J (2017a) Simultaneous silencing of GhFAD2-1 and GhFATB enhances the quality of cottonseed oil with high oleic acid. J Plant Physiol 215:132–139

Liu Q, Wu M, Zhang B, Shrestha P, Petrie J et al (2017b) Genetic enhancement of palmitic acid accumulation in cottonseed oil through RNAi down-regulation of ghKAS 2 encoding β-ketoacyl-ACP synthase II (KASII). Plant Biotechnol J 15:132–143

Lusas EW, Jividen GM (1987) Glandless cottonseed: a review of the first 25 years of processing and utilization research. J Am Oil Chem Soc 64:839–854

Mayorga H, Gonzales J, Menchu JF, Rolz C (1975) Preparation of a low free gossypol cottonseed flour by dry and continuous processing. J Food Sci 40:1270–1274

McMichael SC (1959) Hopi cotton, a source of cottonseed free of gossypol pigments. Agron J 51:630–630

McMichael SC (1960) Combined effects of glandless genes gl2 and gl3 on pigment glands in the cotton plant. Agron J 52:385–386

Mendoza MS, Dubreucq B, Miquel M, Caboche M, Lepiniec L (2005) LEAFY COTYLEDON 2 activation is sufficient to trigger the accumulation of oil and seed-specific mRNAs in Arabidopsis leaves. FEBS Lett 579:4666–4670

Mu J, Tan H, Zheng Q, Fu F, Liang Y et al (2008) LEAFY COTYLEDON1 is a key regulator of fatty acid biosynthesis in Arabidopsis. Plant Physiol 148:1042–1054

Ohlrogge JB, Jaworski JG (1997) Regulation of fatty acid synthesis. Annu Rev Plant Biol 48:109–136

Palle SR, Campbell LM, Pandeya D, Puckhaber L, Tollack LK et al (2013) RNAi-mediated ultra-low gossypol cottonseed trait: performance of transgenic lines under field conditions. Plant Biotechnol J 11:296–304

Patel M, Jung S, Moore K, Powell G, Ainsworth C et al (2004) High-oleate peanut mutants result from a MITE insertion into the FAD2 gene. Theor Appl Genet 108:1492–1502

Paterson AH, Wendel JF, Gundlach H, Guo H, Jenkins J et al (2012) Repeated polyploidization of *Gossypium* genomes and the evolution of spinnable cotton fibres. Nature 492:423–427

Pham AT, Lee JD, Shannon JG, Bilyeu KD (2010) Mutant alleles of FAD2-1A and FAD2-1B combine to produce soybeans with the high oleic acid seed oil trait. BMC Plant Biol 10:195

Pirtle IL, Kongcharoensuntorn W, Nampaisansuk M, Knesek JE, Chapman KD et al (2001) Molecular cloning and functional expression of the gene for a cotton Δ-12 fatty acid desaturase (FAD2). BBA-Gene Struct Exp 1522:122–129

Pouvreau B, Baud S, Vernoud V, Morin V, Py C et al (2011) Duplicate maize Wrinkled1 transcription factors activate target genes involved in seed oil biosynthesis. Plant Physiol 156:674–686

Puckhaber LS, Dowd MK, Stipanovic RD, Howell CR (2002) Toxicity of (+)-and (−)-gossypol to the plant pathogen, *Rhizoctonia solani*. J Agric Food Chem 50:7017–7021

Quampah A, Liu HY, Xu HM, Li JR, Wu JG et al (2012) Mapping of quantitative trait loci for oil content in cottonseed kernel. J Genet 91:289–295

Rathore KS, Sundaram S, Sunilkumar G, Campbell LM, Puckhaber L et al (2012) Ultra-low gossypol cottonseed: generational stability of the seed-specific, RNAi-mediated phenotype and resumption of terpenoid profile following seed germination. Plant Biotechnol J 10:174–183

Reverdatto S, Beilinson V, Nielsen NC (1999) A multisubunit acetyl coenzyme A carboxylase from soybean. Plant Physiol 119:961–978

Romano GB, Scheffler JA (2008) Lowering seed gossypol content in glanded cotton (*Gossypium hirsutum* L.) lines. Plant Breed 127:619–624

Rong J, Feltus FA, Waghmare VN, Pierce GJ, Chee PW et al (2007) Meta-analysis of polyploid cotton QTL shows unequal contributions of subgenomes to a complex network of genes and gene clusters implicated in lint fiber development. Genetics 176:2577–2588

Said JI, Lin Z, Zhang X, Song M, Zhang J (2013) A comprehensive meta QTL analysis for fiber quality, yield, yield related and morphological traits, drought tolerance, and disease resistance in tetraploid cotton. BMC Genomics 14:776

Said JI, Knapka JA, Song M, Zhang J (2015a) Cotton QTLdb: a cotton QTL database for QTL analysis, visualization, and comparison between *Gossypium hirsutum* and *G. hirsutum* × *G. barbadense* populations. Mol Gen Genomics 290:1615–1625

Said JI, Song M, Wang H, Lin Z, Zhang X et al (2015b) A comparative meta-analysis of QTL between intraspecific *Gossypium hirsutum* and interspecific *G. hirsutum* × *G. barbadense* populations. Mol Gen Genomics 290:1003–1025

Santos JEP, Villasenor M, Robinson PH, DePeters EJ, Holmberg CA (2003) Type of cottonseed and level of gossypol in diets of lactating dairy cows: plasma gossypol, health, and reproductive performance. J Dairy Sci 86:892–905

Sasaki Y, Nagano Y (2004) Plant acetyl-CoA carboxylase: structure, biosynthesis, regulation, and gene manipulation for plant breeding. Biosci Biotech Bioch 68:1175–1184

Shang L, Abduweli A, Wang Y, Hua J (2016) Genetic analysis and QTL mapping of oil content and seed index using two recombinant inbred lines and two backcross populations in Upland cotton. Plant Breed 135:224–231

Shang X, Cheng C, Ding J, Guo W (2017) Identification of candidate genes from the SAD gene family in cotton for determination of cottonseed oil composition. Mol Gen Genomics 292:173–186

Shockey J, Dowd M, Mack B, Gilbert M, Scheffler B et al (2017) Naturally occurring high oleic acid cottonseed oil: identification and functional analysis of a mutant allele of *Gossypium barbadense* fatty acid desaturase-2. Planta 245:611–622

Singh S, Singh VV, Choudhary AD (2010) Combining ability estimates for oil content, yield components, and fibre quality traits in cotton (*G. hirsutum*) using an 8× 8 diallel mating design. Trop Subtrop Agroecosyst 12:161–166

Song XL, Zhang TZ (2007) Identification of quantitative trait loci controlling seed physical and nutrient traits in cotton. Seed Sci Res 17:243–251

Stipanovic RD, Lopez JD, Dowd MK, Puckhaber LS, Duke SE (2006) Effect of racemic and (+)- and (−)-gossypol on the survival and development of *Helicoverpa zea* larvae. J Chem Ecol 32:959–968

Sturtevant D, Horn P, Kennedy C, Hinze L, Percy R et al (2017) Lipid metabolites in seeds of diverse *Gossypium* accessions: molecular identification of a high oleic mutant allele. Planta 245:595–610

Su Y, Liang W, Liu Z, Wang Y, Zhao Y et al (2017) Overexpression of GhDof1 improved salt and cold tolerance and seed oil content in *Gossypium hirsutum*. J Plant Physiol 218:222–234

Sunilkumar G, Campbell LM, Puckhaber L, Stipanovic RD, Rathore KS (2006) Engineering cottonseed for use in human nutrition by tissue-specific reduction of toxic gossypol. PNAS 103:18054–18059

Suzuki M, Wang HH-Y, McCarty DR (2007) Repression of the LEAFY COTYLEDON 1/B3 regulatory network in plant embryo development by VP1/ABSCISIC ACID INSENSITIVE 3-LIKE B3 genes. Plant Physiol 143:902–911

Turner JH, Ramey HH, Worley S (1976) Relationship of yield, seed quality, and fiber properties in Upland cotton. Crop Sci 16:578–580

Vroh Bi I, Baudoin J, Hau B et al (1999) Development of high-gossypol cotton plants with low-gossypol seeds using trispecies bridge crosses and in vitro culture of seed embryos. Euphytica 106:243–251

Wang H, Guo J, Lambert KN, Lin Y (2007a) Developmental control of Arabidopsis seed oil biosynthesis. Planta 226:773–783

Wang HW, Zhang B, Hao YJ, Huang J, Tian AG et al (2007b) The soybean Dof-type transcription factor genes, GmDof4 and GmDof11, enhance lipid content in the seeds of transgenic Arabidopsis plants. Plant J 52:716–729.

Wang N, Ma J, Pei W, Wu M, Li H et al (2017) A genome-wide analysis of the lysophosphatidate acyltransferase (LPAAT) gene family in cotton: organization, expression, sequence variation, and association with seed oil content and fiber quality. BMC Genomics 18:218

Weiss SB, Kennedy EP, Kiyasu JY (1960) The enzymatic synthesis of triglycerides. J Biol Chem 235:40–44

Wendel JF, Cronn RC (2003) Polyploidy and the evolutionary history of cotton. Adv Agron 78:139–186

Wendel JF, Grover CE (2015) Taxonomy and evolution of the cotton genus, *Gossypium*. 25-44. *In*: Fang DD, Percy RG, Cotton., Amer. Soc. Agron., Madison, WI

Wu J, Jenkins JN, McCarty JC, Thaxton P (2009) Seed trait evaluation of *Gossypium barbadense* L. chromosomes/arms in a *G. hirsutum* L. background. Euphytica 167:371–380

Xu Z, Li J, Guo X, Jin S, Zhang X (2016) Metabolic engineering of cottonseed oil biosynthesis pathway via RNA interference. Sci Rep 6:33342

Yeap WC, Lee FC, Shabari Shan DK, Musa H, Appleton DR et al (2017) WRI 1-1, ABI 5, NF-YA 3, and NF-YC 2 increase oil biosynthesis in coordination with hormonal signaling during fruit development in oil palm. Plant J 91:97–113

Yu XH, Rawat R, Shanklin J (2011) Characterization and analysis of the cotton cyclopropane fatty acid synthase family and their contribution to cyclopropane fatty acid synthesis. BMC Plant Biol 11:97

Yu J, Yu S, Fan S, Song M, Zhai H et al (2012) Mapping quantitative trait loci for cottonseed oil, protein, and gossypol content in a *Gossypium hirsutum*× *Gossypium barbadense* backcross inbred line population. Euphytica 187:191–201

Yu J, Jung S, Cheng CH, Ficklin SP, Lee T et al (2013) CottonGen: a genomics, genetics, and breeding database for cotton research. Nucleic Acids Res 42:D1229–D1236

Yuan Y, Wang X, Wang L, Xing H, Wang Q et al (2018) Genome-wide association study identifies candidate genes related to seed oil composition and protein content in *Gossypium hirsutum* L. Front Plant Sci 9:1359

Zhang D, Pirtle IL, Park SJ, Nampaisansuk M, Neogi P et al (2009) Identification and expression of a new delta-12 fatty acid desaturase (FAD2-4) gene in Upland cotton and its functional expression in yeast and *Arabidopsis thaliana* plants. Plant Physiol Biochem 47:462–471

Zhang J, Yu J, Pei W, Li X, Said J et al (2015a) Genetic analysis of Verticillium wilt resistance in a backcross inbred line population and a meta-analysis of quantitative trait loci for disease resistance in cotton. BMC Genomics 16:577

Zhang T, Hu Y, Jiang W, Fang L, Guan X et al (2015b) Sequencing of allotetraploid cotton (*Gossypium hirsutum* L. acc. TM-1) provides a resource for fiber improvement. Nat Biotechnol 33:531–537

Zhang Y, Maximova SN, Guiltinan MJ (2015c) Characterization of a stearoyl-acyl carrier protein desaturase gene family from chocolate tree, *Theobroma cacao* L. Front Plant Sci 6:239–239

Zhao Y, Huang Y, Wang Y, Cui Y, Liu Z et al (2018a) RNA interference of GhPEPC2 enhanced seed oil accumulation and salt tolerance in Upland cotton. Plant Sci 271:52–61

Zhao Y, Liu Z, Wang X, Wang Y, Hua J (2018b) Molecular characterization and expression analysis of GhWRI1 in Upland cotton. J Plant Biol 61:186–197

Zhao Y, Wang Y, Huang Y, Cui Y, Hua J (2018c) Gene network of oil accumulation reveals expression profiles in developing embryos and fatty acid composition in Upland cotton. J Plant Physiol 228:101–112

Chapter 4
Olive-Tree Genome Sequencing: Towards a Better Understanding of Oil Biosynthesis

Mehtap Aydin, Huseyin Tombuloglu, Pilar Hernandez, Gabriel Dorado, and Turgay Unver

Contents

M. Aydin
Genetics and Bioengineering Department, Yeditepe University, Istanbul, Turkey

H. Tombuloglu (✉)
Department of Genetics Research, Institute for Research and Medical Consultations (IRMC), Imam Abdulrahman Bin Faisal University, Dammam, Saudi Arabia
e-mail: htoglu@iua.edu.sa

P. Hernandez
Institute for Sustainable Agriculture (IAS-CSIC), Consejo Superior de Investigaciones Científicas (CSIC), Córdoba, Spain

G. Dorado
Dep. Bioquímica y Biología Molecular, Campus Rabanales C6-1-E17, Campus de Excelencia Internacional Agroalimentario (ceiA3), Universidad de Córdoba, Córdoba, Spain

T. Unver
Ficus Biotechnology, Ostim Teknopark, Ankara, Turkey

H. Tombuloglu et al. (eds.), *Oil Crop Genomics*,
https://doi.org/10.1007/978-3-030-70420-9_4

4.1 Introduction

Olive (*Olea europaea* L.) is an industrial plant of economic importance, especially cultivated by countries in the Mediterranean Basin (~75 of world's total production of olive oil). According to the FAOSTAT (2018) data, the olive-tree cultivation is carried out in 42 countries. The top five olive-producing ones are: Spain (9.8 million tonnes), Italy (1.8 million t), Morocco (1.5 million t), Turkey (1.5 million t), and Greece (1.07 million t) (Fig. 4.1) http://www.fao.org/faostat/en/#home. As an example of the recent expansion of this crop, state-supported olive-tree planting in Saudi Arabia is carried out especially in the northern regions of the country (i.e., Al-Jouf), where the climatic conditions are appropriate. The planting of mainly the Picual Spanish cultivar began in 2007, reaching ~52,000 acres, with a production of ~20 thousand, as reported by Index Mundi for 2020 https://www.indexmundi.com/agriculture/?country=sa&commodity=olive-oil&graph=production).

Besides the economic importance of olive-tree cultivation, the benefits of its products for human health have been revealed. It has been shown that olive oil has a protective effect especially against cardiovascular, nervous system diseases and cancer, due to its rich content of monounsaturated fatty acids and minor antioxidant components, like phenolic ones (Ruiz-Gutierrez et al. 1998; Visioli and Galli 1998). Indeed, the polyphenolic components (hydroxytyrosol, tyrosol, secoiridoids, lignans and squalenes) (Owen et al. 2004) that make up the composition of olive oil have protective effects against diseases such as breast, colon, stomach, and blood cancer. In vitro studies have revealed that olive oil extracts have an antigenotoxic

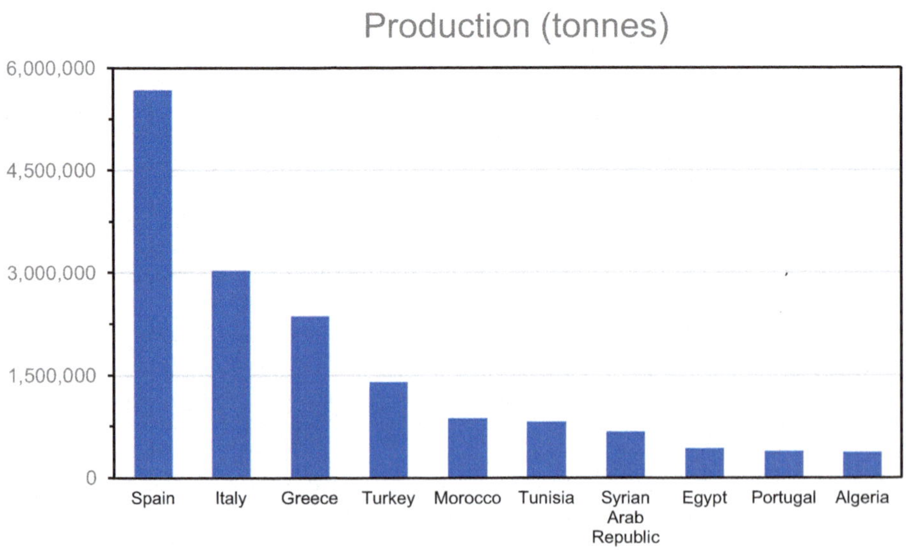

Fig. 4.1 Top 10 olive producers in the world in 1994–2018 (FAOSTAT 2018; http://www.fao.org/faostat/en/#home)

effect on DNA. That protects against colorectal carcinogenesis and decreases the spread (metastasis) of cancer cells, depending on dose and time (Gill et al. 2005). In another study, olive (fruit) extracts decreased colon cancer cell proliferation and lead them to death by apoptosis (Juan et al. 2006; Isik et al. 2012). Studies on olive oil are not only limited to their benefits for human health. Molecular analyses are essential for increasing the content of essential oils and other components that give the olive oil its special organoleptic characteristics, like aroma or antioxidative potential. That allows producing higher-quality olive oil by plant breeding (Angerosa et al. 2001; Perez et al. 2003; Luaces et al. 2008; Sanchez-Ortiz et al. 2012). As an example, studies have revealed that the lipoxygenase pathway involves the lipoxygenase (LOX) enzyme, which contributes to the aroma of olive oil (Olias et al. 1993). Likewise, a series of esters, together with 6 carbon aldehydes and alcohols, has been found to contribute to the aroma of olive oil (Morales et al. 2008; Angerosa et al. 2000). The length of the alkyl chains varies between 3 and 22 carbon atoms. They are used as an energy source in the body.

4.2 Content of Olive Oil

Triacylglycerols or triglycerides (TAG; Fig. 4.2) constitute the largest part of olive oil (~98%). Most of them (~72%) are monounsaturated fatty acids, like oleic (18:1; 55 to 83%) and palmitoleic (16:1; 0.3 to 3.5%) acids. Saturated ones include palmitic (16:0; 7.5 to 20%) and stearic (18:0; 0.5 to 5%) acids. Polyunsaturated fatty acids (PUFA) in olive oil are linoleic (18:2; 3.5 to 21%) and linolenic (18:3; 0 to 1.5%) acids. These rates depend on the cultivar of olive tree, climate conditions, harvesting time of olive fruits (ripening stage), storage, and olive-oil extraction methodology in the mill (Tsimidou et al. 2003; Beltran et al. 2004; Unver et al. 2017), as shown by The Olive Oil Source (http://www.oliveoilsource.com/page/chemical-characteristics).

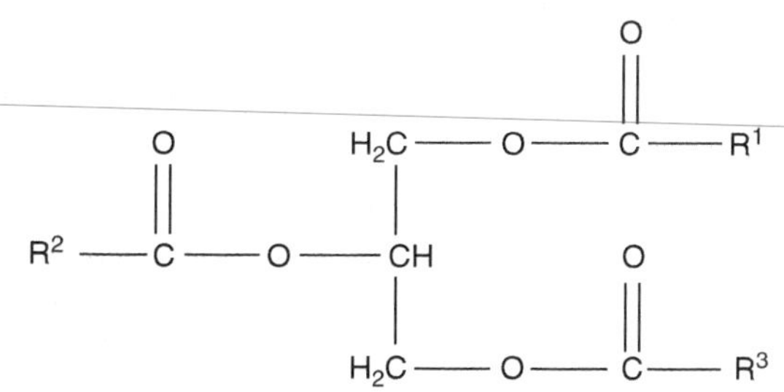

Fig. 4.2 General scheme of triglycerides. R^1, R^2, and R^3 represent long alkyl chains

4.3 Olive-Oil Biosynthesis

Olive oil biosynthesis involves key genes and enzymes. The synthesis of fatty acids in olives is initiated by the formation of palmitate and stearate in plastids. The latter is then saturated by the stearyl-ACP D9 desaturase enzyme. Oleate is either (i) unsaturated during lipid synthesis in the plastid or (ii) transferred from such organelles to the endoplasmic reticulum, where it is turned into unsaturated, being further included in the Kennedy pathway for lipid formation (Gunstone and Harwood 2007). It is also included in acyl-CoA-independent mechanisms, such as phospholipid: diacylglycerol acyltransferase (PDAT) or diacylglycerol: diacylglycerol transacylase (DGTA) (Stobart et al. 1997; Dahlqvist et al. 2000; Hernandez et al. 2008). Apart from this mechanism, it is also estimated that there may be others that may affect the abundance of intermediate molecules. That is responsible for the different concentrations of fatty acids in olive oils from different olive-tree cultivars and ripening stage. The latter has been studied in the olive tree, as well as other species. For instance, carambola or star fish (*Averrhoa carambola*) fruits collected in different months exhibit such variations. Thus, their oil composition changes involve palmitic, palmitoleic, stearic, oleic, cis-vaccenic (18:1c11), linoleic, and alpha-linolenic (18:3) acids (Fatima et al. 2012). Such authors further conducted high-throughput transcriptome sequencing on four different sea buckthorn (*Hippophae rhamnoides*) varieties having different oil efficiency. This way, they identified candidate genes for oil biosynthesis (Fig. 4.3).

4.4 Genome Sequencing and Analyses

4.4.1 Genome Sequencing and Assembly

There are two main olive tree types, mainly grown in the Mediterranean Basin and Near East: the cultivated type (*Olea europaea* L. subsp. *Europaea* var. europaea) and the wild type (*Olea europaea* subsp. *Europaea* var. *sylvestris*). Both have 46 chromosomes (2n). The genome of the domesticated olive cultivar 'Farga' was released in 2016, with a total of 543 Gbp of raw DNA sequence. The genome assembly resulted in 1.31 Gbp (Cruz et al. 2016). The genome of oleaster was shotgun-sequenced with a coverage of 220×, generating 515.7 Gbp of data (Unver et al. 2017). Assembling of the sequence reads resulted in 1.48 Gbp of draft genome, with a quality-assembled genome (N50) of 228 kbp. These findings are in line with the genome-size estimations carried out using flow cytometer and *k*-mer analyses (~1.46 Gbp). The sequence reads longer than 1 kbp (~572 Mbp) were anchored into 23 linkage groups, corresponding to a newly generated genetic map.

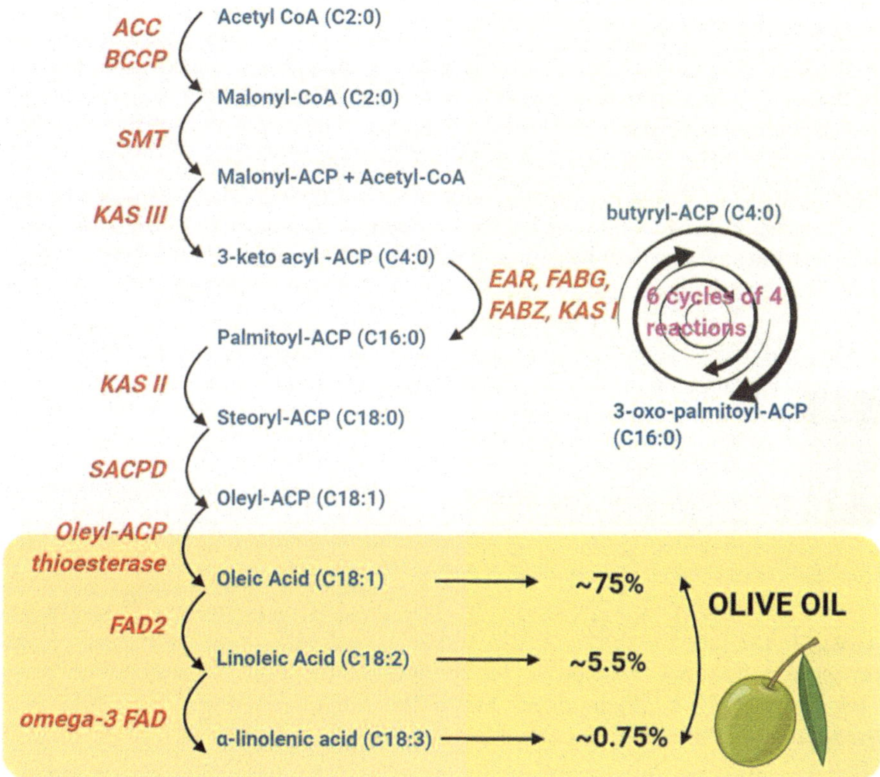

Fig. 4.3 Fatty acid biosynthesis pathway of olive oil. The final product includes ~75% of oleic acid, ~5.5% of linoleic acid, and ~ 0.75% of α-linolenic acid

4.4.2 Genome Annotation

The genome annotation of the oleaster genome was carried out using different methods, including ab initio and homology-based predictions, as well as transcriptome mapping. It was found that a large proportion of the wild olive-tree genome is composed of repetitive DNA, including transposable elements (TE; ~43%) and interspersed repeats (~51%). Long terminal repeats (LTR) are the most abundant type of TE (~40.3% of genome), followed by DNA-type TE (~4.6% of genome). The number of predicted protein-encoding genes was 50,684. Among those, 47,124 genes (93%) were confirmed by transcriptome (RNA-seq) analyses (Unver et al. 2017). The number of protein-coding genes was 56,349 in the domesticated cultivar of olive tree (cv. Farga) (Cruz et al. 2016). It was found that the wild type cultivar of olive tree harbors 31,245 genes which were located on the anchored pseudochromosomes. For the annotation of non-coding RNA (ncRNA), the data from ~90 × 10^6 sRNA reads obtained from different tissues (leaf, stem, fruit and pedicel) were used.

The results revealed that oleaster harbors 498 conserved miRNA families and 125 novel miRNAs. It was predicted that the identified miRNA target 9842 transcripts, of which 7849 are unique. The targeted genes are mainly associated with transcription factors (4606), stress-response (1937), and metabolism (630) (Unver et al. 2017).

The functional prediction of protein-encoding genes of wild olive-tree by Gene Ontology (GO) and Kyoto Encyclopedia of Genes and Genomes (KEGG) led to the annotation of 72.42% and 50.14% of all genes, respectively. The metabolic pathway annotations of oleaster and selected plants, such as oil crops like sesame (*Sesamum indicum*) and soybean or soya (*Glycine max*), the close relatives of wild olive (bladderwort, monkey flower and European ash-tree), and a tree reference-genome (poplar), were analyzed. Thus, oleaster genes are mostly functional in protein folding, sorting, and degradation (4263); biosynthesis of secondary metabolites (2236); carbohydrate metabolism (1905); and lipid metabolism (811) (Unver et al. 2017).

4.4.3 Olive-Genome Evolution

The asterids (Asteridae) are the largest subgroup of the flowering plants, consisting of about 100,000 species, grouped into 110 families in 17 orders. The olive tree belongs to the Lamiales order (~23,810 species, including mint, basil, sesame) and the Oleaceae family (~688 species http://www.plantlist.org/1.1/browse/A/Oleaceae (Wortley et al. 2005; Zhang et al. 2020). The Lamiales order is one of the most diverse and the largest group of asterids. Their diversification mostly depends on whole genome duplications (WGD), which in turn strongly contribute to genome structure and complexity of species (Van de Peer et al. 2017). Oleasters are found in two gene pools, corresponding to the Mediterranean Basin and the East and the West of Near Eastern, respectively. Cultivated olives show a geographical differentiation between the East, Middle, and West of the Mediterranean Basin. However, a dense mixture has occurred in the western part of these gene pools. It has been discussed whether these gene pools are domesticated or not (Gros-Balthazard et al. 2019).

Interestingly, analyses of the wild olive-tree genome revealed multiple signatures of paleopolyploidy events. The distributions of synonymous substitutions per synonymous site (K_S) for the whole paranome (set of all duplicated genes in the genome) showed two certain peaks of duplicates at K_S values around 0.25 and 0.75, respectively (Fig. 4.4). The estimations pointed out two rounds of WGD events, dated around 26–30 million years ago (Mya) and 57–63 Mya, respectively. Similarly, the genome of European ash (*Fraxinus excelsior*), a close relative of olive tree and a member of Oleaceae family, has two WGDs (Sollars et al. 2017). The fourfold synonymous third-codon transversion rates (4DTv) analysis confirmed this finding (Unver et al. 2017). Contrarily, sesame, a member of the same Lamiales order than olive tree, has only one WGD in its genome, suggesting the divergence of Oleaceae (wild olive and ash) and sesame. In addition, the two rounds of WGD in olive and

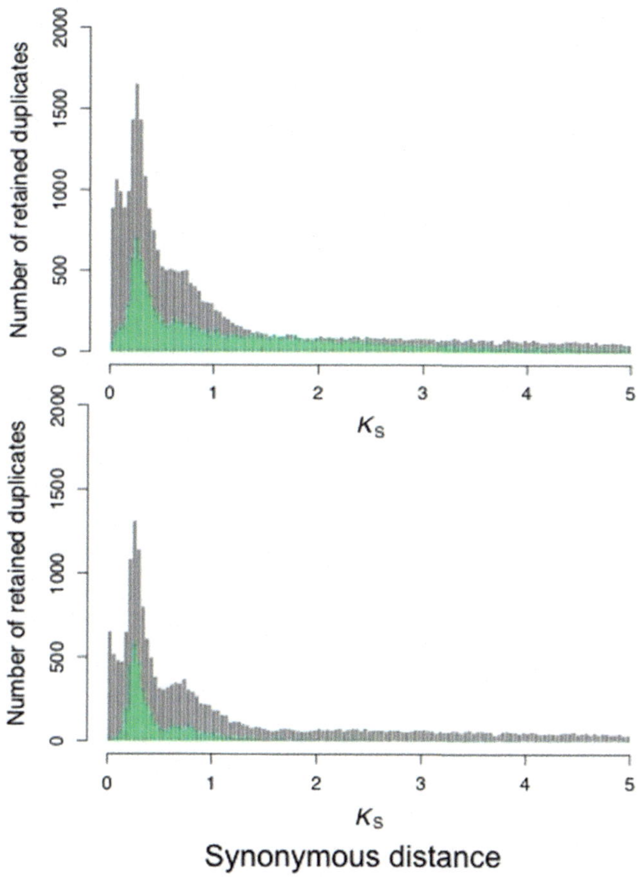

Fig. 4.4 Distributions of synonymous substitutions per synonymous site (K_S) for the whole paranome showed two certain peaks of duplicates at K_S values around 0.25 and 0.75, respectively (Modified from Unver et al. 2017)

ash tree are specific to Oleaceae and occur independently of the WGD in the lineage leading to sesame (Unver et al. 2017).

The possible activation of TE and their effects on the olive-tree genome during domestication has been investigated by Jiménez-Ruiz et al. (2020). For this purpose, the genome of Picual cultivar (the most cultivated Spanish variety) was sequenced and used as the reference genotype. A total of 40 additional cultivated genotypes and 10 oleaster genotypes were sequenced, and their phylogenetic relationships were examined. Besides, DNA from samples which existed in the Roman prehistory in Baetica (southern Spain) were also sequenced. Excavations covered most of the invasions from first and second centuries BC and the first century after Christ; they also covered the present Granada and Jaén provinces of Andalusia, which are the world's most important olive-oil production regions. Analyses of these DNA sequences may shed light not only on the origin of the olive-trees found in the South of Spain but on human history as well (Jiménez-Ruiz et al. 2020).

Although the olive tree is fertile, such crop has been traditionally bred clonally. That is due to the slow growth and long life of these trees. Indeed, the olive tree has a long juvenile period. Thus, it has been reported that only 20% of seedlings reach flowering 12–13 years after seed germination (Santos-Antunes et al. 2005). That may explain the small genetic differences found between commercial cultivars. In spite of such difficulty, ~2000 olive-tree cultivars have been obtained by sexual reproduction and grown for generations, probably with high phenotypic variability, albeit at a limited commercial scale (Gros-Balthazard et al. 2019). Recently, large-scale commercial breeding through sexual reproduction (crossing and selection) has been carried out, in order to increase variability and generate new varieties like Sikitita (patented as Chiquitita in USA) for high-density hedgerow and intensive cultivation (Rallo et al. 2008, 2011).

Phenotypic variability associated with domestication is produced at a shorter time than the natural variability associated with speciation. It should be explained whether the original kinds were dependent on their former variability or whether there are genetic mechanisms increasing variability during the domestication process. Although no clear answer has been given to this question, both mechanisms are likely involved. This way, TE may play a role in the production of natural or human-driven variability (Jiménez-Ruiz et al. 2020).

4.4.4 Role of Key Genes in Oil Biosynthesis

Olive oil is traditionally obtained by the squeezing of the olive fruit. In other words, virgin olive oil is a natural fruit juice obtained by mechanical processes, not involving the use of organic solvents, heat, etc. The olives consist of oil (20–30%), cellulose (17%), carbohydrate (4%), protein (2%), micro-nutrients (0.1%), and water (46,9–56,9%). Polyols and oligosaccharides are synthesized on the leaves of the olive tree. Long-chain fatty acids are synthesized and disintegrated from the carbon source (Unver et al. 2017). A recent study conducted on 113 cultivars from five germplasm collections of olive trees revealed that the fruit fresh weight trait was mainly under genetic control. The oleic/(palmitic + linoleic) acid ratio was regulated by the environment and genotype, having the former the major effect on oil content (Mousavi et al. 2019).

Fatty-acid biosynthesis is one of the most important steps of TAG biosynthesis. It includes the biosynthesis of saturated fatty acids, their elongation and degradation, performed through the activity of many genes encoding the corresponding enzymes. TAG are synthesized from glycerol esterified with three fatty acids. Although unsaturated fatty acid (PUFA) pathways are common in plants, an important group of genes was highlighted when comparing oleaster and sesame. The contraction of genes encoding catabolic enzymes may be responsible for accumulation of different fatty acids. The gene counts for linoleic acid biosynthesis revealed 20 for oleaster and 164 for sesame, showing a striking difference between the species (Hernandez et al. 2019). Fatty acids can be unsaturated via desaturase enzymes, like stearoyl-acyl-carrier-protein (stearoyl-ACP). PUFA (especially, linoleic, and linolenic acids) play key roles in plant metabolism. In addition, they are particularly important for signaling pathways (Hernandez et al. 2019). Previous studies have described the genes involved in oil biosynthesis. They were grouped according to enzyme codes and identities. Among 308 pathways identified, comparison of some of them including Ca^{2+}- transport ATPase (K01537) and acyl-CoA oxidase (K00232) revealed that they are more represented in the oleaster genome than in other plants. Paranoid ortholog analyses were performed on sesame and oleaster genes, in order to understand the evolution of oil biosynthesis between the latter and other high oil-producing plants. Thus, among 2327 oil biosynthesis genes in oleaster, 2025 seem to have homologs in sesame (Unver et al. 2017; Gros-Balthazard et al. 2019) (Fig. 4.5).

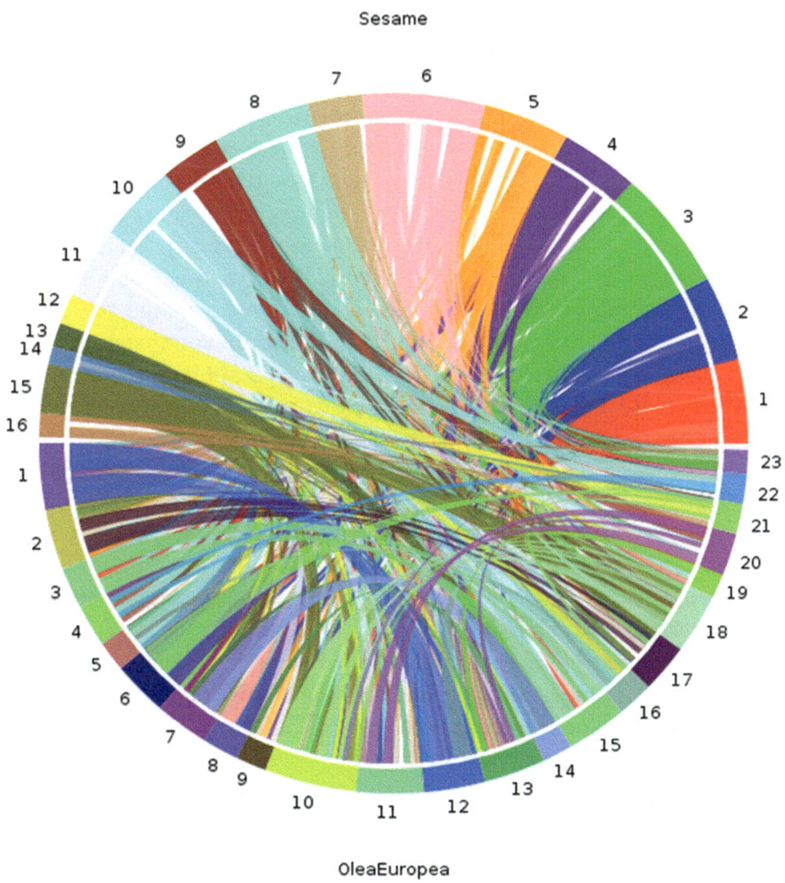

Fig. 4.5 Syntenic relationships of wild olive-tree and sesame chromosomes

Functional analyses of gene expression of tissues obtained from underripe and ripe fruits showed that duplicated oleaster fatty-acid desaturase genes (*FAD2*, *FAD2-1*, *FAD2-2*, *FAD2-4* and *FAD2-5*) showed downregulation on the fruit tissues during the maturation phase (Unver et al. 2017). In addition, the same study revealed that some key functional genes in the PUFA pathway have been expanded by WGD and/or segmental duplications, such as the ones encoding enoyl-ACP reductase (EAR), β-ketoacyl-ACP synthase II (KASII), β-ketoacyl-ACP reductase (FabG), acyl carrier protein (ACP)-hydrolase/thioesterase (ACPTE), and stearoyl-acyl carrier protein desaturase (SACPD).

4.4.5 Analyses of Repetitive Sequences

Large genomes are full of repetitive sequences in plants. Although some repeating sequences seem to be non-functional, others play relevant roles in species evolution. Such biological mechanisms should have appeared as a result of production of extra copies of a sequence. Thus, repetitions are generated. They can split by wide ranges, generating millions of copies, varying between one to thousands of bases. Some repetitive families may represent a significant portion of genomic DNA. Repetition families do not significantly exist in all large genomes. However, many low-repetitive classes may represent most genomic DNA (Cruz et al. 2016).

The importance of the olive-tree derived products (table olives and olive oil) has been acknowledged for a long time. That is due to their organoleptic, cooking, and health benefits. Yet, its genome was mostly unknown, and has only been recently sequenced. The olive tree has a middle-sized haploid genome of 1.4–1.5 Gbp (Contento et al. 2002). Homology-based and transcriptome-mapping approaches have been used to study oleaster TE. The former approach involves repetitive sequencing with RepeatProteinMask and RepeatMasker to identify the repeat element boundaries and family relationships among sequences. RepeatModeler (http://www.repeatmasker.org/RepeatModeler) was used with two ab initio repeat-prediction programs (RECON and RepeatScout). The results showed that 51% of the olive-tree genome is comprised of repetitive sequences. The ratio of TE and interspersed repeats in the assembled genome is about 43%. LTR were the most common elements, taking ~40.3% of the genome. DNA-type TE follow it, with 4.6% of the genome (Unver et al. 2017).

Genome sequence of the Picual cultivar revealed a high genetic diversity, driven by activation of TE (Jiménez-Ruiz et al. 2020). The same study indicated an expansion on TE families in domesticated olive trees, about 5000 years ago, and their insertion near to some critical genes. The latter are related to important traits, such as reproduction, photosynthesis, seed development, and oil production. Therefore, the expansion and activity of TE in domesticated olive trees may have affected their genetic diversity. In addition, and interestingly, there is a small interfering-RNA (siRNA), derived from a TE-rich region, targeting the fatty-acid desaturase 2 (*FAD2*) mRNA, which suppresses its expression during oil production. Such suppression of gene expression leads to accumulation of exceptionally high levels of oleic acid in olive trees (Unver et al. 2017).

4.4.6 Analyses of miRNA

Micro-RNA (miRNA) is an RNA with a length range of 20–24 bases. They exist in eukaryotes and do not encode proteins. Such small RNA molecules have regulatory functions, modulating growth and stress reactions of plants. It has been previously shown that they play critical roles in the regulation of gene expression, for instance during growth, florescence, signal transmission, protein degradation, and biotic and abiotic stresses (Eldem et al. 2013). The first catalogue of sRNA from olive trees was generated by taking advantage of high-throughput pyrosequencing and RNA hybridization technologies. Different olive-tree cultivars were analyzed in order to understand the function of miRNA during growth and development (Donaire et al. 2011). Together with, 93,526,915 raw readings were produced from six transcriptome libraries of olive tree. One hundred and thirty-five miRNA which were previously known and 38 new miRNA belonging to 22 families were detected (Yanik et al. 2013). On the other hand, annotation of ncRNAs in six different wild olive-tree libraries resulted in 90 millions of small-RNA reads. The analyses revealed 498 conserved miRNA families and 125 novel miRNAs. The prediction of miRNA targets exhibited 29,842 miRNA–target pairs. Among those, 7849 target genes were unique. These target genes were associated with transcription factors (4606), stress-response (1937), and metabolism (630) (Unver et al. 2017). Interestingly, transfected synthetic olive-tree miRNA showed functional homology to human miR34a (*hsa-miR34A*), which is strongly associated with critical biological events. They include development, differentiation, inflammation, apoptosis, and carcinogenesis. The transfection resulted in reduced protein expression of *hsa*-miR34a mRNA targets. Moreover, increased apoptosis and decreased proliferation in different tumor cells (THP-1 monocytoids and Human Jurkat E6-1 lymphoid cells) have been observed. These findings revealed the development of plant-derived tumor-suppressing small RNA, with functional homology to *hsa*-miRNA (Minutolo et al. 2018).These findings highlight the critical roles of miRNA in olive-tree physiology, including growth and metabolism.

4.5 Future Perspectives

The olive tree is one of the main cultural, characteristic, and economically important plants of the Mediterranean Basin. The virgin olive oil obtained from its fruits is a high-value product in terms of its benefits in nutrition, health, cosmetics, etc. Recent advances in sequencing technologies open a new era for plant biotechnology. Indeed, they have been used for sequencing the genome and transcriptomes of different olive-tree varieties (Farga and Picual) and oleasters. In particular, many transcriptome libraries have been constructed to understand the high-value products derived from olive trees, the domestication of this species, and its evolution. Genomics and transcriptomics data displayed some key genes that are effective in oil metabolism, disease resistance, stress responses, etc. Further studies may focus on the functional identification of selected genes that are important to breeders, using modern biotechnological approaches, such as cloning, CRISPR/Cas9, etc.

References

Angerosa F, Mostallino R, Basti C, Vito R (2000) Virgin olive oil odour notes: their relationships with the volatile compound from the lipoxigenase pathway and secoiridoid compounds. Food Chem 68:283–287

Angerosa F, Mostallino R, Basti C, Vito R (2001) Influence of malaxation temperature and time on the quality of virgin olive oils. Food Chem 72:19–28

Beltran G, Rio CD, Sanchez S, Martinez L (2004) Influence of harvest date and crop yield on the fatty acid composition of virgin olive oils from Cv. Picual. J Agric Food Chem 52:3434–3440

Contento A, Ceccarelli M, Gelati M, Maggini F, Baldoni L, Cionini P (2002) Diversity of Olea genotypes and the origin of cultivated olives. Theor Appl Genet 104(8):1229–1238

Cruz F, Julca I, Gómez-Garrido J, Loska D, Marcet-Houben M, Cano E et al (2016) Genome sequence of the olive tree, Olea europaea. GigaScience 5(1):s13742–s13016

Dahlqvist A, Stahl U, Lenman M, Banas A, Lee M, Sandager L, Ronne H, Stymne S (2000) Phospholipid:diacylglycerol acyl-transferase: an enzyme that catalyzes the acyl-CoA-independent formation of triacylglycerol in yeast and plants. Proc Natl Acad Sci U S A 97:6487–6492

Donaire L, Pedrola L, de la Rosa R, Llave C (2011) High-throughput sequencing of RNA silencing-associated small RNAs in olive (*Olea europaea* L.). PLoS One 6(11):e27916

Eldem V, Okay S, Unver T (2013) Plant MicroRNAs: new players in functional genomics. Turk J Agric For 37:1–21

Fatima T, Snyder CL, Schroeder WR, Cram D, Datla R, Wishart D, Weselake RJ, Krishna P (2012) Fatty acid composition of developing sea buckthorn (*Hippophae rhamnoides* L.) berry and the transcriptome of the mature seed. PLoS One 7(4):e34099

FAOSTAT F (2018) Crop statistics. https://www.fao.org/statistics/en/

Gill I, Boyd A, McDermott E, McCann M, Servili M, Selvaggini R (2005) Potential anti-cancer effects of virgin olive oil phenols on colourectal carcinogenesis models in vitro. Int J Cancer 117(1):1–7

Gros-Balthazard M, Besnard G, Sarah G, Holtz Y, Leclercq J, Santoni S, Wegman D, Glemin S, Khadari B (2019) Evolutionary transcriptomics reveals the origins of olives and the genomic changes associated with their domestication. Plant J 100:143–157

Gunstone FD, Harwood JL (2007) Occurrence and characterisation of oils and fats. In The lipid handbook with CD-ROM (pp. 51–156). CRC press

Hernández ML, Guschina IA, Martínez-Rivas JM, Mancha M, Harwood JL (2008) The utilization and desaturation of oleate and linoleate during glycerolipid biosynthesis in olive (*Olea europaea* L.) callus cultures. J Exp Bot 59(9):2425–2435

Hernández ML, Sicardo MD, Arjona PM, Rivas Jose MM (2019) Specialized functions of olive *FAD2* gene family members related to fruit development and the abiotic stress response. Plant Cell Physiol 61:427–441

Isik S, Karagoz A, Karaman S, Nergiz C (2012) Proliferative and apoptotic effects of olive extracts on cell lines and healthy human cells. Food Chem 134:29–36

Jiménez-Ruiz J, Ramírez-Tejero JA, Fernández-Pozo N, Leyva-Pérez MDLO, Yan H, Rosa RDL et al (2020) Transposon activation is a major driver in the genome evolution of cultivated olive trees (*Olea europaea* L.). Plant Genome 13(1):e20010

Juan ME, Wenzel U, Ruiz-Gutierrez V, Daniel H, Planas JM (2006) Olive fruit extracts inhibit proliferation and induce apoptosis in HT-29 human colon cancer cells. J Nutr 136(10):2553–2557

Luaces P, Perez AG, Sanz C (2008) Effect of the blanching process and olive fruit temperature at milling on the biosynthesis of olive oil aroma. Eur Food Res Technol 224:11–17

Minutolo A, Potestà M, Gismondi A, Pirrò S, Cirilli M, Gattabria F et al (2018) Olea europaea small RNA with functional homology to human miR34a in cross-kingdom interaction of anti-tumoral response. Sci Rep 8(1):1–14

Morales MT, Aparicio R, Rios JJ (2008) Dynamic headspace gas chromatographic method for determining volatiles in virgin olive oil. J Chromatogr 668:455–462

Mousavi S, de la Rosa R, Moukhli A, El Riachy M, Mariotti R, Torres M et al (2019) Plasticity of fruit and oil traits in olive among different environments. Sci Rep 9(1):1–13

Olias JM, Perez AG, Rios JJ, Sanz C (1993) Aroma of virgin olive oil: biogenesis of the green odor notes. J Agric Food Chem 41:2368–2373

Owen RW, Haubner R, Wurtele G, Hull E, Spiegelhalder B, Bartsch H (2004) Olives and olive oil in cancer prevention. Eur J Cancer Prev 13(4):319–326

Perez AG, Luaces P, Rios JJ, Garcia JM, Sanz C (2003) Modification of volatile compound profile of virgin olive oil due to hot-water treatment of olive fruit. J Agric Food Chem 51:6544–6549

Rallo L, Barranco D, DeLaRosa R, León L (2008) 'Chiquitita' olive. Hortic Sci 43:529–531

Rallo L, Barranco D, DeLaRosa R, León L (2011) Advances in the joint UCO-IFAPA olive breeding program (JOBP). Acta Hortic 924:283–290

Ruiz-Gutierrez V, Muriana FJG, Villar J (1998) Virgin olive oil and cardiovascular diseases. Plasma lipid profile and lipid composition of human erythrocyte membrane. Grasas Aceites 49:9–29

Sanchez-Ortiz A, Romero-Segura C, Gazda VE, Graham IA, Sanz C, Perez AG (2012) Factors limiting the synthesis of virgin olive oil volatile esters. J Agric Food Chem 60:1300–1307

Santos-Antunes F, Leon L, De la Rosa R, Alvarado J, Mohedo A, Trujillo I, Rallo L (2005) The length of the juvenile period in olive as influenced by vigor of the seedlings and the precocity of the parents. Hortic Sci 40:1213–1215

Sollars ES et al (2017) Genome sequence and genetic diversity of European ash trees. Nature 541:212–216

Stobart K, Mancha M, Lenman M, Dahlqvist A, Stymne S (1997) Triacylglycerols are synthesized and utilized by trans-acylation reactions in microsomal preparations of developing safflower (*Carthamus tinctorius* L.) seeds. Planta 203:58–66

Tsimidou M, Blekas G, Boskou D (2003) Olive oil. In: Caballero B (ed) Encyclopedia of food sciences and nutrition, 2nd edn. Academic, Cambridge, MA, pp 4252–4260

Unver T, Wu Z, Sterck L, Turktas M, Lohaus R, Li Z et al (2017) Genome of wild olive and the evolution of oil biosynthesis. Proc Natl Acad Sci U S A 114:44

Van de Peer Y, Mizrachi E, Marchal K (2017) The evolutionary significance of polyploidy. Nat Rev Genet 18(7):411–424

Visioli F, Galli C (1998) Olive oil phenols and their potential effects on human health. J Agric Food Chem 46:4292–4296

Wortley AH, Rudall PJ, Harris DJ, Scotland RW (2005) How much data are needed to resolve a difficult phylogeny?: case study in Lamiales. Syst Biol 54(5):697–709

Yanik H, Turktas M, Dündar E, Hernandez P, Dorado G, Unver T (2013) Genome-wide identification of alternate bearing-associated miRNAin the olive tree (*Olea europaea* L.). BMC Plant Biol 13:10

Zhang C, Zhang T, Luebert F, Xiang Y, Huang CH, Hu Y et al (2020) Asterid phylogenomics/phylotranscriptomics uncover morphological evolutionary histories and support phylogenetic placement for numerous whole genome duplications. Mol Biol Evol. https://doi.org/10.1093/molbev/msaa160

Chapter 5
Translational Genomics of Cucurbit Oil Seeds

Cecilia McGregor and Geoffrey Meru

Contents

Abbreviations

chr Chromosome
EST Expressed sequence tags
FAD2 Omega-6 fatty acid desaturase-2
KP Kernel percentage

C. McGregor (✉)
Institute of Plant Breeding, Genetics & Genomics and Department of Horticulture, University of Georgia, Athens, GA, USA
e-mail: cmcgre1@uga.edu

G. Meru
Department of Horticulture, Tropical Research & Education Center, University of Florida, IFAS, Homestead, FL, USA

H. Tombuloglu et al. (eds.), *Oil Crop Genomics*,
https://doi.org/10.1007/978-3-030-70420-9_5

lin Linoleic acid
ole Oleic acid
pal Palmitic acid
PVE Phenotypic variation explained
QTL Quantitative trait loci
sc Seed coat
SI Seed index
SOP Seed oil percentage
ss Seed size
ste Stearic acid
SYI Seed yield index

5.1 Introduction

The Cucurbitaceae family includes 95 genera and between 950 and 980 different species (Schaefer and Renner 2011). The most economically important crops are cucumber (*Cucumis sativus* L.), melon (*Cucumis melo* L.), pumpkins and squashes (*Cucurbita* spp.), and watermelon (*Citrullus lanatus* (Thunb.) Matsum. & Nakai) (Table 5.1). These crops are mainly produced for consumption of their mature or immature fruit. However, in certain parts of the world, edible seed is an important part of production. According to the Food and Agricultural Organization (FAOSTAT 2020), 953,605 tonnes of cucurbit seed was produced on 1,840,829 ha of land (Table 5.1) in 2018. This represents an increase of 66% in production and 50% in area harvested since 1994 (FAOSTAT 2020). In 2018 approximately 92% of production took place in Africa with 61% taking place in Nigeria (Table 5.2). However, cucurbit seed production numbers are probably underestimated since Zhang (1996) reported that approximately 200,000 tonnes of edible watermelon (*Citrullus lanatus*) seed was produced in China, compared to the 43,012 tonnes reported for all cucurbit seed by FAOSTAT (2020).

Cucurbit seeds are used in various ways, including as snacks, as extracted oil, and as cosmetics and are ground to a flour for use in soups and stews (Vermaak et al. 2011). Recently there has also been an increased interest in utilizing cucurbit seed for biofuel production (Giwa and Akanbi 2020). There is a large variability in the

Table 5.1 Worldwide cucurbit production in 2018 (FAOSTAT 2020)

Crop	Species	Area harvested (ha)	Production (tonnes)
Melons	*Cucumis melo*	1,047,283	27,349,214
Cucurbit seed	Various	1,840,829	953,605
Cucumbers and gherkins	*Cucumis sativus*	1,984,518	75,219,440
Pumpkins, squash, and gourds	*Cucurbita* spp.	2,042,955	27,643,932
Watermelons	*Citrullus lanatus*	3,241,239	103,931,337

Table 5.2 Worldwide annual cucurbit seed production by region and top 10 producing countries (1998–2018) (FAOSTAT 2020)

Region	Country	Average annual production (tonnes)	
Africa			704,775
	Nigeria	466,161	
	Sudan	58,864	
	Democratic Republic of Congo	54,104	
	Cameroon	53,997	
	Sudan (former)	44,929	
	Central African Republic	32,009	
	South Sudan	22,020	
	Chad	20,249	
Americas			9
Asia			51,650
	China, mainland	31,558	
	Iran (Islamic Republic of)	20,090	
Europe			4,850

amount of oil in different cucurbit species (Earle and Jones 1962), but on average 50% oil and 35% protein have been reported in seed kernels (Jacks et al. 1972).

5.2 Genomic Resources for Cucurbitaceae

Cucurbit genomes are relatively small compared to many other important crops (Arumuganathan and Earle 1991) which facilitates genome sequencing. Draft genomes for *Cucumis melo* (Garcia-Mas et al. 2012; Huang et al. 2009), *C. sativus* (Li et al. 2019, 2011; Qi et al. 2013; Yang et al. 2012), *Cucurbita maxima* (Sun et al. 2017), *C. moschata* (Sun et al. 2017), *C. pepo* (Montero-Pau et al. 2018), *C. argyrosperma* (Barrera-Redondo et al. 2019), *Citrullus lanatus* (Guo et al. 2013; Guo et al. 2019; Wu et al. 2019), *Lagenaria siceraria* (Wu et al. 2017), and *Benicia hispida* (Xie et al. 2019) are available at the Cucurbit Genomics Database (CuGenDB; http://cucurbitgenomics.org/) (Zheng et al. 2018) (Table 5.3). In addition to the annotated genomes, CuGenDB also includes transcriptome and expressed sequence tags (ESTs), genetic maps, and comparative genomics tools (e.g., SyntenyViewer) (Zheng et al. 2018). Many high-quality genetic maps (Gonzalo and Monforte 2017) have been developed for melon (Diaz et al. 2011; Pereira et al. 2018), cucumber (Wang et al. 2020a; Weng 2017), *Cucurbita* spp. (Esteras et al. 2012; Montero-Pau et al. 2017), and watermelon (Sandlin et al. 2012), and quantitative trait loci (QTL) have been identified for a large number of disease and fruit traits (Gonzalo and Monforte 2017; Pan et al. 2020). Despite all the available cucurbit genomic resources, relatively little research has been done to utilize these resources for translational genomics of cucurbit seed oil traits. Although seeds from many different

Table 5.3 Cucurbit draft genomes available publicly from the Cucurbit Genomics Database (http://cucurbitgenomics.org/)

Species	Genome size (Mb)	Crop	Cultivar/line	Citation
Cucumis melo ($2n = 2x = 24$)	454[a]	Melon	DHL92	Garcia-Mas et al. (2012)
Cucumis sativus ($2n = 2x = 14$)	367[a]	Cucumber	Chinese long, 9930	Huang et al. (2009) Li et al. (2011) Li et al. (2019)
Cucumis sativus ($2n = 2x = 14$)	367[a]	Cucumber	GY14	Yang et al. (2012)
Cucumis sativus ($2n = 2x = 14$)	367[a]	Cucumber	PI 183967	Qi et al. (2013)
Cucurbita maxima ($2n = 2x = 40$)	386.8[b]	Pumpkin/ squash	Rimu	Sun et al. (2017)
Cucurbita moschata ($2n = 2x = 40$)	372[b]	Pumpkin/ squash	Rifu	Sun et al. (2017)
Cucurbita pepo ($2n = 2x = 40$)	502[a]	Zucchini	MU-CU-16	Montero-Pau et al. (2018)
Cucurbita argyrosperma ($2n = 2x = 40$)	238[c]	Silver-seed gourd	–	Barrera-Redondo et al. (2019)
Citrullus lanatus ($2n = 2x = 22$)	435[a]	Watermelon	97,103	Guo et al. (2013) Guo et al. (2019)
Citrullus lanatus ($2n = 2x = 22$)	435[a]	Watermelon	Charleston gray	Wu et al. (2019)
Lagenaria siceraria $2n = 2x = 22$	334[d]	Bottle gourd	USVL1VR-Ls	Wu et al. (2017)
Benicia hispida	913[e]	Wax gourd	B227	Xie et al. (2019)

[a]Arumuganathan and Earle (1991)
[b]Sun et al. (2017)
[c]Barrera-Redondo et al. (2019)
[d]Achigan-Dako et al. (2008b)
[e]Xie et al. (2019)

cucurbit species are used as sources of oil, we will focus here on *Cucurbita* and *Citrullus* since most of the genomic research associated with cucurbit seed oil have been carried out for these genera (Table 5.4).

5.3 Cucurbita

The *Cucurbita* genus constitutes a major crop cultivated worldwide for edible flesh and seeds, as well as ornamental purposes (Cutler and Whitaker 1961). *Cucurbita* consists of 12 or 13 species (Nee 1990), 5 among which (*C. pepo*, *C. moschata*, *C. maxima*, *C. argyrosperma*, and *C. ficifolia*) are widely cultivated around the

Table 5.4 Chromosomal (Chr) positions (bp) of the QTL associated with seed size traits and the corresponding phenotypic variance explained (PVE) in two independent *C. maxima* F_2 populations

Trait	Name	Chr(LG)	PVE (%)	Position (bp)	Left flanking marker	References
Seed width	CuQ1	LG2	29.68	–	E15M51a	Tan et al. (2013)
Seed width	CuQ2	LG3	14.67	–	E20M54b	Tan et al. (2013)
Seed width	CuQ3	LG3	7.26	–	E21M53b	Tan et al. (2013)
Seed width	CuQ4	LG4	2.87	–	E20M53b	Tan et al. (2013)
Seed length	SL4-1	LG4	12.6	7,653,955	Marker93772	Wang et al. (2020a, b)
Seed length	SL6-1	LG6	38.6	2,517,576	Marker157285	Wang et al. (2020a, b)
Seed length	SL17-1	LG17	11.1	585,908	Marker421215	Wang et al. (2020a, b)
Seed length	SL18-1	LG18	7	7,718,862	Marker477731	Wang et al. (2020a, b)
Seed width	SW4-1	LG4	6.9	7,653,955	Marker93772	Wang et al. (2020a, b)
Seed width	SW5-1	LG5	10.9	4,546,137	Marker137649	Wang et al. (2020a, b)
Seed width	SW6-1	LG6	28.9	1,815,380	Marker154765	Wang et al. (2020a, b)
Seed width	SW8-1	LG8	13	2,504,270	Marker200328	Wang et al. (2020a, b)
100-seed weight	HSW6-1	LG6	13.2	3,588,530	Marker161001	Wang et al. (2020a, b)
100-seed weight	HSW17-1	LG17	17.2	2,047,202	Marker424340	Wang et al. (2020a, b)

world (Whitaker 1947; Moya-Hernández et al. 2018). *Cucurbita* is native to the Americas, with the majority of wild species occurring south of Mexico City and extending towards the Mexico-Guatemala border (Whitaker and Robinson 1986). The crop was distributed transcontinentally by voyagers in the sixteenth century (Whitaker 1947). Evidence supports *Cucurbita* as a New World plant and one of the most important crops pre-Colombian time (Cutler and Whitaker 1961). Although modern cultivation of *Cucurbita* is primarily for flesh consumption, it is likely that the original use of the crop was for seed consumption because the flesh of wild species is bitter (Cutler and Whitaker 1961; Paris 2016; Robinson and Decker-Walters 1997). Today, seeds of various *Cucurbita* species are a significant source of income and nutrition worldwide (Baxter et al. 2012; Fruhwirth and Hermetter 2007; Meru et al. 2018, 2019; Nakić et al. 2006). Records of commercial production of *C. pepo* pumpkins for seed consumption and oil production date back to the seventeenth century in Styria, Austria (Teppner 2004). Originally the landraces grown during this period had hulled seeds which necessitated manual de-hulling prior to use. However, towards the end of the nineteenth century, a spontaneous mutation resulted in the occurrence of hull-less seed phenotype (Teppner 2004). This trait allowed rapid expansion of seed pumpkin cultivation in Austria and across Europe because it eliminated the need for de-hulling prior to snacking or oil production (Lelley et al. 2009; Loy 2004; Teppner 2004). Currently, more than 18,000 ha of hull-less seed pumpkins are under cultivation in Austria (Lelley et al. 2009). In North America, seeds are primarily used as snack food in trail mixes with various nuts, seeds, dried fruit, as well as an ingredient in breakfast cereal and bread (Baxter et al. 2012; Loy 2004). Other minor uses include as seed oil which can be purchased by bottle or as formulated capsules in health-food stores (Stevenson et al. 2007). Pumpkin seeds used in North America are derived from cultivars of *C. maxima*, *C. pepo*, and *C. argyrosperma* (Loy 2004), which are grown primarily for flesh consumption and must be manually de-hulled prior to use (Lelley et al. 2009; Loy 2004). Seeds of *C. maxima* are primarily produced in Oregon or imported from China, while those of *C. argyrosperma* are imported from Central America (Loy 2004). In the United States, breeding efforts are focused towards development of high seed yield hull-less pumpkin cultivars (Loy 2004; Paris 2015), as well as cultivars that combine hull-less trait and superior flesh quality (Meru and Fu 2018).

5.3.1 Major Nutritional Components of Cucurbita *Seeds*

5.3.1.1 Seed Oil and Fatty Acid Composition

Cucurbita seeds are an energy-rich food source, as evidenced by their high seed oil content (40–50% of total seed weight) (Lazos 1986; Meru et al. 2018). However, seed oil content is species and genotype dependent. Various studies have reported wide variation in seed oil content across *Cucurbita* species, including *C. pepo* (29.3–51.0%) (Applequist et al. 2006; El-Adawy and Taha 2001a; El-Adawy and Taha 2001b; Meru et al. 2018), *C. moschata* (29.1–43.3%) (Datta and Lal 1977; Tsuyuki et al. 1985), *C. maxima* (10.9–42.3%) (Applequist et al. 2006; Stevenson et al. 2007; Tsuyuki et al. 1985), *C. argyrosperma* (36–40.1%) (Applequist et al. 2006), and *C. ficifolia* (43.5%) (Bernardo-Gil and Cardoso-Lopes 2004). Another important determinant of seed oil content is the seed phenotype, which influences the proportion of the kernel relative to the hull, in the seed. Seed oil content in *Cucurbita* and many oil crops is a function of kernel oil percentage, kernel percentage, and hull components (Leon et al. 1995; Meru et al. 2018; Song and Zhang 2007; Yan et al. 2009). Since there is insignificant amount of oil in the hulls, genotypes with hulled seeds typically accumulate lesser oil than their less-hulled counterparts of similar genotype (Meru et al. 2018). In *C. pepo*, hull-less seeds (seed oil content = 44.6%) have been shown to accumulate more oil than hulled seeds (seed oil content = 35.91%) (Meru et al. 2018). Similar conclusions have been reported for watermelon, where normal seeded accessions with thicker seed coats have less seed oil content than their egusi counterparts, which have thinner seed coats (Prothro et al. 2012b).

Palmitic ($C_{16:0}$), stearic ($C_{18:0}$), oleic ($C_{18:1}$), and linoleic ($C_{18:2}$) fatty acids are the main components of the oil, with the latter most prevalent (Bavec et al. 2007; Meru et al. 2018). Variation in fatty acid composition is genotype dependent in *Cucurbita*, with a wide range for palmitic (9.5–14.5%), stearic (3.1–7.65%), oleic (18.42–46.9%), and linoleic (35.38–64.05%) acid (Meru et al. 2018; Murkovic et al. 1996; Nakić et al. 2006). In addition, low field temperatures during seed filling stage can influence conversion of oleic acid to linoleic acid (Lelley et al. 2009). The degree of unsaturation [oleic acid and linoleic acid, (78.6–86.1%)] in *Cucurbita* is similar to that of soybean (84.4%) and sunflower (88.6%) (Baboli and Kordi 2010) and may contribute towards a reduced risk of arteriosclerosis and heart-related ailments (Wassom et al. 2008).

5.3.1.2 Seed Protein

Cucurbita seed is rich in protein (35% w/w), with albumins and globulins making up approximately 60% of the crude protein (Bavec et al. 2007; Fruhwirth and Hermetter 2007). As a result, *Cucurbita* seed flour and meal are used to supplement protein levels in human and animal diets, respectively (Lazos 1992). Seed protein content in *Cucurbita* is genotype dependent (17.3–44.4%) (Achu et al. 2005; Alfawaz 2004; Ardabili et al. 2011; Akwaowo et al. 2000; El-Soukkary 2001; Idouraine et al. 1996; Meru et al. 2018; Younis et al. 2000) and is generally higher in semi-hulled seeds (seed protein content = 26.13%) than in hull-less seeds (22.63%) (Meru et al. 2018).

5.3.1.3 Antioxidants and Minerals

Cucurbita seed is an important source for antioxidants (tocopherols and tocotrienols), which are linked to reduced risk of gastric, breast, lung, and colorectal cancer (Lelley et al. 2009; Nesaretnam et al. 2007; Stevenson et al. 2007). Phytosterols present in *Cucurbita* seed serve to limit cholesterol uptake and are beneficial for treatment of enlarged prostate (benign prostate hyperplasia) (Fruhwirth and Hermetter 2007; Thompson and Grundy 2005). Carotenoids (lutein, B-carotene, violaxanthin, luteoxanthin, auroxanthin epimers, lutein epoxide, flavoxanthin, and chrysanthemaxanthin) in *Cucurbita* seed may reduce the development risk of age-related macular degeneration (Lelley et al. 2009). *Cucurbita* seeds are also an excellent source of the elements K, P, and Fe (Loy 2004).

5.3.2 Biology and Genetics of the Hull-Less Seed Trait

The hull-less seed trait (Fig. 5.1) in *Cucurbita* is conferred by a single recessive allele, designated *n* or *h*, with other potential gene modifiers (Whitaker and Robinson 1986). It first occurred in the late nineteenth century in Austria. A major dominant gene is responsible for the wild-type seed coat (Fig. 5.1) (Lelley et al. 2009). A hulled *Cucurbita* seed has five layers, namely: (1) epidermis (outer layer), (2) hypodermis (composed of small densely packed cells), (3) sclerenchyma (1–2 tiers of thickened cells), (4) aerenchyma (1–3 tiers of thickened cells), and (5) the chlorenchyma (parenchyma rich in intercellular spaces) (Teppner 2004). The *n* allele results in a reduction of the amount of lignin and cellulose in the hypodermis,

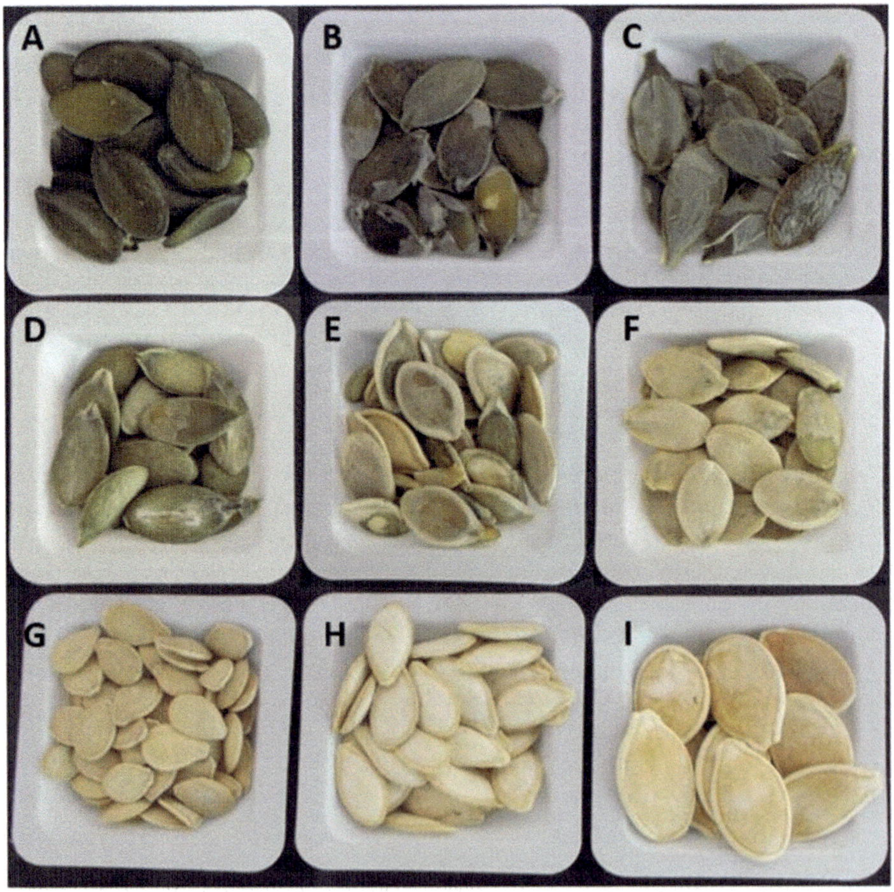

Fig. 5.1 Phenotypic variation of *Cucurbita* seed coat, where (**a–f**) represent hull-less *C. pepo* seeds of Beppo, Kakai, Styrian, Naked Bear, Triple Treat, and Baby Bear cultivars, while G–I represent hulled seeds of Sweet Dumpling (*C. pepo*), Waltham Butternut (*C. moschata*), and Big Max (*C. maxima*) cultivars

sclerenchyma, and parenchyma tissues of the seed coat (Fruhwirth and Hermetter 2007). Many inheritance studies have shown that F_2 populations derived from a cross between hulled and hull-less parents yield 75% hulled and 25% hull-less progeny (reviewed by Lelley et al. 2009). However, variation exists for lignin accumulated in the testa, resulting in a range of hull-less phenotypes (Fig. 5.1). Variation in hull-less phenotypes has been reported among collections of *C. pepo* cultivars (open pollinated and hybrids), as well as plant introductions from germplasm collections around the world (Meru and Fu 2018; Murovec 2015; Teppner 2004).

5.3.3 *Considerations for* Cucurbita *Seed Pumpkin Breeding*

5.3.3.1 Seed Yield and Yield Components

The harvest index (proportion of biological yield converted into reproductive biomass) can be improved to increase seed yield by growing bush habit/compact plants that yield heavy fruit load (Lelley et al. 2009). However, cultivars with compact growth habit should also be bred for multiple branching to lessen photosynthate competition among fruits, which is a function of location on the plant (Paris 2015). Compact plants promote efficient field management because they allow uniform canopy growth and fruit maturation (Paris 2015). Compact plants offer higher plant density, which in turn increases seed yield per ha. However, at very high planting density (>24,000 plants/ ha), fruit number per plant can be adversely affected, while a range between 18,000 and 24,000 plants/ ha is optimum for compact plants (Lelley et al. 2009).

Another important yield parameter is seed index (SI = seed dry biomass/total fruit biomass) and seed yield index (SYI = seed dry biomass/fruit fresh weight) (Lelley et al. 2009). Genetic improvement of SI and SYI increases assimilates partitioned to the seeds while reducing the pericarp tissue to seeds ratio. Seed size (seed weight, seed width, and seed length) is also an important determinant for seed yield. Generally, larger seeds are preferred for snacking; however, they are not required for pumpkin seeds harvested mechanically for oil production (Lelley et al. 2009). In *Cucurbita*, seed size is positively correlated with fruit size, but does not progressively increase in fruit larger than 3 kg. Similarly, seed yield per fruit is positively correlated with fruit size, but the relationship is not linear in fruit larger than 3 kg (Lelley et al. 2009). A trial comparing seed yield between two hull-less seed pumpkin cultivars, Beppo (fruit weight = 2.79 kg) and Naked Bear (fruit weight = 0.59 kg), found no significant difference in the number of seeds per fruit (Meru and Fu 2018), despite a fivefold difference in fruit size (Fig. 5.2). It is therefore noteworthy that SI and SYI can be increased rapidly in smaller fruits (0.5 kg – 1.5 kg) (Lelley et al. 2009). Breeders can also exploit heterosis to increase SI and SYI; however, the yield obtained must be significantly higher to justify the cost of hybrid production (Lelley et al. 2009).

Fig. 5.2 Fruits of Beppo (**a**) and Naked Bear (**b**) hull-less pumpkin cultivars (*Cucurbita pepo*) harvested in summer of 2018 at the University of Florida Tropical Research and Education Center

5.3.3.2 Enhancement of *Cucurbita* Seed Nutritive Value

As previously highlighted, *Cucurbita* seed is rich in beneficial nutrients for human consumption, including seed oil and seed protein, unsaturated fatty acids, antioxidants, and mineral elements. The natural variation in these traits across *Cucurbita* species provides an opportunity for breeders to develop enhanced nutrition for specific target markets. This is important as the market for pumpkin seeds expands and the demand for healthy foods increases, especially in North America (Meru et al. 2018). Phenotypic relationships among these traits can provide a breeder useful insights on the best selection strategies, particularly for tandem improvement of traits. The negative correlation between seed oil and protein content in *Cucurbita* limits the extent to which the two traits can be simultaneously improved (Jan and Seiler 2007; Meru et al. 2018). However, breeders can exploit the positive correlation between seed size and seed oil to improve the latter by selecting larger seeds in segregating populations. Conversely, indirect selection for higher seed protein content may be achieved by selecting smaller seeds (Meru et al. 2018). Indirect selection for seed oil and protein content would not only be cost-effective but would also allow rapid and non-destructive phenotyping for the traits. A strong negative correlation between linoleic and oleic acids (−0.96–0.98) presents a challenge for breeders seeking to improve the two traits tandemly (Meru et al. 2018; Murkovic et al. 1996). The relationship between the two fatty acids may be explained by the fact that they share a common pathway in which conversion of oleic acid into linoleic acid is catalyzed by omega-6 fatty acid desaturase-2 (FAD2) enzyme (Bachlava et al. 2009). Breeders may consider developing high oleic-low linoleic acid *Cucurbita* seed lines by inducing functional mutation in FAD2 locus, as in the case of peanut (Yu et al. 2008). *Cucurbita* seed oil from high oleic-low linoleic acid lines may find application in culinary frying due to improved smoke point. The positive correlation between palmitic and stearic acids (Meru et al. 2018) can be exploited to develop *Cucurbita* lines with high saturated fats for industrial development of solid or semisolid fats without harmful chemical processes such as hydrogenation or transesterification (Ascherio and Willett 1997; Panthee et al. 2005).

5.3.4 Opportunities for Marker-Assisted Selection in Cucurbita *Seed Pumpkin*

Extensive studies have been conducted to elucidate the morphological and biological mechanisms underlying phenotypic variation of economically important traits in *Cucurbita* seed. However, there are far too few studies undertaken to identify molecular mechanisms underlying these traits in *Cucurbita*. Availability of such information would be useful in accelerating genetic gains in the breeding program for superior *Cucurbita* seed pumpkin cultivars. Recent efforts have focused on identifying quantitative trait loci (QTL) underlying seed size variation in *C. maxima* (Table 5.1) (reviewed by Guo et al. 2020). Tan et al. (2013) identified four seed width QTL in an F_2 population between Indian large-grain ‘0515-1’ and the small-grain ‘0460-1-1’ *C. maxima* pumpkin lines. The phenotypic variation explained (PVE) by these QTL ranged between 2.87% and 29.68%. Using a different F_2 population (parental lines 2013-12′ and ‘9-6’), Wang et al. (2020b) identified 10 QTL associated with seed length, seed width, and seed weight (Table 5.1). Among these, QTL SL6-1 (38.6%), SW6-1 (28.9%), and HSW17-1 (17.2%) explained the largest variation for seed length, seed width, and seed weight, respectively. Markers linked to these seed size loci may be used for marker-assisted selection (MAS) for larger seed size in *C. maxima*. Although no QTL have been identified for seed size traits in *C. pepo*, a recent RNA seq study of ‘Sweet Reba’ fruit and seed transcriptome revealed insights into the genes involved in seed and embryo development (Wyatt et al. 2015). These genes will serve as an invaluable resource towards elucidating the genetic mechanism underlying seed development traits in *C. pepo*.

Hull-less seed trait is easily distinguishable phenotypically. However, having molecular markers tightly linked to the trait would allow rapid MAS at the seedling stage or through seed-based genotyping (Meru et al. 2013). Using an F2 population between Lady Godiva oil-pumpkin and Bianco Friulano crookneck cultivar, Gong et al. (2008) mapped the hull-less mutation as a morphological trait on LGp9 (chromosome 12) of C. pepo genome. Later, Kaur (2016) validated four SSR markers (CMTm261, CMTp182, CMTm47 and CMTp257) linked to the hull-less trait for MAS in segregating C. pepo populations.

5.4 Citrullus

Citrullus seed is an important source of oil and protein in Africa and Asia, especially for subsistence farmers (Mahla et al. 2014; National Research Council 2006; Zhang 1996). The exact *Citrullus* species used is often difficult to discern.

Historically there has been inconsistency in species classification used in published research, which continues to some degree into the present. A review of literature shows that *Citrullus* species used as oilseeds have been described as *C. vulgaris*, *C. edulis*, *C. colocynthis*, *Cucurbita citrullus* (Giwa and Akanbi 2020), *C. lanatus* subsp. *mucosospermus* (Achigan-Dako et al. 2008b), and *Colocynthis citrullus* (Bankole et al. 2005), among others.

Here we will use the taxonomic classification by Chomicki and Renner (2015) that describes the genus *Citrullus* as containing seven species including the well-known, sweet watermelon *C. lanatus* (previously *C. lanatus* subsp. *lanatus* or *C. vulgaris*) and *C. amarus* (previously *C. lanatus* var. *citroides*) that is an important source of disease resistance in watermelon breeding (Branham et al. 2019; Branham et al. 2020; McGregor 2011; Ren et al. 2020). *C. ecirrhosus* Cogn., *C. rehmii* De Winter, and *C. naudinianus* Urschler s.n. (M) are all native to Southern Africa, with the latter the only dioecious species (Paris 2015; Renner et al. 2017). *C. mucosospermus* (previously *C. lanatus* subsp. *mucosospermus)*, the closest relative to *C. lanatus*, is native to West Africa and is grown for seeds production, while *C. colocynthis* (L.) Schrad is cultivated for seed oil and medicinal uses (Renner et al. 2017). Molecular genetic studies of seed oil traits in *Citrullus* have been limited to *C. lanatus* and *C. mucosospermus* although *C. colocynthis* is also economically important as an oil seed.

5.4.1 Seed Coat Types

Citrullus seed displays a large amount of phenotypic variation in seed size and color (Fig. 5.3) and can generally be divided into two coat types: normal and egusi. Egusi watermelon seed has a modified seed coat that includes a fleshy outer mucilaginous layer, clearly visible when seed is fresh (Fig. 5.3b) (Gusmini et al. 2004). This layer becomes desiccated when the seed dries and the dry seed resembles normal seed types, albeit with a very thin, papery seed coat (Fig. 5.3c). The mucilaginous layer reappears when the seed is rehydrated. The thin seed coat makes it much easier to remove the hull of egusi seed than that of normal seed. This is an important characteristic since seeds are usually manually de-hulled by grasping the seed with both hands and twisting to break the seed coat and release the kernel (Giwa and Akanbi 2020). Egusi seed is usually large, yellowish, sometimes with a dark edge (Fig. 5.3a). Seed with the normal seed coat has a greater variety of sizes and colors, including white, black, green, yellow, tan, and dotted (Fig. 5.3a). The word "egusi" originates from the Nigerian yoruba and igbo languages meaning "melon" (Adebayo and Yusuf 2015) and can refer to seed from several different cucurbit species including *Citrullus colocynthis*, *Cucumeropsis mannii*, *Lagenaria siceraria*, *Cucurbita maxima*, *Cucurbita moschata*, and *Cucumis sativus* (Achigan-Dako et al. 2008a, b; Giwa and Akanbi 2020). Here we will use the term "egusi" to refer exclusively to *C. mucosospermus* seed that has an outer mucilaginous layer. *Citrullus* genotypes with the outer mucilaginous layer have been classified as *C. mucosospermus*;

Fig. 5.3 (**a**) Phenotypic variation of *Citrullus* seed. The two top rows are egusi seed types. (**b**) The normal seed coat type (left) and the egusi seed coat type with the fleshy outer mucilaginous layer in fresh seed (right)(Photo by Lucky Paudel). (**c**) Normal seed coat type (left) and the dry egusi seed coat type (right)

however, there are some genotypes that are classified as *C. mucosospermus* that do not have the egusi seed phenotype (Paudel et al. 2019b).

The egusi seed trait is controlled by a single recessive gene, *eg* (Gusmini et al. 2004; Prothro et al. 2012b). This locus was mapped to chromosome (chr) 6 of the watermelon genome (Fig. 5.4) (Paudel et al. 2019b; Prothro et al. 2012b). Several high-throughput molecular markers for trait selection have been published (Paudel et al. 2019b), and Luan et al. (2019) suggested *Cla007520*, a member of the CPP protein family as a candidate gene for the trait. No difference in expression levels of

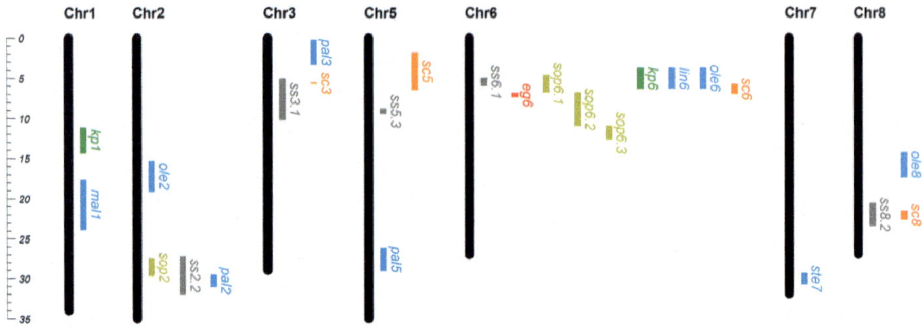

Fig. 5.4 QTL associated with seed oil percentage (*sop*) (Prothro et al. 2012b), the egusi locus (*eg*) (Luan et al. 2019; Paudel et al. 2019a; Prothro et al. 2012b), kernel percentage (*kp*) (Meru and McGregor 2013), linoleic acid (*lin*) (Meru and McGregor 2014), oleic acid (*ole*) (Meru and McGregor 2014), stearic acid (*ste*) (Meru and McGregor 2014), palmitic acid (*pal*) (Meru and McGregor 2014), seed size (*ss*) (Li et al. 2018; Meru and McGregor 2013; Prothro et al. 2012a; Ren et al. 2014), and seed coat color (sc)(Li et al. 2020; Paudel et al. 2019b) in *Citrullus*. Seed size loci named according to Guo et al. (2020)

Cla007520 between the parental WI-1 (normal seed) and PI 186490 (egusi) was observed, but a 3 bp deletion leading to loss of a serine residue in PI 186490 is thought to be a causal mutation (Luan et al. 2019). The modification of the seed coat due to the *eg* locus is important because it has a direct effect on oil content and ease of processing of *Citrullus* seed.

5.4.2 *Seed Oil Percentage (SOP)*

Seed oil percentage is measured as the lipid content of seed as a percentage of dry weight. Jarret and Levy (2012) found that SOP of egusi seed (avg = 35.6%) was significantly higher than that of *C. lanatus* (avg = 23.2%) and *C. amarus* (avg = 22.6) (Fig. 5.5). Prothro et al. (2012b) mapped QTL for SOP in a *C. lanatus* (PI 279461; SOP = 25.2%) x *C. mucosospermus* (PI 560023; SOP = 40.6%) F_2 population and found that ~83% of the PVE for SOP in the population was explained by a QTL (*sop6.2*) that overlaps with the *eg* locus (Fig. 5.4). There was also an interaction between the *eg* locus and a minor SOP QTL on Chr 2 (*sop2*), but the effect of the interaction was small (0.7%). Two additional QTL (*sop6.1* and *sop6.3*) were detected close to *sop6.2*, and further analysis showed that *sop6.1* was associated with SOP of individuals in the population with normal seed types. *Sop6.1* overlaps with a seed size QTL (*ss6.1*) previously identified in watermelon (Prothro et al. 2012a). Significant negative correlations have been observed between SOP and seed size parameters, while significant positive correlations have been observed between SOP and kernel percentage (Jarret and Levy 2012; Meru and McGregor 2013).

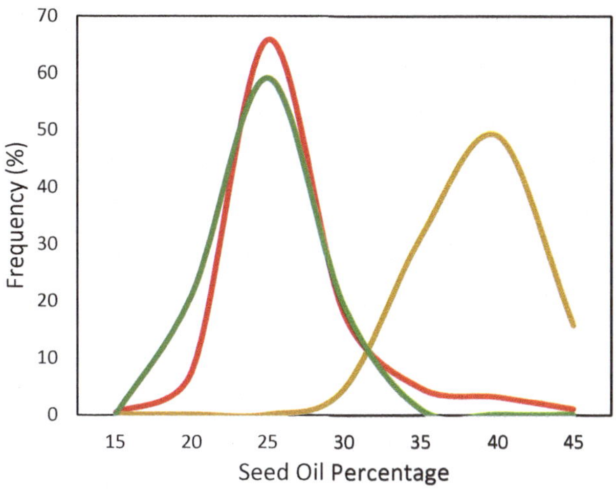

Fig. 5.5 Frequency distribution of seed oil percentage of *C. lanatus* (red, $N = 321$), *C. amarus* (green, $N = 110$), and egusi (yellow, $N = 45$) accessions in the USDA germplasm collection. (Based on data from Jarret and Levy 2012)

5.4.3 Kernel Percentage (KP)

The kernel percentage is the percentage of the overall seed weight that is due to the kernel. This characteristic is also sometimes measured as the hull/kernel ratio. Meru and McGregor (2013) observed correlations of 82% and 76% between SOP and KP in normal and egusi seed types, respectively. These observations were in a single segregating population, but is supported by results from Jarret and Levy (2012) that observed hull/kernel ratio of 0.49 for egusi seed and 1.37 for *C. lanatus* seed and an 82% negative correlation between SOP and hull/kernel ratio of 24 *Citrullus* accessions from 3 different species. In egusi seed, the mucilaginous layer dries to a very thin seed coat that leads to a higher KP. The thin seed coat means that a larger percentage of the overall seed weight is due to the kernel, leading to the pleiotropic effect of the *eg* locus on seed coat type, SOP, and KP. Meru and McGregor (2013) split the F_2 progeny from the *C. lanatus* x *C. mucosospermus* population used by Prothro et al. (2012b) into normal and egusi subpopulations and mapped KP in the subpopulations separately. In the normal seed subpopulation, a KP QTL (*kp6*, PVE = 22%) overlapped with a locus associated with a seed size locus on chr 6 (*ss6.1*) (Fig. 5.4) (Guo et al. 2020; Li et al. 2018; Prothro et al. 2012a). For the egusi seed, a QTL associated with the KP (*kp1*, PVE = 34%) was identified on chr 1 (Meru and McGregor 2013). No known seed size QTL have been described at this location; however, except for the Meru and McGregor (2013) study, no other study has attempted to map seed size for egusi seed. Meru and McGregor (2013) reported significant negative correlations between KP and seed size traits (weight and width) in both the normal and egusi subpopulations, but only in the normal subpopulation for seed length. These results suggest that in addition to the pleiotropic effect of the *eg* locus, seed size is an important parameter associated with KP and therefore SOP.

5.4.4 Seed Size (SS)

Seed size is an important trait for *Citrullus* edible seed, not only due to its role in consumer preference but also due to its negative correlation with KP and SOP (Meru and McGregor 2013). Producers prefer larger seed since increases in seed size, particularly seed width, are translated to higher yields. Generally, a seed width larger than 10 mm is desirable (Zhang 1996). Guo et al. (2020) recently published an extensive review on seed size QTL in cucurbits, so here we will limit our discussion to only the most important QTL in *Citrullus*, especially those that colocalize with other important seed oil traits. Prothro et al. (2012a) were the first to describe QTL associated with seed size in watermelon, and a number of the QTL have since been confirmed in other studies [see Guo et al. (2020) for review]. The QTL with the largest effect on seed size was *ss6.1*, and this locus has since been fine mapped by Li et al. (2018). The authors also identified *Cla009291*, *Cla009310*, and *Cla009301* as candidate genes and proposed that a SNP leading to a premature stop codon in the first exon of *Cla0009310* is the causal mutation. They also developed a marker assay, caps5_S6, for selection of this SNP.

ss6.1 colocalizes with QTL for SOP (*ss6.*1), oleic acid (*ole6*), linoleic acid (*lin6*), and seed coat color (*sc6*) (Meru and McGregor 2013, 2014; Prothro et al. 2012b) (Fig. 5.4). The colocalization on *sc6* with *ss6.1* is thought to be due to linkage rather than pleiotropy (Paudel et al. 2019a; Poole et al. 1941). Even though large seed sizes are often preferred for edible seed, the negative correlation of seed size with SOP means that if seed can be mechanically de-hulled, the optimal seed size for oil yield might be smaller seed.

5.4.5 Fatty Acid Composition

In a study of 30 *C. amarus*, 33 *C. lanatus*, and 33 *C. mucosospermus* accessions from the USDA germplasm collection, Jarret and Levy (2012) found that seed extracts contained an average of between 45.37 and 73% linoleic acid, 7.98 and 33.95% oleic acid, 5.03 and 13.84% stearic acid, and 9.68 and 14.38% palmitic acid (Fig. 5.6). This is generally in line with what has been found in other studies (Sultana and Ashraf 2019). The fatty acid profile of watermelon is thus high in linoleic acid, but low in oleic acid, although considerable variation exists in the germplasm (Fig. 5.6).

The only mapping study for *Citrullus* fatty acid content was done by Meru and McGregor (2014) in the same *C. lanatus* x *C. mucosospermus* population used for

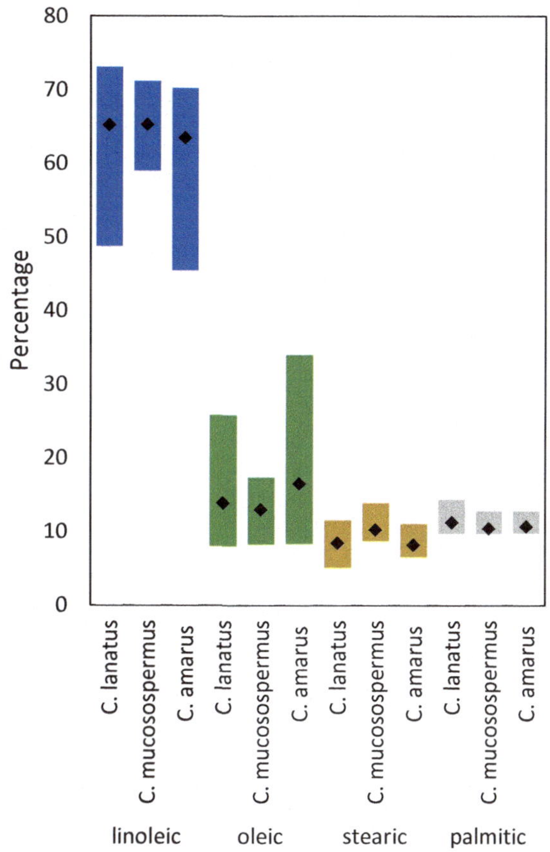

Fig. 5.6 Percentage of primary fatty acids in seed extracts of *C. lanatus* (*N* = 33), *C. mucosospermus* (*N* = 33), and *C amarus* (*N* = 30). The bars represent the range and the diamond the average. (Based on data from Jarret and Levy 2012)

mapping the *eg* locus, SOP (Paudel et al. 2019b; Prothro et al. 2012b), and KP (Meru and McGregor 2013) traits. QTL were identified for linoleic acid on chr 6 (*lin6*; PVE = 21.5%); for oleic acid on chr 2 (*ole2*; PVE = 10.75), 6 (*ole6*; PVE = 17.9%), and 8 (*ole8*; PVE = 13.5%); for stearic acid on chr 7 (*ste7*; PVE = 10.2%); and for palmitic acid on chrs 2 (*pal2*; PVE = 7.6%), 3 (*pal3*; PVE = 10.7%), and 5 (*pal5*; PVE = 12.7%) (Fig. 5.4). *lin6* and *ole6* colocalized as can be expected from the shared metabolic pathway. These QTL also colocalized with *ss6.1* (Meru and McGregor 2013; Prothro et al. 2012a), KP, and SOP loci. In addition, *pal2* colocalize with *ss2.2* and *SOP2* loci. Meru and McGregor (2014) also identified a number of candidate genes for these QTL, including *Cla013264* (fatty acid elongase) for *pal2*; *Cla008157* (omega-3 fatty acid desaturase), *Cla008263* (fatty acid elongase), and *Cla002633* (acyl-ACP thioesterase) for *pal3*; *Cla009335* (acyl carrier protein) for *ole6* and *lin6*; *Cla010780* (acyl-ACP desaturase) for *ste7*; and *Cla013862* (acyl-ACP desaturase) for *ole8*.

5.4.6 Seed Coat Color

Seed coat color plays an important role in consumer preference for seed consumption. Some consumers prefer red seed, while others might prefer seed with light coat color, but dark margins (Jensen et al. 2011; Zhang 1996). The seed coat characteristics are often used to classify seed like *Serewe* which has brown smooth seed coats or *Fombou* which has red and brown stripes (Jensen et al. 2011; Sanusi 2015). Poole et al. (1941) described a model for inheritance of watermelon seed coat color based on the interaction of four genes, *R*, *T*, *W*, and the modifier *D*. Together these genes code for black (*RTWD*), dotted black (*RTWd*), green (*rTW*), tan (*RtW*), clump (*RTw*), red (*rtW*), white tan-tipped (*Rtw*), and white pink-tipped (*rtw*) seeds. Paudel et al. (2019a) mapped the location of the *R* (*sc3*), *W* (*sc6*), and *D* (*sc8*) genes as well as a locus they named T^1 (*sc5*), since the inheritance did not fit that of the *T* locus described by Poole et al. (1941) (Fig. 5.4). Li et al. (2020) fine mapped the *sc3* locus and identified a polyphenol oxidase (PPO) gene, *Cla019481*, as a candidate gene. PPO is involved in melanin biosynthesis, and the proposed causal polymorphism is a frameshift due to single nucleotide insertion that leads to a premature stop codon in the gene. Due to the wide variation in seed coat colors, it is expected that many additional loci are still to be discovered.

5.5 Conclusion

Cucurbits are important oil seed crops with significant genomic resources. However genomic studies on oil seed traits are limited and have not fully exploited the abundance of genomic resources available for these crops. Candidate genes have been identified for some key traits including the egusi seed trait in *Citrullus*, and molecular markers are available for several other traits including the hull-less seed trait in *Cucurbita*. However, there is an urgent need to identify additional loci for molecular breeding and candidate genes as potential targets for gene editing to accelerate breeding for oil traits in cucurbits.

References

Achu MB, Fokou E, Tchiégang C, Fotso M, Tchouanguep FM (2005) Nutritive value of some Cucurbitaceae oilseeds from different regions in Cameroon. Af J Biotechnol 4:1329–1334

Achigan-Dako EG, Fagbemissi R, Avohou HT, Vodouhe RS, Coulibaly O, Ahanchede A (2008a) Importance and practices of Egusi crops (*Citrullus lanatus* (Thunb.) Matsum. & Nakai, *Cucumeropsis mannii* Naudin and *Lagenaria siceraria* (Molina) Standl. cv. ' Aklamkpa ') in sociolinguistic areas in Benin. Biotechnol Agron Soc Environ 12:393–403

Achigan-Dako EG, Fuchs J, Ahanchede A, Blattner FR (2008b) Flow cytometric analysis in *Lagenaria siceraria* (*Cucurbitaceae*) indicates correlation of genome size with usage types and growing elevation. Plant Syst Evol 276(1):9

Adebayo AA, Yusuf KA (2015) Analele universit. Eftimie Murgu Resita 22:11–22
Alfawaz MA (2004) Chemical composition and oil characteristics of pumpkin (*Cucurbita maxima*) seed kernels. Res Bult 129:5–18
Akwaowo EU, Ndon BA, Etuk EU (2000) Minerals and antinutrients in fluted pumpkin (*Telfairia occidentalis* Hook f.). Food Chem 70:235–240
Applequist WL, Avula B, Schaneberg BT, Wang YH, Khan IA (2006) Comparative fatty acid content of seeds of four *Cucurbita* species grown in a common (shared) garden. J Food Compos Anal 19:606–611
Ardabili A, Farhoosh R, Khodaparast MH (2011) Chemical composition and physicochemical properties of pumpkin seeds (*Cucurbita pepo* subsp. pepo var. Styriaka) grown in Iran. J Agric Sci Technol 13:1053–1063
Arumuganathan K, Earle E (1991) Nuclear DNA content of some important plant species. Plant Mol Biol Report 9(3):208–218
Ascherio A, Willett WC (1997) Health effects of trans-fatty acids. Am J Clin Nutr 66:1006–1010
Baboli ZM, Kordi A (2010) Characteristics and composition of watermelon seed oil and solvent extraction parameters effects. J Am Oil Chem Soc 87:667–671
Bachlava E, Dewey RE, Burton JW, Cardinal AJ (2009) Mapping candidate genes for oleate biosynthesis and their association with unsaturated fatty acid seed content in soybean. Mol Breed 23:337–347
Bankole SA, Osho A, Joda AO, Enikuomehin OA (2005) Effect of drying method on the quality and storability of 'egusi' melon seeds (*Colocynthis citrullus* L.). Afr J Biotechnol 4:799–803
Barrera-Redondo J, Ibarra-Laclette E, Vázquez-Lobo A, Gutiérrez-Guerrero YT, Sánchez de la Vega G, Piñero D et al (2019) The genome of *Cucurbita argyrosperma* (silver-seed gourd) reveals faster rates of protein-coding gene and long noncoding RNA turnover and neofunctionalization within *Cucurbita*. Mol Plant 12(4):506–520
Bavec F, Mlakar SG, Rozman Č, Bavec M (2007) Oil pumpkins: niche for organic producers. In: Janick J, Whipkey A (eds) Issues in new crops and new uses. ASHS Press, Alexandria, pp 185–189
Baxter GG, Murphy K, Paech A (2012) The potential to produce pumpkin seed for processing in North East Victoria. Rural Industr Develop Corp 11(145):5–36
Bernardo-Gil MG, Cardoso-Lopes LM (2004) Supercritical fluid extraction of Cucurbita ficifolia seed oil. Eur Food Res Technol 219:593–597
Branham SE, Levi A, Wechter WP (2019) QTL mapping identifies novel source of resistance to Fusarium wilt race 1 in *Citrullus amarus*. Plant Dis 103(5):984–989
Branham SE, Patrick Wechter W, Ling K-S, Chanda B, Massey L, Zhao G et al (2020) QTL mapping of resistance to *Fusarium oxysporum* f. sp. *niveum* race 2 and papaya ringspot virus in *Citrullus amarus*. Theor Appl Genet 133(2):677–687
Chomicki G, Renner SS (2015) Watermelon origin solved with molecular phylogenetics including Linnaean material: another example of museomics. New Phytol 205:526–532
Cutler HC, Whitaker TW (1961) History and distribution of the cultivated cucurbits in the Americas. Am Antiq 26:469–485
Datta N, Lal BM (1977) Distribution of oil in different anatomical parts of some cucurbit kernels. Assoc Food Sci 14:24–25
Diaz A, Fergany M, Formisano G, Ziarsolo P, Blanca J, Fei Z et al (2011) A consensus linkage map for molecular markers and Quantitative Trait Loci associated with economically important traits in melon (*Cucumis melo* L.). BMC Plant Biol 11(1):111
Earle FR, Jones Q (1962) Analyses of seed samples from 113 plant families. Econ Bot 16(4):221–250
El-Adawy TA, Taha KM (2001a) Characteristics and composition of watermelon, pumpkin, and paprika seed oils and flours. J Agric Food Chem 49:1253–1259
El-Adawy TA, Taha KM (2001b) Characteristics and composition of different seed oils and flours. Food Chem 74:47–54

El-Soukkary FAH (2001) Evaluation of pumpkin seed products for bread fortification. Plant Food Human Nutr 56:365–384
Esteras C, Gómez P, Monforte AJ, Blanca J, Vicente-Dólera N, Roig C et al (2012) High-throughput SNP genotyping in *Cucurbita pepo* for map construction and quantitative trait loci mapping. BMC Genomics 13(1):80
FAOSTAT. Crop Production 2020 [cited 2020 14 July]. Available from: http://www.fao.org/faostat/en/#data/QC
Fruhwirth GO, Hermetter A (2007) Seeds and oil of the Styrian oil pumpkin: components and biological activities. Eur J Lipid Sci Technol 109:1128–1140
Garcia-Mas J, Benjak A, Sanseverino W, Bourgeois M, Mir G, González VM et al (2012) The genome of melon (*Cucumis melo* L.). Proc Natl Acad Sci 109(29):11872–11877
Giwa SO, Akanbi TO (2020) A review on food uses and the prospect of egusi melon for biodiesel production. BioEnergy Research
Gong L, Stift G, Kofler R, Pachner M, Lelley T (2008) Microsatellites for the Genus *Cucurbita* and an SSR-based genetic linkage map of *Cucurbita pepo* L. Theor Appl Genet 117:37–48
Gonzalo MJ, Monforte AJ (2017) Genetic mapping of complex traits in cucurbits. In: Grumet R, Katzir N, Garcia-Mas J (eds) Genetics and genomics of cucurbitaceae. Springer, Cham, pp 269–290
Guo S, Zhang J, Sun H, Salse J, Lucas WJ, Zhang H et al (2013) The draft genome of watermelon (*Citrullus lanatus*) and resequencing of 20 diverse accessions. Nat Genet 45:51–58
Guo S, Zhao S, Sun H, Wang X, Wu S, Lin T et al (2019) Resequencing of 414 cultivated and wild watermelon accessions identifies selection for fruit quality traits. Nat Genet 51(11):1616–1623
Guo Y, Gao M, Liang X, Xu M, Liu X, Zhang Y et al (2020) Quantitative trait loci for seed size variation in cucurbits – a review. Front Plant Sci 11:304
Gusmini G, Wehner TC, Jarret RL (2004) Inheritance of egusi seed type in watermelon. J Hered 95(3):268–270
Huang S, Li R, Zhang Z, Li L, Gu X, Fan W et al (2009) The genome of the cucumber, *Cucumis sativus* L. Nat Genet 41(12):1275–1281
Idouraine A, Kohlhepp EA, Weber CW (1996) Nutrient constituents from eight lines of naked seed squash (*Cucurbita pepo* L.). J Agric Food Chem 44:721–724
Jacks TJ, Hensarling TP, Yatsu LY (1972) Cucurbit seeds: I. Characterizations and uses of oils and proteins. A review. Econ Bot 26(2):135–141
Jan C, Seiler G (2007) Sunflower. In: Singh RJ (ed) Genetic resources, chromosome engineering and crop improvement. CRC Press, New York, p 137
Jarret RL, Levy IJ (2012) Oil and fatty acid contents in seed of *Citrullus lanatus* Schrad. J Agric Food Chem 60(20):5199–5204
Jensen BD, Touré FM, Ag Hamattal M, Touré FA, Nantoumé AD (2011) Watermelons in the sand of Sahara: cultivation and use of indigenous landraces in the Tombouctou region of Mali. Ethnobot Res Appl 9:151–162
Kaur N (2016) Genetic analysis of important economic traits and validation of molecular markers linked to hull-less seed trait in pumpkin (*Cucurbita pepo* subsp. pepo var. styriaca) [dissertation]. Punjab Agricultural University, Ludhiana
Lazos ES (1986) Nutritional, fatty acid, and oil characteristics of pumpkin and melon seeds. J Food Sci 4:83–87
Lazos ES (1992) Certain functional properties of defatted pumpkin seed flour. Plant Food Human Nutr 42:257–273
Lelley T, Loy BL, Murkovic M (2009) Hull-Less oil seed pumpkin. In: Vollmann J, Rajcan I (eds) Oil crops, handbook of plant breeding. Springer, New York, pp 469–492
Leon A, Lee M, Rufener G, Berry S, Mowers R (1995) Use of RFLP markers for genetic linkage analysis of oil percentage in sunflower seed. Crop Sci 35:558–564
Li B, Lu X, Gebremeskel H, Zhao S, He N, Yuan P et al (2020) Genetic mapping and discovery of the candidate gene for black seed coat color in watermelon (*Citrullus lanatus*). Front Plant Sci 10:1689

Li N, Shang J, Wang J, Zhou D, Li N, Ma S (2018) Fine mapping and discovery of candidate genes for seed size in watermelon by genome survey sequencing. Sci Rep 8(1):17843

Li Q, Li H, Huang W, Xu Y, Zhou Q, Wang S et al (2019) A chromosome-scale genome assembly of cucumber (*Cucumis sativus* L.). GigaScience 8(6)

Li Z, Zhang Z, Yan P, Huang S, Fei Z, Lin K (2011) RNA-Seq improves annotation of protein-coding genes in the cucumber genome. BMC Genomics 12(1):540

Loy JB (2004) Morpho-physiological aspects of productivity and quality in squash and pumpkins (*Cucurbita* spp.). Crit Rev Plant Sci 23:337–363

Luan F, Fan C, Sun L, Cui H, Amanullah S, Tang L et al (2019) Genetic mapping reveals a candidate gene for egusi seed in watermelon. Euphytica 215(11):182

Mahla HR, Singh JP, Roy MM (2014) Seed purpose watermelon in arid zone. Jodhpur, Central Arid Zone Research Institute

McGregor CE (2011) *Citrullus lanatus* germplasm of Southern Africa. Isr J Plant Sci 60(4):403–413

Meru G, Fu Y (2018) Yield and horticultural performance of naked-seed pumpkin in south Florida. Electronic Data Information Source. HS1323

Meru G, Fu Y, Leyva D, Sarnoski P, Yagiz Y (2018) Phenotypic relationships among oil, protein, fatty acid composition and seed size traits in *Cucurbita pepo*. Sci Hortic 233:47–53

Meru G, Leyva D, Michael V, Dorval M, Mainviel R, Fu Y (2019) Genetic variation among *Cucurbita pepo* accessions varying in seed nutrition and seed size. Am J Plant Sci 10:1536–1547

Meru G, McDowell D, Waters V, Seibel A, Davis J, McGregor C (2013) A non-destructive genotyping system from a single seed for marker-assisted selection in watermelon. Genet Mol Res 12:702–709

Meru G, McGregor C (2013) Genetic mapping of seed traits correlated with seed oil percentage in watermelon. HortScience 48(8):955–959

Meru G, McGregor C (2014) Quantitative trait loci and candidate genes associated with fatty acid content of watermelon seed. J Am Soc Hortic Sci 139(4):433–441

Montero-Pau J, Blanca J, Bombarely A, Ziarsolo P, Esteras C, Martí-Gómez C et al (2018) *De novo* assembly of the zucchini genome reveals a whole-genome duplication associated with the origin of the *Cucurbita* genus. Plant Biotechnol J 16(6):1161–1171

Montero-Pau J, Blanca J, Esteras C, Martínez-Pérez EM, Gómez P, Monforte AJ et al (2017) An SNP-based saturated genetic map and QTL analysis of fruit-related traits in zucchini using genotyping-by-sequencing. BMC Genomics 18(1):94

Moya-Hernández A, Bosquez-Molina E, Serrato-Díaz A, Blancas-Flores G, Alarcón-Aguilar FJ (2018) Analysis of genetic diversity of *Cucurbita ficifolia* Bouché from different regions of Mexico, using AFLP markers and study of its hypoglycemic effect in mice. South Afr J Bot 116:110–115

Murkovic M, Hilerbrand A, Winkler J, Leitner E, Pfannhauser W (1996) Variability of fatty acid content in pumpkin seeds (*Cucurbita pepo* L.). Z Lebensm Unters Forsch 203:216–219

Murovec J (2015) Phenotypic and genetic diversity in pumpkin accessions with mutated seed coats. HortScience 50:211–217

National Research Council (2006) Lost crops of Africa: Volume II: Vegetables. National Academies Press, Washington, D.C

Nakić SN, Rade D, Skevin D, Strucelj D, Mokrovcak Z, Bartolic M (2006) Chemical characteristics of oils from naked and husk seeds of *Cucurbita pepo* L. Eur J Lipid Sci Technol 108:963–943

Nee M (1990) The domestication of Cucurbita (Cucurbitaceae). Econ Bot 44(3):56–68

Nesaretnam K, Gomez PA, Selvaduray KR, Razak GA (2007) Tocotrienol levels in adipose tissue of benign and malignant breast lumps in patients in Malaysia. Asia Pac J Clin Nutr 16:498–504

Pan Y, Wang Y, McGregor C, Liu S, Luan F, Gao M et al (2020) Genetic architecture of fruit size and shape variation in cucurbits: a comparative perspective. Theor Appl Genet 133(1):1–21

Panthee D, Pantalone V, West D, Saxton A, Sams C (2005) Quantitative trait loci for seed protein and oil concentration, and seed size in soybean. Crop Sci 45:2015–2022

Paris HS (2015) Origin and emergence of the sweet dessert watermelon, Citrullus lanatus. Ann Bot 116(2):133–148

Paris HS (2016) Germplasm enhancement of *Cucurbita pepo* (pumpkin, squash, gourd: Cucurbitaceae): progress and challenges. Euphytica 208:415–438

Paudel L, Clevenger J, McGregor C (2019b) Chromosomal locations and interactions of four loci associated with seed coat color in watermelon. Front Plant Sci 10:788

Paudel L, Clevenger J, McGregor C (2019a) Refining of the egusi locus in watermelon using KASP assays. Sci Hortic 257:108665

Pereira L, Ruggieri V, Pérez S, Alexiou KG, Fernández M, Jahrmann T et al (2018) QTL mapping of melon fruit quality traits using a high-density GBS-based genetic map. BMC Plant Biol 18(1):324

Poole CF, Grimball PC, Porter DR (1941) Inheritance of seed characters in watermelon. J Agric Res 63:433–456

Prothro J, Sandlin K, Abdel-Haleem H, Bachlava E, White W, Knapp S et al (2012a) Main and epistatic quantitative trait loci associated with seed size in watermelon. J Am Soc Hortic Sci 137(6):452–457

Prothro J, Sandlin K, Gill R, Bachlava E, White V, Knapp S et al (2012b) Mapping of the egusi seed trait locus (*eg*) and quantitative trait loci associated with seed oil percentage in watermelon. J Am Soc Hortic Sci 137(5):311–315

Qi J, Liu X, Shen D, Miao H, Xie B, Li X et al (2013) A genomic variation map provides insights into the genetic basis of cucumber domestication and diversity. Nat Genet 45(12):1510–1515

Ren R, Xu J, Zhang M, Liu G, Yao X, Zhu L et al (2020) Identification and molecular mapping of a gummy stem blight resistance gene in wild watermelon (*Citrullus amarus*) germplasm PI 189225. Plant Dis 104(1):16–24

Ren Y, McGregor C, Zhang Y, Gong G, Zhang H, Guo S et al (2014) An integrated genetic map based on four mapping populations and quantitative trait loci associated with economically important traits in watermelon (*Citrullus lanatus*). BMC Plant Biol 14(1):33

Renner SS, Sousa A, Chomicki G (2017) Chromosome numbers, Sudanese wild forms, and classification of the watermelon genus *Citrullus*, with 50 names allocated to seven biological species. Taxon 66(6):1393–1405

Robinson RW, Decker-Walters DS (1997) Cucurbits. CAB International, New York

Sandlin KC, Prothro JM, Heesacker AF, Khalilian N, Okashah R, Xiang W et al (2012) Comparative mapping in watermelon [*Citrullus lanatus* (Thunb.) Matsum. et Nakai]. Theor Appl Genet 125:1603–1618

Sanusi SM (2015) Profit efficiency of Egusi melon (*Colocynthis citrullus* var. *lanatus*) production in Bida local government area of Niger state, Nigeria. Indian J Econ Dev 11:543–552

Schaefer H, Renner SS (2011) Cucurbitaceae. In: Kubitzki K (ed) The families and genera of vascular plants, Vol 10, Sapindales, Cucurbitales, Myrtaceae. Springer, Berlin, pp 112–174

Song XL, Zhang TZ (2007) Identification of quantitative trait loci controlling seed physical and nutrient traits in cotton. Seed Sci Res 17:243–252

Stevenson DG, Eller FJ, Wang L, Jane JL, Wang T, Inglett GE (2007) Oil and tocopherol content and composition of pumpkin seed oil in 12 cultivars. J Agri Food Chem 55:4005–4013

Sultana B, Ashraf R (2019) Watermelon (*Citrullus lanatus*) oil. In: Ramadan MF (ed) Fruit oils: chemistry and functionality. Springer International Publishing, Cham, pp 741–756

Sun H, Wu S, Zhang G, Jiao C, Guo S, Ren Y et al (2017) Karyotype stability and unbiased fractionation in the paleo-allotetraploid *Cucurbita* genomes. Mol Plant 10(10):1293–1306

Tan XZ, Ge Y, Xu WL, Cui CS, Qu SP (2013) Construction of genetic linkage map and QTL analysis for seed width in pumpkin (*Cucurbita maxima*). Acta Bot Borealioccidentalia Sin 33:697–702

Teppner H (2004) Notes on *Lagenaria* and *Cucurbita* (Cucurbitaceae) review and new contributions. Phyton 44:245–308

Thompson GR, Grundy SM (2005) History and development of plant sterol and stanol esters for cholesterol-lowering purposes. Amer J Cardiol 96:3D–9D

Tsuyuki H, Itoh S, Yamagata K (1985) Lipid and triacylglycerol compositions of total lipids in pumpkin seeds. Nippon Shokuhin Kogyo Gakkaishi 32:7–15

Vermaak I, Kamatou GPP, Komane-Mofokeng B, Viljoen AM, Beckett K (2011) African seed oils of commercial importance — cosmetic applications. S Afr J Bot 77(4):920–933

Wang Y, Bo K, Gu X, Pan J, Li Y, Chen J et al (2020a) Molecularly tagged genes and quantitative trait loci in cucumber with recommendations for QTL nomenclature. Hortic Res 7(1):3

Wang Y, Wang C, Han H, Luo Y, Wang Z, Yan C (2020b) Construction of a high-density genetic map and analysis of seed-related traits using specific length amplified fragment sequencing for *Cucurbita maxima*. Front Plant Sci 10:1782

Weng Y (2017) The cucumber genome. In: Grumet R, Katzir N, Garcia-Mas J (eds) Genetics and genomics of Cucurbitaceae. Springer, Cham, pp 183–197

Wassom JJ, Mikkelineni V, Bohn MO, Rocheford TR (2008) QTL for fatty acid composition of maize kernel oil in Illinois high oil B73 backcross-derived lines. Crop Sci 48:69–78

Whitaker T (1947) American origin of the cultivated cucurbits. Ann Mo Bot Gard 34:101–111

Whitaker T, Robinson R (1986) Squash breeding. In: Baset M (ed) Breeding vegetable crops. AVI Publishing Company Inc, Connecticut, pp 209–242

Wu S, Shamimuzzaman M, Sun H, Salse J, Sui X, Wilder A et al (2017) The bottle gourd genome provides insights into *Cucurbitaceae* evolution and facilitates mapping of a papaya ring-spot virus resistance locus. Plant J 92(5):963–975

Wu S, Wang X, Reddy U, Sun H, Bao K, Gao L et al (2019) Genome of 'Charleston gray', the principal American watermelon cultivar, and genetic characterization of 1,365 accessions in the U.S. National Plant Germplasm System watermelon collection. Plant Biotechnol J 17:2246–2258

Wyatt LE, Strickler SR, Mueller LA, Mazourek M (2015) An acorn squash (*Cucurbita pepo* subsp ovifera) fruit and seed transcriptome as a resource for the study of fruit traits in Cucurbita. Hortic Res 2:14070

Xie D, Xu Y, Wang J, Liu W, Zhou Q, Luo S et al (2019) The wax gourd genomes offer insights into the genetic diversity and ancestral cucurbit karyotype. Nat Commun 10(1):5158

Yan X, Li J, Fu F, Jin M, Chen L, Liu L (2009) Co-location of seed oil content, seed hull content and seed coat color QTL in three different environments in *Brassica napus* L. Euphytica 170:355–364

Yang L, Koo D-H, Li Y, Zhang X, Luan F, Havey MJ et al (2012) Chromosome rearrangements during domestication of cucumber as revealed by high-density genetic mapping and draft genome assembly. Plant J 71(6):895–906

Younis YMH, Ghirmay S, Al-Shihry SS (2000) African *Cucurbita pepo* L.: Properties of seed and variability in fatty acid composition of seed oil. Phytochem 54:71–75

Yu S, Pan L, Yang Q, Min P, Ren Z, Zhang H (2008) Comparison of the 12 fatty acid desaturase gene between high-oleic and normal-oleic peanut genotypes. J Genet Genomics 35:679–685

Zhang J (1996) Breeding and production of watermelon for edible seed in China. Cucurbit Genet Coop Rep 19:66–67

Zheng Y, Wu S, Bai Y, Sun H, Jiao C, Guo S et al (2018) Cucurbit genomics database (CuGenDB): a central portal for comparative and functional genomics of cucurbit crops. Nucleic Acids Res 47(D1):D1128–D1D36

Chapter 6
Genome Sequence of Oil Palm

Amal Mahmoud

Contents

6.1 Introduction

Oil palm belongs to the genus *Elaeis* and family *Arecaceae*. There are two species of the oil palm: *Elaeis guineensis* (African oil palm) and *Elaeis oleifera* (American oil palm) (Chan et al. 2017).

African oil palm is the predominant producer of oil worldwide, and palm oil accounts for 33% of vegetable oil and 45% of all edible oil produced, worldwide (Singh et al. 2013b). The global production of palm oil in 2018 was approximately 71 million tons (http://www.fao.org/faostat/en/#data/QC). Palm oil's unique composition makes it versatile for use in food manufacturing and in the chemical, cosmetic, and pharmaceutical industries (Pantzaris 1997). It is highly suitable for deep frying as well, for its lower content of polyunsaturated linoleic acid and higher levels of saturated fatty acids make it less susceptible to oxidation. Palm oil also contains high levels of natural antioxidants such as tocopherols, tocotrienols, and carotenoids (Mayes et al. 2008).

The importance of oil palm has resulted in interest to sequence its transcriptomes and genome. The genomes of *pisifera* and *dura* forms of oil palm are available, the genome size is 1.8 Gb for *pisifera* form and 1.701 Gb for *dura* from (Jin et al. 2016; Singh et al. 2013b). Genome-wide association analysis (GWAS) was used to

A. Mahmoud (✉)
Department of Biology, College of Science, Imam Abdulrahman Bin Faisal University, Dammam, Saudi Arabia
e-mail: amhsalem@iau.edu.sa

H. Tombuloglu et al. (eds.), *Oil Crop Genomics*,
https://doi.org/10.1007/978-3-030-70420-9_6

identify single-nucleotide polymorphisms (SNPs) and genes, associated with fatty acid content and vitamin E in oil palm trees (Luo et al. 2020; Xia et al. 2019b). However, microsatellite or simple sequence repeats (SSR) markers are favored over other genetic markers because of their many desirable attributes, which include hypervariability, wide genomic distribution, codominant inheritance, a multiallelic nature, and chromosome-specific location and is easily assayed using PCR (Powell et al. 1996; Zaki et al. 2012). Most oil palm studies have focused on quantitative trait loci (QTL) mapping for yield-related traits (e.g., traits of bunch number, fresh fruit bunch yield, oil yield, oil-to-bunch content, and fatty acid composition) (Zhang et al. 2018).

Although African oil palm (*Elaeis guineensis*) is the highest-yielding oil crop per unit area worldwide, its oil is considered unhealthy for human consumption due to its high palmitic acid content (C16:0). Palmitic acid (16:0) is the major fatty acid in the oil (50%) in the mesocarp, in contrast, lauric acid (12:0) is the major fatty acid (50%) in kernel oil (Xia et al. 2019a). In order to facilitate breeding for fatty acid content in oil palm, GWAS was used to identify and validate SNP markers and underlying candidate genes associated with fatty acid content. SNPs-based GWAS resulted in 62 SNP markers, significantly associated with fatty acid composition (Xia et al. 2019b).

Comparative genomics and transcriptomic analyses focused on disease resistance genes (R genes) in order to predict candidates for breeding of pathogen tolerance/resistance. R genes may be used to identify and treat infected plants as they are expressed in early stages of the oil palm (Rosli et al. 2018).

The present work describes the sequence analysis of oil palm genome, focusing on the identified genes, regulating oil composition, and molecular markers, linked to fatty acid content and disease resistance genes.

6.2 Oil Palm Genome Sequence

The publication of the completed oil palm genome sequence in 2013 (Singh et al. 2013b) has facilitated the identification of genes involved in the regulation of important agronomic traits, such as oil composition and disease tolerance (Rosli et al. 2018).

Singh et al. (2013b) reported the 1.8 Gb genome sequence of the African oil palm *Elaeis guineensis* (AVROS, *pisifera* fruit form). A total of 1.535 Gb of assembled sequence and transcriptome data were used to predict at least 34,802 genes, including oil biosynthesis genes and homologues of WRINKLED1 (WRI1), and other transcriptional regulators.

Jin et al. (2016) sequenced the genome of elite *dura* palm (1.701 Gb), and 36,105 genes were predicted, 75.8% of them predicted protein-coding genes showing significant sequence similarity to known genes deposited in the public databases. Further, genes for noncoding RNAs in the *dura* palm draft genome were identified: 636 genes for tRNAs and 1182 genes for rRNAs.

6.2.1 Oil Palm Databases

To explore, retrieve, and analyze the available large expressed sequence tags (ESTs) and whole genome sequence databases of oil palm, the following available databases were developed to manage and share oil palm sequencing data (Table 6.1).

First, the Malaysian Oil Palm Genome Program (MyOPGP) sequences are available for download at the Genomsawit website (http://genomsawit.mpob.gov.my/). The available oil palm genome sequences in MyOPGP were submitted to NCBI with BioProject accession PRJNA217845 (*E. guineensis*) and PRJNA217846 (*E. oleifera*) (Low et al. 2014). Oil palm genome data available in MyOPGP are classified as genome sequence data, gene prediction data, transcriptome data, GeneThresher sequences (genomic sequences generated from the hypomethylated or gene-rich regions of the oil palm genome), SNP variants, sequences, and mapping locations (linkage groups).

Furthermore, PalmXplore is a public-domain archive of predicted oil palm (*Elaeis guineensis*) genes (http://palmxplore.mpob.gov.my/palmXplore/). Databases in PalmXplore include (1) predicted genes and their genomic coordinates; (2) annotations derived from external databases (e.g., Pfam, Gene Ontology, and Kyoto Encyclopedia of Genes and Genomes); and (3) information about genes related to important traits, such as those involved in fatty acid biosynthesis and disease resistance.

In addition, several oil palm databases are available at NCBI (http://www.ncbi.nlm.nih.gov), including databases of genome sequences, proteins, genes, gene expression omnibus (GEO), pathways, substances, and compounds.

Further, a web application, pSatdb (https://ssr.icar.gov.in/index.php), is a microsatellite database of oil palm, which is used for identifying SSRs based on repeat motif type, repeat type, and primer details (Kalyana et al. 2019).

Moreover, different bioinformatics databases were used by Ling et al. (2016) to analyze vitamin E biosynthetic genes from the oil palm. SDSC Biology Workbench

Table 6.1 Oil palm databases

Database		Website
Oil palm databases		
1	PalmXplore	http://palmxplore.mpob.gov.my/palmXplore/
2	NCBI	https://www.ncbi.nlm.nih.gov/
3	Genomsawit	http://genomsawit.mpob.gov.my/
4	OpSatdb	https://ssr.icar.gov.in/index.php
General sequence databases		
5	ExPASy Proteomics tools	http://cn.expasy.org/tools/protscale.html
6	ProtParam tool	http://web.expasy.org/protparam/
7	SignalP and ChloroP	http://www.cbs.dtu.dk/services/SignalP
		http://www.cbs.dtu.dk/services/ChloroP/
8	SMART	http://smart.embl-heidelberg.de/
9	Phylogeny.fr	http://www.phylogeny.fr/version2_cgi/simple_phylogeny.cgi

tools (http://seqtool.sdsc.edu) were used for sequence alignment and characterization of the deduced proteins. Amino acid composition and isoelectric point were performed using ExPASy Proteomics tools (http://cn.expasy.org/tools/protscale.html). The physical and chemical characteristics of all deduced amino acid sequences were analyzed using the ProtParam tool (http://web.expasy.org/protparam/). In addition to this, the signal peptide targeting location of the deduced proteins was predicted using SignalP and ChloroP (http://www.cbs.dtu.dk/services). WoLF PSORT (http://wolfpsort.org/) was used to perform the subcellular localization prediction, and the Simple Modular Architectural Research Tool (SMART) database was used to analyze the protein domain. Phylogenetic and molecular evolutionary analyses were performed using the Phylogeny.fr web services (http://www.phylogeny.fr/version2_cgi/simple_phylogeny.cgi).

6.2.2 Molecular Markers in Oil Palm

Molecular markers play an important role in genetic selection strategies employed for the breeding of oil palm (e.g., variety identification and genetic diversity studies). The availability of ESTs and whole genome databases enabled the development of a database of microsatellite markers for the oil palm genome.

Palm oil is rich in vitamins, including vitamin E. GWAS analysis was used to identify SNPs linked with vitamin E in a diversity panel of 161 *E. guineensis* trees (Luo et al. 2020). Forty-seven SNP markers were significantly associated with the variation of tocopherol and tocotrienol levels (chemical constituents of vitamin E) (Luo et al. 2020).

The SSR markers were used in oil palm for studying genetic diversity (Hayati et al. 2004) and construction of linkage maps (Billotte et al. 2005), QTL mapping (Jeennor and Volkaert 2014), and association mapping (Babu et al. 2017). Singh et al. (2008) exploited oil palm ESTs sequence database and found 145 SSRs in 136 unique ESTs, and few of them were used for genetic diversity studies.

The availability of the sequence data, facilitated mapping of QTL to resolve the genetic basis of several complex traits in oil palm, including yield (Rance et al. 2001; Billotte et al. 2010), fatty acid composition (Singh et al. 2009; Montoya et al. 2013), sex ratio (Ukoskit et al. 2014), and embryogenesis (Ting et al. 2013). Although association mapping has been validated to be a reliable method for identifying trait-associated markers, for marker-assisted selection, this method has rarely been applied in the oil palm. Association-mapping of marker data with phenotypic data of eight oil yield-related traits resulted in the identification of 11 significant QTLs (Babu et al. 2017).

The different forms of oil palm fruit is governed by the *SHELL* gene for regulating shell thickness, distinguishing the oil palm genotype as *dura*, *pisifera*, and *tenera* fruits. The *dura* genotype, for a thick-shelled fruit, is the dominant homozygous Sh/Sh, whereas the *pisifera* genotype, for a shell-less fruit, is the recessive homozygous sh/sh genotype (Corley and Tinker 2003). The *tenera* genotype has a thin-shelled fruit that has 30% more mesocarp and oil production than *dura* and *pisifera*, and is generally produced as a hybrid from the cross between the *dura* and *pisifera* (heterozygous Sh/sh). A genetic marker for shell thickness could be used to distinguish the *dura*, *tenera*, and *pisifera* plants in the nursery, long before they are field planted (Singh et al. 2013a). Babu et al. (2017) identified one cleaved amplified polymorphic site (CAPS) marker, for the differentiation of oil palm fruit type, which produced two alleles (280 and 250 bp) in *dura* genotypes, three alleles in *tenera* genotypes (550, 280, and 250 bp) and one allele in *pisifera* genotypes (550 bp). The CAPS marker will facilitate selection and timely distribution of desirable high-yielding *tenera* sprouts to the farmers instead of waiting for 4 ± 5 years to distinguish fruit forms after being field planted.

6.2.3 Identification of Oil Palm Genes

The identification of genes associated with traits of interest, such as those for fatty acid biosynthesis and disease resistance, would be extremely helpful in accelerating the breeding and selection of oil palm.

Singh et al. (2013a) described the mapping and identification of the *SHELL* gene, which is responsible for the different fruit forms. Oil palm breeding involves crossing *dura* and *pisifera* palms to produce *tenera* progeny with higher oil yield. Oil yield is controlled by variant alleles of the type-II MADS-box gene, *SHELL*, that impact the presence and thickness of the shell, surrounding the fruit kernel, and is a homologue of *SEEDSTICK*, which is responsible for ovule and seed development in Arabidopsis (Chan et al. 2017; Singh et al. 2020; Zhang et al. 2018).

Singh et al. (2014) identified the *VIRESCENS* (*VIR*) gene. *VIR* determines the fruit exocarp color, which is an indicator of ripeness.

Identification of disease resistance genes can help to improve screening for disease resistance/tolerance for major oil palm pathogens. Chan et al. (2017) identified 210 candidate genes, for disease-resistance in the oil palm tree, grouped in to six classes based on their protein domain structures. Eight genes were unique to the oil palm tree and are potentially involved in oil palm–specific interactions with pathogen *Avr* gene products.

Zhang et al. (2018) found that the *EgGDSL* gene was highly expressed in oil palm trees with high oil content than in those with low oil content, demonstrating that the transcription level of *EgGDSL* correlated with the amount of oil accumulation. The gene may be valuable for engineering fatty acid metabolism in crop-improvement programs and for marker-assisted breeding.

Xia et al. (2019b) investigated SNPs associated with fatty acid content in the oil palm and identified dozens of candidate genes involved in fatty acid biosynthesis and metabolic pathways. *EgFatB1* gene, highly and specifically expressed in the mesocarp, was validated using Arabidopsis transformation; overexpression of EgFatB1 resulted in increased saturated-fat content in Arabidopsis seeds. Xia et al. (2019a) found the acyl-ACP thioesterase B (*FatB*) gene, involved in fatty acid biosynthesis, is associated with high palmitic acid content in the mesocarp. Overexpression of this gene caused a significant increase in palmitic acid content.

Luo et al. (2020) found that the *EgHGGT* gene (homogentisate geranylgeranyl transferase), involved in the biosynthesis of tocotrienols, was highly expressed in the mesocarp than in other tissues. Induced overexpression of the gene in Arabidopsis caused a significant increase in vitamin E content and production of α-tocotrienols compared to that of wild Arabidopsis.

6.2.4 Genetic Diversity

It was noticed that selection for genetic improvement caused a loss of genetic diversity in the oil palm. Genetic variation was studied using microsatellite markers in nine crosses (*Dura* × *Pisifera*) of *Elaeis guineensis* Jacq. obtained from various commercial companies in Malaysia, France, Costa Rica, and Colombia. A significant reduction in allele number was observed in this variety, compared to the number of alleles reported for wild oil palm populations (Arias et al. 2012). The genetically improved *dura* and *pisifera* palms used in producing the hybrid *tenera* in Southeast Asia showed the lowest number of SNPs and reduced genetic diversity, compared with that of *tenera* palms collected in Africa (Jin et al. 2016). It is well known that the average oil yield of oil palms in Southeast Asia is much higher than that in Africa because of the extensive selective breeding for oil yield in Southeast Asia (Corley and Tinker 2015). Therefore, these form-specific alleles in the elite *dura* and *pisifera* palms in Southeast Asia may be useful in improving the production performances of oil palms in Africa.

The coding regions represents around 7% in the oil palm genome, and about 2% of the total SNPs were found to be present in these regions; the remaining SNPs (~98%) were found in noncoding regions. The average Ka/Ks ratio (nonsynonymous SNPs/synonymous SNPs) in oil palms was 1.4 (Jin et al. 2016), which is among the highest of all plants reported so far (1.31–1.61 in soybean, 1.2 in rice, and 0.83 in *Arabidopsis thaliana*) (Initiative 2000; Goff et al. 2002; Lam et al. 2010). The average Ka/Ks ratio of R genes was 1.7, which is much higher than that (1.4) of all genes in the oil palm genome, suggesting a strong positive selection of R genes in palms (Jin et al. 2016).

Although the number of R genes in oil palms is less than that in other plants (566 R genes/*dura* genome), they are more variable (Goff et al. 2002; Jin et al. 2016; Paterson et al. 2009; Schnable et al. 2009). Extreme diversity and rapid evolution of R genes enhance the resistance of the oil palm against challenges from diverse pathogens (Jin et al. 2016).

Jin et al. (2016) found two transition SNPs (SNP1 G_{169}→A and SNP2 A_{248}→T) in the *SHELL* allele sequence of the oil palm. In SNP1, the substitution of “G” was present in all genotypes, whereas the substitution “A” was present in a single *dura* genotype. In SNP2, the substitution A was present only in *dura* genotypes, whereas substitution “T” was present only in *pisifera* genotypes. In *tenera*, the *SHELL* allele sequences were found to contain either A or T substitutions. These results indicate that SNP2 contributes to variation in the fruit forms of *dura*, *pisifera*, and *tenera* genotypes. Similarly, Singh et al. (2013a) found two independent mutations in the MADS-box transcription factor of the *SHELL* gene. They also found one SNP in the 28th or 30th codon of the oil palm genome, which impairs the normal DNA binding of the *SHELL* gene, leading to a shell-less phenotype.

6.3 Conclusion

Availability of genetic and genomic resources of oil palm can potentially improve the nutritional value of palm oil for human consumption. The availability of molecular markers associated with agronomic traits allow breeders to rapidly identify desired traits. SNPs markers associated with fatty acid content are useful for selecting low palmitic acid genotypes during breeding and subsequently improving the nutritional value of palm oil. Moreover, the study of gene affecting the relative palmitic acid content, allowing us to screen for low palmitic acid content.

References

Arias D, Montoya C, Rey L, Romero H (2012) Genetic similarity among commercial oil palm materials based on microsatellite markers. Agron Colomb 30(2):188–195

Babu BK, Mathur RK, Kumar PN, Ramajayam D, Ravichandran G, Venu MVB, Babu SS (2017) Development, identification and validation of CAPS marker for SHELL trait which governs dura, pisifera and tenera fruit forms in oil palm (*Elaeis guineensis* Jacq.). PLoS One 12(2):1–16

Billotte N, Marseillac N, Risterucci A-M, Adon B, Brottier P, Baurens F-C, Singh R, Herrán A, Asmady H, Billot C, Amblard P, Durand-Gasselin T, Courtois B, Asmono D, Cheah SC, Rohde W, Ritter E, Charrier A (2005) Microsatellite-based high density linkage map in oil palm (Elaeis guineensis Jacq.). Theor Appl Genet 110(4):754–765

Billotte N, Jourjon MF, Marseillac N, Berger A, Flori A, Asmady H, Adon B, Singh R, Nouy B, Potier F, Cheah SC, Rohde W, Ritter E, Courtois B, Charrier A, Mangin B (2010) QTL detection by multi-parent linkage mapping in oil palm (Elaeis guineensis Jacq.). Theor Appl Genet 120:1673–1687

Chan K-L, Tatarinova TV, Rosli R, Amiruddin N, Azizi N, Halim MAA, Sanusi NSNM, Jayanthi N, Ponomarenko P, Triska M, Solovyev V, Firdaus-Raih M, Sambanthamurthi R, Murphy D, Low E-TL (2017) Evidence-based gene models for structural and functional annotations of the oil palm genome. Biol Direct 12(21):1–23

Corley RHV, Tinker PB (2003) The oil palm, 4th edn. Wiley-Blackwell, Malden

Corley RHV, Tinker PHB (2015) The oil palm, 5th edn. Wiley-Blackwell, Malden

Goff SA, Ricke D, Lan TH, Presting G, Wang R, Dunn M, Glazebrook J, Sessions A, Oeller P, Varma H, Hadley D, Hutchison D, Martin C, Katagiri F, Lange BM, Moughamer T, Xia Y, Budworth P, Zhong J, Miguel T, Paszkowski U, Zhang S, Colbert M, Sun W-L, Chen L, Cooper B, Park S, Wood TC, Mao L, Quail P, Wing R, Dean R, Yu Y, Zharkikh A, Shen R, Sahasrabudhe S, Thomas A, Cannings R, Gutin A, Pruss D, Reid J, Tavtigian S, Mitchell J, Eldredge G, Scholl T, Miller RM, Bhatnagar S, Adey N, Rubano T, Tusneem N, Robinson

R, Feldhaus J, Macalma T, Oliphant A, Briggs S (2002) A draft sequence of the rice genome (Oryza Sativa L. Ssp. Japonica). Science 296(5565):92–100

Hayati A, Wickneswari R, Maizura I, Rajanaidu N (2004) Genetic diversity of oil palm (Elaeis guineensis Jacq.) germplasm collections from Africa: implications for improvement and conservation of genetic resources. Theor Appl Genet 108:1274–1284

Initiative AG (2000) Analysis of the genome sequence of the flowering plant Arabidopsis thaliana. Nature 408:796–815

Jeennor S, Volkaert H (2014) Mapping of quantitative trait loci (QTLs) for oil yield using SSRs and gene-based markers in African oil palm (Elaeis guineensis Jacq.). Tree Genet Genomes 10:1–14

Jin J, Lee M, Bai B, Sun Y, Qu J, Rahmadsyah, Alfiko Y, Lim CH, Suwanto A, Sugiharti M, Wong L, Ye J, Chua N-H, Yue GH (2016) Draft genome sequence of an elite Dura palm and whole-genome patterns of DNA variation in oil palm. DNA Res 23(6):527–533

Kalyana BB, Mary RKL, Sarika S, Mathur RK, Naveen KP, Ravichandran G, Anitha P, Bhagya HP (2019) Development and validation of whole genome-wide and genic microsatellite markers in oil palm (*Elaeis guineensis* Jacq.): first microsatellite database (OpSatdb). Sci Rep 9:1899

Lam H-M, Xu X, Liu X, Chen W, Yang G, Wong F-L, Li M-W, He W, Qin N, Wang B, Li J, Jian M, Wang J, Shao G, Wang J, Sun SS-M, Zhang G (2010) Resequencing of 31 wild and cultivated soybean genomes identifies patterns of genetic diversity and selection. Nat Genet 42:1053–1059

Ling KS, Abdullah SNA, Ling HC, Amiruddin MD (2016) Molecular cloning, gene expression profiling and in silico sequence analysis of vitamin E biosynthetic genes from the oil palm. Plant Gene 5:100–108

Low E-TL, Rosli R, Jayanthi N, Halim MAA, Azizi N, Chan K-L, Maqbool NJ, Maclean P, Brauning R, McCulloch A, Moraga R, Ong-Abdullah M, Singh R (2014) Analyses of hypomethylated oil palm gene space. PLoS One 9(1):1–15

Luo T, Xia W, Gong S, Li ASMZ, Liu R, Dou Y, Tang W, Fan H, Zhang C, Xiao Y (2020) Identifying vitamin E biosynthesis genes in *Elaeis guineensis* by genome-wide association study. J Agric Food Chem 68:678–685

Mayes S, Hafeez F, Price Z, MacDonald D, Billotte N, Roberts J (2008) Molecular research in oil palm, the key oil crop for the future. In: Moore PH, Ming R (eds) Genomics of tropical crop plants. Springer, New York

Montoya C, Lopes R, Flori A, Cros D, Cuellar T, Summo M (2013) Quantitative trait loci (QTLs) analysis of palm oil fatty acid composition in an interspecific pseudo-backcross from Elaeis oleifera (H.B.K.) Cortés and oil palm (Elaeis guineensis Jacq.). Tree Genet Genomes 9:1207–1225

Pantzaris TP (1997) Pocketbook of palm oil uses. Palm Oil Research Institute of Malaysia, Kuala Lumpur

Paterson AH, Bowers JE, Bruggmann R, Paterson AH, Bowers JE, Bruggmann R, Dubchak I, Grimwood J, Gundlach H, Haberer G, Hellsten U, Mitros T, Poliakov A, Schmutz J, Spannag M, Tang H, Wang X, Wicker T, Bharti AK, Chapman J, Feltus FA, Gowik U, Grigoriev IV, Lyons E, Maher CA, Martis M, Narechania A, Otillar RP, Penning BW, Salamov AA, Wang Y, Zhang L, Carpita NC, Freeling M, Gingle AR, Hash CT, Keller B, Klein P, Kresovich S, McCann MC, Ming R, Peterson DG, Rahman MUR, Ware D, Westhoff P, Mayer KFX, Messing J, Rokhsar DS (2009) The Sorghum bicolor genome and the diversification of grasses. Nature 457:551–556

Powell W, Morgante M, Andre C, Hanafey M, Vogel J, Tingey S, Rafalski A (1996) The comparison of RFLP, RAPD, AFLP and SSR (microsatellite) markers for germplasm analysis. Mol Breed 2:225–238

Rance KA, Mayes S, Price Z, Jack PL, Corley RHV (2001) Quantitative trait loci for yield components in oil palm (Elaeis guineensis Jacq.). Theor Appl Genet 103:1302–1310

Rosli R, Amiruddin N, Halim MAA, Chan P-L, Chan K-L, Azizi N, Morris PE, Low E-TL, Ong-Abdullah M, Sambanthamurthi R, Singh R, Murphy DJ (2018) Comparative genomic and transcriptomic analysis of selected fatty acid biosynthesis genes and CNL disease resistance genes in oil palm. PLoS One 13(4):1–17

Schnable PS, Ware D, Fulton RS, Stein JC, Wei F, Pasternak S, Liang C, Zhang J, Fulton L, Graves TA, Minx P, Reily AD, Courtney L, Kruchowski SS, Tomlinson C, Strong C, Delehaunty K, Fronick C, Courtney B, Rock SM, Belter E, Du F, Kim K, Abbott RM, Cotton M, Levy A, Marchetto P, Ochoa K, Jackson SM, Gillam B, Chen W, Yan L, Higginbotham J, Cardenas M, Waligorski J, Applebaum E, Phelps L, Falcone J, Kanchi K, Thane T, Scimone A, Thane N, Henke J, Wang T, Ruppert J, Shah N, Rotter K, Hodges J, Ingenthron E, Cordes M, Kohlberg S, Sgro J, Delgado B, Mead K, Chinwalla A, Leonard S, Crouse K, Collura K, Kudrna D, Currie J, He R, Angelova A, Rajasekar S, Mueller T, Lomeli R, Scara G, Ko A, Delaney K, Wissotski M, Lopez G, Campos D, Braidotti M, Ashley E, Golser W, Kim H, Lee S, Lin J, Dujmic Z, Kim W, Talag J, Zuccolo A, Fan C, Sebastian A, Kramer M, Spiege L, Nascimento L, Zutavern T, Miller B, Ambroise C, Muller S, Spooner W, Narechania A, Ren L, Wei S, Kumari S, Faga B, Levy MJ, McMahan L, Buren PV, Vaughn MW, Ying K, Yeh C-T, Emrich SJ, Jia Y, Kalyanaraman A, Hsia A-P, Barbazuk WB, Baucom RS, Brutnell TP, Carpita NC, Chaparro C, Chia J-M, Deragon J-M, Estill JC, Fu Y, Jeddeloh JA, Han Y, Lee H, Li P, Lisch DR, Liu S, Liu Z, Nagel DH, McCann MC, SanMigue P, Myers AM, Nettleton D, Nguyen J, Penning BW, Ponnala L, Schneider KL, Schwartz DC, Sharma A, Soderlund C, Springer NM, Sun Q, Wang H, Waterman M, Westerman R, Wolfgruber TK, Yang L, Yu Y, Zhang L, Zhou S, Zhu Q, Bennetzen JL, Dawe RK, Jiang J, Jiang N, Presting GG, Wessler SR, Aluru S, Martienssen RA, Clifton SW, McCombie WR, Wing RA, Wilson RK (2009) The B73 maize genome: complexity, diversity, and dynamics. Science 326:1112–1115

Singh R, Zaki N, Ting N-C, Rosli R, Tan S-G, Low E-T, Ithnin M, Cheah S-C (2008) Exploiting an oil palm EST database for the development of gene-derived and their exploitation for assessment of genetic diversity. Biologia 63:227–235

Singh R, Tan SG, Panandam JM, Rahimah AR, Ooi LCL, Low E-TL, Sharma M, Jansen J, Cheah S-C (2009) Mapping quantitative trait loci (QTLs) for fatty acid composition in an interspecific cross of oil palm. BMC Plant Biol 9(114):1–19

Singh R, Low EL, Ooi LC-L, Ong-Abdullah M, Ting N-C, Nagappan J, Nookiah R, Amiruddin MD, Rosli R, Abdul Manaf MA, Chan K-L, Halim MA, Azizi N, Lakey N, Smith SW, Budiman MA, Hogan M, Bacher B, Brunt AV, Wang C, Ordway JM, Sambanthamurthi R, Martienssen RA (2013a) The oil palm SHELL gene controls oil yield and encodes a homologue of SEEDSTICK. Nature 500:244–246

Singh R, Ong-Abdullah M, Low EL, Manaf MAA, Rosli R, Nookiah R, Ooi LC-L, Ooi S–E, Chan K-L, Halim MA, Azizi N, Nagappan J, Bacher B, Lakey N, Smith SW, He D, Hogan M, Budiman MA, Lee EK, DeSalle R, Kudrna D, Goicoechea JL, Wing RA, Wilson RK, Fulton RS, Ordway JM, Martienssen RA, Sambanthamurthi R (2013b) Oil palm genome sequence reveals divergence of interfertile species in old and new worlds. Nature 500:335–339

Singh R, Low E-TL, Ooi LC-L, Ong-Abdullah M, Nookiah R, Ting N-C, Singh R, Low ETL, Ooi LC-L, Ong-Abdullah M, Nookiah R, Ting N-C, Marjuni M, Chan P-L, Ithnin M, Abdul Manaf MA, Nagappan J, Chan K-L, Rosli R, Halim MA, Azizi N, Budiman MA, Lakey N, Bacher B, Brunt AV, Wang C, Hogan M, He D, MacDonald JD, Smith SW, Ordway JM, Martienssen RA, Sambanthamurthi R (2014) The oil palm VIRESCENS gene controls fruit colour and encodes a R2R3- MYB. Nat Commun 5:1–8

Singh R, Low E-TL, Ooi LC-L, Ong-Abdullah M, Ting N-C, Nookiah R, Ting N-C, Marjuni M, Chan P-L, Ithnin M, Abdul Manaf MA, Nagappan J, Chan K-L, Rosli R, Halim MA, Azizi N, Budiman MA, Lakey N, Bacher B, Brunt AV, Wang C, Hogan M, He D, MacDonald JD, Smith SW, Ordway JM, Martienssen RA, Sambanthamurthi R (2020) Variation for heterodimerization and nuclear localization among known and novel oil palm SHELL alleles. New Phytol 226(2):426–440

Ting NC, Jansen J, Nagappan J, Ishak Z, Chin CW, Tan SG, Cheah S-C, Singh R (2013) Identification of QTLs associated with callogenesis and embryogenesis in oil palm using genetic linkage maps improved with SSR markers. PLoS One 8(1):1–16

Ukoskit K, Chanroj V, Bhusudsawang G, Pipatchartlearnwong K, Tangphatsornruang S, Tragoonrung S (2014) Oil palm (Elaeis guineensis Jacq.) linkage map, and quantitative trait locus analysis for sex ratio and related traits. Mol Breed 33:415–424

Xia W, Luo T, Dou Y, Zhang W, Mason AS, Huang D, Huang X, Tang W, Wang J, Zhang C, Xiao Y (2019a) Identification and validation of candidate genes involved in fatty acid content in oil palm by genome- wide association analysis. Front Plant Sci 10:1263

Xia W, Luo T, Zhang W, Mason AS, Huang D, Huang X, Tang W, Dou Y, Zhang C, Xiao Y (2019b) Development of high-density SNP markers and their application in evaluating genetic diversity and population structure in Elaeis guineensis. Front Plant Sci 10:130

Zaki NM, Singh R, Rosli R, Ismail I (2012) *Elaeis oleifera* genomic-SSR markers: exploitation in oil palm germplasm diversity and cross-amplification in arecaceae. Int J Mol Sci 13:4069–4088

Zhang Y, Bai B, Lee M, Alfiko Y, Suwanto A, Yue GH (2018) Cloning and characterization of *EgGDSL*, a gene associated with oil content in oil palm. Sci Rep 8:1–11

Chapter 7
Argane Genetics and Genomics

Hassan Ghazal, Oussama Badad, Houcine Zaid, Tatiana Tatusova, Stacy Pirro, Slimane Khayi, Fatima Gaboun, Kamal Aberkani, Aissam El Finti, Mary Kinsel, Abdelaziz Zahidi, Naima Ait Aabd, Jamila Mouhaddab, Fouad Msanda, Abdellah Idrissi Azami, Rachid Mentag, and Abdelhamid El Mousadik

Contents

H. Ghazal (✉)
National Center for Scientific and Technological Research (CNRST), Rabat, Morocco

School of Medicine, University Mohammed VI for Health Sciences, Casablanca, Morocco
e-mail: hassan.ghazal@fulbrightmail.org

O. Badad · H. Zaid
Faculty of Sciences, University Mohammed V, Rabat, Morocco

T. Tatusova
National Center for Biotechnology Information, National Institutes of Health, Bethesda, MD, USA

S. Pirro
Iridian Genome, Bethesda, MD, USA

S. Khayi · F. Gaboun · R. Mentag
CRRA-Rabat, National Institute for Agricultural Research (INRA), Rabat, Morocco

K. Aberkani
Polydisciplinary Faculty of Nador, University Mohammed Premier, Oujda, Morocco

A. El Finti · A. Zahidi · J. Mouhaddab · F. Msanda · A. El Mousadik
Laboratory of Biotechnology and Valorization of Natural Resources (LBVRN), Faculty of Sciences, University Ibn Zohr, Agadir, Morocco

M. Kinsel
Department of Chemistry & Biochemistry, Southern Illinois University in Carbondale, Carbondale, IL, USA

N. Ait Aabd
CRRA-Agadir, National Institute for Agricultural Research (INRA), Rabat, Morocco

A. Idrissi Azami
School of Medicine, University Mohammed VI for Health Sciences, Casablanca, Morocco

H. Tombuloglu et al. (eds.), *Oil Crop Genomics*,
https://doi.org/10.1007/978-3-030-70420-9_7

7.1 Introduction

Argane trees (*Argania spinosa* L. Skeels) have been here for millions of years. Today they only grow on a narrow semidesert strip between the Atlantic coast of Morocco and the Atlas mountains (Nouaim et al. 2002). The *Argania spinosa*, better known as the Argane tree, is the "tree of life" in the midwestern area of Morocco of which it is endemic. Believed to have taken roots in Argana village, 80 *km* North of the city of Agadir (M'Hirit et al. 1998), this tree is well known for its "kernels" of healing oil or liquid gold obtained from its fruits (Charrouf and Guillaume 2008). The *Argania spinosa* tree is suitable for harsh conditions, heat survival, drought, and poor soil (Nouaim et al. 2002). It is less familiar outside of Morocco as it only grows in the country's Southwest, approximately between the cities of Essaouira and Agadir, in an area covering 900,000 *ha* and has been given UNESCO protection as a biosphere reserve (Arganeraie Biosphere Reserve) in 1998 ("UNESCO-MAB Biosphere Reserve Directory" 2018). The *Argania spinosa* is an endangered species that plays a vital role in resisting the ever-creeping Sahara Desert (Nouaim et al. 2002). There are around 21 million trees in the region where the Argane trees grow which play a vital role in the food chain and the environment, although their numbers are decreasing. The thorny trees that can attain heights of 8–10 m (Msanda et al. 2005) last longer than the olive tree and do not require cultivation (Msanda et al. 2005). The Argan trunk is frequently twisted and gnarled; the fruit has a dark, fleshy surface like an olive tree, but larger and rounder; within is a nut with an incredibly hard shell that contains one, two, or three almond-shaped kernels in turn (Nouaim et al. 2002).

According to the International Development Research Center (IDRC), Argane trees help three million Moroccans, around 10 *percent* of the country's population, who use husks as firewood, animal fodder berries, and pipes to produce the precious oil (El Monfalouti et al. 2010). Argane oil, extracted from the almond of the fruit of *Argania spinosa*, plays a very large role in the local economy including wood to provide lumber for construction, heating, and charcoal, the leaves and pulp of the fruit are used for animal feed, and the seeds provide the Argane oil (Charrouf and Guillaume 2008).

During the eighteenth century, the distribution area of the Argane forest significantly decreased. Moreover, between 1970 and 2007, about 44 *percent* of the forest was lost (Msanda et al. 2005). Though multiple factors exist, desertification and overgrazing are the major pressures on the Argane forest. Hence the management and conservation of Argane Forest's remaining genetic resources are urgent priorities (Msanda et al. 2005). Several studies have been carried out in recent decades to assess the genetic diversity of the Argane tree using morphological, chemical, biochemical, standard molecular marker techniques and omics, both with the intention to characterize the genetic diversity of Argane trees, address ecological and preservation issues, and characterize the oil composition and metabolic production pathways. The Argane tree still appears in its wild and spontaneous state within its natural ecosystem, in the form of trees and shrubs with different architectural, phenological, vigor, and diverse fruit shapes (Ait Aabd 2013). This variability is seen even within the same locality and under similar ecogeographic conditions.

7.2 Argane Genetics

Molecular markers have been used to characterize the genetic structure of the Argane tree. Work on genetic diversity is moderately limited; nonetheless, it underwent an important evolution either in the markers used or in the size of the populations evaluated. Several markers have allowed a partial description of the genetic variability in samples reared in nurseries from seedlings originating from different origins of the Argane tree, in particular by 10 allozymic loci (El Mousadik and Petit 1996), 12 loci derived from chloroplast (cp) DNA *RFLP*-PCR (El Mousadik and Petit 1996), and fragments amplified by RAPD (Bani-Aameur and Benlahbil 2004). A genetic diversity assessment of the chloroplast genome from 95 seedlings from 19 populations using 12 *PCR*-RFLP markers showed a high level of genetic differentiation between populations ($G_{ST} = 0.60$). This study also showed a strong geographical structuring of the studied populations with allozymes (El Mousadik and Petit 1996).

Bani-Aameur and Benlahbil (2004) used eight RAPD markers to identify the profiles of three provenances of Argane and their progeny (5 months) as well as the discriminating power between the genotypes (mother tree) and their progeny. This analysis allowed the identification of 54 loci fingerprints which enabled to identify genotype profiles of parents.

Similarly, Majourhat et al. (2008) used 19 RAPD markers to characterize and study the genetic relationships between 38 genotypes of Argane tree with three morphotypes: oval, rounded, and fusiform, according to fruit phenotype. The markers generated 146 fragments, 140 of which are polymorphic. The results also showed great genetic variability within the accessions of the tested Argane trees and a very high rate of polymorphism. Thus, the comparative analysis of the fruit shape with the results of the RAPD analysis did not produce any significant correlation. This result shows that information based on the fruit shape was not sufficient for genetic discrimination. The estimation of genetic differentiation parameter has not been made.

Later, new approaches based on the study of genetic variability from samples taken directly from trees and not from their progeny (seedlings from seedlings) were undertaken. The latter strategy made it possible to better estimate the genetic structuring of trees in situ according to different ecogeographical situations. Thus, the first work representing 86 loci from 9 ISSR primers (dominant markers revealed by mass) revealed a genetic differentiation (G_{ST}) of the order of 0.22, thus illustrating a rather limited gene flow between 5 populations belonging to the most representative region of the Argane tree (Ait Aabd et al. 2015).

Future studies should expand the sampling areas to cover all provinces of the natural distribution of Argane trees, especially the inclusion of the pre-Saharan extreme zones and the two relict sites of Oued Grou (near the city of Rabat) and Beni Snassen (North-East of Morocco) to see the effect of biogeographic isolation on the diversity and genetic structure of the Argane tree.

Accordingly, Mouhaddab et al. (2017), using the same types of markers, were able to evaluate the G_{ST} between 24 populations (240 trees in total) representing both isolated provenances (relics from the North and Saharan provenances from Southern Morocco), the genetic differentiation coefficient found was very high ($G_{ST} = 0.44$). G_{ST} was similar between ISSR and 60 shrubs of Argane planted in 4 different sites in Tunisia ($G_{ST} = 0.18$), and this value increased from plantations introduced into Tunisia ($G_{ST} = 0.17$) (Louati et al. 2019).

In addition, microsatellites were transferred to the Argane tree from two species belonging to the Sapotaceae family and were tested by Majourhat et al. (2008). These markers were applied to characterize the genetic structure of 38 genotypes of Argane tree from the Essaouira region according to fruit shape. However, these markers, considered as codominant, appeared invalid according to the ploidy level ($2n = 2x = 24$) of the Argane tree (Ait Aabd 2013). This requires the development of SSR markers from the Argane genome.

Recently, 11 SSR markers have been developed from the Argane genome using high-throughput sequencing (El Bahloul et al. 2014). These markers have been tested on 150 genotypes of the Argane tree, are proven to be informative, and have the ability to detect genotype polymorphism. On the other hand, and for the first time, AFLP markers have been introduced (Pakhrou et al. 2017). This technique has been applied to study the genetic structuring of 13 natural origins of the Argane tree in the region of Essaouira and the Anti-Atlas. The use of 4 pairs of primers gave rise to 477 loci, a number that testifies the effectiveness of these markers for the study of genetic diversity. The results also showed a high level of genetic differentiation between populations ($G_{ST} = 0.22$) and strong stand structuring. On the other hand, the microsatellites revealed weak genetic differentiation ($G_{ST} = 0.06$) between 24 provenances recording very high allelic richness (172 alleles with an average of 43 per locus) (Mouhaddab et al. 2017). The low differentiation obtained reflects a very important allelic mixing between stands. In contrast to other studies via ISSR and AFLP where differentiation between populations is high, the value recorded by the microsatellites (SSR) reflects the effect of common alleles compared to those that are stand specific but rare due to their low frequencies. It can also be noted that the dominant markers revealed in loci group (RAPD, ISSR, and AFLP) do not follow the same trend in terms of genetic differentiation as SSRs that are codominant with a huge polymorphism per loci. This observation clearly illustrates the risk to be taken if one bases oneself only on a single category of markers to estimate the genetic structuring in the Argane tree and in particular on the strategy to be taken for the construction of core collections. Indeed, Mouhaddab et al. (2017) showed in a first initiative that the same populations and the same individuals (240 mother trees/ genotypes) are structured in 2 very different core collections, according to the use of ISSR or SSR. Thus, maximizing the diversity of genotypic data from ISSRs and SSRs allowed the identification of 13 and 96 genotypes representing, respectively, 2.7% and 20% of the whole collection (240 mother trees). A recent study, conducted by Louati et al. (2019) and based on the comparison of 60 trees at the level of the polymorphism of the *delta-6* desaturase (D6D) locus, revealed the existence of ten groups of similar genotypes without respecting the origin or provenance. This result reinforces the mistrust to consider regarding the construction of germplasm (Core Collection) so as not to lose genotypes that go unnoticed based on the polymorphism of dominant markers based on nonspecific primers.

7.3 The *Argania spinosa* Genome

Advanced genetic studies prepared the entrance of *Argania* genomics with the advancement of the genome sequencing strategies. Elucidating the genomic origin of the valuable metabolites of the *A. spinosa*, namely, fatty acids, tocopherols, squalene, sterols, and phenolic compounds, became of great importance for molecular breeding in the ascending Argane oil industry. Given the availability of various sequencing platforms and the ingenuity to combine them for better de novo assemblies, annotations, and saturated genomic maps of high resolution, *A. spinosa* genomics is yet in its early days. In 2015, the first Argania genome individual, named *Amghar*, has been sequenced and raw DNA sequence reads deposited in GenBank (BioSample, SAMN04014715; SRA, SRS1539294; BioProject, PRJNA294096). Up to today, the only genome evidence available is a draft genome providing a draft assembly of the Argane tree genome using the combination of 144 *Gb* Illumina and 7.2 *Gb* PacBio reads (Khayi et al. 2018). This draft genome was assembled by using a hybrid de novo assembly approach combining long and short reads and thus creating a reliable draft reference genome for the *Argania spinosa* species consisting of 75,327 scaffolds with an N50 of 49.916 k*b*, totaling 671 *Mb* which is close to the estimated genome size using *k*-mer frequency. The assembly tends to be of an acceptable quality taking into account the below criteria of discrimination: a *k*-mer frequency study estimating the genome size through jellyfish analysis and a mapping of 94 *percent* of Illumina reads back toward assembly; thus from a total of 1440 Benchmarking Universal Single-Copy Orthologs (BUSCO) genes, 1291 genes (89%) were complete (1179 in single copy and 112 duplicated), 62 genes (4.3%) were represented partially, while 87 genes (6%) were missing from the assembly (Khayi et al. 2018).

The Sapotaceae family consists of approximately 50 genera and 1100 species distributed in the tropical regions with some exception, in particular, *A. Spinosa*, in the Midwestern portion of Morocco (Vaghani 2003). Previous biogeography and phylogenetic analyses of *Sapotaceae* species based on various elements, including several chloroplast genes (Duangjai et al. 2006), the chloroplast *ndhF* gene combined with morphological data (Swenson and Anderberg 2005), and the chloroplast *trnH*-psbA regions coupled with nuclear internal transcribed spacers (ITS), were identified (Stride et al. 2014). Such phylogenetic studies, inferred using few genetic markers, strongly suggested that *Argania*'s satellite genus be included in *Sideroxylon*'s group, thus revisiting *A. spinosa* phylogenetic status. Therefore, further genetic markers are required to explain this debatable phylogenetic revision.

Phylogenetic analysis using 69 chloroplast (cp) protein-coding genes shared by 11 members of Ericales and 1 species of Lamiales showed that the Sapotaceae family was clustered between *S. wightianum* and *P. campechiana* from the related Theaceae and Primulaceae families (Khayi et al. 2019, 2020). These aforementioned results were in concordance with a cladistic study of the largely tropical family Sapotaceae based on both morphological and molecular data (cp gene ndhF) (Stride et al. 2014). Furthermore, the resulting trees depicted that *A. spinosa* and *Sideroxylon mascatense* clustered together and belong to the genus Sideroxylon that is in discrepancy with the studies founded on combined loci (chloroplast, mitochondrial, and nuclear) (Larson et al. 2020), which positioned the *Sapotaceae* near to the Ebenaceae rather than the Primulaceae family. For the Asterids clade, inconsistent topology has been stated between the nuclear genome and plastome phylogenies and could be clarified by considering several evolutionary processes such as hybridization, horizontal gene transfer, and gene gain and loss (Stull et al. 2020).

The complete genome composition of chloroplast will help to elucidate the phylogenetic landscape of the Argane tree within *Sapotaceae* and will also be useful for further studies on the preservation and breeding of this essential medicinal and culinary plant. Thus, the full chloroplast (cp) genome of *A. spinosa* was sequenced, assembled, and analyzed against another two members of the Sapotaceae family. The *Argania* chloroplast genome cp is 158,848 *bp* long with an estimated *GC* percentage of 36.8 *percent*. The cp genome possesses a characteristic quadripartite yet circular structure consisting of a pair of 25.945 *bp* inverted regions (IR) in length separating small single-copy (SSC) and large single-copy (LSC) regions of 18.591 and 88.367 *bp*, respectively. The *Argania* chloroplast genome's annotation indicated the presence of 130 genes, namely, 85 protein-coding genes, 8 genes for ribosomal RNA (rRNA), and 37 genes for transfer RNA (tRNA) (Khayi et al. 2019, 2020).

Several botanical, agronomic, chemical, and socioeconomic studies have been carried out on the Argane tree. There is, however, little information concerning the Argane tree genomics. The draft genome assembly is the first step toward a global and integrative omics strategy for exhaustive characterization of the Argane tree. Throughout the field of Argane science, omics studies including genomics, transcriptomics, metabolomics, epigenomics, and metagenomics are becoming hot topics with a shared cause: understanding the organization, the functioning, and the evolution of the plant precious genetic material. This will require answers to identify genomic-scale biological problems. To this end and for the need to coordinate the increasing interests of the plant genomics community for this precious plant and integrated Argane omics analysis, the International Argane Genome Consortium (IAGC) and a dedicated resource website have been created (http://www.arganome.org/).

7.4 *Argania spinosa* Metabolomics

Plant metabolites have an immense chemical variety, with each plant having its own diverse selection of metabolites. The diversity poses methodological problems for both the simultaneous sampling of several metabolites and the quantitative study of the identified metabolites. We are only starting to grasp the functions of such metabolites, many of which are implicated in adaptation to different ecological niches and others having useful applications, including pharmaceutical. Spectacular developments in plant metabolomics, along with the assistance of systems biology, give fresh opportunities to investigate the remarkable complexities of plant biochemical ability. Genomics tools could be paired with metabolic profiling to classify essential genes that could be optimized for better crop plant growth and oil production.

Metabolites variation is what makes Argane oil special (Khallouki et al. 2017). Diverse compounds including essential oils, fatty acids, triacylglycerols, flavonoids, monophenols, phenolic acids, cinnamic acids, saponins, triterpenes, phytosterols, ubiquinone, melatonin, aminophenols, vitamin E, and several secondary metabolites have been described (Khallouki et al. 2003). As reported by Khallouki et al. (2003), linoleic acid, preceded by unsaturated oleic acid, dominates the fatty acids in Argane oil. Palmitic acid predominates alongside stearic acid for the saturated fatty acids. Moreover, the fraction of Argane oil sterol consists of four derived sterols: stigmastane, consisting of 7-sterols and composed of spinasterol and schottenol (Charrouf and Guillaume 2008), *stigmasta-8,22-diene-3*-beta-ol (Matthäus et al. 2010), but also campesterol a 5-sterol (Hilali et al. 2005). Argane oil contains higher squalene levels (greater than 3.2 *g/kg*) (Khallouki et al. 2003). Scarcely known structures in this oil include derivatives of lupane, ursane, and oleanane that include β-amyrin, butyrospermol, and tirucallol. Argane oil also contains tocopherols or vitamin E which is an essential compound. In addition, γ-form is the prevailing one. In Argane oil, characterized by glycosylated ferulic acid, polyphenolics are present in very low concentrations in combination with syringic acid, vanillic acid, tyrosol, vanillin, and *p*-hydroxybenzaldehyde. Virgin Argane oil also represents a great source of the antioxidant coenzyme Q10 (CoQ10) and melatonin (El Monfalouti et al. 2010). Resorcinol and camphor were also classified as a major compound among the volatile compounds (Tahrouch et al. 1998). Certain compounds include *endo*-borneol and 2-(4-methylcyclohex-3-enyl) propan-2-ol (Matthäus et al. 2010).

The Argane fruit fraction which is unsaponifiable also includes triterpenoids, for instance, erythrodiol, lupeol, α- and β-amyrin, taraxasterol, betulinaldehyde, and betulin (Charrouf et al. 1990). The Argane oil also contains pentacyclic triterpenic free acids. The main triterpene acid followed by oleanolic acid and maslinic acid together is the ursolic acid (Guinda et al. 2011).

Very small sterols in the flesh (pulp) include schottenol and spinasterol (Charrouf et al. 1992). Additional structures have been established such as gallic acid, protocatechuic acid, rhamnetin-O-rutinoside, isorhoifolin, hesperidin, hyperoside, isoquercetin, naringenin-7-O-glucoside, quercetin-3-O-arabinoside, naringenin, quercetin, luteolin, and an unspecified procyanidin dimer (Charrouf and Guillaume 2008; Charrouf et al. 2007). Many polyphenolic compounds were characterized from unripe Argane fruits. That was expressed by catechins, flavonoids, procyanidins, free phenolic acids, and glycosides of phenolic acid. Some identified polyphenolics have also been described, and these comprise catechol, resorcinol, 4-hydroxybenzyl alcohol, vanillin, tyrosol, p-hydroxybenzoic acid, vanillyl alcohol, 3,4-dihydroxybenzyl alcohol, 4-hydroxy-3-methoxyphenethylalcohol, methyl-3,4-benzoate, vanillic acid, hydroxytyrosol, 3,4-dihydroxybenzoic acid (protocatechuic acid), syringic acid, epicatechin, and catechin (Mojica et al. 2016).

Additional studies have found secondary metabolites, namely, arganimides and argaminolics, of which arganimide A was initially identified in *Argania* (Khallouki et al. 2017). Eventually, polysaccharides, particularly hemicellulose, are also among the Argane oil's principal compounds. These are metabolites of pentoses, hexoses, and glucuronic acid (Aboughe-Angone et al. 2008). Another fascinating piece of research performed by Khallouki et al. (2015) first found new molecules in Argane oil, namely, epicatechin-(4β→8)-catechin dimer (procyanidin B1), p-coumaric acid glycoside, epicatechin-(4β→8)-epicatechin dimer (procyanidin B2), caffeic acid glycoside, epicatechin-(4β→8)-epicatechin-(4β→8)-epicatechin trimmer (procyanidin C1), p-hydroxybenzaldehyde, ferulic acid glycoside, vanillic acid, sinapic acid glycoside, p-coumaric acid, ferulic acid, sinapic acid, rutin pentoside, quercetin glycopentoside, 4,4′-dihydroxy-3,3′-imino-di-benzoic acid, quercetin-3-O-rhamnogalactoside, quercetin glycohydroxybenzoate, quercetin glycocaffeate, quercetin glycosinapate, quercetin glycoferulate, and quercetin glycocoumarate.

There are other known molecules not described in this review. However, the discovery of new metabolites is one of the main concerns in the Argane oil research, and this largely relies on the nature of methods used for oil extraction and for metabolite profiling and analysis. With advancements in separation and spectrometry technologies (nuclear magnetic resonance (NMR), gas chromatography–mass spectrometry (GC-MS), liquid chromatography–mass spectrometry (LC-MS), etc.) resulting in high sensitivity, high resolution, high throughput, and wide dynamic range, metabolomics can be applied to answer many Argane biochemical questions and improve metabolite profiling. We have been using oil extract from *Argania* seeds on both NMR and GC-MS protocols. Our preliminary analysis suggested the existence of potentially new metabolites. Further studies using various types of oil extracts and combining metabolomics profiling methods are ongoing.

7.5 Perspectives and Prospective Impact of the *Argania* Genomics

The *A. spinosa* genetics and genomics will be useful for assessing biodiversity leading to efficient conservation of this endangered endemic tree. Whole-genome sequencing shows an organism's whole DNA and offers the most thorough analysis of genetic and epigenetic variations. An integrated omics strategy will lead to a better understanding of the complex biological, physiological, and molecular processes in Argane trees and will enable the characterization of relevant developmental and metabolic genes, such as those involved in the early and late rooting process, which is an unresolved problem facing the cultivation and domestication of this crop and the production of argan oil. In addition, the genome will not only allow genome-assisted cultivar breeding but will also provide a deeper understanding of cosmetics and the pharmacological applications of essential metabolic pathways along with their underlying genes.

Particularly, immediate work will be focused on the use of predictive tools and transcriptome analysis to structurally annotate the genome. Future work will perform functional gene annotation and gathering evidence on genome duplication and evolution of comparative genomes. An accurate annotation depends heavily on the transcriptomic data, analysis, and study. The methylomes, genomic variation, metabolome, and analysis of Argane oil biosynthesis pathways as well as the tree's microbiome are all in the program.

Given the fact that Argane oil is an important resource, yet it is not the only product of this tree, further studies of this plant are required to prevent the prominent danger that lies within the Argane tree ecosystems, yet another reason to promote genomic studies to breed this crop. Additionally, artificial intelligence (AI) proved its efficiency giving more perspective to plant genomic studies and transcriptomics prediction. For instance, recent studies on the maize genome to predict regulatory regions using machine learning (ML) have been achieved (Mejía-Guerra and Buckler 2019). Other studies were done to predict the macronutrients deficiencies in tomato (Ghosal et al. 2018) and reveal the phenotypes of stress in different plants (Do et al. 2018). ML can further be used to deepen our understanding to the *Argania spinosa* as well as revealing the biosynthesis pathways of all its products or predicting the genomics regions related to each product or even outlining the perfect conditions that might lead to better yields.

7.6 Conclusion

The *Argania* tree is renowned for its outstanding resistance to drought and for its good adaptability to different types of soil, but mostly for its nutritional qualities for both humans and livestock. The genetics and genomics tools are efficient tools for the acceleration of breeding programs to identify and select traits/varieties that confer further tolerance and higher fruit and oil yield and quality. Unfortunately, taking full advantage of omics applications faces many challenges in Morocco such as lack of core facilities and advanced bioinformatics skills. However, it is imperative to enhance the Argane genome evolution and genome-derived biotechnology in breeding and agriculture. This is the mission of the established International Argane Genome Consortium to promote Argane genomics and its applications.

References

Aboughe-Angone S, Nguema-Ona E, Ghosh P, Lerouge P, Ishii T, Ray B, Driouich A (2008) Cell Wall carbohydrates from fruit pulp of *Argania Spinosa*: structural analysis of pectin and xyloglucan polysaccharides. Carbohydr Res 343(1):67–72. https://doi.org/10.1016/j.carres.2007.10.018

Ait Aabd N (2013) Apport Des Marqueurs Phénotypiques et Moléculaires Pour l'analyse de La Variabilité Génétique de l'arganier Présélection Pour Le Rendement En Huile. Université Ibn Zohr, Agadir. http://www.congresarganier.ma/pdf/ait-aabd-these.pdf

Ait Aabd N, Msanda F, El Mousadik A (2015) Genetic diversity of the endangered argan tree (Argania Spinosa L.) (Sapotaceae) revealed by ISSR analysis. Basic Res J Agric Sci Rev 4:176–186. http://www.basicresearchjournals.org

Bani-Aameur F, Benlahbil S (2004) Variation in RAPD markers of *Argania Spinosa* trees and their progenies. For Genet 11(January):337–342

Charrouf Z, Guillaume D (2008) Argan oil: occurrence, composition and impact on human health. Eur J Lipid Sci Technol 110(7):632–636. https://doi.org/10.1002/ejlt.200700220

Charrouf Z, Fkih-tetouani S, Rouessac F (1990) Occurrence of erythrodiol in *Argania spinosa*. Albirunya 6135-6138

Charrouf Z, Wieruszeski JM, Fkih-Tetouani S, Leroy Y, Charrouf M, Fournet B (1992) Triterpenoid saponins from *Argania spinosa*. Phytochemistry 31:2079–2086

Charrouf Z, Hilali M, Jauregui O, Soufiaoui M, Guillaume D (2007) Separation and characterization of phenolic compounds in Argan fruit pulp using liquid chromatography-negative electrospray ionization tandem mass spectroscopy. Food Chem 100(4):1398–1401. https://doi.org/10.1016/j.foodchem.2005.11.031

Do H, Than K, Larmande P (2018) Evaluating named-entity recognition approaches in plant molecular biology. BioRxiv Cold Spring Harbor Laboratory. https://doi.org/10.1101/360966

Duangjai S, Wallnofer B, Rosabelle S, Jerome M, Mark WC (2006) Generic delimitation and relationships in Ebenaceae Sensu Lato: evidence from six plastid DNA regions. Am J Bot 93(12):1808–1827. https://doi.org/10.3732/ajb.93.12.1808

El Bahloul Y, Dauchot N, Machtoun I, Gaboun F, Cutsem VP (2014) Development and characterization of microsatellite loci for the Moroccan endemic endangered species *Argania Spinosa* (Sapotaceae). Appl Plant Sci 2(4):1300071. https://doi.org/10.3732/apps.1300071

El Monfalouti H, Guillaume D, Denhez C, Charrouf Z (2010) Therapeutic potential of argan oil: a review. J Pharm Pharmacol. Blackwell Publishing Ltd https://doi.org/10.1111/j.2042-7158.2010.01190.x

EL Mousadik A, Petit RJ (1996) Chloroplast DNA phylogeography of the argan tree of Morocco. Mol Ecol 5(4):547–555. https://doi.org/10.1111/j.1365-294x.1996.tb00346.x

Ghosal S, Blystone D, Singh AK, Ganapathysubramanian B, Singh A, Sarkar S (2018) An explainable deep machine vision framework for plant stress phenotyping. Proc Natl Acad Sci U S A 115(18):4613–4618. https://doi.org/10.1073/pnas.1716999115

Guinda A, Rada M, Delgado T, Castellano JM (2011) Pentacyclic triterpenic acids from *Argania Spinosa*. Eur J Lipid Sci Technol 113(2):231–237. https://doi.org/10.1002/ejlt.201000342

Hilali M, Charrouf Z, El Aziz Soulhi A, Hachimi L, Guillaume D (2005) Influence of origin and extraction method on argan oil physico-chemical characteristics and composition. J Agric Food Chem 53(6):2081–2087

Khallouki F, Younos C, Soulimani R, Oster T, Charrouf Z, Spiegelhalder B, Bartsch H, Owen RW (2003) Consumption of argan oil (Morocco) with its unique profile of fatty acids, tocopherols, squalene, sterols and phenolic compounds should confer valuable cancer chemopreventive effects. Eur J Cancer Prev 12(1):67–75. https://doi.org/10.1097/00008469-200302000-00011

Khallouki F, Haubner R, Ricarte I (2015) Identification of polyphenolic compounds in the flesh of argan (Morocco) fruits. Food Chem 179:191–198

Khallouki F, Voggel J, Breuer A, Klika KD, Ulrich CM, Owen RW (2017) Comparison of the major polyphenols in mature argan fruits from two regions of Morocco. Food Chem 221(April):1034–1040. https://doi.org/10.1016/j.foodchem.2016.11.058

Khayi S, Azza NE, Gaboun F, Pirro S, Badad O, Claros MG, Lightfoot DA, Unver T, Chaouni B, Merrouch R, Rahim B, Essayeh S, Ganoudi M, Abdelwahd R, Diria G, Alaoui MM, Labhilili M, Iraqi D, Mouhaddab J, Sedrati H, Memari M, Hamamouch N, de Dios AJ, Boukhatem N, Mrabet R, Dahan R, Legssyer A, Khalfaoui M, Badraoui M, Van de Peer Y, Tatusova T, El Mousadik A, Mentag R, Ghazal H (2018) First draft genome assembly of the argane tree (*Argania Spinosa*). F1000Research 7(1310). https://doi.org/10.12688/f1000research.15719.1

Khayi S, Gaboun F, Pirro S, Tatusova T, El Mousadik A, Ghazal H, Mentag R (2019) Chloroplast genome assembly of *Argania spinosa*: insights into phylogeny and simple sequences repeats. Proceeding of the 5th International Congress of Argane, p63, 10–11 December, 2019, Agadir, Morocco.

Khayi S, Gaboun F, Pirro S, Tatusova T, El Mousadik A, Ghazal H, Mentag R (2020) Complete chloroplast genome of *Argania spinosa*: structural organization and phylogenetic relationships in Sapotaceae. Plants 9(10):1354

Larson DA, Walker JF, Vargas OM, Smith SA (2020) A consensus Phylogenomic approach highlights paleopolyploid and rapid radiation in the history of Ericales. Am J Bot 107(5):773–789. https://doi.org/10.1002/ajb2.1469

Louati M, Khouja A, Ben Abdelkrim A, Salhi Hannachi A, Baraket G (2019) Adaptation of *Argania Spinosa L.* in Northern Tunisia: soil analysis and morphological traits variability. Sci Hortic 255(September):220–230. https://doi.org/10.1016/j.scienta.2019.05.035

M'Hirit O, Benzyane M, Benchekroun F, El Yousfu S.M, Bendaanoun M (1998) L'arganier - Une Espèce Fruitière-Forestière à Usages Multiples. Vol. 1 Belgique

Majourhat K, Jabbar Y, Hafidi A, Martínez-Gómez P (2008) Molecular characterization and genetic relationships among most common identified morphotypes of critically endangered rare moroccan species *Argania Spinosa* (Sapotaceae) using RAPD and SSR markers. Ann For Sci 65(8). https://doi.org/10.1051/forest:2008069

Matthäus B, Guillaume D, Gharby S, Haddad A, Harhar H, Charrouf Z (2010) Effect of processing on the quality of edible argan oil. Food Chem 120(2):426–432. https://doi.org/10.1016/j.foodchem.2009.10.023

Mejía-Guerra MK, Buckler ES (2019) A K-Mer grammar analysis to uncover maize regulatory architecture. BMC Plant Biol 19(1):103. https://doi.org/10.1186/s12870-019-1693-2

Mojica MA, León A, Rojas-Sepúlveda AM, Marquina S, Mendieta-Serrano MA, Salas-Vidal E, Villarreal ML, Alvarez L (2016) Aryldihydronaphthalene-type lignans from Bursera Fagaroides Var. Fagaroides and their antimitotic mechanism of action. RSC Adv 6(6). https://doi.org/10.1039/c5ra23516b

Mouhaddab J, Ait Aabd N, Msanda F, Filali-Maltouf A, Belkadi B, Ferradouss A, El Modafar C, Koraichi SI, Ghazal H, El Mousadik A (2017) Assessing genetic diversity and constructing a core collection of an endangered Moroccan endemic tree [*Argania spinosa* (L.) Skeels]. Moroccan J Biol 13:1–2

Msanda F, El Aboudi A, Peltier JP (2005) Synthèse Biodiversité et Biogéographie de l'arganeraie Marocaine. Cahiers Agricultures 14. https://revues.cirad.fr/index.php/cahiers-agricultures/article/view/30528

Nouaim R, Mangin G, Breuil M.C, Chaussod R (2002) The Argan tree (*Argania Spinosa*) in Morocco: propagation by seeds, cuttings and in-vitro techniques. Agrofor Syst 54 (1): 71–81. https://doi.org/10.1023/A:1014236025396

Pakhrou O, Medraoui L, Yatrib C, Alami M, Filali-maltouf A, Belkadi B (2017) Assessment of genetic diversity and population structure of an endemic Moroccan tree (*Argania Spinosa L.*) based in IRAP and ISSR markers and implications for conservation. Physiol Mol Biol Plants 23(3):651–661. https://doi.org/10.1007/s12298-017-0446-7

Stride G, Nylinder S, Swenson U (2014) Revisiting the biogeography of Sideroxylon (Sapotaceae) and an evaluation of the taxonomic status of Argania and Spiniluma. Aust Syst Bot 27(2):104. https://doi.org/10.1071/SB14010

Stull GW, Soltis PS, Soltis DE, Gitzendanner MA, Smith SA (2020) Nuclear Phylogenomic analyses of Asterids conflict with Plastome trees and support novel relationships among major lineages. Am J Bot 107(5):790–805. https://doi.org/10.1002/ajb2.1468

Swenson U, Anderberg AA (2005) Phylogeny, character evolution, and classification of Sapotaceae (Ericales). Cladistics 21(2):101–130. https://doi.org/10.1111/j.1096-0031.2005.00056.x

Tahrouch S, Rapior S, Bessière JM, Andary C (1998) Les Substances Volatiles de *Argania Spinosa* (Sapotaceae). Acta Botanica Gallica 145(4):259–263. https://doi.org/10.1080/12538078.1998.10516305

UNESCO-MAB Biosphere Reserve Directory (2018). http://www.unesco.org/mabdb/br/brdir/directory/database.asp

Vaghani S.N (2003) Fruits of tropical climates | fruits of the Sapotaceae. Encyclopedia of food sciences and nutrition; Elsevier:2790–2800. ISBN 978-0-12-227055-0

Chapter 8
On "The Most Useful" Oleaginous Seeds: *Linum usitatissimum* L., A Genomic View with Emphasis on Important Flax Seed Storage Compounds

Lucija Markulin, Yuliia Makhno, Samantha Drouet, Sara Zare, Sumaira Anjum, Duangjai Tungmunnithum, Mohammad R. Sabzalian, Bilal Haider Abbasi, Eric Lainé, Hanna Levchuk, and Christophe Hano

Contents

Lucija Markulin and Yuliia Makhno have equal contribution of the first authors.
Hanna Levchuk and Christophe Hano corresponding authors.
These authors have equal contribution of the senior authors.

L. Markulin · S. Drouet · E. Lainé · C. Hano (✉)
Laboratoire de Biologie des Ligneux et des Grandes Cultures, INRAE USC1328, Orleans University, Orléans Cedex 2, France
e-mail: eric.laine@univ-orleans.fr; hano@univ-orleans.fr

Y. Makhno · H. Levchuk (✉)
Flax Breeding Lab, Institute of Oilseed Crops of the National Academy of Agricultural Sciences of Ukraine, Zaporizhzhya region, Ukraine

S. Zare · M. R. Sabzalian
Department of Agronomy and Plant Breeding, College of Agriculture, Isfahan University of Technology (IUT), Isfahan, Iran
e-mail: sara.zare@ag.iut.ac.ir; sabzalian@iut.ac.ir

H. Tombuloglu et al. (eds.), *Oil Crop Genomics*,
https://doi.org/10.1007/978-3-030-70420-9_8

8.1 Introduction

Linum usitatissimum L. (common flax) is one of the oldest domesticated plants. Its popularity has risen again in recent decades since it was characterized as a functional food. The Latin meaning of the name *usitatissimum* ("the most useful") reflects its importance. Flax and linseed are both terms used to reference this crop. In Americas terms are often used interchangeably; however, flax refers to a plant grown for human consumption and linseed for a plant grown for industrial use. Contrarily, in Europe, the term flax is used to describe *Linum* grown for fiber and linseed for flax grown for oil production (Diederichsen and Richards 2003).

Flax is unique because it was domesticated for two very different phenotypes, namely, phenotype with an unbranched stem for the development of long linen fibers (flax) and a small shorter bushy phenotype for high seed yield (linseed). Flax has a long and rich tradition of use. It is an important source of fiber. Its seed is a rich source of both the health-promoting oil, namely, omega 3, α-linolenic acid (ALA), and lignans, in particular secoisolariciresinol diglucoside (SDG) (Fig. 8.1). The major seed coat-derived lignan, SDG, has been reported to suppress the

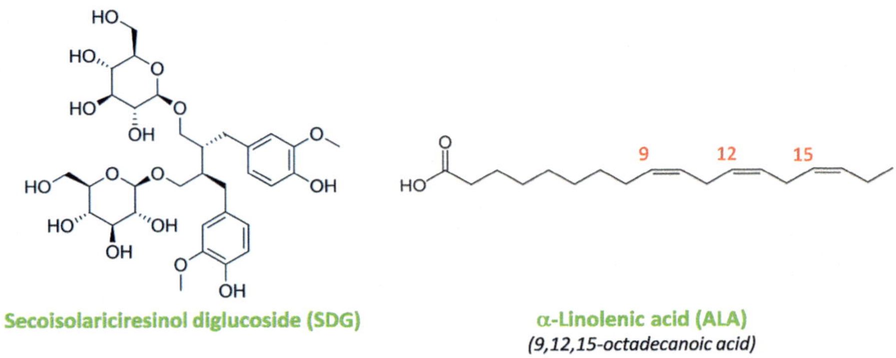

Fig. 8.1 Chemical structure of the health-promoting lignan secoisolariciresinol diglucoside (SDG) and omega-3 fatty acid α-linolenic acid (ALA) accumulated in high amounts in flax seeds

S. Anjum
Department of Biotechnology, Kinnaird College for Women, Lahore, Pakistan
e-mail: sumaira.anjum@kinnaird.edu.pk

D. Tungmunnithum
Department of Pharmaceutical Botany, Faculty of Pharmacy, Mahidol University, Bangkok, Thailand
e-mail: duangjai.tun@mahidol.ac.th

B. H. Abbasi
Department of Biotechnology, Quaid-i-Azam University, Islamabad, Pakistan
e-mail: bhabbasi@qau.edu.pk

development of atherosclerosis (Prasad 2009), reduce incidence rates of type 1 diabetes, delay development of type 2 diabetes progression (Hano et al. 2013; Prasad and Dhar 2016), and act preventively against the development of some hormone-dependent cancers (Lainé et al. 2009). Its seed is also rich in storage proteins such as lectins that may contribute to the specific adaptation of flax plants to various abiotic stress factors (Levchuk et al. 2013). In the present chapter, a genomic view with emphasis on these important flax seed storage compounds is provided.

8.2 A Short History of Flax and Its Usages

The oldest wild flax fibers used by humans have been discovered in Dzudzuana Cave in Georgia (Kvavadze et al. 2009). They date back to the Paleolithic era (30,000 years ago) and have likely been used to make baskets and clothing (Kvavadze et al. 2009). Evidence of both linseed and flax fiber usage also appear in the early Neolithic area excavation sites (McCorriston 1997). Some of the oldest flax seed excavation sites are found in the Fertile Crescent in Syria, Turkey, and Iran (Vaisey-Genser and Morris 2003). Seeds from the wild *Linum* species found at Tell Mureybit (North of Euphrates and East of Aleppo) date to the second half of the ninth-millennium B.C.E (van Zeist and Bakker-Heeres 1975). The earliest evidence of domesticated flax found at Jericho date between 9500 and 9000 B.C.E (Hopf 1983). The archaeobotanical evidence of cultivated flax has also been found at the location of Tell Ramad (20 km southwest of Damascus) where nearly 300 flax seeds have been recovered and dated circa 6200 years B.C.E (van Zeist and Bakker-Heeres 1975). Remains of linen textile have been found in Nahal Hemar Cave in Judean desert (seventh-millennium B.C.E) (Schick 1988). More evidence of domesticated flax found in Syria (van Zeist and Bakker-Heeres 1975), Iran, and Iraq (Helbaek 1969) date to the middle of the fifth-millennium B.C.E. Flax use was particularly popular in ancient Egypt during the time of the Pharaohs where flax's oils were used to embalm the dead and linen was used to wrap mummies (Baumann 1960). In Europe, flax fibers have been found near the Swiss lakes indicating production of linen dating back to 5000 years B.C.E (Condra 2008). Early conquests and trade spread domesticated flax through the Mediterranean and Asia, while later colonization brought cultivated flax all the way to North America and Australia (Vaisey-Genser and Morris 2003).

Today, flax is one of the most important oil and fiber crops alongside oilseed rape, sunflower, cotton, and soybean. More than 800 thousand tons of flax fiber and tow and more than 2.8 million tons of linseed have been produced per year between 2004 and 2014 (FAOSTATS). Leading countries in flax production are Canada, India, Russia, China, Kazakhstan, and France.

Flax is a multipurpose crop that can be produced for multitude reasons, aimed for human consumption to industrial use. In early history flax has been used as an oil source and its fibers were used to make tools and clothing. The use of flax was first limited by the development of the use of wool and later of cotton. The industrial processing of flax started in the mid-eighteenth century but declined by the mid-twentieth century after the popularization of the cotton and synthetic alternatives. Interest in flax has been renewed in the recent decade since it was described as a functional food. Functional food is a food that provides beneficial effects on one or more target functions in the body beyond basic nutritional value. It leads to an improved state of health and/or reduces the risk of certain diseases (Contor 2001).

Flax seed is recommended in addition to human diet because of its polyunsaturated fatty acids, lignans (predominantly SDG), and other phenolics. Polyunsaturated acids ALA and (linoleic acid) LA reduce the risk of cancer and cardiovascular disease (Herchi et al. 2014). Additionally, conversion of ALA to DHA (docosahexaenoic acid) showed beneficial effect on the brain in rats (Sugasini and Lokesh 2015).

While there are many benefits to consumption of flax oil, high amount of omega fatty acids makes flax oil highly susceptible to oxidation. Only flax varieties with low modified ALA content are suitable for human consumption. Additionally, flax oil is important in linoleum production. It is used as a carrier in oil paint and printing ink and in varnishes, as a protective coating of wood (Singh et al. 2011).

Flaxseed dietary fibers are particularly beneficial in weight control. They regulate appetite, lower nutrient absorption by forming a viscous gel, and reduce fat absorption and lower LDL-cholesterol (Kristensen and Jensen 2010; Ibrügger et al. 2012; Kristensen et al. 2012).

Additionally, potential for antimicrobial activity of seedcake (residue of oil extraction from seed) due to the presence of phenolic compounds has been investigated against some multidrug-resistant bacterial strains with promising results (Zuk et al. 2014). Both seedcake and whole flax seed have been used as animal feed. Chickens fed by flax seed oil gave more nutritional eggs (Coorey et al. 2015).

Flax also found its way to the NASA. Space exploration requires astronauts to perform actions outside a spacecraft. During the spacewalk astronauts are exposed to a large amount of radiation from which they are protected only by their suit. Additionally, protocols require 100% hyperoxia prior to leaving the spacecraft. Flaxseed consumption has shown to mitigate lung damage caused by radiation and hyperoxia (Christofidou-Solomidou et al. 2011; Pietrofesa et al. 2014). Dr. Christofidou-Solomidou and researchers at the Perelman School of Medicine are conducting a pilot study for NASA in hope to provide preventive measures against lung damage caused by space exploration (Penn Medicine 2011; Pietrofesa et al. 2014).

8.3 Phylogeny of Linaceae Family

Linum usitatissimum L. is a flowering plant of the genus *Linum* of the family Linaceae (Fig. 8.2). The family Linaceae is a part of the order Malpighiales. Its 14 genera and around 250 species (Dressler et al. 2014) are divided into 2 subfamilies. Eight genera including the genus *Linum* make the subfamily Linoideae that consists of annual and perennial, predominantly temperate plants. Subfamily Hugonioideae

Kingdom: Plantae
Division: Tracheophyta
Sub-division: Spermatophytina
Class: Magnoliopsida
Sub-class: Rosidae
Order: Malpighiales
Family: Linaceae
Subfamily: Linoideae
Genus: *Linum*
- **Section**: ***Linum***
  - Species: *L. usitatissimum, L. bienne, L. perenne, L. grandiflorum*
- **Section**: ***Dasylinum***
  - Species: *L. hirsutum*
- **Section**: ***Linopsis***
  - Species: *L. corymbulosum*
- **Section**: ***Cathartolinum***
  - Species: *L. catharticum*
- **Section**: ***Syllinum***
  - Species: *L. flavum, L. album*

Fig. 8.2 Taxonomic hierarchy of the genus *Linum*, its sections, and some of its species

consists of six genera (the largest genus Hugonia) characterized by tropical trees, shrubs, and lianas (Dressler et al. 2014; APG 2016). Two subfamilies are monophyletic sister lineages (McDill and Simpson 2011).

The genus *Linum* consists of around 250 species (Rogers 1982; Dressler et al. 2014) found primarily in temperate and subtropical regions (McDill et al. 2009). Due to the high morphological diversity of the genus *Linum*, botanists have proposed several divisions based on different characteristics. Traditionally used classification divides the genus *Linum* into the five sections that form two major clades: *Linum* and *Dasylinum* (blue flowering clade) and *Cathartolinum* (monotypic; *L. catharticum*), *Syllinum*, and *Linopsis* (yellow flowering clade) (Tutin et al. 1968; Rogers 1982; McDill et al. 2009).

Most of wild *Linum* species (most species of blue flowering clade and yellow-flowered clade) are characterized by the phenomenon of heterostyly (Fig. 8.3).

Heterostyly is the presence in the plant population of two types (floral morphs), which differ in height of pistils and stamens – some flowers have short stamens and long pistils (long-styled morph) and other flowers have short pistils and long stamens (short-styled morph). The seeds in these plants formed only after cross-pollination between different morphs. According to this classification, cross-pollinated species are called heterostylous species and self-pollination species are called homostylous one.

The genus *Linum* is very interesting from the standpoint study of pollination mechanisms. The members of this genus are present as self-pollinated species (homostyled species) as well as cross-pollinated species (heterostyled species).

Fig. 8.3 The phenotype of the different floral morphs of heterostyled species *Linum* (*Linum perenne L.*). (**a**) The phenotype of short-styled (SS) floral morph; (**b**) the phenotype of long-styled (LS) floral morph; (**c** and **d**) the phenotype of pistils (1) and stamens (2) of SS (**c**) and LS (**d**) of floral morphs

Among these species, different types of self-incompatibility are observed. Several studies have been carried out on the breeding system of *Linum* species (Dulberger 1974, 1987). Darwin (1877), one of the first researchers of this species, revealed the existence of distyly in several species such as *L. pubescens* Banks & Sol., *L. grandiflorum* Desf., *L. mucronatum* Bertol., *L. flavum* L., *L. perenne* L., *L. austriacum* L., and *L. maritimum* L. Genus *Linum* has also historically been important for the study of heterostyly, which was recognized by Darwin in a European blue flax, *Linum perenne* (Darwin 1863).

Section *Linum* consists of about 50 species that can be divided into 2 groups based on morphology of their sepals and stigmas (heterostylous (1) or homostylous (2), respectively): (1) *L. perenne* group (*L. austriacum*, *L. lewisii*, *L. pallescens*, and *L. perenne*) and (2) a homostylous group (*L. bienne*, *L. decumbens*, *L. grandiflorum*, *L. marginale*, *L. narbonense*, and *L. usitatissimum)* (McDill et al. 2009).

L. usitatissimum is cultivated flax not found in the wild used as a source of fibers and oils. Other flax species have also found their purpose, e.g., as ornamental plants (e.g., *L. grandiflorum)* or as sources of medically important compounds (e.g., *L. album)* (McDill et al. 2009).

Section *Linum* is paraphyletic (because of *L. stelleroides*) in relation to section *Dasylinum* which is monophyletic and nested inside of the section *Linum* (McDill et al. 2009). Sections *Cathartolinum* and *Syllinum* (monophyletic) are nested inside section *Linopsis* (para- or polyphyletic) (McDill et al. 2009). Taxonomic separation in two clades and five sections (McDill et al. 2009) is supported by lignan accumulation analysis in seed and aerial parts of different species of the genus *Linum* (Schmidt et al. 2010, 2012). These results showed that blue flowering clade accumulates arylnaphthalene/aryldihydronaphthalene (AN/ADN) lignans and yellow flowering clade accumulates aryltetralin (AT) lignans (Schmidt et al. 2010, 2012). Specialization to high accumulation of either AN/AND or AT lignans probably appeared after segregation of two clades in common ancestor of each clade but before section separation and remained dominant biosynthetic pathway through diversification of the clades (Schmidt et al. 2010).

8.4 The Flax Genome

The origin of cultivated flax to this date remains inconclusive. Several authors proposed theories for the center of flax origin, most notable ones being multiple independent domestication events and single domestication event origin. Due to the wide biogeographical range of flax, theory of multiple centers of domestication includes Central Asia, Mediterranean, Ethiopia, Fertile Crescent (Vavilov 1992), as well as European-Siberian region (Zeven and de Wet 1982) as possible centers of origin. On the other side, Allaby et al.'s (2005) analysis of stearoyl-ACP desaturase II alleles from 30 accessions of cultivated and pale flax (*L. angustifolium* Huds.) suggests a single domestication origin and proposes Near Eastern and European germplasms for further origin studies. Flax appears to be domesticated first for its oil and not for the fiber production (Allaby et al. 2005).

There are two proposed flax progenitors: *L. angustifolium* Huds. and *L. bienne* Mill. Some authors believe they are the same species (Tutin et al. 1968), and some believe they are separate species. Muravenko et al.'s (2003) study based on chromosomal and molecular markers suggests *L. angustifolium* as the wild ancestor and proposes *L. bienne* to be treated as a subspecies of *L. usitatissimum*. Vromans (2006) suggests that *L. bienne* is more closely related to *L. angustifolium* than to *L. usitatissimum*. Both species like *L. usitatissimum* contain 30 (2n) chromosomes, have homostylous flowers, and are self-pollinated (Hall et al. 2016). They are able to cross-breed and give fertile progeny. They are biennial or perennial (rarely annual), while *L. usitatissimum* is an annual plant. While generally *L. bienne* is stated as a progenitor of cultivated flax in addition to different lengths of annual cycles, *L. angustifolium* has narrow leaves while cultivated flax has larger leaves. Pale flax (*L. angustifolium*) seed is released once it is ripe, while cultivated flax seed stays enclosed in the capsule. *L. usitatissimum* accumulates dibenzylbutyrolactone lignans (DBBL), and *L. bienne* accumulates furofuran (FF)-type lignans (Schmidt et al. 2010).

The *Linum* genus encompasses around 250 diploid species, of which 75 were evaluated for their chromosome numbers. Chromosome numbers varied significantly from $n = 7$ to 43 with predominant $n = 9$ and 15 (Goldblatt 2007; Rice et al. 2014).

In addition, B chromosomes (i.e., supernumerary chromosomes) have been detected in *L. capitatum*, *L. flavum*, and *L. tauricum*, whereas the cultivated (*L. usitatissimum* L.) and wild flax (*L. bienne*) possess the same number of haploid chromosome $n = 15$ (Nosova 2005; Nosova et al. 2005). Polyploidy, an inheritable condition of having more than two homologous sets of chromosomes, is an evolutionary factor that contributes to plant diversification and shapes the karyotypes (Otto and Whitton 2000). The variation in chromosome numbers within the genus *Linum* suggests occurrences in chromosomal alteration in the speciation of this taxon, e.g., polyploidization by full gene duplications, chromosome rearrangements, and potential instances of aneuploid reduction or increase (Ray 1944; Rogers 1982). Evidence in the lineage of cultivated flax and its close relatives has shown that it may have undergone two polyploidization events: an ancient (paleoploidization) one about 30 million years ago (MYA) and a more recent (mesopolyploidization) one 5–9 MYA (Wang et al. 2012; Sveinsson et al. 2014) (Fig. 8.4).

Cultivated flax C-banding karyotypes showed similarities to wild relatives *L. austriacum* L. ($n = 9$) and *L. grandiflorum* Desf. ($n = 8$), suggesting that it may have arisen from an ancient polyploidization event involving one or both species accompanied by subsequent loss and/or rearrangement of chromosomes (Muravenko et al. 2003, 2010). Illumina next-generation transcriptome sequence (NGS) of 11 *Linum* species has revealed, through a sequence analysis of paralog genes, the signature of a paleopolyploidization event unique to the 20–40 MYA flax lineage (You et al. 2018). On the basis of data from the whole-genome shotgun (WGS) sequence of the flax genome (Wang et al. 2012), a more recent whole-genome duplication (WGD) event was proposed to occur around 5–9 MYA and refined 3.7 MYA based on the Ks of gene pairs in the sorted flax genome (You et al. 2018). These findings indicate that flax has undergone paleopolyploidization and mesopolyploidization, followed by rearrangements and deletions or a fusion of chromosome arms from its ancient progenitor with a haploid chromosome number of 8 (You et al. 2018).

8.5 Genomics Considerations About Flax

Self-pollinated, diploid, and with a relatively small genome released (Wang et al. 2012; You et al. 2018) are key features which make flax an ideal crop for breeding and genetic studies. Flax has an average genome size of 750 Mb with a genetic potential estimated to encode *ca* 43,000 genes (Wang et al. 2012). The availability of this sequence information is an excellent opening tool for developing breeding strategies.

Flax genome contains a large number of duplicated genes which could lead to rapid genomic changes (Wang et al. 2012). Flax has undergone whole-genome

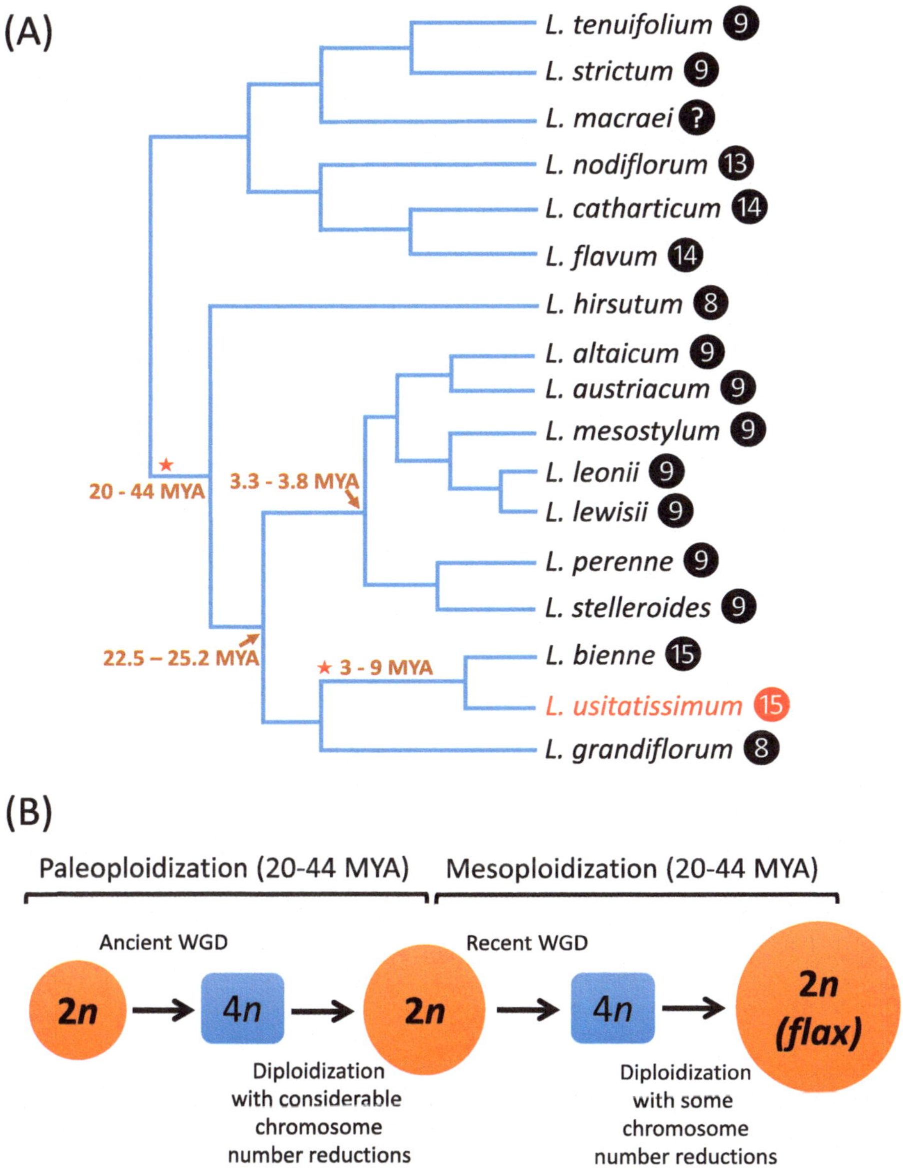

Fig. 8.4 (**a**) Cladogram of flax (*L. usitatissimum*) and other related *Linum* species showing haploid chromosome numbers inside circles. The approximate placement of the whole-genome duplication (WGD) events occurring in the common ancestor of cultivated (*L. usitatissimum*) and wild (*L. bienne*) flax is indicated by stars. (Tree drawn based on the integration of data from McDill et al. (2009), Wang et al. (2012), Sveinsson et al. (2014), Fu et al. (2016), and You et al. (2018), and redraw from You and Cloutier (2019)). (**b**) Proposed evolutionary history of flax. MYA, million years ago; WGD, whole-genome duplication. (Adapted from You and Cloutier (2019))

duplication events, one about 30 MYA and a more recent one 5–9 MYA (Wang et al. 2012; Sveinsson et al. 2014). It has been shown that, in genome progeny, retrotransposons can cause heritable genomic modifications, resulting in increased phenotypic differences and plant development under multiple environmental factors (Schneeberger and Cullis 1991; Chen et al. 2005; Johnson et al. 2011). This natural genome plasticity may be valuable for further domestication and selection of essential genes controlling important agronomical traits such as oil and lignan contents. For example, association mapping has been used to identify important regions in the flax genome related to yield (Soto-Cerda et al. 2014).

Besides, CRISPR (Clustered Regularly Interspaced Short Palindromic Repeats)/Cas9 (CRISPR-associated system 9) has been recently used successfully to modify a flax gene (Sauer et al. 2016). Virus-induced gene silencing has been successful to modify a flax gene expression in seed coat (Hano et al. 2020). Latest advancements in gene editing-based technology together with the recently released genome sequences open up different possibilities for flax modifications aimed at improving yield.

8.5.1 Flaxseed α-Linolenic Acid (ALA)

L. usitatissimum seed is an excellent oil source. The total oil content in flaxseed goes up to 45% (Łukaszewicz et al. 2004; Bozan and Temelli 2008), but the lipid profile varies between flax varieties. Flax oil mainly consists of 18-carbon chain fatty acids, primarily ALA (18:3n-3). ALA generally makes around half of the fatty acid content; however, it can make as low as 2% (var. Linola) and as high as 73% (var. La Estanzuela 117) of total oil content depending on the flax variety (Łukaszewicz et al. 2004). Next most abundant fatty acids are LA (18:2n-6) and oleic acids (18:1n-9) (Łukaszewicz et al. 2004). Small quantities of stearic acid (18:0), palmitic acid (16:0) (Łukaszewicz et al. 2004; Bozan and Temelli 2008), and arachidic acid (20:0, aka eicosenoic acid) can be detected (Łukaszewicz et al. 2004) as well as some 22- and 24-carbon chain fatty acids (Westcott and Muir 2003). Cool growing conditions increase the oil content in the plant and lower the protein contents.

The main interest is on ALA and linoleic acid which are essential in the human diet. ALA is a precursor of eicosapentaenoic acid (EPA) and DHA whose positive influence on the human immune system is well documented (Simopoulos 1999; Kew et al. 2004). However, the rate of this conversion in humans remains controversial. Up to 35% of dietary ALA is catabolized in CO_2 for energy in first 24 h (Burdge et al. 2002; Burdge and Wootton 2002), and only a small portion is converted to EPA and DHA. While humans have the ability to convert ALA to longer-chained n-3 fatty acids, this conversion is limited and often depends on the ratio of other fatty acids present. Studies showed that ALA intake can increase EPA and DPA levels while conversion to DHA happens only with lower LA ratio (Harper et al. 2006; Wood et al. 2015).

As fatty acids are prone to oxidation, only flax varieties with specific lipid profile are suitable for the production of edible oils (e.g., Linola) (Łukaszewicz et al. 2004). Even cold extraction doesn't provide a lot of protection as the natural antioxidants that protect fatty acids from oxidation are not extracted together with the oil (Łukaszewicz et al. 2004). To prolong its shell life, flax oil comes in dark bottles and is sometimes supplemented with vitamin E, a fat-soluble antioxidant (Łukaszewicz et al. 2004). Because of the high ALA content, flax oil is unsuitable for cooking. To bypass high oxidation instability of ALA, varieties such as Linola (Solin) have been developed (Dribnenki et al. 2003). They are deficient in an enzyme desaturase that converts LA to ALA and have low ALA content (<5%) and accumulate LA (~70%) increasing the overall oxidative stability of the flax oil (Hall et al. 2016).

Unsaponifiable matter makes a small part of the flax oil but contains antioxidant molecules that help prevent lipid oxidation such as sterols and tocopherols (Hall et al. 2016). In addition, lignans are also speculated to help prevent lipid oxidation (Kitts et al. 1999; Hano et al. 2017; Garros et al. 2018). Oil is mainly present in the cotyledons (~ 74%), then in endosperm and seed coat (~ 21.6%), and 4.4% in embryo axis (Dorrell 1970). Lignans as well as flavonols present in the seed coat could potentially act as antioxidants to protect against UV, or protect against peroxidation during seedling phase. They could also be useful as natural antioxidant added at permissive level in flax oil (Hano et al. 2017).

ALA content is a major target for improving flax seed economic aspects. Some analyses have focused on gene identification as well as expression analysis involved in the production of ALA (Venglat et al. 2011; Rajwade et al. 2014). Some desaturase genes show significant variation associated with the production of ALA (Fig. 8.5) (Rajwade et al. 2014).

The gene expression of *FATTY ACID DESATURASE 2 (FAD2)*, *FAD3A*, and *FAD3B* has been shown to be highly correlated with ALA accumulation during flax seed development (Fig. 8.6) (Rajwade et al. 2014), thus suggesting that multiple genes and even gene families are required for ALA production (Rajwade et al. 2014).

Interestingly, a number of mucilage-related genes are highly expressed in the seed coat during flax seed development (Venglat et al. 2011). These genes, including *RHAMNOSE SYNTHASE*, a member of a gene family required for mucilage biosynthesis, have been also proposed as important targets for potentially increasing oil contents (Western et al. 2004; Shi et al. 2012; Venglat et al. 2011). Other expression analysis studies have already identified key genes involved in oil production, and further expression modification approaches could help to provide tailored oil content that would be beneficial to the production of healthy oils or oils used for specific industrial settings.

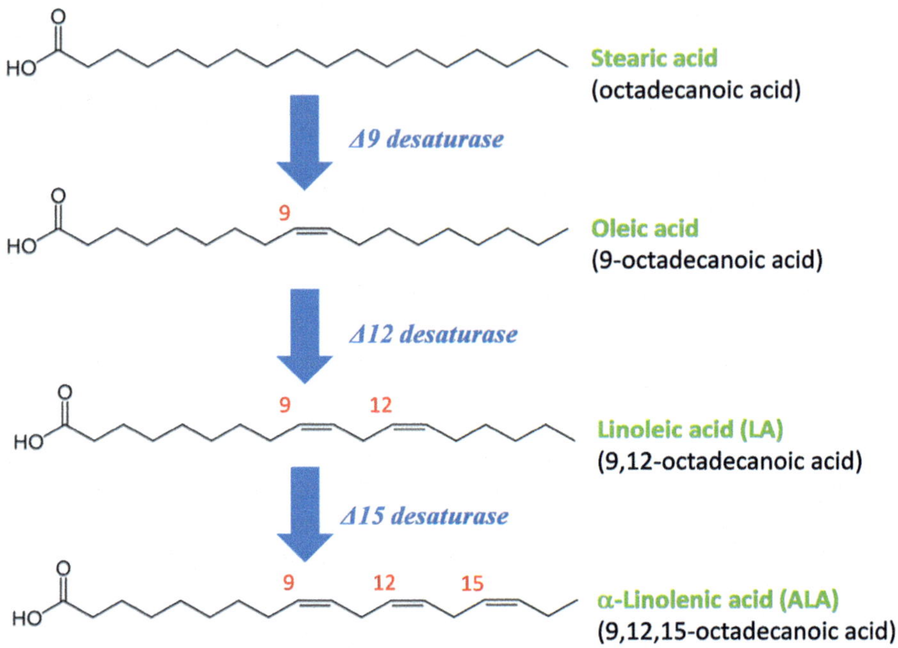

Fig. 8.5 Desaturase actions leading to α-linolenic acid

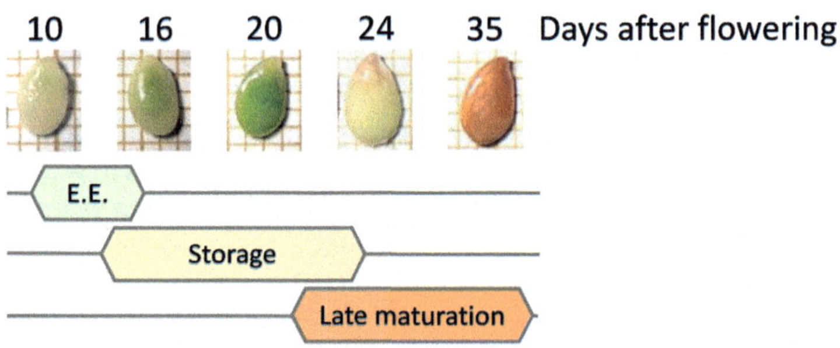

Fig. 8.6 The different phases of flax seed maturation. (Adapted from Gutierrez et al. (2006)) E.E.: early embryo development

8.5.2 Flaxseed Storage Proteins

The protein content in flaxseed has been reported to range from 10.5% (Bajpai et al. 1985) to 31% (Salunkhe and Desai 1986), but protein values are generally well above 36%. Differences can be attributed to both genetics and environment. As in other oilseeds, a negative correlation is observed between the oil and protein contents of the seeds (Naqvi et al. 1987; Bajpai et al. 1985). However, the relationship between protein and oil contents in the same populations varies from year to year and from location to location (Oomah and Mazza 1993).

Because of their abundance and economic importance, seed storage proteins were among the earliest of all proteins to be characterized. The detailed study of seed storage proteins dates from the turn of the century, when Osborne (1924) classified them into groups on the basis of their extraction and solubility in water (albumins), dilute saline (globulins), alcohol-water mixtures (prolamins), and dilute acid or alkali (glutelins). The major seed storage proteins include albumins, globulins, and prolamins.

The information about the fractional composition of flax proteins is extremely contradictory. So, some researchers indicate that in flax seeds the predominant fraction are globulins. They are at least 40–50% of the total amount of proteins (Pleshkov 1968; Boechko 1985). According to Barbashov, the flax seed protein complex is represented by water-, salt-, and alkali-soluble proteins. This predominant component is water-soluble proteins, and the proportion of alkali-soluble proteins is up to 10% (Barbashov 2005). According to Sammour and Poliakov and colleagues, the flax seeds contain all four protein fractions; however, the amount of albumin and globulins is predominant (Sammour 1999; Poliakov et al. 2011). A similar dynamic is observed in annual wild species of the section *Linum* (*L. angustifolium*, *L. bienne*, *L. grandiflorum*, and *L. hispanicum*); however, they have a significant predominance of prolamins, which is associated with a possible associated tolerance to abiotic stress factors (Poliakov et al. 2011).

The seed storage proteins (SSPs) are characterized by polymorphism and at the same time have biological specificity; therefore, they can be used as markers of species and varietal identification. The main method for this identification is polyacrylamide gel electrophoresis (PAGE). Makhno and colleagues developed a technique for separating the flaxseed proteins using the native PAGE. On the basis of the analysis of various varieties and lines of flax, a standard electrophoretic spectrum was proposed, which contains five main zones of electrophoretic mobility of the proteins. The first three zones (I, II, III) are constant (non-polymorphic), and zones IV and V are variable (polymorphic) parts of the spectrum (Makhno et al. 2010).

The protein profile of wild flax species significantly differed from cultivated flax in the number of zones, the manifestation, and mobility of protein components. Homostylous species (*L. bienne, L. angustifolium, L. hispanicum*) and heterostylous species (*L. thracicum, L. austriacum, L. hirsutum*) present some differences with cultivated flax in terms of protein distribution in these different zones. Both their quantity and intensity were in the non-polymorphic zones I and II whereas variable components were also observed in zones III, IV and V. Each wild species had several types of protein spectrum. This fact on our opinion is due to heterostyly of these species (Makhno and Polyakova 2007).

Seed storage proteins provide the necessary resources for seed germination and the first stages of plant ontogenesis (Wakasa and Takaiwa 2013). Some these proteins (lectins) perform the function of protecting the young plant from biotic and abiotic stress factors (Jiang et al. 2010).

Lectins are carbohydrate-binding proteins that specifically recognize diverse sugar structures and mediate a variety of biological processes (Lis and Sharon 1998; Vijayan and Chandra 1999).

Levchuk and colleagues have established the physiological role of lectins in vital living processes at the initial stages of ontogeny of the flax seed: flax seed contains two lectin groups with different localization and physiological functions. The first group is located in the fat-free residue and is released into the environment during the seed germination, promoting the formation of the microflora, and the second group is located in the fat layer. It combines with lipase into a lectin-enzyme functional complex, which provides an effective immobilization of auxiliary material for seed germination (Levchuk 2012). Also, these research group (Levchuk et al. 2013) found that in the course of adaptation of plants resistant to hyperthermia, osmotic stress, and salinity, a considerable rise in the coefficient of activity of membrane lectins occurs, whereas the adaptation to hypothermia elevated the coefficient of activity of cell wall lectins. As to total soluble lectins, the adaptation of flax plants was associated with the changes in the range of their carbohydrate specificity. For example, following adaptation to various stresses such as hyperthermia, lectins have been shown to bind to glucose and glucosamine; for osmotic stress, to mannose and xylose; response to salinity, to galactose, glucose and glucosamine; in the case of cold resistance, total soluble lectins recognize lactose and fructose. It was concluded that lectins may participate in specific adaptation of flax plants to various abiotic stress factors.

8.5.3 Flaxseed Lignan SDG

Lignans are a major group of secondary metabolites widely spread in the plant kingdom. Lignan diversity in *Linum* species was investigated by Schmidt et al. (2010, 2012). Analyses of aerial parts of 41 members of *Linum* species show 64 different lignans that are divided into 2 major groups, aryltetralin group (AT) (sections Syllinum, Linopsis, and Cathartolinum) and aryldihydronaphthalene/arylnaphthalene (ADN/AN) (sections Linum and Dasylinum). Some members of Linum section contain high quantities of dibenzylbutyrolactones (e.g., *Linum usitatissimum*), while furofuranolignans are only found in *L. bienne* (Schmidt et al. 2010). Seed analyses of 20 *Linum* species show similar distribution as in aerial parts. Section Syllinum contains AT lignans, while Linum and Dasylinum generally contain ADN/AN lignans. Secoisolariciresinol diglucoside is found in all but one investigated species of Linum and Dasylinum. On the contrary, no SDG is detected in section Syllinum. Most species contained only 8R,8′R-SDG, while 8S,8′S-SDG is the predominant form in only three species (*L. usitatissimum*, *L. decumbens*, and *L. narbonense*) (Schmidt et al. 2012).

Unlike lignin, they are not present in all plants. Most model plants used so far for functional genetic studies (e.g., *Arabidopsis thaliana* and *Nicotiana tabacum*) have poor or null lignan contents. *L. usitatissimum* shows the highest content of SDG in

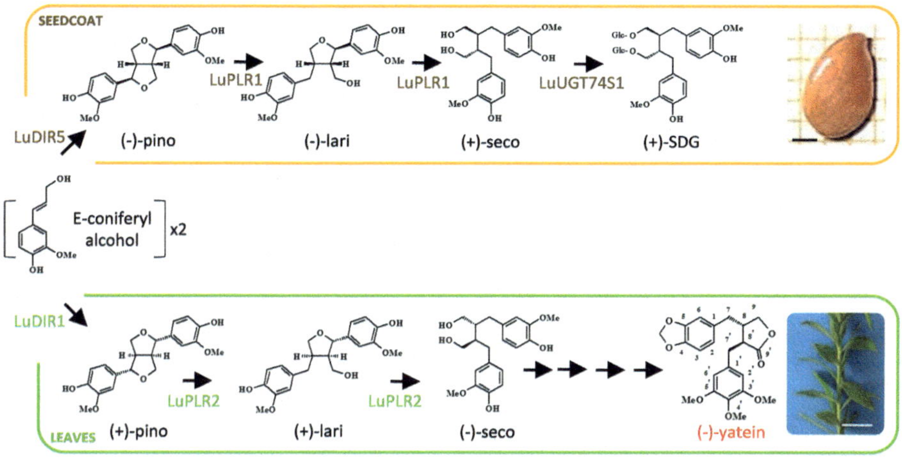

Fig. 8.7 Lignan biosynthetic pathway in flax (*Linum usitatissimum L.*) [Adapted from Ford et al. 2001, Ghose et al. 2014, and Dalisay et al. 2015]. After LuDIR5 or LuDIR6 control over stereoselective coupling of two *E*-coniferyl alcohol moieties to afford (−)-pinoresinol, LuPLR1 (pinoresinol-lariciresinol reductase (1) converts this into the major lignan (+)-secoisolariciresinol, which initially accumulates in the seed coat in its diglucosylated form SDG by LuUGT74S1, and is then converted into higher molecular weight (HMW) oligomers. In contrast, LuDIR1 engenders control over (+)-pinoresinol biosynthesis. The latter is then acted on by LuPLR2 (pinoresinol lariciresinol reductase (2) to afford the (−)-secoisolariciresinol enantiomer which is then subsequently metabolized into other lignans, such as (−)-yatein

the seed, more than 60–700 times in any other analyzed plant foods (Thompson et al. 1991; Johnsson et al. 2000; Eliasson et al. 2003; Renouard et al. 2010; Corbin et al. 2015).

The lignan biosynthesis has been characterized, and the proposed pathway is confirmed and widely accepted (Ford et al. 2001). Lignans, neolignans, and lignin share a common phenylpropanoid precursor, namely, the coniferyl alcohol, and both lignin formation and the first step in monolignol polymerization are considered to arise via bimolecular phenoxy radical coupling (Fang et al. 2013; Fig. 8.7).

A major difference in the biosynthesis of these compounds results from the stereoselective and the stereospecific coupling in the formation of optically active pinoresinol, whereas optically inactive lignin is formed in non-stereospecific process (Gang et al. 1999). The synthesis starts with coupling of two coniferyl alcohols by an oxidase (laccase or peroxidase) with the help of dirigent protein (DIR) to form pinoresinol (Davin et al. 1997; Dalisay et al. 2015; Corbin et al. 2018). Dirigent proteins may have a role in the formation of different pinoresinol enantiomers by holding coniferyl alcohols in a specific orientation during coupling. Pinoresinol is then reduced to secoisolariciresinol via lariciresinol in a two-step reduction by a pinoresinol-lariciresinol reductase (PLR). *Linum usitatissimum*, expresses at least two PLRs with opposite enantiospecificity (Hemmati et al. 2010; Corbin et al. 2017). Depending on the plant species and organ, secoisolariciresinol can be further oxidized in matairesinol by secoisolariciresinol dehydrogenase (SDH) as was shown in Xia et al. (2001), or secoisolariciresinol can be glycosylated to secoisolariciresinol diglucoside (SDG) with secoisolariciresinol glycosyltransferase as is often the case in seeds of *L. usitatissimum* (Ghose et al. 2014).

Transgenic plants with the knocked-down *LuPLR1* gene by RNAi produce seeds devoid of lignan (Renouard et al. 2014). There are speculations that there are other isozymes that have not been identified yet (Markulin et al. 2019a).

The *LuPLR1* gene expression is mainly localized in the seed coat during all developmental stages of flax seed (Hano et al. 2006a; Fang et al. 2016). Abscisic acid (ABA), a key growth regulator, has a strong relationship with the gene expression and SDG biosynthesis (Renouard et al. 2012). ABA accumulation starts in developmental stage 2 (WS2; around DAF10–16, Fig. 8.5) and reaches its maximum in developmental stage 4 (around DAF24, Fig. 8.5; Renouard et al. 2012). Exogenous application of ABA to seeds in WS2 causes an increase of *LuPLR1* gene expression and SECO quantity, while inhibition of ABA biosynthesis causes their decrease (Renouard et al. 2012). The effect of the ABA is realized through at least two cis regulation elements present in the gene promoter (Corbin et al. 2013a). The two plant phytohormones, ABA and gibberellin (GA), showed opposite influence on the lignan biosynthesis (Corbin et al. 2013a). The role of the embryo in expression control is still not fully understood. Ca^{2+} ions are universal secondary messengers involved in wide variety of processes in the plant (Tuteja and Mahajan 2007). Ca^{2+} showed to be an important factor of the *LuPLR1* gene expression and lignans biosynthesis mediated by the ABA. Treatments that inhibit influx of Ca^{2+} from extracellular space or intracellular storages as well as disruption of calmodulin (CaM)-like) proteins and Ca^{2+}-dependent protein kinases (CDPK) activity have shown a role for the MYB2 *cis*-acting element in the transduction of the signal. The ABA-responsive elements, ABRE and MYB2, could play a role in the signal arrived from the cellular entrance of Ca^{2+}, possibly through the action of the CaM, LuCML15b (Markulin et al. 2019b). The signal was transduced by action of CDPK, which could be involved by either direct interaction with cis-acting elements or a part of the phosphorylation cascaded (Markulin et al. 2019b).

Interestingly, another negative effector of the *LuPLR1* gene expression and SECO synthesis is LuERA1 protein (enhanced response to ABA) (Corbin et al. 2013b). It is a β-subunit of the heterodimeric protein called protein farnesyltransferase. It makes a post-translation modification by transferring C15 farnesyl unit to the C-terminal end cysteine of the target protein. *LuERA1* is expressed in roots, leaves, and cell suspension but predominantly in the stem and flax seed. The strongest expression is in the seed coat which corresponds to *LuPLR1* gene expression and lignan biosynthesis (Corbin et al. 2013b). The expression overlaps of *LuPLR1* and *LuERA1* enable LuERA1 to modulate gene expression of the *LuPLR1*. LuERA1 probably affects lignan synthesis through negative regulation of ABA signaling. *A. thaliana era1* mutant shows enhanced sensitivity to ABA, while inhibition of farnesyltransferase positively affects lignan biosynthesis (Corbin et al. 2013b). LuERA1 action is probably mediated through ABRE (ABA-responsive element). Indeed, the mutation of ABRE abolishes the positive effect of the farnesyltransferase inhibitor on the *LuPLR1* gene (Corbin et al. 2013b). The LuERA1 could possibly inhibit *LuPLR1* expression through farnesylation of the ABI3 (ABA-insensitive 3) transcription factor involved in the ABA signaling (Brady et al. 2003; Corbin et al. 2013b). In *A. thaliana*, ERA1 and farnesylation events have been also reported to control seed size, oil content, as well as fatty acid profile (Dutilleul et al. 2016), making this gene an attractive candidate to control ALA and SDG contents but also drought tolerance.

Note that the *LuPLR1* gene shows weak activation in the cell suspension treated with *Botrytis cinerea* or *Fusarium oxysporum* between 8 and 48 h post-elicitation (Hano et al. 2006b, 2008). A transcription factor LuWRKY36 controlling lignan biosynthesis in response to ABA and *F. oxysporum* elicitation has been identified (Markulin et al. 2019c) and could be a promising target to increase SDG accumulation in flax seeds.

8.6 Conclusions

While great progress has been made in last few decades in establishing the target genes for development and amelioration of linseed used in agriculture, still much remains to be accomplished. Initial studies have shown that flax is highly adaptable to modern molecular genetic engineering techniques. Additionally, future investigations in flax genomics will greatly benefit from reduced sequencing costs and new approaches. There are still many areas where flax biology can be investigated to improve our understanding of this ancient crop with the goal of increasing yields and quality. Many important gene targets involved in key processes, such as ALA, storage proteins, and SDG biosynthesis during seed development, have been established through transcriptome analysis. However, significant gaps still exist to definitely connect some specific genes to these pathways and/or obtain a comprehensive view of their regulatory mechanisms. The transcriptomic analysis of the flax seed development is still underdeveloped. To date, there are still far more papers about the health-modulating effects of many flaxseed compounds than about the expression regulation of the genes involved in the biosynthesis of ALA, proteins, and/or SDG. Addressing these questions could potentially contribute to a better understanding of flax seed production and the identification of novel gene targets to further enhance flax resistance to various abiotic stresses as well as its nutritional and health-promoting aspects of flax seeds.

Acknowledgments Part of this work was supported by Loire Valley Regional Council (Acti-LIN) and Ligue contre le Cancer (IMPACT, Cell4LiFE).

References

Allaby RG, Peterson GW, Merriwether DA, Fu YB (2005) Evidence of the domestication history of flax (*Linum usitatissimum* L.) from genetic diversity of the sad2 locus. Theor Appl Genet 112:58–65

APG (2016) An update of the Angiosperm Phylogeny Group classification for the orders and families of flowering plants: APG IV. Bot J Linn Soc 181:1–20

Bajpai M, Pandey S, Vasishtha AK (1985) Spectrum of variability of characteristics and composition of the oils from different genetic varieties of linseed. J Am Oil Chem Soc 62:628

Barbashov AV (2005) Group composition of the protein complex of flax seeds of modern varieties. Food Technol 4:S.71–S.72

Baumann BB (1960) The botanical aspects of ancient Egyptian embalming and burial 1. Econ Bot 14:84–104

Boechko FF (1985) Biological chemistry Higher school, p 658

Bozan B, Temelli F (2008) Chemical composition and oxidative stability of flax, safflower and poppy seed and seed oils. Bioresour Technol 99:6354–6359

Brady SM, Sarkar SF, Bonetta D, McCourt P (2003) The ABSCISIC ACID INSENSITIVE 3 (ABI3) gene is modulated by farnesylation and is involved in auxin signaling and lateral root development in Arabidopsis. Plant J 34:67–75

Burdge GC, Wootton SA (2002) Conversion of a-linolenic acid to eicosapentaenoic, docosapentaenoic and docosahexaenoic acids in young women. Br J Nutr 88:411–420

Burdge GC, Jones AE, Wootton SA (2002) Eicosapentaenoic and docosapentaenoic acids are the principal products of a-linolenic acid metabolism in young men*. Br J Nutr 88:355–363

Chen Y, Schneeberger RG, Cullis CA (2005) A site-specific insertion sequence in flax genotrophs induced by environment. New Phytol 167:171–180

Christofidou-Solomidou M, Tyagi S, Tan K-S, Hagan S, Pietrofesa R, Dukes F, Arguiri E, Heitjan DF, Solomides CC, Cengel KA (2011) Dietary flaxseed administered post thoracic radiation treatment improves survival and mitigates radiation-induced pneumonopathy in mice. BMC Cancer 11:269

Condra J (2008) The Greenwood encyclopedia of clothing through world history. Greenwood Press, Westport

Contor L (2001) Functional food science in Europe. Nutr Metab Cardiovasc Dis 11(4 Suppl):20

Coorey R, Novinda A, Williams H, Jayasena V (2015) Omega-3 fatty acid profile of eggs from laying hens fed diets supplemented with chia, fish oil, and flaxseed. J Food Sci 80:S180–S187

Corbin C, Decourtil C, Marosevic D, Bailly M, Lopez T, Renouard S, Doussot J, Dutilleul C, Auguin D, Giglioli-Guivarc'h N et al (2013a) Role of protein farnesylation events in the ABA-mediated regulation of the Pinoresinol-Lariciresinol Reductase 1 (LuPLR1) gene expression and lignan biosynthesis in flax (Linum usitatissimum L.). Plant Physiol Biochem 72:96–111

Corbin C, Renouard S, Lopez T, Lamblin F, Lainé E, Hano C (2013b) Identification and characterization of cis-acting elements involved in the regulation of ABA- and/or GA-mediated LuPLR1 gene expression and lignan biosynthesis in flax (Linum usitatissimum L.) cell cultures. J Plant Physiol 170:516–522

Corbin C, Fidel T, Leclerc EA, Barakzoy E, Sagot N, Falguiéres A, Renouard S, Blondeau J-P, Ferroud C, Doussot J et al (2015) Development and validation of an efficient ultrasound assisted extraction of phenolic compounds from flax (Linum usitatissimum L.) seeds. Ultrason Sonochem 26:176–185

Corbin C, Drouet S, Mateljak I, Markulin L, Decourtil C, Renouard S, Lopez T, Doussot J, Lamblin F, Auguin D et al (2017) Functional characterization of the pinoresinol–lariciresinol reductase-2 gene reveals its roles in yatein biosynthesis and flax defense response. Planta 246:405. https://doi.org/10.1007/s00425-017-2701-0

Corbin C, Drouet S, Markulin L, Auguin D, Lainé É, Davin LB, Cort JR, Lewis NG, Hano C (2018) A genome-wide analysis of the flax (*Linum usitatissimum* L.) dirigent protein family: from gene identification and evolution to differential regulation. Plant Mol Biol 97(1–2):73–101

Dalisay DS, Kim KW, Lee C, Yang H, Rübel O, Bowen BP, Davin LB, Lewis NG (2015) Dirigent protein-mediated lignan and cyanogenic glucoside formation in flax seed: integrated omics and MALDI mass spectrometry imaging. J Nat Prod 78:1231–1242

Darwin C (1863) On the existence of two forms, and on their reciprocal sexual relations, in several species of the genus *Linum*. J Proc Linn Soc Lond Bot 7:69–83

Darwin C (1877) The different forms of flowers on plants of the same species. Murray, London

Davin LB, Wang HB, Crowell AL, Bedgar DL, Martin DM, Sarkanen S, Lewis NG (1997) Stereoselective bimolecular phenoxy radical coupling by an auxiliary (dirigent) protein without an active center. Science 275:362–366

Diederichsen A, Richards K (2003) Cultivated flax and the genus Linum L. In: Muir AD, Westcott ND (eds) Flax genus Linum. Routledge, London, pp 22–54

Dorrell D (1970) Distribution of fatty acids within seed of flax. Can J Plant Sci 50:71–75

Dressler S, Repplinger M, Bayer C (2014) Linaceae. In: Kubitzki K (ed) Fam. genera Vasc. plants. Vol. 11 Flowering plants. Eudicots. Springer, Berlin/Heidelberg, pp 237–246

Dribnenki JCP, McEachern SF, Chen Y, Green AG, Rashid KY (2003) Linola™ 2047 low linolenic flax. Can J Plant Sci 83:81–83

Dulberger R (1974) Structural dimorphism of stigmatic papillae in distylous *Linum* species. Am J Bot 61:238–243

Dulberger R (1987) Fine structure and cytochemistry of the stigma surface and incompatibility in some distylous *Linum* species. Ann Bot 59:203–217

Dutilleul C, Ribeiro I, Blanc N, Nezames CD, Deng XW, Zglobicki P, Palacio Barrera AM, Atehortùa L, Courtois M, Labas V, Giglioli-Guivarc'h N, Ducos E (2016) ASG2 is a farnesylated DWD protein that acts as ABA negative regulator in Arabidopsis. Plant Cell Environ 39(1):185–198

Eliasson C, Kamal-Eldin A, Andersson R, Aman P (2003) High-performance liquid chromatographic analysis of secoisolariciresinol diglucoside and hydroxycinnamic acid glucosides in flaxseed by alkaline extraction. J Chromatogr A 1012:151–159

Fang J, Ramsay A, Paetz C, Tatsis EC, Renouard S, Hano C, Grand E, Fliniaux O, Roscher A, Mesnard F, Schneider B (2013) Concentration kinetics of secoisolariciresinol diglucoside and its biosynthetic precursor coniferin in developing flaxseed. Phytochem Anal 24(1):41–46

Fang J, Ramsay A, Renouard S, Hano C, Lamblin F, Chabbert B, Mesnard F, Schneider B (2016) Laser microdissection and spatiotemporal pinoresinol-lariciresinol reductase gene expression assign the cell layer-specific accumulation of secoisolariciresinol diglucoside in flaxseed coats. Front Plant Sci 7:1743

Ford JD, Huang KS, Wang HB, Davin LB, Lewis NG (2001) Biosynthetic pathway to the cancer chemopreventive secoisolariciresinol diglucoside-hydroxymethyl glutaryl ester-linked lignan oligomers in flax (Linum usitatissimum) seed. J Nat Prod 64:1388–1397

Fu YB, Dong Y, Yang M-H (2016) Multiplexed shotgun sequencing reveals congruent three genome phylogenetic signals for four botanical sections of the flax genus *Linum*. Mol Phylogenet Evol 101:122–132

Gang DR, Kasahara H, Xia ZZ-Q, Vander MK, Bauw G, Boerjan W, Van Montagu M, Davin LB, Lewis NG, Vander Mijnsbrugge K et al (1999) Evolution of plant defense mechanisms. J Biol Chem 274:7516–7527

Garros L, Drouet S, Corbin C, Decourtil C, Fidel T, Lebas de Lacour J, Leclerc EA, Renouard S, Tungmunnithum D, Doussot J, Abbasi BH, Maunit B, Lainé E, Fliniaux O, Mesnard F, Hano C (2018) Insight into the influence of cultivar type, cultivation year, and site on the lignans and related phenolic profiles, and the health-promoting antioxidant potential of flax (*Linum usitatissimum* L.) seeds. Molecules 23(10):2636

Ghose K, Selvaraj K, Mccallum J, Kirby CW, Sweeney-Nixon M, Cloutier SJ, Deyholos M, Datla R, Fofana B (2014) Identification and functional characterization of a flax UDP-glycosyltransferase glucosylating secoisolariciresinol (SECO) into secoisolariciresinol monoglucoside (SMG) and diglucoside (SDG). BMC Plant Biol 14:1–17

Goldblatt P (2007) The index to plant chromosome numbers: past and future. Taxon 56:984–986

Gutierrez L, Conejero G, Castelain M, Guénin S, Verdeil JL, Thomasset B, Van Wuytswinkel O (2006) Identification of new gene expression regulators specifically expressed during plant seed maturation. J Exp Bot 57:1919–1932

Hall LM, Booker H, Siloto RMP, Jhala AJ (2016) Flax (Linum usitatissimum L.). In: McKeon TA, Hayes DG, Hildebrand DF, Weselake RJ (eds) Industrial Oil Crop. Elsevier Inc, New York, pp 157–194

Hano C, Addi M, Bensaddek L, Crônier D, Baltora-Rosset S, Doussot J, Maury S, Mesnard F, Chabbert B, Hawkins S et al (2006a) Differential accumulation of monolignol-derived compounds in elicited flax (Linum usitatissimum) cell suspension cultures. Planta 223:975–989

Hano C, Martin I, Fliniaux O, Legrand B, Gutierrez L, Arroo RRJ, Mesnard F, Lamblin F, Lainé E (2006b) Pinoresinol-lariciresinol reductase gene expression and secoisolariciresinol diglucoside accumulation in developing flax (Linum usitatissimum) seeds. Planta 224:1291–1301

Hano C, Addi M, Fliniaux O, Bensaddek L, Duverger E, Mesnard F, Lamblin F, Lainé E (2008) Molecular characterization of cell death induced by a compatible interaction between Fusarium oxysporum f. sp. linii and flax (Linum usitatissimum) cells. Plant Physiol Biochem 46:590–600

Hano C, Renouard S, Molinié R, Corbin C, Barakzoy E, Doussot J, Lamblin F, Lainé E (2013) Flaxseed (Linum usitatissimum L.) extract as well as (+)-secoisolariciresinol diglucoside and its mammalian derivatives are potent inhibitors of Î±-amylase activity. Bioorg Med Chem Lett 23:3007–3012

Hano C, Corbin C, Drouet S, Quéro A, Rombaut N, Savoire R, Molinié R, Thomasset B, Mesnard F, Lainé E (2017) The lignan (+)-secoisolariciresinol extracted from flax hulls is an effective protectant of linseed oil and its emulsion against oxidative damage. Eur J Lipid Sci Technol 118:1600219

Hano C, Drouet S, Lainé E (2020) Virus-induced gene silencing (VIGS) in flax (*Linum usitatissimum* L.) seed coat: description of an effective procedure using the transparent testa 2 gene as a selectable marker. In: Virus-induced gene silencing in plants. Humana, New York, pp 233–242

Harper CR, Edwards MJ, DeFilippis AP, DeFilipis AP, Jacobson TA (2006) Flaxseed oil increases the plasma concentrations of cardioprotective (n-3) fatty acids in humans. J Nutr 136:83–87

Helbaek H (1969) Plant collecting, dry farming and irrigation agriculture in prehistoric Deh Luran. In: Hole F, Flannery KV, Neely JF (eds) Prehistory human ecology Deh Luran plain. University of Michigan, Ann Arbor, pp 383–426

Hemmati S, von Heimendahl CBI, Klaes M, Alfermann AW, Schmidt TJ, Fuss E (2010) Pinoresinol-lariciresinol reductases with opposite enantiospecificity determine the enantiomeric composition of lignans in the different organs of Linum usitatissimum L. Planta Med 76:928–934

Herchi W, Arráez-Román D, Trabelsi H, Bouali I, Boukhchina S, Kallel H, Segura-Carretero A, Fernández-Gutierrez A (2014) Phenolic compounds in flaxseed: a review of their properties and analytical methods. An overview of the last decade. J Oleo Sci 63:7–14

Hopf M (1983) Jericho plant remains. In: Kenyon K, Holland T (eds) Excavitation at Jericho. British School of Archaeology in Jerusalem, London, pp 576–621

Ibrügger S, Kristensen M, Skau Mikkelsen M, Astrup A (2012) Flaxseed dietary fiber supplements for suppression of appetite and food intake. Appetite 58:490–495

Jiang SY, Ma Z, Ramachandran S (2010) Evolutionary history and stress regulation of the lectin superfamily in higher plants. BMC Evol Biol 10:79

Johnson C, Moss T, Cullis C (2011) Environmentally induced heritable changes in flax. J Vis Exp:pii: 47:e2332

Johnsson P, Kamal-Eldin A, Lundgren LN, Aman P (2000) HPLC method for analysis of secoisolariciresinol diglucoside in flaxseeds. J Agric Food Chem 48:5216–5219

Kew S, Mesa MD, Tricon S, Buckley R, Minihane AM, Yaqoob P (2004) Effects of oils rich in eicosapentaenoic and docosahexaenoic acids on immune cell composition and function in healthy humans. Am J Clin Nutr 79:674–681

Kitts DD, Yuan YV, Wijewickreme AN, Thompson LU (1999) Antioxidant activity of the flaxseed lignan secoisolariciresinol diglycoside and its mammalian lignan metabolites enterodiol and enterolactone. Mol Cell Biochem 202:91–100

Kristensen M, Jensen MG (2010) Dietary fibres in the regulation of appetite and food intake. Importance of viscosity. Appetite 56:65–70

Kristensen M, Jensen MG, Aarestrup J, Petersen KE, Søndergaard L, Mikkelsen MS, Astrup A (2012) Flaxseed dietary fibers lower cholesterol and increase fecal fat excretion, but magnitude of effect depend on food type. Nutr Metab (Lond) 9:1–8

Kvavadze E, Bar-Yosef O, Belfer-Cohen A, Boaretto E, Jakeli N, Matskevich Z, Meshveliani T (2009) 30,000-year-old wild flax fibers. Science 325:1359

Lainé E, Hano C, Lamblin F (2009) Lignans. In: Chemoprevention of cancer and DNA damage by dietary factors. Wiley-VCH, Weinheim, pp 555–577

Levchuk HM (2012) The spectrum of physiological functions of lectin-like proteins in *Linum humile* mill. seeds. Curr Issues Biol Environ Chem: Electronic Version [electronic resource], Zaporizhzhya National University 4(2):67–76

Levchuk HM, Voitovich EN, Lyakh VA (2013) Lectins of oil-seed flax plants exposed to abiotic stress. Russ J Plant Physiol 60(1):77–83

Lis H, Sharon N (1998) Lectins: carbohydrate-specific proteins that mediate cellular recognition. Chem Rev 98:637–674

Łukaszewicz M, Szopa J, Krasowska A (2004) Susceptibility of lipids from different flax cultivars to peroxidation and its lowering by added antioxidants. Food Chem 88:225–231

Makhno Y, Polyakova I (2007) Comparative study of the storage linseed proteins spectrum in the seeds of wild Linum species. Sci Tech Bull Inst Oilseed Crops UAAS – Zaporizhzhya 12:45–49

Makhno Y, Polyakova I, Lyakh V (2010) Determination of the storage linseed proteins polymorphism in seeds by PAGE. In: Kunakh VA (ed) Factors of experimental evolution of organisms. National Academy of Sciences of Ukraine, Society of geneticists and breeders, Logos, pp 397–401

Markulin L, Corbin C, Renouard S, Drouet S, Gutierrez L, Mateljak I, Auguin D, Hano C, Fuss E, Lainé E (2019a) Pinoresinol–lariciresinol reductases, key to the lignan synthesis in plants. Planta 249(6):1695–1714

Markulin L, Drouet S, Corbin C, Decourtil C, Garros L, Renouard S, Lopez T, Mongelard G, Gutierrez L, Auguin D, Lainé E, Hano C (2019b) The control exerted by ABA on lignan biosynthesis in flax (Linum usitatissimum L.) is modulated by a Ca2+ signal transduction involving the calmodulin-like LuCML15b. J Plant Physiol 236:74–87

Markulin L, Corbin C, Renouard S, Drouet S, Durpoix C, Mathieu C, Lopez T, Auguin D, Hano C, Lainé E (2019c) Characterization of LuWRKY36, a flax transcription factor promoting secoisolariciresinol biosynthesis in response to *Fusarium oxysporum* elicitors in *Linum usitatissimum* L. hairy roots. Planta 250(1):347–366

McCorriston J (1997) The fiber revolution – textile extensification, alienation, and social stratification in ancient Mesopotamia. Curr Anthropol 38:517–549

McDill JR, Simpson BB (2011) Molecular phylogenetics of Linaceae with complete generic sampling and data from two plastid genes. Bot J Linn Soc 165:64–83

McDill J, Repplinger M, Simpson BB, Kadereit JW (2009) The phylogeny of Linum and Linaceae subfamily Linoideae, with implications for their systematics, biogeography, and evolution of heterostyly. Syst Bot 34:386–405

Muravenko OV, Lemesh VA, Samatadze TE, Amosova AV, Grushetskaia ZE, Popov KV, Semenova OI, Khotyleva LV, Zelenin AV (2003) Genome comparisons of three closely related flax species and their hybrids with chromosomal and molecular markers. Russ J Genet 39:414–421

Muravenko OV, Bol'sheva NL, Yurkevich OY et al (2010) Karyogenomics of species of the genus *Linum* L. Russ J Genet 46:1339–1342

Naqvi PA, Rat M, Vasishtha AK (1987) Association of different components of seed and oil in linseed. Indian J Agric Sci 57(4):231–236

Nosova IV (2005) A study of the Linum species, sect. Syllinum and Dasylinum. In: Karyology, karyosystematics and molecular phylogeny, St. Petersburg, Russia pp 73–74

Nosova IV, Yu Semenova O, Samatadze TE, Amosova AV, Bolsheva NL, Zelenin AV, Muravenko OV (2005) Investigation of karyotype structure and mapping of ribosomal genes on chromosomes of wild Linum species by FISH. Biol Membr 22:244–248

Oomah BD, Mazza G (1993) Flaxseed proteins – a review. Food Chem 48:109–114

Osborne TB (1924) The vegetable proteins. Longmans, Green and Company, London

Otto SP, Whitton J (2000) Polyploid incidence and evolution. Annu Rev Genet 34:401–437

Penn Medicine (2011) Penn study shows an ancient crop effective in protecting against a 21st century hazard – PR News. www.pennmedicine.org/news/news-releases/2011/august/penn-study-shows-an-ancient-cr

Pietrofesa RA, Solomides CC, Christofidou-Solomidou M (2014) Flaxseed mitigates acute oxidative lung damage in a mouse model of repeated radiation and hyperoxia exposure associated with space exploration. J Pulm Respir Med 4:1000215. https://doi.org/10.4172/2161-105X.1000215

Pleshkov BP (1968) Workshop on plant biochemistry, p 183

Poliakov VA, Levchuk AN, Lyakh VA (2011) Study of protein complex in oil flax seeds. Bull Zaporozhye Natl Univ 2:23–28

Prasad K (2009) Flaxseed and cardiovascular health. J Cardiovasc Pharmacol 54:369–377

Prasad K, Dhar A (2016) Flaxseed and diabetes. Curr Pharm Des 22:141–144

Rajwade AV, Kadoo NY, Borikar SP, Harsulkar AM, Ghorpade PB, Gupta VS (2014) Differential transcriptional activity of SAD, FAD2 and FAD3 desaturase genes in developing seeds of linseed contributes to varietal variation in the a-linolenic acid content. Phytochemistry 98:41–53

Ray C (1944) Cytological studies on the flax genus, Linum. Am J Bot 31:241–248

Renouard S, Hano C, Corbin C, Fliniaux O, Lopez T, Montguillon J, Barakzoy E, Mesnard F, Lamblin F, Lainé E (2010) Cellulase-assisted release of secoisolariciresinol from extracts of flax (Linum usitatissimum) hulls and whole seeds. Food Chem 122:679–687

Renouard S, Corbin C, Lopez T, Montguillon J, Gutierrez L, Lamblin F, Lainé E, Hano C (2012) Abscisic acid regulates pinoresinol-lariciresinol reductase gene expression and secoisolariciresinol accumulation in developing flax (Linum usitatissimum L.) seeds. Planta 235:85–98

Renouard S, Tribalatc M-A, Lamblin F, Mongelard G, Fliniaux O, Corbin C, Marosevic D, Pilard S, Demailly H, Gutierrez L et al (2014) RNAi-mediated pinoresinol lariciresinol reductase gene silencing in flax (Linum usitatissimum L.) seed coat: consequences on lignans and neolignans accumulation. J Plant Physiol 171:1372–1377

Rice A, Glick L, Abadi S, Einhorn M, Kopelman NM, Salman-Minkov A, Mayzel J, Chay O, Mayrose I (2014) The chromosome counts database (CCDB) – a community resource of plant chromosome numbers. New Phytol 206:19–26

Rogers CM (1982) The systematics of Linum sect. Linopsis (Linaceae). Plant Syst Evol 140:225–234

Salunkhe DK, Desai BB (1986) Linseed, niger, and cottonseed. In: Postharvest biotechnology of oilseeds. CRC Press, Boca Raton, pp 171–186

Sammour RH (1999) Proteins of linseed (Linum usitatissimum L.), extraction and characterization by electrophoresis. Bot Bull Acad Sinica 40:121–126

Sauer NJ, Narvaez-Vasquez J, Mozoruk J, Miller RB, Warburg ZJ, Woodward MJ, Mihiret YA, Lincoln TA, Segami RE, Sanders SL, Walker KA, Beetham PR, Schopke CR, Gocal GFW (2016) Oligonucleotide-mediated genome editing provides precision and function to engineered nucleases and antibiotics in plants. Plant Physiol 170:1917–1928

Schick T (1988) Nahal Hemar cave: cordage, basketry and fabrics. Atiqot 38:31–43

Schmidt TJ, Hemmati S, Klaes M, Konuklugil B, Mohagheghzadeh A, Ionkova I, Fuss E, Wilhelm Alfermann A (2010) Lignans in flowering aerial parts of Linum species–chemodiversity in the light of systematics and phylogeny. Phytochemistry 71:1714–1728

Schmidt TJ, Klaes M, Sendker J (2012) Lignans in seeds of Linum species. Phytochemistry 82:89–99

Schneeberger RG, Cullis CA (1991) Specific DNA alterations associated with the environmental induction of heritable changes in flax. Genetics 128:619–630

Shi L, Katavic V, Yu Y, Kunst L, Haughn G (2012) Arabidopsis glabra2 mutant seeds deficient in mucilage biosynthesis produce more oil. Plant J 69:37–46

Simopoulos AP (1999) Essential fatty acids in health and chronic disease. Am Soc Clin Nutr 70:560S–569S

Singh KK, Mridula D, Rehal J, Barnwal P (2011) Flaxseed: a potential source of food, feed and fiber. Crit Rev Food Sci Nutr 51:210–222

Soto-Cerda BJ, Duquid S, Booker H, Rowland G, Diederichsen A, Cloutier S (2014) Genomic regions underlying agronomic traits in linseed (Linum usitatissimum L.) as revealed by association mapping. J Integr Plant Biol 56:75–87

Sugasini D, Lokesh BR (2015) Rats given linseed oil in microemulsion forms enriches the brain synaptic membrane with docosahexaenoic acid and enhances the neurotransmitter levels in the brain. Nutr Neurosci 18:87–96

Sveinsson S, Mcdill J, Wong GKS, Li J, Li X, Deyholos MK, Cronk QCB (2014) Phylogenetic pinpointing of a paleopolyploidy event within the flax genus (Linum) using transcriptomics. Ann Bot 113:753–761

Thompson LU, Robb P, Serraino M, Cheung F (1991) Mammalian lignan production from various foods. Nutr Cancer 16:43–52

Tuteja N, Mahajan S (2007) Calcium signaling network in plants an overview. Plant Signal Behav 2:79–85

Tutin TG, Heywood VH, Burges NA, Moore DM, Valentine DH, Walters SM, WD A (eds) (1968) Flora Europaea – volume 2: Rosaceae to Umbelliferae. Cambridge University Press, Cambridge

Vaisey-Genser M, Morris DH (2003) Introduction. History of the cultivation and use of flaxseed. In: Muir AD, Westcott ND (eds) Flax genus Linum. Routledge, London, pp 1–21

van Zeist W, Bakker-Heeres JAH (1975) Evidence for linseed cultivation before 6000 BC. J Archaeol Sci 2:215–219

Vavilov NI (1992) Origin and geography of cultivated plants. Cambridge University Press, New York

Venglat P, Xiang D, Qiu S, Stone SL, Tibiche C, Cram D, Alting-Mees M, Nowak J, Cloutier S, Deyholos M et al (2011) Gene expression analysis of flax seed development. BMC Plant Biol 11:74

Vijayan M, Chandra N (1999) Lectins. Curr Opin Struct Biol 9:707–714

Vromans J (2006) Molecular genetic studies in flax (Linum usitatissimum L.). Wageningen University, Wageningen

Wakasa Y, Takaiwa F (2013) Seed storage proteins (Chapter in Brenner's Encyclopedia of genetics (2nd ed))

Wang Z, Hobson N, Galindo L, Zhu S, Shi D, McDill J, Yang L, Hawkins S, Neutelings G, Datla R et al (2012) The genome of flax (Linum usitatissimum) assembled de novo from short shotgun sequence reads. Plant J 72:461–473

Westcott ND, Muir AD (2003) Chemical studies on the constituents of Linum spp. In: Muir AD, Westcott ND (eds) Flax genus Linum. Routledge, London, pp 55–73

Western TL, Young DS, Dean GH, Tan WL, Samuels AL, Haughn GW (2004) MUCILAGE-MODIFIED4 encodes a putative pectin biosynthetic enzyme developmentally regulated by APETALA2, TRANSPARENT TESTA GLABRA1, and GLABRA2 in the Arabidopsis seed coat. Plant Physiol 134:296–306

Wood K, Mantzioris E, Gibson R, Ramsden C, Muhlhausler B (2015) The effect of modifying dietary LA and ALA intakes on omega-3 long chain polyunsaturated fatty acid (n-3 LCPUFA) status in human adults: a systematic review and commentary. Prostaglandins Leukot Essent Fat Acids 95:47–55

Xia Z-Q, Costa MA, Pélissier HC, Davin LB, Lewis NG (2001) Secoisolariciresinol dehydrogenase purification, cloning, and functional expression. Implications for human health and protection. J Biol Chem 276:12614–12623

You FM, Cloutier S (2019) Assembly of the flax genome into chromosomes. In: Genetics and genomics of Linum. Springer, Cham, pp 73–87

You FM, Xiao J, Li P, Yao Z, Jia G, He L, Zhu T, Luo M-C, Wang X, Deyholos MK, Cloutier S (2018) Chromosome-scale pseudomolecules refined by optical, physical, and genetic maps in flax. Plant J 95:371–384

Zeven A, de Wet J (1982) Dictionary of cultivated plants and their regions of diversity, 2nd edn. Centre for Agricultural Publishing and Documentation, Wageningen

Zuk M, Dorotkiewicz-Jach A, Drulis-Kawa Z, Arendt M, Kulma A, Szopa J (2014) Bactericidal activities of GM flax seedcake extract on pathogenic bacteria clinical strains. BMC Biotechnol 14:70

Part II
Oil Crop Genomics

Chapter 9
Coconut Genomics

H. D. D. Bandupriya and S. A. C. N. Perera

Contents

9.1 Introduction

Coconut is a perennial tropical palm belonging to family Arecaceae. It is a multipurpose tree species providing a multitude of products including food and beverage, fiber, and wood among many others. Coconut is economically and socially important and blended into the lifestyles of millions of people living in over 80 countries in the humid tropics. Ideally, coconut flourishes in the tropical belt between 200 North and South of the equator, up to an altitude of about 1200 m from the sea level.

H. D. D. Bandupriya (✉)
Department of Plant Sciences, Faculty of Science, University of Colombo,
Colombo, Sri Lanka
e-mail: dbandupriya@pts.cmb.ac.lk

S. A. C. N. Perera
Department of Agricultural Biology, Faculty of Agriculture, University of Peradeniya,
Peradeniya, Sri Lanka

H. Tombuloglu et al. (eds.), *Oil Crop Genomics*,
https://doi.org/10.1007/978-3-030-70420-9_9

Bright sunshine, temperatures of 27–30 °C, and well-aerated deep soils are the best growing conditions for coconut.

World coconut extent in 2018 was recorded to be 12,355,183 ha (FAOSTAT 2018) including both small to medium holdings and plantations of larger acreages producing a total yield of 62,268,105 tonnes (FAOSTAT 2018). The majority of the coconut plantations are located in South and Southeast Asia. Countries such as Indonesia, the Philippines, India, and Sri Lanka and Brazil in South America are the leading coconut producers. With its diverse uses that provide raw material to small, medium, and large-scale industries, coconut has been termed as the "tree of life" in coconut-growing countries and has been identified as a tree species having the potential for alleviating poverty in the rural community in the Asia and Pacific region of the world.

The major and the most traditional produce of coconut is the coconut oil which accounts for 1.8% of the global vegetable oil industry. Coconut oil contains saturated fatty acids resulting in a relatively high melting point and some level of resistance to rancidity (Marina et al. 2009). Due to high levels of digestibility, resulting from the abundance of medium and short-chain fatty acids, coconut oil is used in certain infant milk formulas as a source of fat (Banzon 1997). The relatively novel form of coconut oil, "virgin coconut oil," is produced from fresh coconut kernel by extracting under low heat so that the natural vitamins and enzymes are preserved intact (Marikkar et al. 2007). Virgin coconut oil has gained much attention for its pharmaceutical properties that are being uncovered. Despite being a saturated fatty acid, coconut oil is heavily used as cooking oil in many of the Asian countries. Coconut oil and dried kernel are heavily used in confectionery industries to produce a variety of products ranging from biscuits, ice cream, hard-butter, and many more. In addition to being used in food preparations, coconut oil is also used as a raw material in non-food industries such as detergents, soap, cosmetics, etc. The "coconut milk" which is extracted from fresh coconut kernel is used in Asian cooking creating a demand for fresh coconuts in countries such as Sri Lanka, Thailand, southern parts of India, etc. Coconut milk is also processed as a substitute for cows' milk for the community having lactose intolerance. Coconut sugar, jaggery, vinegar, toddy, and hard liquor are products obtained by tapping the immature inflorescence for coconut sap. The most recent coconut product that has gained global attention is the natural health drink obtained from the "coconut water" which is actually the liquid endosperm in tender or mature coconut.

9.2 Botany and Genetics of Coconut

Coconut (*Cocos nucifera* L.) is a monocot palm, and it is the only species in the genus *Cocos*. Further, there are no records on any wild relatives or even closely related species of coconut. The diploid genome of coconut consists of 16 pairs of chromosomes ($2n = 2 \times = 32$). Coconut possesses a fibrous root system and an unbranched stem, which produces a prominent root bole in tall varieties, a crown consisting of about 30–40 fronds, and inflorescences maturing to fruit bunches stemming out from each leaf axis. The coconut inflorescence is monoecious producing female flowers at the basal ends and an abundance of male flowers at the distal ends of spikelets that emanate from the peduncle of the inflorescence. The emission of inflorescences continues throughout the economic lifespan of coconut once the palms attain the reproductive stage at an average of 12–14 inflorescences per palm per annum.

Genetics of natural pollination is an interesting feature in coconut: the tall varieties being naturally cross-pollinated due to nonoverlapping of the male and female phases of the same inflorescence and between successive inflorescences while the dwarf coconuts being naturally self-pollinated due to overlapping of the male and female phases of the same inflorescence. Coconuts do not record any genetic incompatibilities and also are amenable to both artificial cross- and self-pollinations making it easier for the manipulation of pollination in genetic improvement.

9.2.1 Genetic Resources of Coconut

Distinct phenotypes of coconuts in the world are believed to be at different evolutionary and domestication stages. However, two main groups of global genetic resources have been distinguished as tall and dwarf coconuts based on the stature of the palm. The tall coconuts, as the name refers, are tall in stature, while the dwarfs are shorter. Besides this deviation in height of the palm, the tall coconuts are naturally cross-pollinating, while the dwarf coconuts are predominantly self-pollinating. Included in these two broad types of coconuts are many other varietal groups, forms, morphotypes, or variants in different coconut-growing countries, and the nomenclature of such groups is not globally standardized.

Different countries do have their own coconut classifications but keeping common to the two main types of Tall (*Typica*) and Dwarf (*Nana*) coconuts which are usually referred to as the two main varieties. However, the nomenclature of different morphotypes within tall coconuts is not uniform across classifications. Some countries do name them as varieties, while some include them as forms or deviants within varieties. For example, in Sri Lanka, within the tall variety, the distinct phenotypes are identified forms within variety Typica as Bodiri, Porapol (Liyanage 1958), etc. The dwarf coconuts are fairly uniform apart from the fruit color which can be green, yellow, orange, or brown, and they are denoted as such across different countries. Besides, in certain countries a third coconut type that displays mixed characteristics of tall and dwarf coconuts has been recorded: King coconut in Sri Lanka (Liyanage 1958), Gangabondam in India, and Compact dwarfs in the Pacific Islands (Bourdeix et al. 2005) being examples.

A concerted effort led and guided by the International Coconut Genetic Resources Network (COGENT) in collaboration with Bioversity International for the conservation of global coconut germplasm has taken place in the latter part of the twentieth century continuing to the early years of the twenty-first century. This resulted in

Table 9.1 Distinct features of cultivated coconut in the world

Cultivar	Specific characteristics (relative to other two cultivars)
Tall	Tall stature, presence of a prominent root bole, hardy palms, longer juvenile period (about 6–7 years), larger crown and fruits, no lag periods in nut production, relatively smaller number of nuts per bunch, higher kernel and copra amount per nut, tolerant to abiotic stresses, cross-pollinating breeding behavior
Dwarf	Short stature, lacking a root bole, fragile looking palms, shorter juvenile period (about 2–3 years), smaller crown and smaller fruits, lag periods in nut production, higher number of nuts per bunch, low kernel and copra productivity, susceptible to abiotic stresses, predominantly self-pollinating breeding behavior
Hybrids	Intermediate stature, presence of a root bole, intermediate juvenile period (about 3–4 years), intermediate crown, larger fruits, intermediate number of nuts per bunch, no lag periods in nut production, higher number of nuts per bunch relative to talls, intermediate responses to abiotic stresses

the collection and conservation of germplasm in national as well as international ex situ field gene banks and a somewhat standard nomenclature of genetic resources. Accordingly, the nomenclature for tall coconuts included an area or country name, while the dwarf coconuts included the country, area, or a specific name along with the color form. Under this nomenclature, Sri Lanka Tall Ambakelle, West Coast Tall, and San Ramon Tall are some examples for tall coconuts, while Malayan Yellow Dwarf, Chowghat Green Dwarf, and Sri Lanka Brown Dwarf are some examples for dwarf coconut accessions.

The tall coconuts in the world are the main commercially important type of coconut. The dwarf coconuts are gaining popularity as the beverage type in addition to their role as a parent in coconut hybridization. Therefore, the cultivated coconuts form three groups as natural talls, dwarfs, and both natural and genetically improved hybrid coconuts having distinct features (Table 9.1).

9.2.2 *Origin and Domestication of Coconut*

After many suggestions and theories, currently, two regions in the world, namely, the islands in Southeast Asia and the South Indian subcontinent, have been reported to be the localities of origin of tall coconuts (Gunn et al. 2011). The dwarf coconuts have been hypothesized to have evolved from tall coconuts resulting from a reduction in allelic composition followed by subsequent mutations, as they represent only a subset of alleles that are present in tall coconuts and are divergent in their phenotypes (Perera et al. 2001).

Following its origin, coconut has been dispersed throughout the oceanic tropical regions including African and American continents. The early dissemination of coconut has been hypothesized to be naturally by floating of coconuts with the sea currents facilitated by the fibrous outer cover of nuts. In addition, it is reported that early sea voyagers carried coconuts as food and beverage in their vessels and dumped the remaining in their destinies introducing coconut across shores. Later on, in the colonization era, human-mediated transfer of coconuts across countries has taken place resulting in its wide-scale cultivation in the tropics.

9.3 DNA-Based Molecular Marker Studies

Starting from the late 1920s for about 80 years, coconut breeding relied solely on conventional approaches. In addition, germplasm characterization, pathogen detection, and all subdisciplines on research on coconut entirely depended on conventional approaches up to about the 1990s. With the elucidation of DNA structure in the 1950s, gradually, the molecular marker technology was developed and by the 1980s this technique became an applied science in plants, animals, humans, and microbes. The application of molecular marker technology in coconut was initiated in the 1990s with the validation of first-generation molecular markers in coconut.

Conventional approaches on coconut breeding have led to much progress in improving and manipulating the genetics of the palm for desirable traits. However, there were inherent constraints associated with such conventional approaches, especially in a perennial tree such as coconut having a long gestation period. The long time period required for a breeding cycle, the very high genotype and environment interaction, and the non-gene-targeted approaches typical to conventional coconut breeding have been instrumental in a longer duration and lack of efficiency in coconut improvement. Another subdiscipline, germplasm characterization, was entirely dependent on morphology and did not reveal the true genetic variation.

The introduction of molecular marker technology paved the way for the detection of DNA-based true genetic variation in coconut germplasm and more gene-targeted approaches which are independent of environmental effects in coconut breeding. The introduction of molecular markers in coconut research was a slow process due to lack of technical know-how, laboratory facilities, and other infrastructures and funding in coconut-growing countries which are mainly from the developing world. However, as of today, much progress has been achieved in using the molecular markers, especially in genetic diversity studies and genome mapping as described below.

9.3.1 Genetic Variation and Diversity Studies

The initial research on using molecular marker technology in coconut was the assessment of genetic diversity in coconut germplasm. Pioneering research on assessing the genetic diversity of coconut using molecular markers was reported in 1995 with the use of polymerase chain reaction (PCR) to amplify copia-like 16R1 repetitive elements in 17 coconut accessions (Rohde et al. 1995). This preliminary study revealed genetic similarity between tall coconuts from the Indian Ocean and East Africa and the clustering of Panama Tall with coconuts from the Pacific and Southeast Asia.

The hybridization-based marker system restriction fragment length polymorphism (RFLP) markers were used by Lebrun et al. (1998) to genotype both tall and dwarf coconuts to reveal separate subgroups on geographical regions, the Far East and the Pacific in one and the South Asia and African coconuts in the second group.

With the introduction of polymerase chain reaction (PCR), random amplified polymorphic DNA (RAPD) markers were commonly used by many research groups for genetic diversity studies (Table 9.2). The results were consistent wherever global collections have been studied in single experiments. However, RAPD is a dominant marker, and there was less transferability of techniques across laboratories and the polymorphism detected was not satisfactory.

Another molecular marker system that was limitedly used in the early days of molecular diversity studies in coconut is amplified fragment length polymorphism (AFLP) which combines the techniques of hybridization and PCR in detecting the genotypes. The Sri Lankan coconut germplasm was characterized using AFLP (Perera et al. 1998), and an international collection was subjected to AFLP analysis by Teulat et al. (2000). Although AFLPs display higher throughput than RFLP or RAPD; the cumbersome procedure, the need for the use of radiolabeling at least at the initial years, and the dominant nature of the marker limited the adoption of

Table 9.2 Some important research on assessing genetic diversity using RAPD markers

Research	Salient findings	References
Analysis of 17 coconut populations from the South Pacific region	High within-population diversity, occurrence of two geographic groups, and low gene migration among populations	Ashburner et al. (1997)
Analysis of ex situ-conserved germplasm in Sri Lanka	Revelation of the narrow genetic base in Sri Lankan germplasm	Everard (1996), Dassanayaka et al. (2003)
Study of East African Tall genotypes	Identification of two major genotypic groups	Duran et al. (1997)
Characterizing coconut germplasm in India	A set of informative markers were identified with 1–16 polymorphic bands per marker	Ratnambal et al. (2001)
Analysis of accessions of indigenous and exotic origin, ex situ-conserved in India	Genetic closeness of intermediate coconuts to dwarfs than the tall coconuts	Upadhyay et al. (2002)
Assessing the genetic divergence among 19 coconut populations	Identification of the genetic distinctness of all and the lack of duplicates in the collection	Daher et al. (2002)
Analysis of 30 ex situ-conserved coconut accessions from Pacific and Nicobar Islands	Identification of distinct genetic variation among the accessions	Sankaran et al. (2012)

AFLPs in a wider scale. Several studies on analyzing the genetic diversity of coconut ISSR markers have also been reported (Duran et al. 1997).

9.3.1.1 Use of Simple Sequence Repeat (SSR) Markers in Diversity Analysis

After the initial attempts of using molecular techniques for diversity studies, the need for a coconut-specific and a codominant marker system to capture the heterozygosity that is common in tall coconuts was reiterated. The invented SSR markers fulfilled the need with the development of coconut-specific SSR markers by several groups (Perera et al. 1999; Rivera et al. 1999) using Sri Lanka Tall and Tagnanan Tall coconut, respectively. Following their development, the use of SSR markers became highly popular, and many research teams used this technology in genetic diversity analysis in local and international coconut collections (Dassanayaka et al. 2003; Perera et al. 2000, 2001, 2003; Meerow et al. 2003; Teulat et al. 2000). These initial SSR analyses revealed important information related to the origin and dissemination of coconut, allelic richness in tall coconuts as opposed to the allelic reduction in dwarfs, narrow or rich genetic basis in certain local collections, and a vast array of other important information. The information derived led to important recommendations for parental identification in breeding programs, develop or revise strategies in germplasm collection and decide on germplasm exchange and importation activities, etc. For example, the introduction of imported exotic material as parents in hybridization projects is a practical application that resulted from the identification of the narrow genetic base of Sri Lankan coconuts.

During the last 10–15 years, the collection of SSR marker pool was enriched and SSR markers still remain to be the choice of markers in many studies leading from diversity analyses. Assessing the tolerance of coconut accessions from Africa, to lethal yellowing (Konan et al. 2007), testing hybridity (Kriswiyanti et al. 2013), determination of the population structure of coconut (Meerow et al. 2003; Kamaral et al. 2016) and genetic characterization and DNA finger printing of novel coconut phenotypes (Kamaral et al. 2017) are a few examples for practical applications of SSR in coconut.

With the extensive use of SSR markers in coconut diversity analysis, an SSR marker kit comprising of 14 SSR loci was developed (Baudouin and Lebrun 2002) to standardize the diversity analysis and to facilitate the comparison of results originating from different experiments. With these developments, the SSR marker system has been the most applied in diversity analysis and related studies among all the marker systems developed to date.

The high-throughput marker system Diversity Array Technology (DArT) has also been validated for coconut by analyzing a collection of Sri Lankan coconut germplasm (Perera and Kilian 2008).

9.3.2 Linkage Mapping and QTL Identification and Association Studies

The application of molecular marker technology in coconut has been advanced into the mapping of the coconut genome. Genome mapping in a perennial natural cross-pollinating species such as coconut is highly challenging. One of the main constraints for linkage mapping is the development of a structured segregating for genome mapping. The usual practice in self-pollinated crops is to develop segregating populations such as F_2s, backcrosses, doubled haploid populations, or even recombinant inbred lines. However, the development of such populations in a highly cross-pollinated perennial palm such as coconut is a daunting task. A mapping population should be sufficiently large to capture sufficient relevels of recombinants to map QTL with an acceptable level accuracy (Perera 2010). However, the development of a sufficiently large population in coconut is also hindered by the low seed output in coconut needing a longer duration to get a population of required size making it more complex by the differences in age of the individuals in the mapping population. Combining progeny from several half-sib families has been identified as an alternative strategy to overcome this problem (Kearsey and Luo 2003).

The need for dense molecular marker coverage is another requirement for mapping QTL of desirable traits on the genome map. Coconut being a large genome with 16 pairs of chromosomes, a considerable number of markers that are polymorphic for the alleles of interesting traits will be essential in mapping QTL.

Despite these challenges, several research teams have been successful in developing linkage maps of coconut. Accordingly, Rohde et al. (1999) used ISTR markers in developing the first genome map in coconut by using a segregating population derived from a cross between East African Tall and Laguna Tall. A segregating population deriving from a cross between Malayan Yellow Dwarf and Laguna Tall was used to construct the second genome map by positioning 382 AFLP, ISTR, RAPD, and ISSR markers (Herran et al. 2000) and mapping germination-related QTL on this linkage map. Ritter et al. (2000) were successful in mapping QTL for girth and leaf production on the same map developed by Herran et al. (2000). A mapping population later developed in the Ivory Coast, using a segregating population developed by a cross between Cameroon Red Dwarf and Rennell Island Tall anchored 280 markers in the framework map. QTL for yield parameters, nut number, bunch, and components of fruits were mapped on this QTL map (Baudouin et al. 2006; Lebrun et al. 2001).

The association mapping technique, which is based on linkage disequilibrium, facilitates mapping traits on the existing populations avoiding the need for segregating populations. This mapping approach would be highly useful in a crop such as coconut for getting rid of the difficulty in developing conventional segregating populations for genome mapping. However, it is essential to have dense marker saturation across the coconut genome for fine mapping of QTL using association mapping to make marker-assisted selection in coconut a reality.

9.4 Genomics

The way genes and genetic data are organized inside the genome and determination of their biological functions are together called genomics. Not only the genomics but also the transcriptomics (the study of the set of genes that transcribed into RNA) and proteomics (the study of the translated proteins) have influenced greatly on the understanding of the complex developmental events in plants. Bioinformatics tools play a significant role in managing the huge data produced through these technologies.

9.4.1 Unraveling the Coconut DNA for Candidate Genes

Studies on the coconut genes offer valuable information on key biological processes, such as embryogenesis, general plant development, fruit development, and fruit nutrition quality. In most cases, the annotated sequences from other species have been used as clues to identity homology genes and their possible roles in coconut. These studies provided useful guides for further improvements by allowing researchers to identify specific environments or developmental stages which show higher gene expression.

9.4.1.1 Somatic Embryogenesis

Development of in vitro protocols for the multiplication of elite coconut palms which show superior characters is one of the major areas coconut researchers are working on. Despite the progress recorded for tissue culture protocol development, embryogenesis rate and plant regeneration remain low. Even though earlier studies were mainly confined to develop methodologies, with the advent of smart molecular biology techniques, researchers were keen to understand the gene diversity that provokes somatic cells to become embryogenic.

Various research have been undertaken to apprehend and interpret the molecular basis of embryogenesis in vitro in coconut. Knowing that kinases play a role in plant embryogenesis, Perez-Nunez et al. (2009) identified a somatic embryogenesis receptor-like kinase (SERK) gene (*CnSERK*) during somatic embryogenesis of coconut plumular explants. Gene amplification was based on the degenerate primers designed on conserved sequences of homology genes in other plant species. Results of the in situ expression analysis allowed the authors to locate the spatiotemporal *CnSERK* expression in meristematic areas of the embryogenic structures. The discovery of the expression of *CnSERK* prior to the formation of somatic embryos which was barely detected in non-embryogenic calli suggested using *CnSERK* expression as a potential marker for detecting the competence of somatic cells to become embryogenic in vitro (Perez-Nunez et al. 2009). A putative cyclin-dependent kinase (CDKA) gene (*CnCDKA*) which shows a possible role in cell cycle control was isolated in a similar manner during somatic embryogenesis of coconut (Montero-Cortés et al. 2010). A progressive increase of *CnCDKA* transcripts shown in the cells which attain embryogenic competence proposed that *CnCDKA* may be exploited as a molecular marker to examine the meristematic capability of in vitro-cultured tissues.

Bandupriya et al. (2013) identified and characterized an *AINTEGUMENTA-like* gene (*CnANT*) in a tall coconut cultivar which encodes typical APETALA2 (AP2) conserved domains and the linker region. Primers designed based on the conserved domain regions of rice and oil palm led the identification of CnANT gene initially. A comparison of this gene, done between tall and dwarf cultivars, revealed one single nucleotide polymorphism in the first exon and one Indel in the first intron which made it possible to distinguish between these two cultivars upon a SSR marker designed based on the Indel. Expression analysis of *CnANT* in different tissue types reported enhanced expression in mature zygotic embryos, embryogenic callus, and embryogenic structures compared to other phases of somatic embryogenesis and vegetative tissues (Bandupriya et al. 2014). *CnANT* over-expression on Arabidopsis contributed to the induction of plant regeneration in hormone-free media suggesting *CnANT*'s role in cell proliferation during in vitro regeneration.

Exploiting the prior knowledge to force research in coconut is more obvious when working with transcriptome data for the identification of possible roles assigned for certain genes. This is observed in a study conducted on expression analysis of 14 genes in 6 different developmental stages of the plant regeneration process starting from plumule explants (Rajesh et al. 2016). *CLAVATA1* (*CLV*) showed higher expression at the initial callogenesis stage, while germin-like protein (GLP), PICKLE (PKL), glutathione S-transferase, WRKY, and WUSCHEL (WUS) showed their higher expression in somatic embryos. The rest of the genes, *APETALA2/ethylene-responsive factor, arabinogalactan protein gene* (*AGP*), *AINTEGUMENTA* (*ANT*), *late embryogenesis-abundant* gene (*LEA*), *SERK*, *mitogen-activated protein kinase* (*MAPK*), *SAUR family protein gene*, and embryogenic cell protein (ECP), were abundantly expressed in the embryogenic calli rather than in the initial callus or somatic embryos.

In an attempt to evaluate and compare the embryogenic potential of calli maintained for a long-term (21 weeks old) with the calli at their initial stage (21 days old) through gene expression analysis, enhanced expression of embryogenesis-related genes such as ECP, glutathione-S-transferase (GST), leafy cotyledon (LEAFY), and wuschel (WUS) was reported in calli even after 21 weeks (Bhavyashree et al. 2015). However, some other genes such as germin-like protein (GLP), SERK, pickle (PKL), and WRKY, which have possible roles in somatic embryogenesis, showed higher expression in initial embryogenic calli stage. Not all the genes were fully cloned from coconut in the current study (Bhavyashree et al. 2015). Nevertheless, it was concluded that coconut plumular calli could be maintained in their embryogenic potential for an extended period. Significant genotypic differences in gene expression patterns of above embryogenesis-related genes were revealed between different coconut cultivars (Bhavyashree et al. 2016).

9.4.1.2 Endosperm Development and Oil Biosynthesis

Coconut endosperm is the primary source of raw material for most products based on coconut. Thus, endosperm development matters on the ultimate impact of the quality of many coconut-based products. Coconut seed is a major reservoir for accumulating a large quantity of oil and stores roughly 63% of the seed weight. Several pieces of research highlighted the health benefits of medium-chain fatty acids (MCFAs) in coconut oil, particularly the lauric acid (Dayrit et al. 2008; Orsasova et al. 2015). This has increased the enthusiasm of researchers to work towards the development of outstanding coconut varieties not only with superior yield characters but also with enhanced oil quality enriched with high MCFAs. For this purpose, exploring coconut oil biosynthesis pathway genes is essential. The understanding of the underlined genetic regulatory mechanisms that govern lipid metabolism, nutrient composition, and fiber content is also important to maintain or enhance endosperm quality (Reynolds et al. 2019).

Cloning of a coding region of an endosperm-related gene was reported first in the 90th decade. Knutzon et al. (1995) cloned the *lysophosphatidyl acyltransferase* gene (*LPAAT*) from a coconut endosperm cDNA library. This enzyme is considered essential to control the metabolism of lysophosphatidic acid into different phosphatidic acids. Expression analysis based on the stable genetic transformation of coconut LPAAT promoter in rice showed the confined expression of LPAAT in rice endosperm (Xu et al. 2010).

Coconut stores lipids in the endosperm as oil bodies. Oleosins are unique structural proteins present in oil bodies of plant parts such as seeds and leaves which store high oil contents. The main role is to maintain the oil body size and stability by forming a monolayer with phospholipids while entrenched in the phospholipid coat (Napier et al. 2001). A full-length seed storage protein cDNA sequence, homologous to oleosin, was recognized and characterized from coconut cDNA libraries prepared from different fruit developmental stages (Li and Fan 2009). This gene was found to be expressed only in the endosperm which they termed as "pulp" on their report. Furthermore, a significant increase of the expression has been reported during pulp development. Recently, three genes coding for three isoforms of oleosins (CnOLE500a, CnOLE500c, and CnOLE300a) have been isolated and characterized in coconut (Vargas et al. 2018). The oleosin gene which had 375 nucleotides coding for 125 amino acids reported by Li and Fan (2009) corresponded to CnOLE500c in the study carried out by Vargas et al. (2018). Higher expressions were recorded for all three isoforms during endosperm development.

The first positive fatty acid biosynthesis regulator recognized in plants was the *WRI1* transcription factor (Cernac and Benning 2004). A functional gene which is an AP2/EREBP domain-containing transcription factor in the WRINKLED1 (WRI1) gene family has been detected from coconut (*CoWRI1)* endosperm (Sun et al. 2017). *WRI1* is an important regulator of oil accumulation in coconut endosperm tissues. The functional analysis explored through heterologous expression of *CoWRI1* in rice and Arabidopsis noticeably enhanced the oil content (Sun et al. 2017).

Coconut oil is rich with MCFAs which are much more beneficial than the long-chain fatty acids common in many vegetable oils. Stearoyl-acyl carrier protein (ACP) desaturase (SAD, EC 1.14.99.6) enzyme plays a vital role in fatty acid biosynthesis in higher plants by catalyzing the desaturation of stearoyl-ACP to oleoyl-ACP. By doing this, SAD plays an influential role in maintaining the saturated fatty acids to unsaturated fatty acids ratio (Lindqvist et al. 1996). A full-length cDNA of a stearoyl-acyl carrier protein desaturase (*CocoFAD*) has been isolated from a cDNA library developed from coconut endosperm (*C. nucifera* L.). *CocoFAD* was found to be expressed at a significant level in 8-month-old endosperm, but the expression was reduced by 50% in the 15-month endosperm (Gao et al. 2014). Heterologous expression of *CocoFAD* in *Saccharomyces cerevisiae* strain INVSc1 increased palmitoleic acid (16:1) and oleic acid (18:1) levels significantly while stearic acid (18:0) is reduced.

Owing to the high energy demand, the actual mechanism by which coconut accumulates MCFAs is an important topic for researchers to work on. A master regulator enzyme responsible for fatty acid chain length regulation in plants identified as β-ketoacyl-ACP synthase (KAS) I was isolated and characterized from coconut (CocoKASI) endosperm (Yuan et al. 2014). A 60% increase in the *CocoKASI* transcript accumulation was observed in young endosperm than in leaves. When *CocoKASI* was overexpressed under *Nicotiana tabacum* seed-specific napin promoter, short-chain and MCFAs were increased (Yuan et al. 2014). Similarly, acyl-acyl carrier protein (ACP) thioesterase (CnFatB3) which was upregulated significantly during fruit development was identified (Yuan et al. 2017). Overexpression of *CnFatB3* under *Nicotiana tabacum* seed-specific napin promoter increased the amounts of lauric acid, myristic acid, palmitic acid, and stearic acid in Arabidopsis by 30-, 80-, 4-, and 2-fold, respectively. The results give insights into the regulation of fatty acid biosynthesis and the accumulation of MCFAs in coconuts. The findings will contribute to the engineering, altering, and accumulation of short-chain and MCFAs not only in coconut but also in other oil crops as well to meet energy demands.

A gene-encoded type 2 diacylglycerol acyltransferase (DGAT2), which is considered as an important enzyme in triacylglycerol (TAG) biosynthesis (Zheng et al. 2019), and a cytoplasmic fructose-1, 6-bisphosphate aldolase (FBA) gene which is a key enzyme function in glycolysis, gluconeogenesis, and the Calvin cycle (Feng et al. 2020) have been identified recently in coconut endosperm (gene names are *CoDGAT2* and *CoFBA1*, respectively). *CoDGAT2* transcripts were abundant in the coconut endosperm approximately 7 months after pollination. Moreover, a significant increase in the linoleic acid concentration and decrease in eicosadienoic acid and arachidic acid were observed when *CoDGAT2* was overexpressed in Arabidopsis under a seed-specific promoter (Zheng et al. 2019). When a similar approach was followed to check the expression of *CoFBA1* in transgenic Arabidopsis, total lipid content was increased up to 18% and component-wise, linolenic acid and eicosanoic acid showed a significant increase (Zheng et al. 2019).

Availability of whole-genome sequences for several Arecaceae crop species such as *C. nucifera*, *E. guineensis*, and *P. dactylifera* has allowed comparison of genes in important biological pathways. As an initiative, fatty acid composition in *C. nucifera*, *E. guineensis* and *P. dactylifera* tissues were analyzed and 806, 840 and 848 lipid-related genes were identified respectively. These genes were categorized into 22 lipid metabolism pathways (Xiao et al. 2019). The majority of these lipid-related genes were highly homologous between the three observed species. Analysis of evolutionary relationships of some of the predominant genes in the lipid and carbohydrate gene metabolism pathways showed a close relationship between two oil-producing species coconut and oil palm, and as expected date palm was different from these two species.

Curd coconut or macapuno coconut is identified based on an abnormal phenotype found in the endosperm of certain tall type coconuts (Islam et al. 2009). The endosperm is characterized by having soft jelly-like meat and thick soft gel-like liquid endosperm which fills the entire cavity (Fig. 9.1). Lack of the activity of α-galactosidase which is one of glycoside hydrolase enzymes in coconut has been pronounced to cause the abnormal endosperm due to the incapacity of hydrolyzing oligogalactomannan. The full-length cDNA of α-D-galactosides gene in the normal coconut and three mutant alleles S137R, R216T, and P250R in the curd coconuts have been identified and characterized through in silico analyses, three-dimensional modeling, and expression analyses (Phoeurk et al. 2018). According to the differential transcription level study, 20–583 times of mutant gene transcript accumulation were observed in curd coconuts than in normal coconuts (Phoeurk et al. 2018). Furthermore, heterologous expression of mutants in *Komagataella phaffii* showed no α-galactosidase activity, while normal α-D-galactosides protein showed α-galactosidase activity in *K. phaffii*.

Aromatic coconut with a pleasant "pandan-like" or "popcorn-like" aroma in the liquid endosperm is a popular trade character identified in dwarf coconut in Thailand. The presence of a volatile compound, 2-acetyl-1-pyrroline (2AP), is the reason for this nice aroma. The gene which plays a vital role in the 2AP biosynthesis was identified in aromatic rice (Bradbury et al. 2005), and the gene was renamed as *amino*

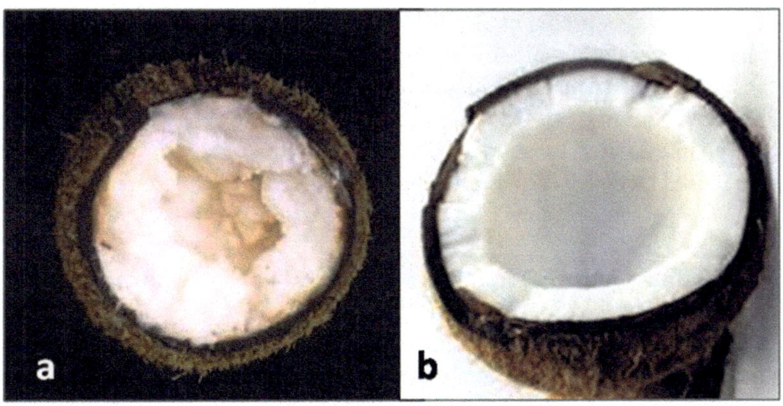

Fig. 9.1 Halved macapuno nut showing (**a**) the distinct endosperm and (**b**) normal nut

aldehyde dehydrogenase (*AMADH*) based on the gene family it belongs to (Arikit et al. 2011). A rice aromatic gene ortholog was identified in coconut (*CnAMADH2*) by Saensuk et al. (2016). Coding sequence comparisons of *CnAMADH2* in both non-aromatic and aromatic palms revealed a single nucleotide change from guanine (G) to cytosine (C), respectively (Saensuk et al. 2016), in the two accessions. The substitution was observed in the 14th exon causing a nonsynonymous amino acid change from alanine (A) to proline (P) at the 442nd position of CnAMADH2. However, this single nucleotide variation was not present in the aromatic yellow dwarf coconut, thus suggesting the presence of an alternative allele of *AMADH* in the yellow variety. Characterization of *CnAMADH2* in yellow dwarf coconut identified a single nucleotide change from G to A in the 9th exon causing an amino acid change in the highly conserved region of the ALDH cysteine (C) active site (Dumhai et al. 2019). Three-dimensional structural modeling outcome revealed structural changes due to the mutation which could reduce *AMADH* activity, thus increasing 2AP. Functional PCR markers that are capable of distinguishing between the aromatic and non-aromatic alleles were developed and validated based on the sequence variation found in yellow dwarf and green dwarf aromatic varieties.

The plant disease resistance (R) genes encoding proteins are characterized by the presence of the cytoplasmic nucleotide-binding site and leucine-rich repeat (NBS-LRR) domains. Recognizing pathogens and activating signal transduction pathways to induce disease resistance are the putative functions of NBS-LRR class genes. In understanding plant-pathogen interactions, the isolation and characterization of the NBS-LRR-encoding genes are of significance for effective disease management. Puch-Hau et al. (2015) isolated cDNAs encoding NBS-type RGAs from coconut palms resistant and susceptible to the devastating lethal yellowing disease and characterized the same. Based on the bioinformatic analysis, 143 sequences which show high homology to NBS-type RGA were reported. Seven different clades included in the non-TIRNBS-LRR subclass were reported based on the phylogenetic analysis. The transcript accumulation pattern of some NBS-type RGAs has been checked in the presence of exogenous salicylic acid (SA), and an increased expression in response to SA has been observed for the majority of the genes tested. As an extension of this study, Puch-Hau et al. (2016) used coconut NBS-type RGAs to explore the evolutionary relationship of these genes in different species of Arecaceae, Poaceae, and Brassicaceae families. Species-specific evolution of resistance gene candidates in coconut palm was revealed in this study. Rachana et al. (2016) amplified and characterized RGAs from Chowghat Green Dwarf using the primers designed on the NBS-LRR domain conserved motifs of the date palm.

9.4.1.3 Database and Genomic Resources Available for Coconut Functional Studies

Genomic resources and data repositories available for coconut studies are relatively limited, and a few online genomic resources are available for coconut and related palm species (Table 9.3). Moreover, open-source online repositories and

Table 9.3 A list of web-based genomic resources available for coconut

Source	URL
The International Coconut Genetic Resources Network	http://www.inibap.org/cogent/
Center for International Cooperation in Agricultural Research for Development (CIRAD) Montpellier France	http://www.cirad.fr
The European Palm Society	http://www.palmsociety.org.uk/
International Coconut Community	http://www.apccsec.org/
Palms of the World Online (Palmweb)	http://www.palmweb.org/
TropGeneDB	http://tropgenedb.cirad.fr/
ICAR-Central Plantation Crops Research Institute India	http://www.cpcri.gov.in

bioinformatic tools have contributed greatly to the rapid growth of omics-based improvements in coconut.

9.4.2 *Transcriptomics*

Even though gene isolation and characterization of individual genes were initiated in coconut, still there are a lot of knowledge gaps in many biological processes. Each plant has its unique genes or processes regulated during its developmental process. To mine "new" or "unique" genes, a global gene sequencing method is recommended; thus, transcriptome approaches are ideal. The complete set of mRNA and non-coding RNA (ncRNA) transcripts produced by a cell or an organism at a given time is collectively called as transcriptomes. Transcriptome profiling or transcriptome analysis is not the assessment of the expression level of a single or few genes but the study of a large set of genes, often an entire genome, simultaneously. This technique has been used in diverse areas of plant and animal research to obtain a global snapshot of the complete mRNA component of the cell at a given time. Without the construction of clonal libraries, direct sequencing of cDNA fragments and characterization are now facilitated due to recent advances in high-throughput sequencing systems integrated with "omics" resources. High-throughput DNA sequencing has been used in *C. nucifera* to provide novel insights into the molecular basis of different biological events and distinguish between sample pairs at different environmental conditions or disease conditions. Moreover, molecular characterization of tissues at different developmental stages has also been carried out. Since the first coconut transcriptome data analysis was recorded some 7 years ago (Fan et al. 2013), available public sequences have increased slowly compared to other major crops.

Coconut cultivations worldwide are affected by a variety of diseases. Enhancing disease resistance in coconut is one of the major goals coconut breeders are trying to achieve. Host-pathogen interactions associated with root wilt disease have been carried out through a transcriptomic study, and this is the first recorded transcriptome analysis in coconut (Rajesh et al. 2017). A total of 59,282 transcripts have been assembled from healthy and diseased samples of Chowghat Green Dwarf (CGD) palm upon RNA-Seq analysis utilizing an Illumina HiSeq 2000 platform. Similarity searches revealed that, out of 59,282 transcripts, 39,665 transcripts had significant hits in either UniProt or date palm proteome databases (Rajesh et al. 2017). Transcripts that were differentially expressed (2718 in number) in diseased samples could be assembled into pathways such as plant-pathogen interactions, cell wall formation, primary and secondary metabolism, hormone biosynthesis, and cellular transport. Verma et al. (2017) reported the first relational transcriptome database for Coconut Root Wilt Disease (CnTDB) which is available on http://webtom.cabgrid.res.in/cntdb/. CnTDB has been developed based on three-tier architecture and contains 285,235 transcripts with all blast information and annotations. CnTDB is considered the first transcriptome database for root wilt disease which is equipped with genic SSRs, SNPs, and indels which could be used as functional markers in future research.

In an attempt to identify disease resistance (R) genes in coconut, 243 resistance gene analog sequences (RGA) were recorded (Rajesh et al. 2015) based on comprehensive bioinformatics analysis of transcriptome data generated through RNA-Seq. These 243 RGAs were grouped into 6 classes based on conserved motif and domain predictions. R genes that are based on leucine-rich repeat (NBS-LRR) class have been clustered into a separate clade in the phylogenetic analysis, due to the presence of two motifs (TIR or CC) in their N-terminal regions (Rajesh et al. 2015).

Aside from the study of host-pathogen interactions, the hunt for potential genes regulating the embryogenesis is another major area studied in coconut using transcriptomic analysis. EST approach for gene identification was particularly useful a decade ago since there were very few coconut sequences available in GenBank. Bandupriya and Dunwell (2015) compared four Expressed Sequence Tag (EST) libraries prepared from the immature embryo, mature embryo, microspore-derived embryo, and mature leaves to recognize embryogenesis-related genes through an abundance analysis. A total of 32,621 putative unigenes were recorded in all 4 libraries, out of which 18,651 had significant sequence similarities to the non-redundant protein database. Of these unigenes, 6694 could be mapped into 139 Kyoto Encyclopedia of Genes and Genomes (KEGG) pathway most of which were included in carbohydrate metabolism, energy metabolism, lipid metabolism, amino acid metabolism, and nucleotide metabolism. The comparison of transcriptomes among four libraries allowed the identification of highly expressed genes such as chitinase, β-1,3-glucanase, ATP synthase CF0 subunit, thaumatin-like protein, and metallothionein-like protein among the transcriptomes generated in embryo-specific tissues compared to leaf tissue.

Recently, Rajesh et al. (2016) reported on the generation of more than 40,000 coconut transcriptome from coconut embryogenic calli using Illumina paired-end sequencing. The study was mainly aimed to identify the molecular events engaged in the somatic embryogenesis pathway as a means of overcoming the constraints associated with coconut somatic embryogenesis. Upon identification of 14 putative genes engaged in somatic embryogenesis, a quantitative real-time PCR (qRT-PCR) analysis was performed using six developmental stages. This study reported that not all the embryogenesis-related genes are expressed at ones in a particular development stage, but they are differentially expressed in different developmental stages. *CLAVATA 1 (CLV1)* was upregulated in initial callogenesis stages, while *SERK*, *mitogen-activated protein kinase* (*MAPK*), *AP2*, *SAUR* family protein (SAUR), embryogenic cell protein (ECP), arabinogalactan protein (AGP), late embryo genesis-abundant protein (LEA), and *AINTEGUMENTA* (*ANT*) were abundantly expressed in the embryogenic callus. Germin-like protein (GLP), glutathione S-transferase (GST), PICKLE (PKL), WUSCHEL (WUS), and WRKY were highly expressed in somatic embryos than in other stages (Rajesh et al. 2016). In addition, small RNA profiles of coconut embryogenic calli (EC) and non-embryogenic calli (NEC) have been generated and analyzed using Illumina Hiseq 2000 platform (Sabana et al. 2020). A total of 110 conserved and 97 novel miRNAs have been identified in the two types of calli. Ten members of the conserved group miRNAs showed differential expression between the two callus types. Among the novel miRNAs, 48 were specific to EC, while 21 were specifically found in NEC. Twenty-eight miRNAs were shared between the two libraries. Such results thus provide an insight into coconut SE at a molecular level, and the information provided could be used to develop and characterize novel molecular markers for SE.

In order to gain insight on the physiological and molecular basis of coconut endosperm development, differentially expressed genes were identified during coconut endosperm development from four suppression subtractive hybridization (SSH) libraries (Liang et al. 2015).

Important information on gene functioning and their involvement in regulatory networks mainly during coconut fruit development were disclosed. Out of the 1272 total clones, 63% showed similarity to known proteins. KEGG pathway analysis revealed that genes such as LPAAT, phospholipase D, acetyl-CoA carboxylase carboxyltransferase beta subunit, 3-hydroxyisobutyryl-CoA hydrolase-like, and pyruvate dehydrogenase E1 β subunit are responsible for fatty acid biosynthesis. In addition, triosephosphate isomerase, cellulose synthase, and glucan 1,3-β-glucosidase were identified as genes related to carbohydrate metabolism. Furthermore, 32 continuously upregulated and 7 downregulated genes have been identified during endosperm development (Liang et al. 2015).

The genome assembly of 'Catigan Green Dwarf' (CATD) coconut, constructed based on the 15X PacBio® and 50X Illumina® MiSeq sequence data (Lantican et al. 2019), has recently been used for genome sequence identification and structure prediction of six MCFA synthesis-related genes using bioinformatics tools (Manohar et al. 2019). Selection of these putative gene candidates [acyl-ACP thioesterase A and B (FatA and FatB); β-ketoacyl-ACP synthases I, II, and III (KasI, KasII, and KasIII); and LPAAT] for this study was based on prior knowledge of the endosperm oil biosynthesis pathway (Bates et al. 2013). KasII and KasIII MCFA genes which have not been reported in coconut before were characterized. Moreover, gene-specific markers based on the coding regions of MCFA genes successfully perceived genetic diversity of 48 coconut accessions.

Fan et al. (2013) reported a collection of 57,304 unigenes (after assembling of the RNA-Seq data) from young leaves and endosperm of Hainan Tall cultivar. In total, 23,168 unigenes were successfully mapped into 215 KEGG pathways. Among these unigenes, 347 were associated with fatty acid synthesis and metabolism. These unigenes were successfully assigned to different steps (fatty acid biosynthesis, unsaturated fatty acid, citrate cycle, fatty acid metabolism, and fatty acid elongation) in the fatty acid biosynthesis pathway. Similar approaches on the characterization of the transcriptome of the Laguna Tall coconut variety [(79,263 transcripts) Punzalan et al. 2019] and a fragrant dwarf coconut variety [(152,809 transcripts) Huang et al. 2014] were reported recently. Subsequently, based on Huang et al.'s (2014) study, the factors necessary for the RNA-directed DNA methylation have been identified in 10 Gb of collected sequence data. The information generated here provided insights on the small RNA-mediated epigenetic control of endosperm maturation during coconut seed development. Similarly, comparative transcriptome analysis between 7-month-old endosperms of Aromatic and non-Aromatic Green Dwarf coconuts led to the identification of gene(s) responsible for the "pandan"-like aroma (Saensuk et al. 2016).

Armero et al. (2017) tried to use oil palm as a reference species for the investigation of comparative genomics in coconut. Improvement of transcriptomic data for the availability of proteomic information which could be useful as a starting point for functional studies evolutionary queries was the objective of the study. A total of 29,366 proteins were generated by this method, and 1246 out of these were derived from new contigs. The functional profile of coconut proteome was identified as similar to rice, while several metabolic pathways were related to secondary metabolism and biotic stress.

9.4.3 *Sequencing of the Coconut Genome: Genome Size Estimation to Whole-Genome Sequencing*

Genetic improvement of a crop species is blessed with the availability of genome sequence information. Being a non-model crop, coconut was not significant enough to enter in the first round of sequencing projects based on Sanger sequencing. Nevertheless, relatively cheap NGS technologies assisted the researchers to unravel the palm genomes such as oil palm, date palm, and coconut (Al-Dous et al. 2016; Al-Mssallem et al. 2013; Singh et al. 2013; Xiao et al. 2019). Initial studies on genome size and ploidy levels of coconut were crucial for the progress of coconut genome sequencing (Gunn et al. 2015).

9.4.3.1 Estimation of Coconut Nuclear DNA Content

The understanding of the DNA content in absolute units (genome size) is a prerequisite in many research areas ranging from evolutionary studies to genome mapping (Galbraith et al. 1997). The genome size of an organism is normally linked and with growth and development and adaptation to environmental changes. Variations in genome size among related species are important in evolutionary studies since these changes may characterize species into different clades depending on the characters such as polyploidy, deletions, rearrangement, and transposon amplifications. In an attempt to estimate the DNA content in coconut using a flow cytometer analysis, the average nuclear DNA content was estimated at 5.6 ± 0.2 pg DNA/2C ($\simeq 5.4 \times 109$ bp) (Sandoval et al. 2003). Different tissues such as immature leaves, plumules extracted from zygotic embryos, and calli were used in the study. Gunn et al. (2015) predicted that the average genome size of coconut would be around 5.96 pg based on a study, done on flow cytometric analysis of 23 coconut varieties which included a mixture of coconut tall, dwarf, and hybrid cultivars. The coconut genome was identified as a large genome that shows intraspecific variations linked to domestication. Significant variations were recorded among tall varieties than that of dwarf varieties (Gunn et al. 2015). A study done by Neto et al. (2016) using 14 coconut accessions, which comprises eight tall and six dwarf cultivars, recorded varying 2C DNA contents ranging from 5.72 to 5.48 pg for talls and from 5.58 to 5.52 pg for dwarfs. The average DNA content of two types of coconut was recorded as 5.59 and 5.55 pg (tall and dwarf accessions, respectively).

9.4.3.2 Whole-Genome Sequencing

In 2017, the first draft genome of coconut (cultivar Hainan Tall) was successfully sequenced using the Illumina HiSeq 2000 platform (Xiao et al. 2017). A total of 419.67 Gb raw reads were generated covering the predicted 2.42 Gb *C. nucifera* genome. The scaffold length generated from the study was 2.2 Gb (N50 = 418 Kb) which covers more than 90% of the estimated genome size. Moreover, evaluation using Benchmarking Universal Single Copy Ortholog (BUSCO) tool showed that 90.8% of the coconut genome was covered by the obtained scaffold sequences and that the genome annotation was complete at 74.1%. The genome contains 28,039 protein-coding genes, and the number is less than that of recorded for palm species such as *Phoenix dactylifera* (PDK30: 28,889), *P. dactylifera* (DPV01: 41,660), and *Elaeis guineensis* (EG5: 34,802). The genome annotation results showed that the major fraction (72.75%) of the genome is composed of transposable elements, of which the largest proportion (92.23%) was the long-terminal repeat elements (LTRs). Analysis of the Bayesian molecular clock showed that the time of divergence between coconut and oil palm was about 46.0 (25.4–83.3) million years ago (Xiao et al. 2017). *C. nucifera* and *E. guineensis* have more recent divergence time than *C. nucifera* and *P. dactylifera* indicating a closer relationship between *C. nucifera* and *E. guineensis*.

The draft genome of a dwarf coconut variety, 'Catigan Green Dwarf' (CATD), developed using Illumina Miseq, PacBio SMRT, and Dovetail Chicago sequencing technologies, was reported recently (Lantican et al. 2019). Even though there were reports on draft genome assemblies of dwarf varieties, Catigan Green Dwarf's draft genome is the first published dwarf genome in coconut. The estimated genome size was 2.15 Gbp, and the estimation was based on the homozygous k-mer peak of the preprocessed Illumina Miseq short reads. Once the genome was enhanced through Chicago sequencing, 2.1 Gbp total genome size resulted. The final assembly consisted of 7998 scaffolds with N50 being 570,487 bp. BUSCO analysis showed that the genome contains 91.8% complete and 3% single-copy orthologs providing evidence to support completeness of the assembled genome. According to the genome statistics reported by Lantican et al. (2019), 34,958 protein-coding genes are present in the coconut genome. These include many genes related to significant traits such as oil biosynthesis, pest/disease resistance, drought tolerance, and other transcriptional regulators.

The alignment of the assembled dwarf coconut genome against the tall genome showed a high degree of synteny as expected (Lantican et al. 2019). Nevertheless, findings on duplicating and widely distributing nature of syntenic orthologs and out-paralog suggest potential duplications of the chromosomal segments, events in transposition, and other genomic modifications in the two genomes. The estimated genome sizes of the two published coconut genomes are higher than the genome sizes of *P. dactylifera* and *E. guineensis* which reported as 660 Mpb (Al-Mssallem et al. 2013) and 1.8 Gpb (Singh et al. 2013), respectively. However, the genome size predicted for Catigan Green Dwarf is considerably lower (by 280 Mbp) than that of Hainan Tall (HAT) coconut variety (Xiao et al. 2019) directing to controversial debate on the evolution of "Dwarf" and "Tall" type coconuts. Variations such as single nucleotide polymorphisms (SNPs) and insertions/deletions (InDels) among the two reference genomes of coconut (Dwarf and Tall) have been characterized as an initiative for developing SNP genotyping arrays and other possible genotyping techniques. Out of 58,503 variants detected in the non-repeat regions between the two genomes, 57,872 (nearly 99%) are SNPs. Major proportion (69.5%) of these SNPs are categorized as single-base transitions (A/G, C/T), while 30.5% are transversion substitutions (A/C, A/T, G/C, T/G). SNP frequency reported in the non-repeat region of the genome was one SNP in every 7881 bp sequence. Six hundred thirty-one InDels are recorded in the coconut genome, categorizing majority (62.12%) into single-base InDels. Next abundant Indel is the di-nucleotide InDels (20.29%).

Genomic resources available as two draft genomes, one for a tall variety and the second for a dwarf variety, will be a rich resource for coconut breeders and geneticists providing immense knowledge to facilitate the genetics and genomics programs. Furthermore, the release of coconut genome sequences will provide the way for rapid identification and functional and structural characterization of many gene families in coconut, through in silico computational studies and expression analysis. This will ultimately assist molecular breeding in *C. nucifera* and accelerate the breeding process towards the development of stress-resilient high-yielding superior coconut varieties.

9.5 Coconut Organelle Genomics

Mitochondrial and plastid genomes are considered important even though they represent only less than 1% of the genes which we see in the nucleus. Organelle genomes encompass maternal inheritance and have high copy numbers in cells. Some essential genes of important pathways such as respiration, energy production, and photosynthesis are encoded in the organelle genomes. According to the reports, agronomically important cytoplasmic male sterility genes are also located in the organelle genomes. With the advancement of high-throughput sequencing technologies, the sequencing of organelle genomes has been facilitated. The information generated on these organelle genomes is of interest to develop plants adapted to adverse environmental conditions, tracking the evolutionary developments, and utilizing them in DNA barcoding studies (Daniell et al. 2016).

9.5.1 Coconut Chloroplast Genome

Huang et al. 2013 reported the first complete chloroplast (cp) genome of dwarf coconut. The coconut cp genome is 154,731 bp in length and considered the smallest cp genome so far recorded for palm species. The coconut cp genome, as recorded for most angiosperms, has the conserved structure and content organization which is denoted by having a single spherical, quadripartite molecule. Similar to what has been recorded for land plants, a pair of inverted repeat regions (IR; 53,110 bp) are separated by a large single copy region (LSC; 84,230 bp) and a small single copy region (SSC; 17,391 bp). The genome contains 84 protein-coding genes, 38 tRNA genes, and 8 rRNA genes which constitute a total gene number of 130. Moreover, four pseudogenes named as pseudo ycf1, rps19, and two copies of ycf15 are also represented by the coconut cp genome. Among these 134 genes, 3 protein-coding genes (ycf2, ndhB, and rps7), 4 rRNA genes (rrn16, rrn23, rrn4.5, and rrn5), and 8 tRNA genes are present in two copies (Huang et al. 2013). Even though general gene configuration, gene content, and repeat structure in coconut cp genome are similar to the cp DNA observed in other palm species, there are some unique features unique to coconut. A remarkable increase of RNA editing sites and pseudogenization of rps19-like genes are a few of those unique characters.

According to a comparison done on two major economically important woody palms, oil palm and date palm of the family Arecaceae, the coconut cp genome has the shortest IRs (53,110 bp) (Huang et al. (2013). All these palms comprise AT-rich cp genomes. Phylogenetic analysis based on 47 chloroplast protein-coding genes of different plant taxa strongly supported the position of *C. nucifera* as the closely related species of *E. guineensis* in monocot subtrees (Huang et al. 2013).

9.5.2 *Coconut Mitochondrial Genome*

Unlike the conserved structures of cp genomes, mitochondrial genomes are unique to each plant in their respective genome size, complex structures, and various processes of RNA editing (Bi et al. 2016). Complete coconut mitochondrial (mt) genome assembly of Oman Local Tall coconut cultivar was reported based on the Roche/454 GSF LX data subjected to a homopolymer correction using the Illumina reads (Aljohi et al. 2016). The genome contains 87 protein-coding genes which encode for 72 proteins (8.62% of mt genome), 9 pseudogenes (0.83% of mt genome), 23 tRNAs (0.46% of mt genome), and 3 ribosomal RNAs (1.51% of mt genome) which constitute a gene content of 11.43% (77,542 bp). The regions derived from chloroplast accounted for 5.07% of the total assembly and included 13 proteins, 2 pseudogenes, and 11 tRNAs. The cp-derived genome sequence length recorded for coconut is opposed to the date palm mt genome where 93.5% of the genome sequence is comprised of cp (chloroplast)-derived regions (Fang et al. 2012). The GC content of the coconut mt genome is 45.5%. Sequence variation analysis shows that the mt genome Ti/Tv ratio is lower than that of nuclear genome.

9.6 Genetic Transformation

Not much effort has been dedicated to improving coconut varieties through genetic transformation. Only preliminary studies (Samosir et al. 1999; Andrade-Torres et al. 2011) have been conducted so far. Unavailability of successful, consistent in vitro plant regeneration protocol mainly hampered the research on genetic transformation. Moreover, the uncertainty of the future of genetically modified (GM) coconut due to the controversial issues existing on the consumer acceptance of GM food in many parts of the world is another bottleneck which hindered the GM research in coconut even though it serves as a precious tool in coconut improvement. Rapid alteration of genetic information for important traits (e.g., disease and pest resistance, tolerance to abiotic stress, boosting the yield levels) is possible through novel genome editing (e.g., CRISPR/Cas9) technology even though considering these technologies under the purview of GM is still uncertain.

9.7 Conclusion

Genetic improvement for desirable traits is identified as vital for improving the productivity of coconut worldwide. Biotechnological tools are immensely important to overcome the shortcomings in traditional approaches. Future research should, therefore, focus on developing improved coconut varieties that can tolerate changing climates, biotic and abiotic stresses, and other environmental effects and at the same time meet the stringent quality criteria insisted on the coconut industry. Availability of the whole genome will be of great use for future research; however, fine attention to eliminate the repeated sequences occurring in the genome is needed. Molecular selection and breeding will be benefited from the continuous enhancement in the availability and accessibility of high-throughput sequencing data that would enable long-term sustainability of coconut production.

References

Al-Dous EK, George B, Al-Mahmoud ME, Al-Jaber MY, Wang H, Salameh YM et al (2016) De novo genome sequencing and comparative genomics of date palm (*Phoenix dactylifera*). Nat Biotechnol 29:521–527

Aljohi HA, Liu W, Lin Q, Zhao Y, Zeng J, Alamer A et al (2016) Complete sequence and analysis of coconut palm (*Cocos nucifera*) mitochondrial genome. PLoS One 11(10):e0163990

Al-Mssallem IS, Hu S, Zhang X, Lin Q, Liu W, Tan J et al (2013) Genome sequence of the date palm Phoenix dactylifera L. Nat Commun 4:2274

Andrade-Torres A, Oropeza C, Sáenz L, González-Estrada T, Ramírez-Benítez JE, Becerril K, Chan JL, Rodríguez-Zapata LC (2011) Transient genetic transformation of embryogenic callus of Cocos nucifera L. Biologia 66(5):790–800

Arikit S, Yoshihashi T, Wanchana S, Uyen TT, Huong NT, Wongpornchai S, Vanavichit A (2011) Deficiency in the amino aldehyde dehydrogenase encoded byGmAMADH2, the homologue of rice Os2AP, enhances 2-acetyl-1-pyrroline biosynthesis in soybeans (*Glycine max* L.). Plant Biotechnol J 9:75–87

Armero A, Baudouin L, Bocs S, This D (2017) Improving transcriptome de novo assembly by using a reference genome of a related species: translational genomics from oil palm to coconut. PLoS One 12(3):e0173300

Ashburner GR, Thompson WK, Halloran GM (1997) RAPD analysis of South Pacific coconut palm populations. Crop Sci 37:992–997

Bandupriya HD, Gibbings JG, Dunwell JM (2013) Isolation and characterization of an *AINTEGUMENTA-like* gene in different coconut (*Cocos nucifera* L.) varieties from Sri Lanka. Tree Genet Genomes 9:813–827

Bandupriya HD, Gibbings JG, Dunwell JM (2014) Over expression of coconut *AINTEGUMENTA-like* gene, CnANT, promotes *in vitro* regeneration in transgenic Arabidopsis. Plant Cell Tissue Org Cult 116(1):67–79

Bandupriya HD, Dunwell JM (2015) Transcriptome analysis for discovering candidate genes involved in embryogenesis in coconut (*Cocos nucifera* L.) through 454 pyrosequencing. J Natl Sci Found Sri Lanka 43(4):319–336

Banzon JA (1997) Industrial coconut research and development. Philipp J Coconut Stud 2:13–17

Bates PD, Stymne S, Ohlrogge J (2013) Biochemical pathways in seed oil synthesis. Curr Opin Plant Biol 16:358–364

Baudouin L, Lebrun P (2002) The development of a microsatellite kit and dedicated software use with coconuts. Burotrop Bull 17:16–20

Baudouin L, Lebrun P, Konan JL, Ritter E, Berger A, Billotte N (2006) QTL analysis of fruit components in the progeny of a Rennell Island tall coconut (*Cocos nucifera* L.) individual. Theor Appl Genet 112:258–268

Bhavyashree U, Lakshmi KJ, Rachana KE, Muralikrishna KS, Sajini KK, Rajesh MK et al (2015) Maintenance of embryogenic potential of calli derived from shoot meristem of West Coast Tall cv. of coconut (*Cocos nucifera* L.). J Plant Crop 43(2):105–116

Bhavyashree U, Rachana KE, Lakshmi KJ, Rajesh MK, Karun A (2016) Comparative gene expression profiling during *in vitro* regeneration in two coconut cultivars. Emirates J Food Agric 28(8):572–580

Bi C, Paterson AH, Wang X, Xu Y, Wu D, Qu Y et al (2016) Analysis of the complete mitochondrial genome sequence of the diploid cotton Gossypium raimondii by comparative genomics approaches. Biomed Res Int 2016:5040598

Bourdeix R, Santos G, Labouisse JP, Baudouin L (2005) Useful definition of terms and nomenclature. In: Batugal P, Ramanatha RV, Oliver J (eds) Coconut genetic resources. IPGRI, Rome, pp 9–10

Bradbury LM, Fitzgerald TL, Henry RJ, Jin Q, Waters DL (2005) The gene for fragrance in rice. Plant Biotechnol J 3:363–370

Cernac A, Benning C (2004) WRINKLED1 encodes an AP2/EREB domain protein involved in the control of storage compound biosynthesis in Arabidopsis. Plant J 40:575–585

Daher RF, Pereira MG, Tupinambá EA, Amaral AT Jr, Aragão WM, Ribeiro FE et al (2002) Assessment of coconut tree genetic divergence by compound sample RAPD marker analysis. Crop Breed Appl Biotechnol 2:431–438

Daniell H, Lin CS, Yu M, Chang WJ (2016) Chloroplast genomes: diversity, evolution, and applications in genetic engineering. Genome Biol 17:134

Dassanayaka PN, Everard JM, Karunanayake EH, Nandadasa HG (2003) Characterization of coconut germplasm by microsatellite markers. Trop Agric Res 15:51–61

Dayrit FM, Buenafe OE, Chainani ET, De Vera IM (2008) Analysis of monoglycerides, diglycerides, sterols, and free fatty acids in coconut (*Cocos nucifera* L.) oil by 31P NMR spectroscopy. J Agric Food Chem 56:5765–5769

Dumhai R, Wanchana S, Saensuk C, Choowongkomon K, Mahatheeranont S, Kraithong T et al (2019) Discovery of a novel *CnAMADH2* allele associated with higher levels of 2- acetyl-1-pyrroline (2AP) in yellow dwarf coconut (*Cocos nucifera* L.). Sci Hortic 243:490–497

Duran Y, Rohde W, Kullaya A, Goikoetxea P, Ritter E (1997) Molecular analysis of east African tall coconut genotypes by DNA marker technology. J Genet Breed 51:279–288

Everard JM (1996) Use of molecular markers for breeding of the coconut palm (*Cocos nucifera* L.). [dissertation]. University of New England, Armidale

Fan H, Xiao Y, Yang Y, Xia W, Mason AS, Xia Z et al (2013) RNA-Seq analysis of *Cocos nucifera*: transcriptome sequencing and de novo assembly for subsequent functional genomics approaches. PLoS One 8(3):e59997

Fang Y, Wu H, Zhang T, Yang M, Yin Y, Pan L et al (2012) A complete sequence and transcriptomic analyses of date palm (Phoenix dactylifera L.) mitochondrial genome. PLoS One 7(5):e37164

FAOSTAT (2018) http://www.fao.org/faostat/en/#data. Accessed 20 Jun 2020

Feng D, Gao L, Zheng Y, Li D, Zhou P (2020) Molecular cloning and function characterisation of a cytoplasmic fructose-1, 6-bisphosphate aldolase gene from coconut (*Cocos nucifera* L.). J Hortic Sci Biotechnol 95:1–9

Galbraith DW, Lambert GM, Macas J, Dolezel J (1997) Analysis of nuclear DNA content and ploidy in higher plants. Curr Protoc Cytom 2(1):7–6

Gao L, Sun R, Liang Y, Zhang M, Zheng Y, Li D (2014) Cloning and functional expression of a cDNA encoding stearoyl-ACP Δ9-desaturase from the endosperm of coconut (*Cocos nucifera* L.). Gene 549:70–76

Gunn BF, Baudouin L, Olsen KM (2011) Independent origins of cultivated coconut (*Cocos nucifera* L.) in the old-world tropics. PLoS One 6(6):e21143

Gunn BF, Baudouin L, Beulé T, Ilbert P, Duperray C, Crisp M et al (2015) Ploidy and domestication are associated with genome size variation in palms. Am J Bot 102(10):1625–1633

Herran A, Estioko L, Becker D, Rodriguez MJ, Rohde W, Ritter E (2000) Linkage mapping and QTL analysis in coconut (*Cocos nucifera* L.). Theor Appl Genet 101:292–300

Huang YY, Matzke AJM, Matzke M (2013) Complete sequence and comparative analysis of the chloroplast genome of coconut palm (*Cocos nucifera*). PLoS One 8(8):e74736

Huang YY, Lee CP, Fu JL, Chang BCH, Matzke AJ, Matzke M (2014) De novo transcriptome sequence assembly from coconut leaves and seeds with a focus on factors involved in RNA directed DNA methylation. G3 4:2147–2157

Islam M, Cedo ML, Namuco LO, Borromeo TH, Aguilar EA (2009) Effect of fruit age on endosperm type and embryo germination of Makapuno coconut. Gene Conserve 8:708–722

Kamaral LC, Dassanayaka PN, Perera KL, Perera SA (2016) SSR markers reveal the population structure of yellow (dwarf) coconuts in Sri Lanka. Tree Genet Genomes 12:116

Kamaral LC, Perera SA, Perera KL, Dassanayaka PN (2017) Characterization of Sri Lanka yellow dwarf coconut (*Cocos nucifera* L.) by DNA fingerprinting with SSR markers. J Natl Sci Found Sri Lanka 45(4):405–412

Kearsey MJ, Luo ZW (2003) Mapping, characterization and deployment of quantitative trait loci. In: Plant molecular breeding. Blackwell Publishing Ltd, Oxford, pp 1–29

Knutzon DS, Lardizabal KD, Nelsen JS, Bleibaum JL, Davies HM, Metz JG (1995) Cloning of a coconut endosperm cDNA encoding a 1-acyl-sn-glycerol-3 phosphate acyltransferase that accepts medium-chain- length substrates. Plant Physiol 109:999–1006

Konan KJ, Konan KJ, Koffi KE, Lebrun P, Sangaré A (2007) Coconut microsatellite gene diversity analysis technology transfer to Cote d' Ivoire. Biotechnology 6(3):383–387

Kriswiyanti E, Temaja IG, Sudana IM, Wirya IG (2013) Genetic variation of coconut tall (*Cocos nucifera* L., Arecaceae) in Bali Indonesia based on microsatellite DNA. J Biol Agric Healthc 3(13):97–101

Lantican DV, Strickler SR, Canama AO, Gardoce RR, Mueller LA, Galvez HF (2019) De novo genome sequence assembly of dwarf coconut (*Cocos nucifera* L. 'Catigan green dwarf') provides insights into genomic variation between coconut types and related palm species. G3 9:6

Lebrun P, N'Cho YP, Seguin M, Grivet L, Baudouin L (1998) Genetic diversity in the coconut (*Cocos nucifera* L.) revealed by restriction fragment length polymorphism (RFLP) markers. Euphytica 101:103–108

Lebrun P, Baudouin L, Bourdeix R, Louis KJ, Barker JH, Aldam C et al (2001) Construction of a linkage map of the Rennell Island tall coconut type (*Cocos nucifera* L.) and QTL analysis for yield characters. Genome 44:962–970

Li DD, Fan YM (2009) Cloning, characterization and expression analysis of an oleosin gene in coconut (*Cocos nucifera* L.) pulp. J Hortic Sci Biotechnol 84:483–488

Liang Y, Yuan Y, Liu T, Mao W, Zheng Y, Li D (2015) Identification and computational annotation of genes differentially expressed in pulp development of *Cocos nucifera* L. by suppression subtractive hybridization. BMC Plant Biol 14:205

Lindqvist Y, Huang W, Schneider G, Shanklin J (1996) Crystal structure of D9 stearoyl-acyl carrier protein desaturase from castor seed and its relationship to other di-iron proteins. EMBO J 15:4081–4092

Liyanage DV (1958) Varieties and forms of coconut palms grown in Ceylon. Ceylon Cocon Q 9:1–10

Manohar AN, Lantican DV, Dancel MP, Cardona DE, Ibarra AC, Gulay CR et al (2019) Genome-guided molecular characterization of oil genes in coconut (*Cocos nucifera* L.). Philipp J Sci 148(S1):153–164

Marikkar JM, Jayasundara JM, Prasadika SA, Jayasingha CV, Premakumara GA (2007) Assessment of stability of virgin coconut oil during deep frying. CORD 23(1):62–70

Marina AM, Che Man YB, Nazimah SAH, Amin I (2009) Chemical properties of virgin coconut oil. J Am Oil Chem Soc 86:301–307

Meerow AW, Wisser RJ, Brown JS, Kuhn DN, Schnell RJ, Broschat TK (2003) Analysis of genetic diversity and population structure within Florida coconut (*Cocos nucifera* L.) germplasm using microsatellite DNA, with special emphasis on the Fiji dwarf cultivar. Theor Appl Genet 106:715–726

Montero-Cortés M, Rodríguez-Paredes F, Burgeff C, Pérez-Nuñez T, Córdova I, Oropeza C et al (2010) Characterisation of a cyclin-dependent kinase (CDKA) gene expressed during somatic embryogenesis of coconut palm. Plant Cell Tissue Organ Cult 102(2):251–258

Napier JA, Beaudoin F, Tatham AS, Alexander LG, Shewry PR (2001) The seed oleosins: structure, properties and biological role. Adv Bot Res 35:111–138

Neto MF, Pereira TN, Geronimo IG, Azevedo AO, Ramos SR, Pereira MG (2016) Coconut genome size determined by flow cytometry: tall versus dwarf types. Genet Mol Res 15:gmr7470

Orsasova J, Misurcova L, Ambrozova JV, Vicha R, Mlcek J (2015) Fatty acids composition of vegetable oils and its contribution to dietary energy intake and dependence of cardiovascular mortality on dietary intake of fatty acids. Int J Mol Sci 16(6):12871–12890

Perera SA (2010) QTL analysis in coconut via genome mapping: principles, requirements and prospects. Cocos 20:57–65

Perera SA, Kilian A (2008) Diversity arrays technology: a high throughput molecular marker system for coconut. J Inst Fundam Stud xix(1)(Special issue):60–64

Perera L, Russell JR, Provan J (1998) Evaluating genetic relationships between indigenous coconut (*Cocos nucifera* L.) accessions from Sri Lanka by means of AFLP profiling. Theor Appl Genet 96:545–550

Perera L, Russell JR, Provan J (1999) Identification and characterization of microsatellites in coconut (*Cocos nucifera* L.) and the analysis of coconut populations in Sri Lanka. Mol Ecol 8:344–346

Perera L, Russell JR, Provan J, Powell W (2000) Use of microsatellite DNA markers to investigate the level of genetic diversity and population genetic structure of coconut (*Cocos nucifera* L.). Genome 43:15–21

Perera L, Russell JR, Provan J, Powell W (2001) Levels and distribution of genetic diversity of coconut (*Cocos nucifera* L., var. Typica form typica) from Sri Lanka assessed by microsatellite markers. Euphytica 122:381–389

Perera L, Russell JR, Provan J, Powell W (2003) Studying the genetic relationships among coconut varieties/populations using microsatellite markers. Euphytica 132:121–128

Perez-Nunez MT, Souza R, Saenz L, Chan JL, Gonzalez T, Zuniga JJ et al (2009) Detection of a SERK-like gene in coconut *in vitro* cultures and analysis of its expression during the formation of embryogenic callus and somatic embryos. Plant Cell Rep 28:11–19

Phoeurk C, Somana J, Sornwatana T, Udompaisarn S, Traewachiwiphak S, Sirichaiyakul P et al (2018) Three novel mutations in α-galactosidase gene involving in galactomannan degradation in endosperm of curd coconut. Phytochemistry 156:33–42

Puch-Hau C, Oropeza-Salín C, Peraza-Echeverría S, Gongora-Paredes M, Córdova-Lara I, Narvaez-Cab M et al (2015) Molecular cloning and characterization of disease-resistance gene candidates of the nucleotide binding site (NBS) type from *Cocos nucifera* L. Physiol Mol Plant Pathol 89:87–96

Puch-Hau C, Oropeza C, Góngora-Paredes M, Córdova I, Tun-Suárez J, Sáenz L (2016) New insights into the evolutionary history of resistance gene candidates in coconut palms and their expression profiles in palms affected by lethal yellowing disease. Genes Genomics 38(9):793–807

Punzalan R, Bautista A, Emmanuel E, Rivera R, Rivera S, Saloma C (2019) Transcriptome of the traditional coconut variety Laguna tall. Philipp J Sci 148(S1):153–164

Rachana KE, Naganeeswaran SA, Fayas TP, Thomas RJ, Rajesh MK (2016) Cloning, characterization and expression analysis of NBS-LRR-type resistance gene analogues (RGAs) in coconut. Acta Bot Croat 75(1):1–10

Rajesh MK, Rachana KE, Naganeeswaran SA, Shafeeq R, Thomas RJ, Shareefa M et al (2015) Identification of expressed resistance gene analog sequences in coconut leaf transcriptome and their evolutionary analysis. Turk J Agric For 39:489–502

Rajesh MK, Fayas TP, Naganeeswaran S, Rachana KE, Bhavyashree U, Sajini KK et al (2016) *De novo* assembly and characterization of global transcriptome of coconut palm (*Cocos nucifera* L.) embryogenic calli using Illumina paired-end sequencing. Protoplasma 253(3):913–928

Rajesh MK, Rachana KE, Kulkarni K, Sahu BB, Thomas RJ, Karun A (2017) Comparative transcriptome profiling of healthy and diseased Chowghat green dwarf coconut palms from root (wilt) disease hot spots. Eur J Plant Pathol 151:1–21

Ratnambal MJ, Khumaran PM, Arunachala V, Niral V, Upadyay A, Parthasarathy VA (2001) Coconut genetic resources and molecular approaches. Indian J Plant Genet Resour 14(2):182–184

Reynolds KB, Cullerne DP, El Tahchy A, Rolland V, Blanchard CL, Wood CC, Singh SP, Petrie JR (2019) Identification of genes involved in lipid biosynthesis through de novo transcriptome assembly from *Cocos nucifera* developing endosperm. Plant Cell Physiol 60(5):945–960

Ritter E, Rodriguez MJ, Herrán A, Estioko L, Becker D, Rohde W (2000) Analysis of quantitative trait loci (QTL) based on linkage maps in coconut (*Cocos nucifera* L.). In: Arencibia A (ed) Plant genetic engineering towards the third millennium. Elsevier, Amsterdam, pp 42–48

Rivera R, Edwards KJ, Barker JH, Arnold GM, Ayad G, Hodgkin T et al (1999) Isolation and characterization of polymorphic microsatellites in *Cocos nucifera* L. Genome 42:668–675

Rohde W, Kullaya A, Rodriguez J, Ritter E (1995) Genome analysis of *Cocos nucifera* L. by PCR amplification of spacer sequences separating a subset of copia-like 16RI repetitive elements. J Genet Breed 49:179–186

Rohde W, Becker D, Kullaya A, Rodriguez MJ, Herrán A, Ritter E (1999) Analysis of coconut germplasm biodiversity by DNA marker technologies and construction of a first genetic linkage map. In: Oropeza C, Verdeil JL, Ashburner GR, Cardega R, Santamaria JM (eds) Current advances in coconut biotechnology. Kluwer Academic Publishers, Dordrecht, pp 99–120

Samosir YMS (1999) Optimisation of somatic embryogenesis in coconut (Cocos nucifera L.). PhD thesis, The University of Queensland, Australia

Sabana AA, Rajesh MK, Antony G (2020) Dynamic changes in the expression pattern of miRNAs and associated target genes during coconut somatic embryogenesis. Planta 251:79

Saensuk C, Wanchana S, Choowongkomon K, Wongpornchai S, Kraithong T, Imsabai W et al (2016) De novo transcriptome assembly and identification of the gene conferring a "pandan-like" aroma in coconut (*Cocos nucifera* L.). Plant Sci 252:324–334

Sandoval A, Hocher V, Verdeil JL (2003) Flow cytometric analysis of the cell cycle in different coconut palm (*Cocos nucifera* L.) tissues cultured *in vitro*. Plant Cell Rep 22:25–31

Sankaran M, Damodaran V, Singh DR, Jai SI, Jerard BA (2012) Characterization and diversity assessment in coconut collections of Pacific Ocean Islands and Nicobar Islands. Afr J Biotechnol 11(97):16320–16329

Singh R, Ong-Abdullah M, Low ET, Manaf MA, Rosli R, Nookiah R et al (2013) Oil palm genome sequence reveals divergence of interfertile species in old and new worlds. Nature 500:335–339

Sun R, Ye R, Gao L, Zhang L, Wang R, Mao T et al (2017) Characterization and ectopic expression of CoWRI1, an AP2/EREBP domain-containing transcription factor from coconut (*Cocos nucifera* L.) endosperm, changes the seeds oil content in transgenic *Arabidopsis thaliana* and Rice (*Oryza sativa* L.). Front Plant Sci 8:63

Teulat B, Aldam C, Trehin R (2000) An analysis of genetic diversity in coconut (*Cocos nucifera*) populations from across the geographic range using sequence-tagged microsatellites (SSRs) and AFLPs. Theor Appl Genet 100:764–771

Upadhyay A, Jose J, Manimekalai R, et al (2002) Managing plant genetic diversity. Proceedings of the International Conference; Kula Lumpur, Malaysia. pp 61–66

Vargas AG, Cabanos CS, Garcia RN, Maruyama N, Tecson Mendoza EM (2018) Cloning, molecular analysis, and developmental expression of 3 oleosin cDNA isoforms in coconut (*Cocos nucifera* L.). J Hortic Sci Biotechnol 93:1–9

Verma SK, Jasrotia RS, Iquebal MA, Jaiswal S, Angadi UB, Rai A et al (2017) Deciphering genes associated with root wilt disease of coconut and development of its transcriptomic database (CnTDB). Physiol Mol Plant Pathol 100:255–263

Xiao Y, Xu P, Fan H, Baudouin L, Xia W, Bocs S, Xu J, Li Q, Guo A, Zhou L, Li J, Wu Y, Ma Z, Armero A, Issali AE, Liu N, Peng M, Yang Y (2017) The genome draft of coconut (Cocos nucifera). Gigascience. 6:1–11

Xiao Y, Xia W, Mason AS, Cao Z, Fan H, Zhang B et al (2019) Genetic control of fatty acid composition in coconut (*Cocos nucifera*), African oil palm (*Elaeis guineensis*), and date palm (*Phoenix dactylifera*). Planta 249:333–350

Xu L, Ye R, Zheng Y, Wang Z, Zhou P, Lin Y et al (2010) Isolation of the endosperm-specific LPAAT gene promoter from coconut (*Cocos nucifera* L.) and its functional analysis in transgenic rice plants. Plant Cell Rep 29:1061–1068

Yuan Y, Chen Y, Yan S, Liang Y, Zheng Y, Dongdong L (2014) Molecular cloning and characterization of an acyl carrier protein thio-esterase gene (CocoFatB1) expressed in the endosperm of coconut (*Cocos nucifera*) and its heterologous expression in nicotiana tabacum to engineer the accumulation of different fatty acids. Funct Plant Biol 41:80–86

Yuan Y, Gao L, Sun R, Yu T, Liang Y, Li D et al (2017) Seed specific expression of an acyl–acyl carrier protein thioesterase CnFatB3 from coconut (*Cocos nucifera* L.) increases the accumulation of medium-chain fatty acids in transgenic Arabidopsis seeds. Sci Hortic 223:5–9

Zheng Y, Jin Y, Yuan Y, Feng D, Chen L, Li D et al (2019) Identification and function analysis of a type 2 diacylglycerol acyltransferase (DGAT2) from the endosperm of coconut (*Cocos nucifera* L.). Gene 702:75–82

Chapter 10
Complete Chloroplast Genome Sequences of Coconut cv. Kopyor Green Dwarf and Comparative Genome Analysis to Oil Palm, Date Palm, Sago Palm, and Miniature Sugar Palm

Annisa Rahmawati, Hugo Alfried Volkaert, Diny Dinarti, Ismail Maskromo, Andi Nadia Nurul Latifa Hatta, and Sudarsono Sudarsono

Contents

A. Rahmawati · D. Dinarti · Andi Nadia Nurul Latifa Hatta · S. Sudarsono (✉)
Plant Molecular Biology Lab., Department of Agronomy and Horticulture, Faculty of Agriculture, Bogor Agricultural University, Bogor, West Java, Indonesia
e-mail: sudarsono_agh@apps.ipb.ac.id

H. A. Volkaert
Center for Agricultural Biotechnology, Kasetsart University, Nakhorn Pathom, Thailand

I. Maskromo
Indonesian Palms Research Institute, Agency for Agricultural Research and Development, Manado, North Sulawesi, Indonesia

H. Tombuloglu et al. (eds.), *Oil Crop Genomics*,
https://doi.org/10.1007/978-3-030-70420-9_10

10.1 Introduction

Coconut (*Cocos nucifera* L.) is one of the essential oil-producing crops, and it is the only existing species belonging to the genus *Cocos* in the Arecaceae family. Coconuts are most abundant in the tropical islands, especially in the coastal regions. The proposed coconut centers of origin are Southeast Asia (Malaysia, Indonesia, and the Philippines) and the islands between the Indian and the Pacific Oceans. In those regions, coconut shows the highest genetic diversity (Purseglove 1968). The coconuts evolved and dispersed by floating in the coral atoll ecosystem and are brought to India and East Africa before they finally dispersed to the American continent and other tropical to subtropical regions (Harries and Clement 2014).

Humans can use almost all parts of the coconut palms. Hence, coconuts are known as "the tree of life" (Esquenazi et al. 2002; Lima et al. 2015). Generally, coconut is used as a food source, fibers, beverage, handicrafts, construction materials, and natural medicines. The phytochemical analysis of the shell fibers (mesocarps) indicated they contain a high polyphenol, such as catechins, epicatechins, tannins, and flavonoids.

Based on the botany and the reproduction patterns, coconuts are mainly categorized into two groups: tall and dwarf types. Tall coconuts are mostly cross-pollinating (allogamous) and genetically heterozygous, and they can grow up to 20–30 m in height. The dwarf coconuts are usually self-pollinating (autogamous) and genetically homozygous (Pan et al. 2017), and they can reach up to 10–15 m in height.

In Indonesia, there is a unique coconut with an abnormal endosperm – locally known as kopyor coconut. Kopyor coconut is a natural mutant having fluffy endosperm, and the endosperm is peeled off from the shell. The other coconut endosperm mutants also exist in other Asian countries, such as Macapuno (Philippines), Makhrao Kathi (Thailand), Dikiri Pol (Sri Lanka), and Thairu Thengai (India) (Novarianto et al. 2014). However, there is a clear difference in the mutant phenotype among kopyor coconut and the other reported Makapuno coconuts. In Makapuno coconuts, the abnormal endosperm is associated with the inactivation of a gene encoding α-galactosidase enzyme. The α-galactosidase enzyme function is associated with galactomannan or oligo-galactomannan accumulation (Phoeurk et al. 2018). However, reports have suggested that the gene encoding α-galactosidase enzyme in kopyor coconut is not a mutant gene. In kopyor coconut, the endosperm abnormality is the result of at least two loci. One of the recessive genes (the K locus) controlled the kopyor phenotype, and the other recessive genes control the quantity of endosperm (Larekeng et al. 2015; Setiawan et al. 2020). For the K locus, the genetic makeup of the locus controlling kopyor endosperm phenotype must be a kkk, and the genotype of the zygotic embryo of a kopyor nut is a kk. The kopyor character is a lethal mutant. Hence, the kopyor nuts could not germinate naturally because of the endosperm's inability to support normal embryo development. Such a phenomenon has posed a problem in providing kopyor coconut seedlings to farmers. Therefore, the kopyor nuts supply is still limited (Maskromo et al. 2015), and the development of more kopyor coconut estate is still necessary.

Three types of kopyor coconut exist in Indonesia: kopyor tall, kopyor dwarf, and kopyor hybrid. Kopyor Green Dwarf is officially released and registered as a national coconut variety (No. SK. Mentan No.3996/Kpts/SR.120/12/2010). Some of Kopyor Green Dwarf coconut characteristics are medium fruit size, green fruit color, approximately 51 cm polar circumference, 42 cm equatorial circumference, and 8.2–8.4% oil content. The main advantages of this variety are its early maturity, first harvest at 48 months after planting, approximately 1.1 kg in average fruit weight, approximately 120–140 of the number of fruits per tree or equal to 1400 fruits per hectare per year, and medium to high endosperm quantity. However, the genetic information of this coconut variety is still limited.

10.2 Chloroplast Genome in Genetic Studies

Chloroplasts are believed to be originated from cyanobacterium through endosymbiosis and evolution over time (Jensen and Leister 2014). Chloroplasts play an essential role in many functions of the plant cells. Chloroplast functions include solar energy capture for the photosynthetic processes, carbon fixation, and stress signal response (Ivanova et al. 2017). The most important responsibility of the plant chloroplasts is the photosynthetic conversion of CO_2 to carbohydrates.

As an organelle for photosynthesis apparatus in plants, chloroplasts have their genomes and genetic systems (Gan et al. 2019). In both angiosperm and gymnosperm, the chloroplast genome (cpDNA) has a conserved quadripartite structure consisting of a large single copy (LSC), a small single copy (SSC), and two identical inverted repeats (IR) copies, respectively. However, the cyanobacterium-derived chloroplast genome has reduced its size during evolution because of the losses and the transfer of genes or sequences into the nuclear genome (Nowack et al. 2016).

Generally, the chloroplast genome's characteristics include a circular double-stranded DNA molecule with size ranges from 120,000 to 160,000 base pairs (bp), a haploid genome, and multiple copies cpDNA in each plant cell. Moreover, the cpDNA could be as many as 1000 copies per cell (Birky 1995; Raubeson and Jansen 2005; Ebert and Peakall 2009; Uthaipaisanwong et al. 2012). The cpDNA contains a highly conserved gene content, gene structure, and a relatively low substitution rate. The cpDNA is uniparentally inherited. Such features of cpDNA indicated that the cpDNA is useful for plant population genetic evaluations, phylogenetic analysis, and evolutionary studies (Rogalski et al. 2015; Yan et al. 2018; Meng et al. 2019).

With the advancement of DNA sequencing technology such as next-generation sequencing (NGS), the whole-genome sequence of various plant species can be carried out (Wang et al. 2018; Lopes de Abreu et al. 2018). Since September 2017, chloroplast genomes of various plant species have been identified and accessed at GenBank Organelle Genome Resources (Nguyen et al. 2020). Some complete chloroplast genomes of important species in the family Arecaceae have been sequenced and published, such as *Elaeis guineensis* (Uthaipasanwong et al. 2012), *Phoenix dactylifera* (Yang et al. 2010; Khan et al. 2018), *Borassus flabellifer* (Sakulsathaporn et al. 2017), and *Cocos nucifera* (Huang et al. 2013; Aljohi et al. 2016). However, the complete chloroplast genome of kopyor coconut has yet been sequenced and characterized. In this study, we report the characterization of a complete chloroplast genome of kopyor coconut cv. Kopyor Green Dwarf and do some comparative genome analysis between Arecaceae species. The analysis will hopefully provide a better understanding of the genetic differences of Arecaceae species. The results of this analysis could be useful in the evolutionary study of palm species and the development of the kopyor coconut breeding program and its conservation strategy.

10.3 DNA Sequencing Technology for Chloroplast Genome Study

Finding the plant genome information can significantly accelerate breeding pipelines, especially to access the agronomically important traits based on molecular bases, such as yield and resistance to biotic and abiotic stresses. Chloroplast genomes offer a solution to the problematical doubt about engineering gene resistance that could escape through pollen dispersal. Pollen with maternal chloroplast inheritance may contain metabolically active chloroplasts. However, the chloroplast DNA is missing during pollen maturation; thus, it is not transmitted to the next generation (Daniell et al. 2005). Engineering foreign genes through chloroplast genome for resistance or tolerance purposes are successfully reported in several studies. Some transgene research utilizes the chloroplast genome to herbicides resistance (Daniell et al. 1998), insect resistance (Kota et al. 1999; De Cosa et al. 2001), disease resistance (DeGray et al. 2001), salt tolerance (Kumar et al. 2004), and drought tolerance (Lee et al. 2003). Chloroplast DNA sequence information is also a useful tool for plant identification or barcoding to detect authentication and establish genetic relationships among plant species.

Plant genome sequence data are generated using DNA sequencing technology. It has been approximately 40 years since discovering the pioneer method for DNA sequencing by Frederick Sanger and colleagues, with "Sanger sequencing" as the "first-generation" DNA sequencing method. The dideoxynucleotide chain-termination technique was developed by Sanger (Sanger et al. 1977). Subsequently, the robustness of the dideoxy chain-termination method has become the central approach for future DNA sequencing. The Sanger sequencing method is also used as validation of the next-generation technique results. However, it is only widely used for smaller-scale projects because it is effectively used for low numbers of targets and also produces reads slightly less than one kilobase (kb) in length. Because of the limitations, a discovery for DNA sequencing has developed, next-generation sequencing (NGS). NGS is divided into second-generation DNA sequencing (SGS) and third-generation sequencing (TGS) (Ambardar et al. 2016). NGS allows DNA and RNA sequencing faster and inexpensive than the Sanger sequencing, longer-reads length, high sample throughput, higher base pair, and mutation resolution data. It also provides multiple genetic loci analyses simultaneously from different samples (Beedanagari and John 2014; Hambuch and Mayfield 2014).

The common types of second-generation sequencing are GS FLX by 454 Life Sciences/Roche Diagnostics; Genome Analyzer, HiSeq, MiSeq, and NextSeq by Illumina, Inc.; SOLiD by ABI; and Ion Torrent by Life Technologies. The basic principle of second-generation sequencing systems is sequencing by synthesis (SBS) or sequencing by ligation (SBL). The three major SBS types are pyrosequencing, sequencing by reversible termination, and sequencing by detecting hydrogen ions. Pål Nyrén and colleagues introduced the pyrosequencing in 1993 and then commercialized it by 454 Life Sciences with technical improvement. The technique is observed on four-enzyme real-time monitoring of DNA synthesis by bioluminescence using a cascade that ends in a detectable light signal (bioluminescence). The detection system is based on the pyrophosphate released when a nucleotide is introduced in the DNA strand (Ronaghi et al. 1998; Ahmadian et al. 2005). Pyrosequencing requires the direct action of DNA polymerase to produce the observable output. The advantages of pyrosequencing are its capability to sequence reads in the read length of 700–1000 bp and organized using natural nucleotides (instead of the heavily modified dNTPs used in the chain-termination techniques) (Nyrén 1987). In contrast, its limitations are relatively expensive, the generation of relatively low bases/run (around 700 Mb), and the main difficulty is homopolymer repeats due to the simultaneous incorporation of the same nucleotide (Gupta and Gupta 2014).

Dr. Jingyue Ju first reported sequencing by reversible terminator technology from Colombia University (Li et al. 2003). Illumina platform is the most famous reversible termination technology with SBS strategies. The procedure of reversible termination sequencing starts with (1) library preparation, (2) immobilizing the sequencing templates and primers on a solid support, (3) primer extension by one base and termination, (4) identifying the color of the fluorophore carried by the extended base to identify the incorporated nucleotide after washing away the unincorporated nucleotides, (5) deletion of the fluorescent tag and the 3'-O blocking group, and (6) washing all chemical remnants and repeating the steps. The whole process can be summarized as an extension–termination–cleavage–extension cycle (Chen et al. 2013). Benefit of the sequencing by reversible terminator-based SBS technique is highly accurate because it can eliminate bias from homopolymeric or repetitive regions in a genome or unbiased genome coverage, hence resulting in a high raw base accuracy of ~99%. The common SBS technologies are Genome Analyzer, MiSeq, NextSeq, and HiSeq that offer cost-effective, high-quality pairwise alignments and high-throughput technology because they can be running many samples and obtain large data.

Ion Torrent platforms commonly use sequencing by detecting hydrogen ions, such as Ion Torrent PGM and Proton (Life Technologies, Carlsbad, CA). This sequencing technique uses origin dNTPs and electronic sensors to detect the release of hydrogen ions as dNTPs are incorporated into a developed DNA strand (Besser et al. 2018). The difference between nucleotides is detected using microwells containing the template strand, and a polymerase is sequentially flooded with each dNTP. Moreover, the detection of homopolymer length can be done using a sensor that detects the pH change magnitude and determines how many nucleotides are incorporated. Limitations on the Ion Torrent platform are mostly insertions and deletions in homopolymer runs, resulting from difficulties in evaluating the magnitude of signal when several dNTPs are incorporated in a cycle. Because only one base is incorporated per cycle, Illumina sequencing can provide more accurate results through repetitive sequences and a homopolymer (Mascher et al. 2013).

Sequencing by ligation is the base sequencing approach of the oligonucleotide ligation and detection (SOLiD) sequencing platform by Applied Biosystems. This sequencing principle depends on DNA ligase's ability to detect and incorporate bases in a very specific location. In sequencing by ligation, the attached DNA fragments are clonally amplified by emulsion PCR (Gupta and Gupta 2014). The benefit of SBL is simply implemented because it can be arranged with off-the-shelf reagents. However, SBL has some limitations of short read lengths, fewer output data, and long time to whole run genome sequencing (Ho et al. 2011). In general, the second-generation sequencing platforms have two main limitations: (i) need additional bioinformatic tools/pipelines to assemble the short read length into the original length template and (ii) clonal amplification makes the probability of PCR bias.

Third-generation or single-molecule real-time (SMRT) sequencing has been developed to solve the limitations of second-generation technology. To date, third-generation sequencing covers five sequencers: Helicos Genetic Analysis System, Pacific Biosciences, and Oxford Nanopore's nanopore sequencing are single-molecule sequencing by incorporating and detecting fluorescently labeled nucleotides (Rusk 2009). The other two platforms are Complete Genomics by Beijing Genomics Institute and GnuBIO by BioRad. Complete Genomics is based on hybridization and ligation. In contrast, GnuBIO is based on microfluidics, which combines PCR amplicons from a genomic sample with nanodroplets containing 1 of ~5000 different hexameric sequencing primers, reducing the number of steps into a single high-throughput library preparation step (Ambardar et al. 2016).

We have sequenced and analyzed the complete chloroplast genome of coconut 'Kopyor Green Dwarf.' The sequence is obtained from fresh young coconut leaf collected from a coconut seedling nursery of the Indonesian Palm Crops Research Institute, Manado, Indonesia. The chloroplast genome (cpDNA) is extracted using a modified CTAB protocol (Allen et al. 2006). High-quality DNA is sequenced using the paired-end sequencing from Illumina Genome Analyzer IIX (GAIIX) platform. The trimmed reads are de novo assembled using Ray software (Boisvert et al. 2010). The chloroplast sequence contigs are then compared with the NCBI database using the BLASTx program with a threshold value of 10^{-3}. The selected contigs are used to construct the complete chloroplast genome structure. The gaps are corrected manually to compose the complete cp genome.

10.4 Kopyor Coconut Chloroplast Genome Annotation

Sequencing technologies have accumulated high-throughput sequence data from various organisms. Genome sequence information provides a better understanding of variation between organisms at the nucleotides level and explains how genes are organized within the genome. An enormous amount of sequence and annotation data has been collected in the genome databases. The genome databases are publicly available, such as the Ensembl Plants portal (http://plants.ensembl.org/index.html) and the NCBI genome portal (http://www.ncbi.nlm.nih.gov/genome/). Some web-based platform for organelle genome annotation can be accessed freely, such as AGORA (Jung et al. 2018), CPGAVAS2 (Shi et al. 2019), GenomeVx (Conant and Wolfe 2008), DOGMA (Wyman et al. 2004), CGAP (Cheng et al. 2013), PGA (Qu et al. 2019), Plann (Huang and Cronk 2015), GeSeq (Tillich et al. 2017), and Verdant (McKain et al. 2017).

The kopyor coconut genome has not been annotated in any research. We have analyzed the Kopyor Green Dwarf chloroplast genome (cpDNA) to know the genome's characteristics. The annotation is curated through the GeSeq (https://chlorobox.mpimp-golm.mpg.de/geseq.html), with the setting of protein, rRNA, tRNA, DNA set at 95%, and *Phoenix dactylifera* (NC_013991.2) used as a reference genome. tRNAscan-SE v2.0.3 program is used as a tRNA annotator with default settings (Lowe and Chan 2016). A circular map of the kopyor chloroplast genome is visualized using OGDRAW (Lohse et al. 2013). GC contents from different regions (LSC, SSC, and IR) are calculated using the GC Content Calculator (https://jamiemcg.github.io/bioinf/gc.html) (Minnullina et al. 2019).

The chloroplast genome (cpDNA) generally has a quadripartite structure with circular double-stranded DNA. The quadripartite structure is composed of two copies of the inverted repeat (IR) region, separated by a large single-copy (LSC) region and a small single-copy (SSC) region. The chloroplast genome contains genes involved in the photosynthesis process, including photosystems I and II, cytochrome bf complex, ATP synthase, and transcription and translation (Cooper 2000). Chloroplast genome content and arrangement are highly conserved among most plants and their closest relatives, algae. Genes in the cpDNA are considered more conserved than genes in the nucleus and mitochondrial genomes. The nucleotide substitution rate in cpDNA is lower than in the nuclear genome (Duchene and Bromham 2013; Smith 2015).

The cpDNA of *C. nucifera* cv. Kopyor Green Dwarf has a quadripartite structure, which is common among cpDNAs of the plant. The circular double-stranded DNA of the *C. nucifera* cv. Kopyor Green Dwarf chloroplast genome is 158,462 bp in length. It has a typical quadripartite structure, and it is composed of a pair of inverted repeats (IRs, 27,276 bp) and separated by large single-copy (LSC, 86,198 bp) and also small single-copy (SSC, 17.712 bp) regions (Fig. 10.1). The chloroplast genome encodes 116 genes, including 80 protein-coding genes, 32 tRNA genes, and 4 rRNA genes (Fig. 10.1, Table 10.1). By comparing the published chloroplast genome sequence of coconut accessions number KF285453 (Huang et al. 2013), the kopyor

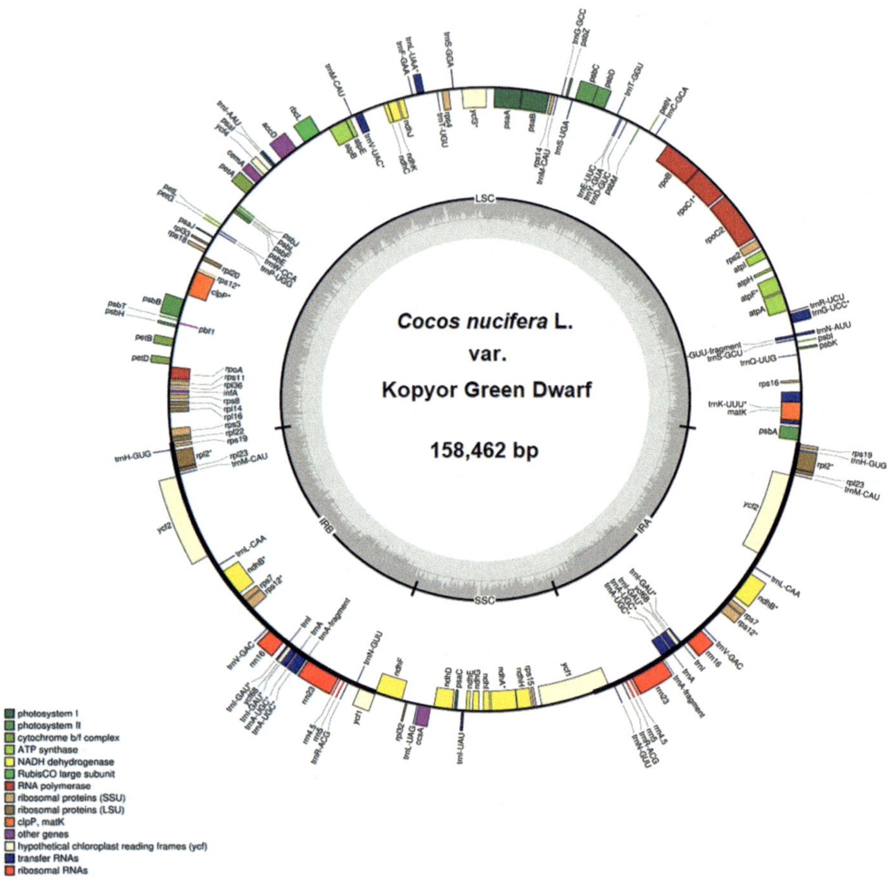

Fig. 10.1 The complete chloroplast genome of *Cocos nucifera* cv. Kopyor Green Dwarf. The genes shown on the outside of the map are transcribed clockwise, while those on the inside are counterclockwise. Gene functions related to processes are coded with colors. *LSC* large single copy, *SSC* small single copy, *IR* inverted repeat

coconut has a larger genome due to a higher number of gene duplications. The kopyor coconut has 21 gene duplications, whereas the other accession only has 15 gene duplications. Moreover, the expansion and contraction of the IR region are hypothesized to explain size differences between cp genomes (Shen et al. 2017).

Twenty-one genes are identified and duplicated in the IR region, including *rps19*, *trnH-GUG*, *rpl2*, *rpl23*, *trnM-CAU*, *ycf2*, *trnL-CAA*, *ndhB*, *rps7*, *rps12*, *trnV-GAC*, *rrn16*, *trnI-GAU*, *ycf68*, *trnA-UGC*, *rrn23*, *rrn4.5*, *rrn5*, *trnR-ACG*, *trnN-GUU*, and *ycf1* (Table 10.2). The *rps12* gene is identified to be trans-spliced. The *rps12* gene is trans-spliced because one of its exons is located in the LSC region and the 3′-tail in IR regions. This case is also found in *P. dactylifera* (Yang et al. 2010), *Cyamopsis tetragonoloba* (Kaila et al. 2017), *Actinidia chinensis* (Yao et al. 2015), *Pongamia pinnata* (Kazakoff et al. 2012), *Allium* (Huo et al. 2019), *Fokienia hodginsii* (Zang

Table 10.1 Features of the chloroplast genome of *Cocos nucifera* cv. Kopyor Green Dwarf

Features	Size or quantity
Length of genomes (bp)	158,462
LSC (bp)	86,198
SSC (bp)	17,712
IR (bp)	27,276
Total GC contents (%)	37.23
LSC GC contents (%)	35.30
SSC GC contents (%)	30.79
IR GC contents (%)	42.38
tRNAs	33
rRNAs	4
Duplicated gene	21
Genes with intron	8
Genes with two introns	2
Genes with an intron	6
tRNA with intron	5

et al. 2019), and *Senecioneae* (Gichira et al. 2019). *ycf1* pseudogene is also reported in *Byrsonima coccolobifolia* and *B. crassifolia* species, which indicate that *ycf1* (5745 bp) is originated from the SSC region before it moves toward the SSC/IRA boundary. It causes a duplication of *ycf1* at its 3′-end in the IRB region, resulting in a *ycf1* pseudogene (1389 bp) (Menezes et al. 2018).

Eight genes are identified to have introns, with the *ycf3*, and *clpP* genes have two introns, while the other six genes (*atpF*, *rpoC1*, *rpl2*, *ndhB*, *rps12*, and *ndhA*) have one intron. Five tRNA genes also possess an intron, i.e., *trnK-UUU*, *trnG-UCC*, *trnV-UAC*, *trnI-GAU*, and *trnA*-UGC. Among those genes, *trnK-UUU* has the largest intron (2626 bp) in which this intron also harbors the *matK* gene (1545 bp). Among the tRNA genes, *trnK-UUU* is the only tRNA gene that contains the maturase (*matK*) gene within its intron. *matK* plays a special role as an essential factor for the splicing of *trnK* intron within which it is contained (Vogel et al. 1997). Although plastid genomes are highly conserved, the *trnK* gene in all *Cuscuta* species is absent. However, it retains the open reading frame (ORF) for *matK* (McNeal et al. 2009). It is proved that *matK* is highly conserved among the plastid genomes, even though *trnK-UUU* is missing. The *matK* gene is also reported to be a highly divergent fragment (Wilson 2004). Therefore it is often used as a marker for inter- and intraspecies phylogenetic analysis (Hardig et al. 2010; Harnelly et al. 2018).

GC content is an important character of a genome because it correlates with the number of microRNA binding sites (Davis et al. 2008), rate of recombination and distribution of genes (Freudenberg et al. 2009), editing RNA organelles (Smith 2009), and regulation of gene expression (Barahimipour et al. 2015). The variation of GC content in monocotyledon plants is between 33.6% and 48.9%, and high GC content is reported to facilitate more complex gene regulation (Smarda et al. 2014). The overall GC content that was observed in the chloroplast genome of *C. nucifera*

Table 10.2 Encoded genes in the chloroplast genome of *Cocos nucifera* cv. Kopyor Green Dwarf

Gene function	Group of genes	Code	List of genes	Number of genes
Self-replication	Small subunit of ribosome	rps	rps16, rps2, rps14, rps4, rps18, rps12[a,b] rps11, rps8, rps3, rps19[a], rps7[a], rps15	15
	rRNA genes	Rrn	rrn16[a], rrn23[a], rrn4.5[a], rrn5[a]	8
	Large subunit of the ribosome	Rpl	rpl33, rpl20, rpl36, rpl14, rpl16, rpl22, rpl2[a], rpl23[a], rpl32	11
	DNA-dependent RNA polymerase	Rpo	rpoC2, rpoC1, rpoB, rpoA	4
	tRNA genes	Trn	trnK-UUU, trnQ-UUG, trnS-GCU, trnN-AUU, trnG-UCC, trnR-UCU, trnC-GCA, trnD-GUC, trnY-GUA, trnE-UUC, trnT-GGU, trnS-UGA, trnG-GCC, trnS-GGA, trnT-UGU, trnL-UAA, trnF-GAA, trnV-UAC, trnM-CAU, trnI-AAU, trnW-CCA, trnP-UGG, trnI-UAU, trnH-GUG[a], trnM-CAU[a], trnL-CAA[a], trnV-GAC[a], trnI-GAU[a], trnA-UGC[a], trnR-ACG[a], trnN-GUU[a], trnL-UAG	40
Photosynthesis	Subunits of ATP synthase	atp	atpA, atpF, atpH, atpI, atpE, atpB	6
	Subunits of NADH dehydrogenase	ndh	ndhJ, ndhK, ndhC, ndhB[a], ndhF, ndhD, ndhE, ndhG, ndhI, ndhA, ndhH	12
	Subunits of cytochrome b/f complex	pet	petN, petA, petL, petG, petB, petD	6
	Subunits of photosystem I	psa	psaB, psaA, psaI, psaJ, psaC	5
	Subunits of photosystem II	psb	psbA, psbK, psbI, psbM, psbD, psbC, psbZ, psbJ, psbL, psbF, psbE, psbB, psbT, psbN, psbH	15
	Subunit of rubisco	rbc	rbcL	1
Other genes	Subunit of acetyl-CoA carboxylase	acc	accD	1
	Envelop membrane protein	cem	cemA	1
	c-Type cytochrome synthesis gene	ccs	ccsA	1
	Protease	clp	clpP	1
	Translational initiation factor	inf	infA	1
	Maturase	mat	matK	1
Unknown function	Conserved open reading frames	ycf	ycf3, ycf4, ycf2[a], ycf68[a], ycf1[a]	8

[a]Two gene copies in the IRs
[b]Gene divided into two independent transcription units

cv. Kopyor Green Dwarf is 37.23% (Table 10.1). GC content in the IR region (42.38%) is higher than in the LSC (35.30%) and SSC region (30.79%).

The IR of *C. nucifera* cv. Kopyor Green Dwarf chloroplast genome is found to have higher GC content than the single-copy regions due to four rRNA genes (*rrn4.5*, *rrn5*, *rrn16*, and *rrn23*). These genes have the highest GC content among other coding regions. The highest GC content could result in the sequence complexity and the stabilization of the entire genome (Kaila et al. 2017). The SSC region has the lowest GC content because of NADH dehydrogenase genes, which have the lowest GC content of any functional groups (Jansen and Ruhlman 2012). GC content among genes in different functional groups of the chloroplast genome is divergent, but the order of the largest to smallest GC content in functional genes is rRNA > tRNA > photosynthetic genes > genetic system genes > NADH (Gao et al. 2009). IR regions containing GC nucleotides greater than LSC and SSC regions are also reported in many chloroplast genomes, such as *Gossypium thurberi* (Talat and Wang 2015), *Cyamopsis tetragonoloba* (Kaila et al. 2017), and *Arachis hypogaea* L. (Wang et al. 2018).

10.5 Codon Usage Analysis in Kopyor Coconut Chloroplast Genome

The relative synonymous codon usage (RSCU) value is obtained from the ratio between the observed codon frequency and the expected codon frequency of the codon synonymous with the same amino acid. The codon usage frequency is calculated only in the CDS region of all protein-coding genes using the CAIcal program (http://genomes.urv.es/CAIcal) (Puigbò et al. 2008), and the RSCU value is calculated manually (Nair et al. 2012). The RSCU value of more than 1.00 is considered an abundant number of codon or high-frequency codon in the genome. In contrast, RSCU values of less than 1.00 are considered fewer codon frequencies (Gun et al. 2018). In order to analyze the codon usage in *C. nucifera* cv. Kopyor Green Dwarf chloroplast genome, RSCU values are determined based on the coding sequences (CDS) of 80 protein-coding genes. All the protein-coding genes are composed of 22,358 codons. The calculation results of RSCU values show that 29 codons have an RSCU value of more than 1.00 (Table 10.3). ATG and TGG for methionine (Met) and tryptophan (Trp), respectively, are given the value of RSCU = 1.00 because a codon encodes both amino acids. Other amino acids, except Met and Trp, are encoded by several synonymous codons.

Leucine (Leu/L) is the most used codon in the *C. nucifera* cv. Kopyor Green Dwarf chloroplast genome (2382 codons or 10.65% of the total codons), whereas cysteine is the least used codon in the genome (280 codons or 1.25% of the total codons). These results have been reported in other species, such as *C. nucifera* with an accession number of KF285453 (Huang et al. 2013), *Ipomoea batatas* (Yan et al. 2015), and *Cyamopsis Tetragonoloba* (Kaila et al. 2017). Based on the RSCU value calculation on the synonymous leucine codon, TTA, TTG, CTT, CTC, CTA, and CTG are 1.88, 1.28, 1.21, 0.43, 0.81, and 0.39, respectively. The result shows that

Table 10.3 Codon usage in the chloroplast genome of *Cocos nucifera* cv. Kopyor Green Dwarf

Amino acid	Codon	Number	RSCU	Amino acid	Codon	Number	RSCU
Phe/F	TTT	859	**1,29**	Ala/A	GCT	561	**1,82**
	TTC	477	0,71		GCC	189	0,61
Leu/L	TTA	748	**1,88**		GCA	349	**1,13**
	TTG	507	**1,28**		GCG	131	0,43
	CTT	480	**1,21**	Tyr/Y	TAT	723	**1,60**
	CTC	170	0,43		TAC	182	0,40
	CTA	322	0,81	His/H	CAT	429	**1,53**
	CTG	155	0,39		CAC	132	0,47
Ile/I	ATT	969	**1,43**	Gln/Q	CAA	613	**1,50**
	ATC	424	0,63		CAG	202	0,50
	ATA	638	0,94	Asn/N	AAT	878	**1,58**
Val/V	GTT	475	**1,43**		AAC	231	0,42
	GTC	167	0,50	Lys/K	AAA	900	**1,50**
	GTA	499	**1,51**		AAG	301	0,50
	GTG	184	0,56	Asp/D	GAT	749	**1,60**
Ser/S	TCT	511	**1,69**		GAC	188	0,40
	TCC	299	0,99	Glu/E	GAA	970	**1,53**
	TCA	379	**1,26**		GAG	299	0,47
	TCG	169	0,56	Cys/C	TGT	212	**1,52**
	AGT	366	**1,21**		TGC	68	0,48
	AGC	88	0,29	Arg/R	CGT	320	**1,37**
Pro/P	CCT	373	**1,57**		CGC	78	0,33
	CCC	185	0,78		CGA	313	**1,33**
	CCA	281	**1,18**		CGG	111	0,47
	CCG	110	0,46		AGA	447	**1,91**
Thr/T	ACT	474	**1,55**		AGG	137	0,58
	ACC	228	0,75	Gly/G	GGT	554	**1,40**
	ACA	392	**1,28**		GGC	138	0,35
	ACG	130	0,42		GGA	640	**1,61**
					GGG	255	0,64

Bold, the RSCU > 1.00; RSCU, relative synonymous codon usage

the chloroplast genome uses the TTA more frequently for leucine translation than other synonymous codons since the value of RSCU for TTA is the highest (1.88). The least used codon is CTG, as shown by its RSCU value (0.39).

ATG is the common initiator codon of most protein-coding genes of basal eudicots, with several alternative start codons also existing. Among *C. nucifera* cv. Kopyor Green Dwarf protein-coding genes, eight genes use alternate start codons, such as AAA in *cemA*, ATT in *petB*, ATC in *rpl16* and *rps16*, ATC in *ndhD*, TAT in *rps12*, ATA in *rpl2*, and GTG in *rps19*. The alternative codons are used as start codons among protein-coding genes, such as ATA, ATC, TTG, and ATT (Wang et al. 2016). It is also identified in the *A. sinensis* chloroplast genome: ATT is used as the start codon for *petB* and *ycf1*; ATC for *rpl16*; ATA for *atpF*; GTG for *rps8*, *psbC*, and *ndhD*; and TTG for *ndhA* and *rpoC1* (Wang et al. 2016). *Oryza sativa* is reported to use ACG and GTG as the start codons for *rpl2* and *rps19*, respectively (Liu and Xue 2004). In maize, AAA is used as a start codon for *atpE* (Krebbers et al. 1982). Meanwhile, ACG is used as initiator codon for *psbL* and *ndhD*, while GTG is used for *rps19*, *psbC*, and *ycf15* genes, in *N. tabacum* (Sugiura et al. 1998).

10.6 Quantity and Distribution of SNPs and InDels in Kopyor Coconut Chloroplast Genome

Single nucleotide polymorphisms (SNPs) and insertion-deletions (InDels) are abundant genetic markers and widespread throughout genome regions. The information on the distribution and quantity of SNPs and InDels would help generate the next genetic markers for evaluating maternal inheritance; identifying species differences; phylogeographic, phylogenetic, gene-specific analysis; and development of future palm breeding programs. The quantity and distribution of SNPs and InDels in Kopyor Green Dwarf chloroplast genome are analyzed by multiple alignment of nine chloroplast genomes from different species: CT Cn (*Cocos nucifera* cv. Kopyor Green Dwarf) as template genome, GU811709.2 (*Phoenix dactylifera*), NC_013991.2 (*P. dactylifera*), FJ212316.3 (*P. dactylifera*), KX028884.1 (*C. nucifera*), KF285453.1 (*C. nucifera*), NC_022417.1 (*C. nucifera*), JF274081.1 (*Elaeis guineensis*), and NC_017602.1 (*E. guineensis*), using MAAFT (Yamada et al. 2016) in Geneious Prime 2019.1.1 version 11 (Kearse et al. 2012). Detection of the number of SNPs and InDels is calculated manually.

A total of 2185 SNPs are identified in the chloroplast genome of *C. nucifera* cv. Kopyor Green Dwarf. The coding regions have a slightly higher number of SNPs than that in the noncoding regions: 1105 (51%) SNPs for coding regions and 1080 (49%) for the noncoding regions (Fig. 10.2). Non-synonymous SNPs (1715) were higher than synonymous ones (470), respectively. The SNP distribution analysis showed that both non-synonymous and synonymous SNPs were present in almost all protein-coding genes. However, no SNPs are found in petG, petN, psbH, psbJ, psbL, psbT, rpl36, and ycf68 (Fig. 10.3). In atpE, infA, ndhB, ndhC, ndhJ, petB, petL, psaI, psbE, psbF, psbK, psbM, psbN, psbZ, rpl16, rpl22, rpl23, rpl33, rps12, and rps15, only non-synonymous SNPs are found, whereas rpl32 gene has only synonymous SNPs. SNPs in atpF, clpP, ndhA, ndhB, rpl2, rpoC1, rps12, and ycf3 genes are detected in their introns.

Nine SNPs (5 non-synonymous, 4 synonymous) are detected in *atpF* intron; 17 SNPs (12 non-synonymous, 5 synonymous) in both introns of *clpP* gene; 20 SNPs (15 non-synonymous, 5 synonymous) in *ndhA* intron; 3 non-synonymous SNPs are detected in *ndhB* intron; 26 SNPs (19 non-synonymous and 7 synonymous) are found in *rpl2* intron; 6 non-synonymous SNPs are found in *rpoC1* intron; 2 non-synonymous SNPs are found in *rps12* intron; 19 SNPs (12 non-synonymous and 7 synonymous) are found in the second intron of *ycf3*.

A total of 444 InDels are identified in the *C. nucifera* cv. Kopyor Green Dwarf chloroplast genome. The number of InDels is higher in the noncoding regions than

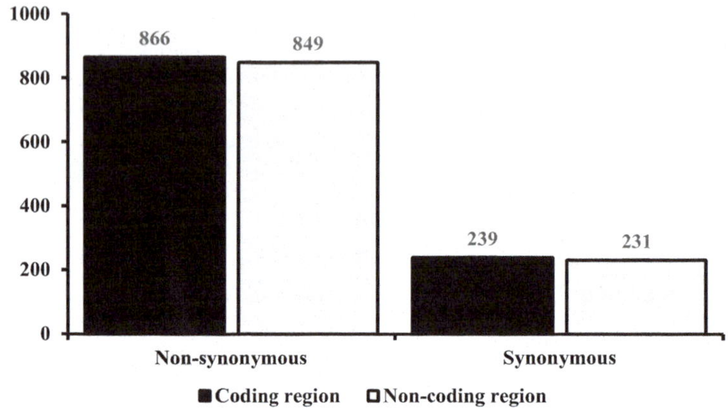

Fig. 10.2 The numbers of single nucleotide polymorphisms (SNPs) in the coding and the noncoding regions of the chloroplast genome of *Cocos nucifera* cv. Kopyor Green Dwarf

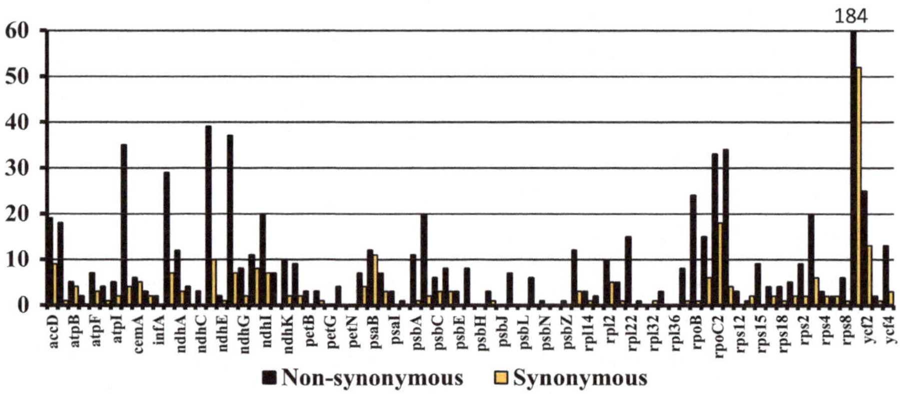

Fig. 10.3 The numbers of non-synonymous and synonymous single nucleotide polymorphisms (SNPs) in each gene in the chloroplast genome of *Cocos nucifera* cv. Kopyor Green Dwarf

in the coding regions. In the noncoding regions, 259 InDels (58% of the total InDels) are detected, whereas, in the coding regions, 185 InDels (42%) are found (Fig. 10.4). These results suggest that the coding regions are more conserved than the noncoding region. InDels are detected in 52 protein-coding genes, i.e., *accD*, *atpA*, *atpB*, *atpF*, *atpH*, *atpI*, *ccsA*, *cemA*, *clpP*, *matK*, *ndhA*, *ndhB*, *ndhC*, *ndhD*, *ndhF*, *ndhG*, *ndhH*, *ndhI*, *ndhJ*, *petA*, *petB*, *psaB*, *psaJ*, *psbB*, *psbD*, *psbE*, *psbF*, *psbI*, *psbK*, *psbM*, *rbcL*, *rpl14*, *rpl16*, *rpl2*, *rpl20*, *rpl22*, *rpoA*, *rpoC1*, *rpoC2*, *rps11*, *rps12*, *rps14*, *rps16*, *rps18*, *rps19*, *rps2*, *rps3*, *rps4*, *rps7*, *ycf1*, *ycf2*, and *ycf4* genes. InDels are found in the intron of *atpF*, *clpP*, *ndhA*, *rpl2*, *rpoC1*, *rps12*, and *ycf3* genes. They are also found between the coding regions, and intergenic spacer (IGS) is located in IGS-*ndhB*, *petG*-IGS, and *psbE-psbF* regions. *ycf1* is identified to have the most variations of InDels than other protein-coding genes (Fig. 10.5). These

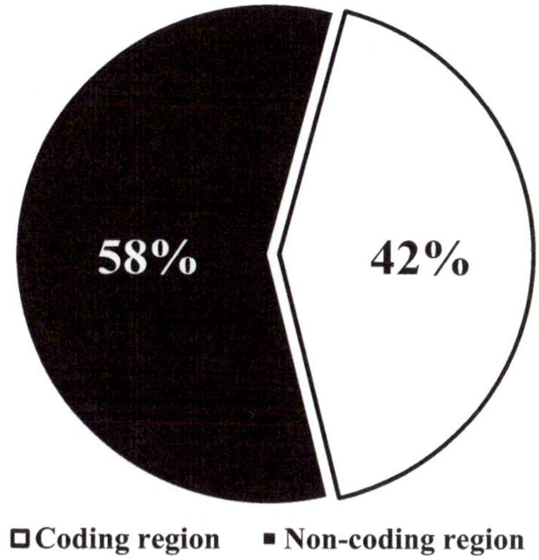

Fig. 10.4 Percentage of insertion-deletion (InDel) in the coding and noncoding regions of the chloroplast genome of *Cocos nucifera* cv. Kopyor Green Dwarf

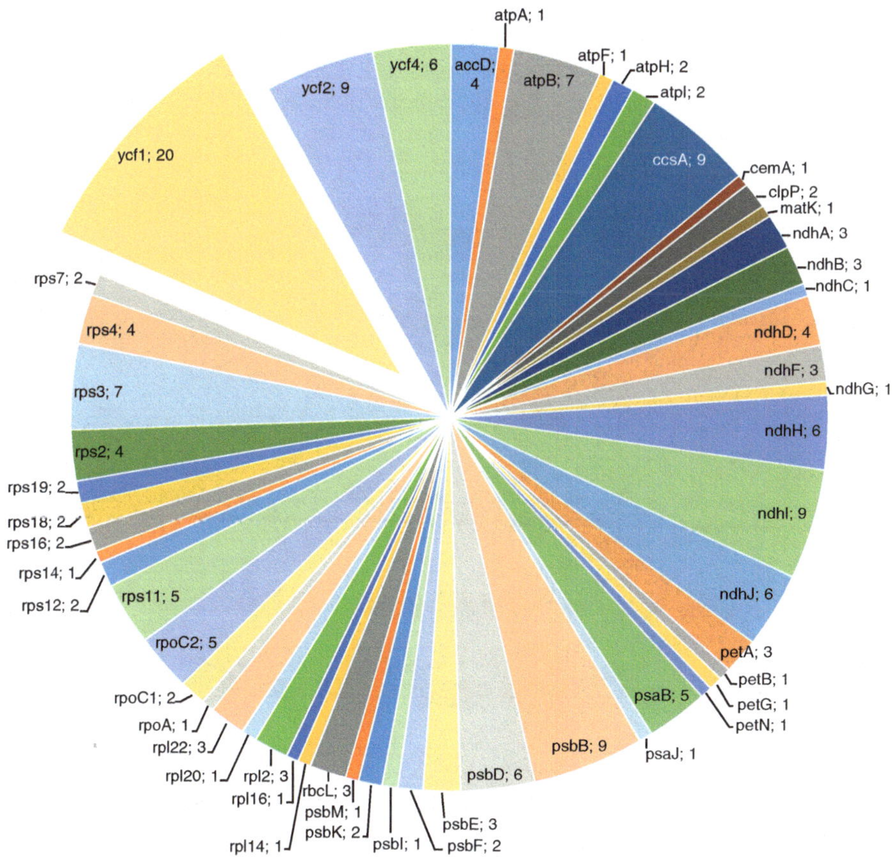

Fig. 10.5 The numbers of insertion-deletions (InDels) in protein-coding genes of the chloroplast genome of *Cocos nucifera* cv. Kopyor Green Dwarf

indicated that *ycf1* has many mutational loci in the chloroplast genome, and the regions are greatly varied.

10.7 Expansion and Contraction of IR Regions of Kopyor Coconut Chloroplast Genome

The existence of the inverted repeat (IR) region greatly impacts the rate of chloroplast genome sequence evolution. Meanwhile, the wide range of expansions and contractions of the IR region is a very common evolutionary event in the evolution of a genome. IR expansion/contraction of those chloroplast genomes is also analyzed using IRscope (Amiryousefi et al. 2018). By using the curated genome annotations, we compare the junction sites of seven selected Arecaceae chloroplast genomes. The comparative analysis of the inverted repeat (IR) regions, large single-copy (LSC) region, and small single-copy (SSC) region among palm chloroplast genomes is purposed to determine the differences in the boundary location of LSC/IR or SSC/IR among genomes. These would indicate the evolution among palm species. Results show that the IR length of several palm species (*C. nucifera*, *E. guineensis*, *P. dactylifera*, *M. warburgii*, and *A. caudata*) is variable 26,553–27,281 bp (Fig. 10.6). The IR regions of *C. nucifera* cv. Kopyor Green

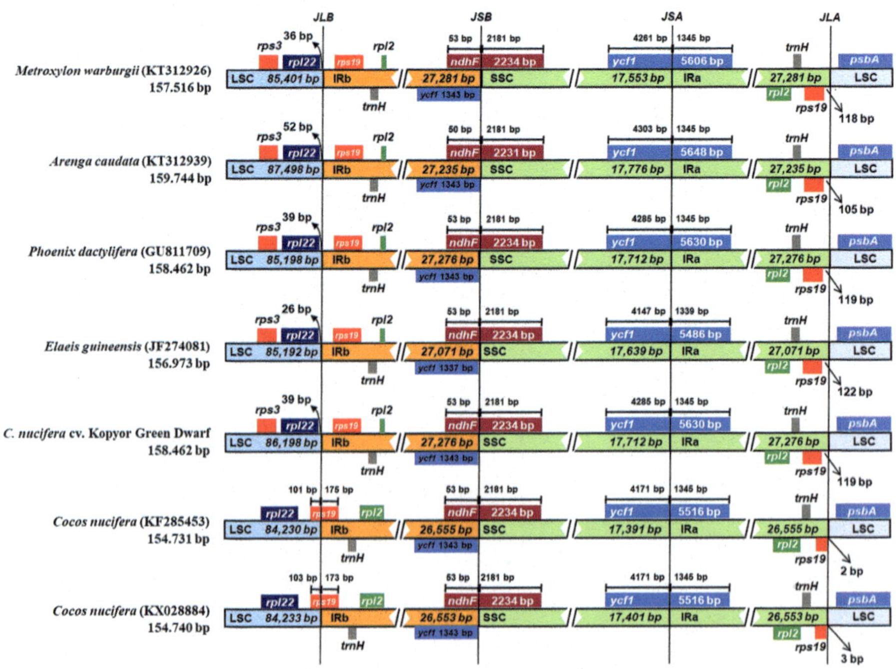

Fig. 10.6 Comparison of IR/LSC and IR/SSC border positions between seven chloroplast genomes, including *Cocos nucifera* cv. Kopyor Green Dwarf, *C. nucifera* (KF285453), *C. nucifera* (KX028884), *Elaeis guineensis* (JF274081), *Phoenix dactylifera* (GU811709), *Metroxylon warburgii* (KT312926), and *Arenga caudata* (KT312939)

Dwarf have the same length as *P. dactylifera* (27,276 bp), whereas the *C. nucifera* cv. Kopyor Green Dwarf is longer than other palms, except *M. warburgii* (26,553–27,235 bp). These indicate that there are expansion and contraction among IRs of palm species chloroplast genomes.

Seven palm species show that the IRA/LSC border is consistently located downstream of *psbA*. The IRB/SSC border is also consistently located within *ndhF*. The location indicates that the border is highly conserved among palm species used in the present study. The IRB/LSC border on five species (including *C. nucifera* cv. Kopyor Green Dwarf) is located in the intergenic spacer between *rpl22* and *rps19* genes. Other *Cocos nucifera* accessions (KF285453 and KX028884) have an IRB/LSC border in the *rps19* gene, which indicates the presence of IR contractions among the *C. nucifera* accessions. The SSC/IRA border is located within the *ycf1* gene of *C. nucifera* cv. Kopyor Green Dwarf, resulting in the *ycf1* gene to become a pseudogene in the IRB region (1343 bp). Not only in the *C. nucifera* cv. Kopyor Green Dwarf, a pseudogene of *ycf1*, is also reported in six other palm species (1337–1343 bp).

10.8 Cross-Species Comparative Chloroplast Genome Analysis

Cross-species sequences comparison is a basic method for identifying conserved, variation regions and functional sequences in the genomes (Frazer et al. 2004). The comparative analysis shows that the coding regions are more highly conserved than the noncoding regions and have higher variability in the intergenic spacer (IGS) regions (Liu et al. 2018). IGS regions are noncoding regions located between two genes. Coding regions are usually less varied because these regions would be fully transcribed into amino acids, and thus the function of the protein-coding genes needs to be maintained. On the contrary, IGS sequences could be more diverse because the noncoding region function is very limited, and IGS could be fully, partially, or not completely transcribed with high mutation rates (Borsch and Quandt 2009; Sakka et al. 2013). The IGS region is most rapidly evolved in single-copy regions (SSC or LSC). Hence, the IGS region could develop genetic markers to analyze interspecific relationships (Korotkova et al. 2014). The IGS regions are used in many types of research, especially for DNA barcoding and genetic marker analysis in *Chimonanthus praecox* (L.), *Typha orientalis*, *Nelumbo nucifera*, *Paeonia suffruticosa*, *Prunus persica*, *Panax bipinnatifidus* (Dong et al. 2012), and Umbelliferae (Degtjareva et al. 2012).

The complete chloroplast genome sequences consisted of *C. nucifera* cv. Kopyor Green Dwarf which was generated in our study, while *C. nucifera* (KF285453), *C. nucifera* (KX028884), *E. guineensis* (JF274081), *P. dactylifera* (GU811709), *Metroxylon warburgii* (KT312926), and *Arenga caudata* (KT312939) were retrieved from NCBI GenBank DNA Database. The chloroplast genomes were used in the subsequent analyses. Genome comparison of seven Palmae chloroplast genomes is performed and visualized using the VISTA software (Frazer et al. 2004) with alignment program as Shuffle-LAGAN and *C. nucifera* cv. Kopyor Green Dwarf as reference. Some chloroplast genome sequences of palm species are aligned to each other and plotted using *C. nucifera* cv. Kopyor Green Dwarf as reference. The comparative analysis of seven palm species chloroplast genomes showed that the coding regions or genic regions are more highly conserved than noncoding regions, such as *ycf2*, *ndhB*, and *rps7*. Meanwhile, the highest polymorphism is observed in the intergenic regions (Fig. 10.7). The IR regions are more conserved compared with LSC or SSC. Some insertion-deletion variations are detected in the chloroplast

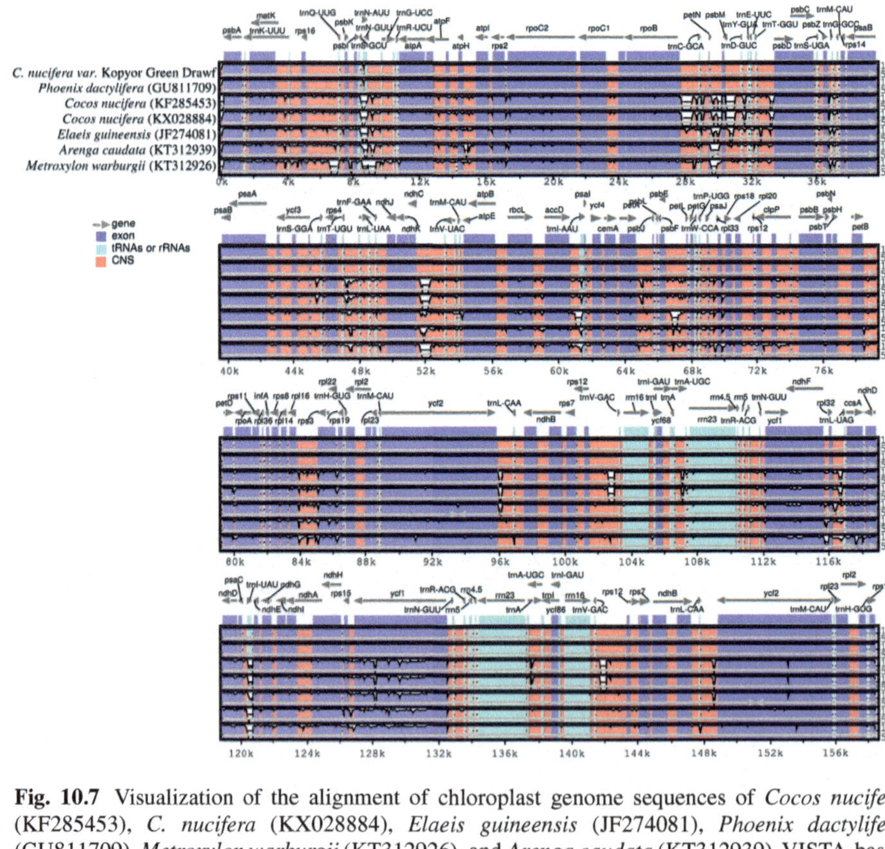

Fig. 10.7 Visualization of the alignment of chloroplast genome sequences of *Cocos nucifera* (KF285453), *C. nucifera* (KX028884), *Elaeis guineensis* (JF274081), *Phoenix dactylifera* (GU811709), *Metroxylon warburgii* (KT312926), and *Arenga caudata* (KT312939). VISTA-based identity plot shows sequence identity among chloroplast genomes and *C. nucifera* cv. Kopyor Green Dwarf chloroplast genome as a reference. A cutoff of 70% identity and 50–100% sequence identity was used for the Y-axis plot. On the X-axis, *B. flabellifer* genes are indicated on top lines, and arrows represent their orientation. Colors distinguish genome regions. CNS, conserved noncoding sequences. Purple, indicates exons; green, indicates tRNA and rRNA

genome region, which are widely distributed in the inter-spacer (ITS) regions, such as *trnN-AUU, trnN-GUU, rpoB-trnC-GCA, petN-psbM, psbM-trnD-GUC, trnT-GGU-psbD, ndhJ-ndhK, ndhC-trnV-UAC, trnI-AAU, psbE-petL, rps12-trnV-GAC, trnI-GAU, trnA-UGC, ndhF-rpl32, rpl32-ccsA, trnI-UAU, psaC-ndhE, rps15-ycf1, ycf1*, and *trnL-ycf2*. Regions with a high variation level can be developed as molecular markers to evaluate the genetic population among varieties or species.

10.9 Cross-Species Comparative Quantity and Distribution of Chloroplast Microsatellites

Microsatellites or simple sequence repeats (SSRs) are repetitive genetic sequences with repeat units widespread in the genomes, usually ranging from one to six nucleotide bases. Chloroplast SSRs (cpSSRs) are abundantly widespread in the LSC, SSC, and IR, distributed in both coding and noncoding regions. In the present study, we identify mono-, di-, and trinucleotide microsatellites for *C. nucifera* cv. Kopyor Green Dwarf, *P. dactylifera* (GU811709), *E. guineensis* (JF274081), *C. nucifera* (KF285453), *C. nucifera* (KX028884), *M. warburgii* (KT312926), and *A. caudata* (KT312939). Repeat sequences (SSR), including mono-, di-, and trinucleotides, are identified within the chloroplast genomes using Phobos version 3.3.12 (http://www.ruhr-uni-bochum.de/ecoevo/cm/cm_phobos.htm) (Mayer 2008) with the search parameters set to ≥8 repeat units for mononucleotide repeats, ≥ 4 repeat units for dinucleotide repeats, and ≥2 repeat units for trinucleotide repeats.

The chloroplast genomes are highly conserved, but the number of SNPs, InDels, and cpSSRs varies among related species. Among those genetic variations, SNP is the most abundant within the genome. SSR variation is higher than InDels. The distribution of repeat sequences and variation sequences among chloroplast genomes enhances the assumption that the mutation rate of noncoding regions is higher compared with coding regions (Niu et al. 2017). LSC and SSC regions contain more SSR than IR regions in angiosperm chloroplast genomes (Huotari and Korpelainen 2012; Yi et al. 2012; Gao et al. 2018). Genetic variations could become a source of genetic material in the genetic analysis that focuses on ecological, systematic, conservation, genetic population, phylogeography, characterization of genetic diversity, and development of molecular markers for phylogenetic studies, DNA fingerprinting, and plant breeding (Vieira et al. 2016; Andrade et al. 2018).

The total number of cpSSRs is variable among the palm species. According to the result of Phobos software, the *C. nucifera* cv. Kopyor Green Dwarf and *P. dactylifera* (GU811709) have the same cpSSRs in their respective cpDNA, i.e., 493 cpSSRs identified. Among the cpSSRs in the intergenic regions, there are 101, 35, and 93 for mono-, di-, and trinucleotide repeats, respectively, while in the genic regions, 79, 28, and 157 cpSSRs are identified, respectively (Figs. 10.8 and 10.9). A total of 479, 464, 462, 438, and 493 cpSSRs are identified in the *E. guineensis* (JF274081), *C. nucifera* (KF285453), *C. nucifera* (KX028884), *M. warburgii*

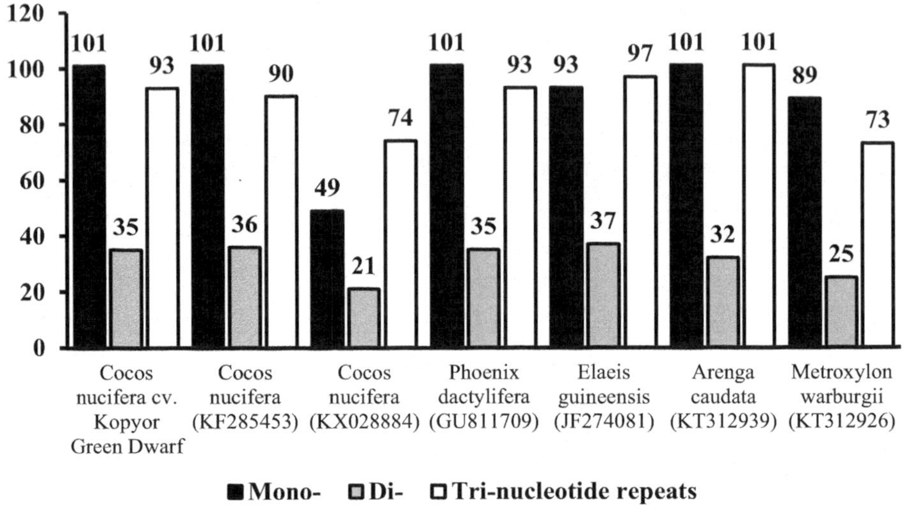

Fig. 10.8 Frequency of cpSSRs in the intergenic regions of the chloroplast genome of *Arenga caudata*, *Cocos nucifera*, *Elaeis guineensis*, *Metroxylon warburgii*, and *Phoenix dactylifera*. The chloroplast genome accession numbers were shown in the graph

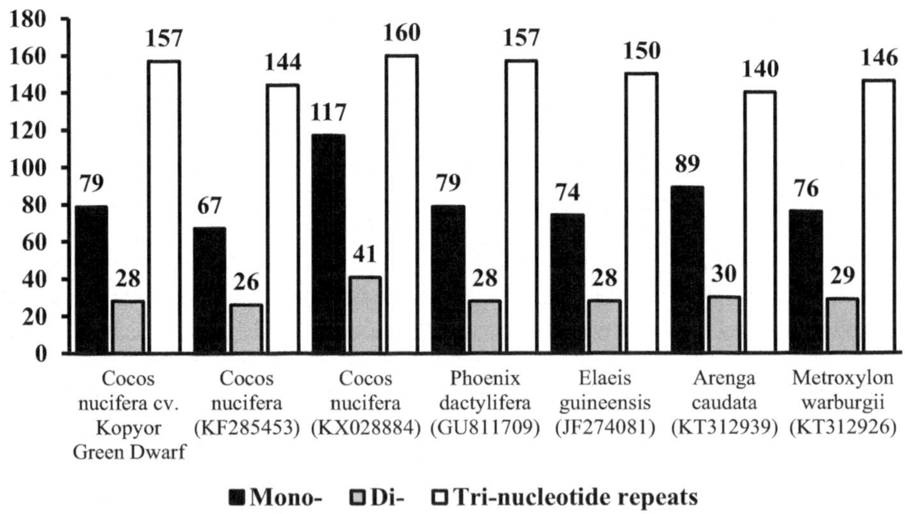

Fig. 10.9 Frequency of cpSSRs in the genic regions of the chloroplast genome of *Arenga caudata*, *Cocos nucifera*, *Elaeis guineensis*, *Metroxylon warburgii*, and *Phoenix dactylifera*. The chloroplast genome accession numbers were shown in the graph

(KT312926), and *A. caudata* (KT312939), respectively. Among those SSR motifs, the trinucleotide is the most common motif found in the genome, followed by mononucleotide. The *clpP*, *ndhA*, *ndhF*, *psbB*, *psbC*, *rpoC2*, *rps14*, *rps19*, *trnI-GAU*, *ycf1*, *ycf2*, and *ycf3* are detected mononucleotide motif in all palm species. Dinucleotide motif is observed in *ndhA*, *ndhB*, *ndhH*, *rpoC1*, *rrn23*, *trnK-UUU*, *ycf1*, and *ycf2* genes from across all palm species analyzed in this study. On the other hand, *atpA*, *atpB*, *atpF*, *atpI*, *clpP*, *ndhA*, *ndhB*, *ndhD*, *ndhF*, *ndhK*, *petB*, *petD*, *psaA*, *psaB*, *psbB*, *psbC*, *psbE*, *rbcL*, *rpl2*, *rpl22*, *rpoB*, *rpoC1*, *rpoC2*, *rps12*, *rps14*, *rps4*, *rrn16*, *rrn23*, *trnI-GAU*, *trnK-UUU*, *trnV-UAC*, *ycf1*, *ycf2*, and *ycf3* genes from across all palm species contain trinucleotide motifs.

10.10 Cross-Species Comparative Phylogenetic Analysis Based on Chloroplast Genome

Phylogenetic analysis is a method of estimating the evolutionary relationships between organisms. In molecular phylogenetic analysis, the sequence of a common gene, protein, or genome can evaluate a species' evolutionary relationship. The evolutionary relationship obtained from the phylogenetic analysis is usually represented as a tree diagram branching called the phylogenetic tree. Phylogenetic analysis was assessed using 39 species of the Arecaceae family, with *Nicotiana tabacum* as an outgroup. All chloroplast genome sequences were aligned using MAFFT (Yamada et al. 2016). The genetic distance models of Tamura-Nei and the neighbor-joining method with 1000 bootstrap replicates are chosen to infer the phylogenetic relationships within Arecaceae in Geneious Prime 2019.1.1 version 11 software.

Thirty-nine complete palm cpDNA sequences available in the NCBI GenBank DNA Database are used to study the phylogenetic relationship among palm species, with cpDNA of *Nicotiana tabacum* as the outgroup. The neighbor-joining tree is reconstructed using the whole accessions. The maternal inheritance and slow rate of substitution in the chloroplast genome have been used for identifying plant phylogeographic research and molecular evolution, such as phylogenetic, phylogenomic, and genomic evolutionary studies (Li and Zheng 2018). The phylogenetic relationship of the Arecaceae family has long been debated. Phylogenomic of chloroplast genomes offers new and in-depth insights into the phylogenetic relationship and the history of diversification among Arecaceae species.

The phylogenetic analysis among the Arecaceae family species has grouped the accessions into five clusters (Fig. 10.10). The first cluster consists of *Dasypogon bromeliifolius*, and the third one is *Nypa fruticans*. Most of the evaluated accessions belong to Group 2 (6 accessions), Group 4 (12 accessions), and Group 5 (19 accessions), respectively. Based on its chloroplast genome, *C. nucifera* cv. Kopyor Green Dwarf belonged to Group 5 and was closely related to the chloroplast genome of four *Phoenix dactylifera* accessions (Fig. 10.10).

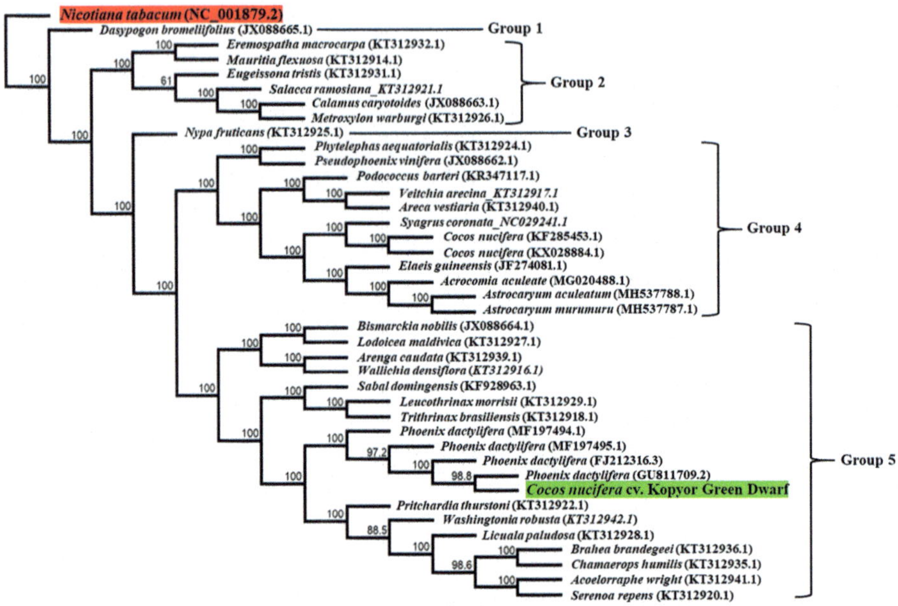

Fig. 10.10 Phylogenetic relationships using the neighbor-joining method of 39 accessions of the Arecaceae family inferred from complete chloroplast genome sequences. The position of *Cocos nucifera* cv. Kopyor Green Dwarf was highlighted green, and *Nicotiana tabacum* (red) was used as an outgroup. The bootstrap support values were obtained using 1000 iterations

10.11 Conclusion

In this chapter, we report and analyze the chloroplast genome sequences of *C. nucifera* cv. Kopyor Green Dwarf, a unique coconut variety from Indonesia, which has an abnormal endosperm (kopyor endosperm). The chloroplast genome of *C. nucifera* cv. Kopyor Green Dwarf has a typical chloroplast genome structure, and it is highly similar to the other chloroplast genomes of the Arecaceae family. According to phylogenetic analysis, *C. nucifera* cv. Kopyor Green Dwarf was closely related to that of *P. dactylifera*. Genetic variations are commonly found in the chloroplast genome, such as single nucleotide polymorphism (SNP), insertion-deletion (InDel), and simple sequences repeat (cpSSR). The trinucleotide cpSSR is the most common motif found in the chloroplast genome of palm species. The number of SNPs is higher than InDels. Repeated sequences, together with the single nucleotide polymorphisms and insertion-deletion variation, are informative sources for developing new molecular markers. The comprehensive data presented in this study provides insight into the evolutionary relationships among the Arecaceae family species. The identification of a whole chloroplast genome of *C. nucifera* cv. Kopyor Green Dwarf might be useful for accelerating the coconut breeding program and further biological studies in palm species.

References

Ahmadian A, Ehn M, Hober S (2005) Pyrosequencing: history, biochemistry and future. Clin Chim Acta 363(1–2):83–94

Aljohi HA, Wanfei L, Qiang L, Yuhui Z, Jingyao Z, Ali A, Alanazi IO, Alawad AO, Al-Sadi AM, Hu S, Yu J (2016) Complete sequence and analysis of coconut palm (*Cocos nucifera*) mitochondrial genome. PLoS One 11(10):e0163990

Allen GC, Flores-Vergara MA, Krasynanski S, Kumar S, Thompson WF (2006) A modified protocol for rapid DNA isolation from plant tissues using cetyltrimethylammonium bromide. Nat Protoc 1(5):2320–2325

Ambardar S, Gupta R, Trakroo D, Lal R, Vakhlu J (2016) High throughput sequencing: an overview of sequencing chemistry. Indian J Microbiol 56(4):394–404

Amiryousefi A, Hyvönen J, Poczai P (2018) IRscope: an on-line program to visualize the junction sites of chloroplast genomes. Bioinformatics 34(17):3030–3031

Andrade MC, Perek M, Pereira FB, Moro M, Tambarussi EV (2018) Quantity, organization, and distribution of chloroplast microsatellites in all species of *Eucalyptus* with available plastome sequence. Crop Breed Appl Biotechnol 18(1):97–102

Barahimipour R, Strenkert D, Neupert J, Schroda M, Merchant SS, Bock R (2015) Dissecting the contributions of GC content and codon usage to gene expression in the model alga *Chlamydomonas reinhardtii*. Plant J 84(4):704–717

Beedanagari S, John K (2014) Next-generation sequencing. In: Wexler P (ed) Encyclopedia of toxicology, 3rd edn. Academic, Oxford, pp 501–503

Besser J, Carleton HA, Smidt PT, Lindsey RL, Trees E (2018) Next-generation sequencing technologies and their application to the study and control of bacterial infections. Clin Microbiol Infect 24(4):335–341

Birky CW (1995) Uniparental inheritance of mitochondrial and chloroplast genes: mechanisms and evolution. Proc Natl Acad Sci U S A 92(25):11331–11338

Boisvert S, Laviolette F, Corbeil J (2010) Ray: simultaneous assembly of reads from a mix of high-throughput sequencing technologies. Journal of Computational Biology 17(11):1519–1533

Borsch T, Quandt D (2009) Mutational dynamics and phylogenetic utility of non-coding chloroplast DNA. Plant Syst Evol 282(3–4):169–199

Chen F, Dong M, Ge M, Zhu L, Ren L, Liu G, Mu R (2013) The history and advances of reversible terminators used in new generations of sequencing technology. Genomics Proteomics Bioinformatics 11(1):34–40

Cheng J, Zeng X, Ren G, Liu Z (2013) CGAP: a new comprehensive platform for the comparative analysis of chloroplast genomes. BMC Bioinformatics 14:95

Conant GC, Wolfe KH (2008) GenomeVx: simple web-based creation of editable circular chromosome maps. Bioinformatics 24(6):861–862

Cooper GM (2000) The cell. In: A molecular approach, 2nd edn. Sinauer Associates, Sunderland

Daniell H, Datta R, Varma S, Gray S, Lee SB (1998) Containment of herbicide resistance through genetic engineering of the chloroplast genome. Nat Biotechnol 16:345–348

Daniell H, Kumar S, Dufourmantel N (2005) Breakthrough in chloroplast genetic engineering of agronomically important crops. Trends Biotechnol 23(5):238–245

Davis N, Biddlecom N, Hecht D, Fogel GB (2008) On the relationship between GC content and the number of predicted microRNA binding sites by MicroInspector. Comput Biol Chem 32(3):222–226

De Cosa B, Moar W, Lee SB, Miller M, Daniell H (2001) Overexpression of the Bt cry2Aa2 operon in chloroplasts leads to formation of insecticidal crystals. Nat Biotechnol 19:71–74

DeGray G, Rajasekaran K, Smith F, Sanford J, Daniell H (2001) Expression of an antimicrobial peptide via the chloroplast genome to control phytopathogenic bacteria and fungi. Plant Physiol 127(3):852–862

Degtjareva GV, Logacheva MD, Samigullin TH, Terentieva EI, Roman CMV (2012) Organization of chloroplast psbA-trnH intergenic spacer in dicotyledonous angiosperms of the family Umbelliferae. Biochemistry (Mosc) 77(9):1056–1064

Dong W, Liu J, Yu J, Wang L, Zhou S (2012) Highly variable chloroplast markers for evaluating plant phylogeny at low taxonomic levels and for DNA barcoding. PLoS One 7(4):e35071

Duchene D, Bromham L (2013) Rates of molecular evolution and diversification in plants: chloroplast substitution rates correlate with species-richness in the Proteaceae. BMC Evol Biol 13(65):1–11

Ebert D, Peakall R (2009) Chloroplast simple sequence repeats (cpSSRs): technical resources and recommendations for expanding cpSSR discovery and applications to a wide array of plant species. Mol Ecol Resour 9(3):673–690

Esquenazi D, Wigg MD, Miranda MM, Rodrigues HM, Tostes JB, Rozental S, da Silva AJ, Alviano CS (2002) Antimicrobial and antiviral activities of polyphenolics from *Cocos nucifera* Linn. (Palmae) husk fiber extract. Res Microbiol 153(10):647–652

Frazer KA, Pachter L, Poliakov A, Rubin EM, Dubchak I (2004) VISTA: computational tools for comparative genomics. Nucleic Acids Res 32:W273–W279

Freudenberg J, Wang M, Yang Y, Li W (2009) Partial correlation analysis indicates causal relationships between GC-content, exon density and recombination rate in the human genome. BMC Bioinformatics 10(Suppl 1):S66

Gan P, Liu F, Li R, Wang S, Luo J (2019) Chloroplasts-beyond energy capture and carbon fixation: tuning of photosynthesis in response to chilling stress. Int J Mol Sci 20(20):5046

Gao L, Yi X, Yang YX, Su YJ, Wang T (2009) Complete chloroplast genome sequence of a tree fern *Alsophila spinulosa*: insights into evolutionary changes in fern chloroplast genomes. BMC Evol Biol 9:1–14

Gao X, Zhang X, Meng H, Li J, Zhang D, Liu C (2018) Comparative chloroplast genomes of *Paris* Sect. *Marmorata*: insights into repeat regions and evolutionary implications. BMC Genomics 19(Suppl 10):878

Gichira AW, Avoga S, Li Z, Hu G, Wang Q, Chen J (2019) Comparative genomics of 11 complete chloroplast genomes of Senecioneae (Asteraceae) species: DNA barcodes and phylogenetics. Bot Stud 60:1–17

Gun L, Yumiao R, Haixian P, Liang Z (2018) Comprehensive analysis and comparison on the codon usage pattern of whole *Mycobacterium tuberculosis* coding genome from different areas. Biomed Res Int 2018:3574976

Gupta AK, Gupta UD (2014) Next generation sequencing and its applications. In: Verma AS, Singh A (eds) Animal biotechnology models in discovery and translation. Academic/Elsevier, pp 345–367. https://doi.org/10.1016/B978-0-12-416002-6.00019-5

Hambuch TM, Mayfield J (2014) Next-generation sequencing. In: Pathobiology of human disease. Academic, San Diego, pp 4131–4139

Hardig TM, Anttila CK, Brunsfeld SJ (2010) A phylogenetic analysis of *Salix* (Salicaceae) based on matK and ribosomal DNA sequence data. J Bot 2010(197696):1–12

Harnelly E, Thomy Z, Fathiya N (2018) Phylogenetic analysis of Dipterocarpaceae in Ketambe Research Station, Gunung Leuser National Park (Sumatra, Indonesia) based on rbcL and matK genes. Biodiversitas 19(3):1074–1080

Harries HC, Clement CR (2014) Long-distance dispersal of the coconut palm by migration within the coral atoll ecosystem. Ann Bot 113(4):565–570

Ho A, Murphy M, Wilson S, Atlas SR, Edwards JS (2011) Sequencing by ligation variation with endonuclease V digestion and deoxyinosine-containing query oligonucleotides. BMC Genomics 12(598):1–8

Huang DI, Cronk QCB (2015) Plann: a command-line application for annotating plastome sequences. Appl Plant Sci 3(8):1500026

Huang YY, Matzke AJM, Matzke M (2013) Complete sequence and comparative analysis of the chloroplast genome of coconut palm. PLoS One 8(8):e74736

Huo Y, Gao L, Liu B, Yang Y, Kong S, Sun Y, Yang Y, Wu X (2019) Complete chloroplast genome sequences of four *Allium* species: comparative and phylogenetic analyses. Sci Rep 9(1):12250

Huotari T, Korpelainen H (2012) Complete chloroplast genome sequence of *Elodea canadensis* and comparative analyses with other monocot plastid genomes. Gene 508(1):96–105

Ivanova Z, Sablok G, Daskalova E, Zahmanova G, Apostolova E, Yahubyan G, Baev V (2017) Chloroplast genome analysis of resurrection tertiary relict *Haberlea rhodopensis* highlights genes important for desiccation stress response. Front Plant Sci 8(204):1–15

Jansen RK, Ruhlman TA (2012) Plastid genomes of seed plants. In: Bock R, Knoop V (eds) Genomics of chloroplast and mitochondria. Springer, London, pp 103–126

Jensen PE, Leister D (2014) Chloroplast evolution, structure and functions. F1000Prime Rep 6(40):1–14

Jung J, Kim JI, Jeong YS, Yi G (2018) AGORA: organellar genome annotation from the amino acid and nucleotide references. Bioinformatics 34:2661–2663

Kaila T, Chaduvla PK, Rawal HC, Saxena S, Tyagi A, Mithra SVA, Solanke AU, Kalia P, Sharma TR, Singh NK, Gaikwad K (2017) Chloroplast genome sequence of cluster bean (*Cyamopsis tetragonoloba* L.): genome structure and comparative analysis. Genes (Basel) 8(212):1–18

Kazakoff SH, Imelfort M, Edwards D, Koehorst J, Biswas B, Batley J, Scott PT, Gresshoff PM (2012) Capturing the biofuel wellhead and powerhouse: the chloroplast and mitochondrial genomes of the leguminous feedstock tree *Pongamia pinnata*. PLoS One 7(12):e51687

Kearse M, Moir R, Wilson A, Havas SS, Cheung M, Sturrock S, Buxton S, Cooper A, Markowitz S, Duran C, Thierer T, Ashton B, Meintjes P, Drummond A (2012) Geneious basic: an integrated and extendable desktop software platform for the organization and analysis of sequence data. Bioinformatics 28(12):1647–1649

Khan AL, Asaf S, Lee I-J, Al-Harrasi A, Al-Rawahi A (2018) First chloroplast genomics study of *Phoenix dactylifera* (var. Naghal and Khanezi): a comparative analysis. PLoS One 13(7):e0200104

Korotkova N, Nauheimer L, Ter-Voskanyan H, Allgaier M, Borsch T (2014) Variability among the most rapidly evolving plastid genomic regions is lineage-specific: implications of pairwise genome comparisons in *Pyrus* (Rosaceae) and other angiosperms for marker choice. PLoS One 9(11):e112998

Kota M, Daniell H, Varma S, Garczynski SF, Gould F, Moa WJ (1999) Overexpression of the *Bacillus thuringiensis* (Bt) Cry2Aa2 protein in chloroplasts confers resistance to plants against susceptible and Bt-resistant insects. Proc Natl Acad Sci U S A 96(5):1840–1845

Krebbers ET, Larrinua IM, McIntosh L, Bogorad L (1982) The maize chloroplast genes for the beta and epsilon subunits of photosynthetic coupling factor CFI are fused. Nucleic Acids Res 10(16):4985–5002

Kumar S, Dhingra A, Daniell H (2004) Plastid-expressed betaine aldehyde dehydrogenase gene in carrot cultured cells, roots, and leaves confer enhanced salt tolerance. Plant Physiol 136(1):2843–2854

Larekeng SI, Maskromo I, Purwito A, Matjiik NA, Sudarsono S (2015) Pollen dispersal and pollination patterns studies in Pati Kopyor coconut using molecular markers. Cord 31(1):46–60

Lee SB, Kwon HB, Kwon SJ, Park SC, Jeong MJ, Han SE, Byun MO, Daniell H (2003) Accumulation of trehalose within transgenic chloroplasts confers drought tolerance. Mol Breed 11:1–13

Li B, Zheng Y (2018) Dynamic evolution and phylogenomic analysis of the chloroplast genome in Schisandraceae. Sci Rep 8(9285):1–11

Li Z, Bai X, Ruparel H, Kim S, Turro NJ, Ju J (2003) A photocleavable fluorescent nucleotide for DNA sequencing and analysis. Proc Natl Acad Sci U S A 100:414–419

Lima EB, Sousa CN, Meneses LN, Ximenes NC, Júnior MAS, Vasconcelos GS, Lima NBC, Patrocínio MCA, Macedo D, Vasconcelos SMM (2015) *Cocos nucifera* (L.) (Arecaceae): a phytochemical and pharmacological review. Braz J Med Biol Res 48(11):953–956

Liu QB, Xue QZ (2004) Codon usage in the chloroplast genome of rice (*Oryza sativa* L. ssp. japonica). Acta Agron Sin 30:1220–1224

Liu HY, Yu Y, Deng YQ, Li J, Huang ZX, Zhou SD (2018) The chloroplast genome of *Lilium henrici*: genome structure and comparative analysis. Molecules 23(6):1276

Lohse M, Drechsel O, Kahlau S, Bock R (2013) OrganellarGenomeDRAW – a suite of tools for generating physical maps of plastid and mitochondrial genomes and visualizing expression data sets. Nucleic Acids Res 41:W575–W581

Lopes de Abreu N, Alves RJV, Cardoso SRS, Bertrand YJK, Sousa F, Hall CF, Pfeil BE, Antonelli A (2018) The use of chloroplast genome sequences to solve phylogenetic incongruences in *Polystachya* Hook (Orchidaceae Juss). PeerJ 6(e4916):1–26

Lowe TM, Chan PP (2016) tRNAs can-SE on-line: search and contextual analysis of transfer RNA genes. Nucleic Acids Res 44(W1):W54–W57

Mascher M, Wu S, Amand PS, Stein N, Poland J (2013) Application of genotyping-by-sequencing on semiconductor sequencing platforms: a comparison of genetic and reference-based marker ordering in barley. PLoS One 8(10):e76925

Maskromo I, Tenda ET, Tulalo MA, Novarianto H, Sukma D, Sukendah S (2015) Keragaman fenotipe dan genetik tiga varietas kelapa genjah kopyor asal Pati Jawa Tengah. J Littri 21(1):1–8. Indonesian

Mayer C (2008) Phobos version 3.3.12. A tandem repeats search tool for complete genomes. http://www.rub.de/spezzoo/cm

McKain MR, Hartsock RH, Wohl MM, Kellogg EA (2017) Verdant: automated annotation, alignment and phylogenetic analysis of whole chloroplast genomes. Bioinformatics 33(1):130–132

McNeal JR, Kuehl JV, Boore JL, Leebens-Mack J, dePamphilis CW (2009) Parallel loss of plastid introns and their maturase in the Genus Cuscuta. PLoS One 4(6):e5982

Menezes APA, Moreira LCR, Buzatti RSO, Nazareno AG, Carlsen M, Lobo FP, Kalapothakis E, Lovato MB (2018) Chloroplast genomes of *Byrsonima* species (Malpighiaceae): comparative analysis and screening of high divergence sequences. Sci Rep 8(1):2210

Meng D, Xiaomei Z, Wenzhen K, Xu Z (2019) Detecting useful genetic markers and reconstructing the phylogeny of an important medicinal resource plant, *Artemisia selengensis*, based on chloroplast genomics. PLoS One 14(2):e0211340

Minnullina L, Pudova D, Shagimardanova E, Shigapova L, Sharipova M, Mardanova A (2019) Comparative genome analysis of *Uropathogenic Morganella morganii* strains. Front Cell Infect Microbiol 9(167):1–14

Nair RR, Nandhini MB, Monalisha E, Murugan K, Sethuraman T, Nagarajan S, Rao NSP, Ganesh D (2012) Synonymous codon usage in the chloroplast genome of *Coffea arabica*. Bioinformation 8(22):1096–1104

Nguyen VB, Giang VNL, Waminal NE, Park HS, Kim NH, Jang W, Lee J, Yang TJ (2020) Comprehensive comparative analysis of chloroplast genomes from seven *Panax* species and development of an authentication system based on species-unique single nucleotide polymorphism markers. J Ginseng Res 44(1):135–144

Niu ZT, Pan JJ, Zhu SY, Li LD, Xue QY, Liu W, Ding X (2017) Comparative analysis of the complete Plastomes of *Apostasia wallichii* and *Neuwiedia singapureana* (Apostasioideae) reveals different evolutionary dynamics of IR/SSC boundary among photosynthetic orchids. Front Plant Sci 8:1713

Novarianto H, Maskromo I, Dinarti D, Sudarsono (2014) Production technology for kopyor coconut seednuts and seedlings in Indonesia. Cord 30(2):31–40

Nowack ECM, Price DC, Bhattacharya D, Singer A, Melkonian M, Grossmana AR (2016) Gene transfers from diverse bacteria compensate for reductive genome evolution in the chromatophore of *Paulinella chromatophora*. Proc Natl Acad Sci U S A 113(43):12214–12219

Nyrén P (1987) Enzymatic method for continuous monitoring of DNA polymerase activity. Anal Biochem 167(2):235–238

Pan K, Wang W, Wang H, Fan H, Wu Y, Tang L (2017) Genetic diversity and differentiation of the Hainan Tall coconut (*Cocos nucifera* L.) as revealed by inter-simple sequence repeat markers. Genet Resour Crop Evol 65(3):1035–1048

Phoeurk C, Somana J, Sornwatana T, Udompaisarn S, Traewachiwiphak S, Sirichaiyakul P, Phongsak T, Arthan D (2018) Three novel mutations in α-galactosidase gene involving in galactomannan degradation in endosperm of curd coconut. Phytochemistry 156:33–42

Puigbò P, Bravo IG, Vallve SG (2008) CAIcal: a combined set of tools to assess codon usage adaptation. Biol Direct 3(38):1–8
Purseglove JW (1968) Origin and distribution of the coconut. Trop Sci 10:191–199
Qu XJ, Moore MJ, Li D-Z, Yi TS (2019) PGA: a software package for rapid, accurate, and flexible batch annotation of plastomes. Plant Methods 15(50):1–12
Raubeson LA, Jansen RK (2005) Chloroplast genomes of plants. In: Henry RJ (ed) Plant diversity and evolution. CABI Publishing, London, pp 45–68
Rogalski M, Vieira LDN, Fraga HP, Guerra MP (2015) Plastid genomics in horticultural species: importance and applications for plant population genetics, evolution, and biotechnology. Front Plant Sci 6(586):1–17
Ronaghi M, Uhlén M, Nyrén P (1998) A sequencing method based on real-time pyrophosphate. Science 281(5375):363–365
Rusk N (2009) Cheap third-generation sequencing. Nat Methods 6(4):244–245
Sakka H, Baraket G, Dkhil SD, Azzouzi SZ, Hannachi AS (2013) Chloroplast DNA analysis in Tunisian date-palm cultivars (Phoenix dactylifera L.): sequence variations and molecular evolution of trnL (UAA) intron and trnL (UAA) trnF (GAA) intergenic spacer. Sci Hortic 164:256–269
Sakulsathaporn A, Wonnapinij P, Vuttipongchaikij S, Apisitwanich S (2017) The complete chloroplast genome sequence of Asian Palmyra palm (*Borassus fabellifer*). BMC Res Notes 10(740):1–7
Sanger F, Nicklen S, Coulson AR (1977) DNA sequencing with chain-terminating inhibitors. Proc Natl Acad Sci U S A 74(12):5463–5467
Setiawan A, Rahayu MS, Maskromo I, Purwito A, Sudarsono (2020) Inheritance pattern of endosperm quantity and Kopyor coconut (*Cocos nucifera* L.) fruit variations. IOP Conf Ser Earth Environ Sci 418(2020):012039
Shen X, Wu M, Liao B, Liu Z, Bai R, Xiao S, Li X, Zhang B, Xu J, Chen S (2017) Complete chloroplast genome sequence and phylogenetic analysis of the medicinal plant *Artemisia annua*. Molecules 22(8):1–14
Shi L, Chen H, Jiang M, Wang L, Wu X, Huang L, Liu C (2019) CPGAVAS2, an integrated plastome sequence annotator and analyzer. Nucleic Acids Res 47:W65–W73
Smarda P, Bureš P, Horová L, Leitch IJ, Mucina L, Pacini E, Tichý L, Grulich V, Rotreklová O (2014) Ecological and evolutionary significance of genomic GC content diversity in monocots. PNAS 111(39):E4096–E4102
Smith DR (2009) Unparalleled GC content in the plastid DNA of Selaginella. Plant Mol Biol 71(6):627–639
Smith DR (2015) Mutation rates in plastid genomes: they are lower than you might think. Genome Biol Evol 7(5):1227–1234
Sugiura M, Hirose T, Sugita M (1998) Evolution and mechanism of translation in chloroplast. Annu Rev Genet 32:437–459
Talat F, Wang K (2015) Comparative bioinformatics analysis of the chloroplast genomes of a wild diploid *Gossypium* and two cultivated allotetraploid species. Iran J Biotechnol 13(3):47–56
Tillich M, Lehwark P, Pellizzer T, Ulbricht-Jones ES, Fischer A, Bock R, Greiner S (2017) GeSeq-versatile and accurate annotation of organelle genomes. Nucleic Acids Res 45(W1):W6–W11
Uthaipaisanwong P, Chanprasert J, Shearman JR, Sangsrakru D, Yoocha T, Jomchai N, Jantasuriyarat C, Tragoonrung S, Tangphatsornruang S (2012) Characterization of the chloroplast genome sequence of oil palm (*Elaeis guineensis* Jacq.). Gene 500(2):172–180
Vieira MLC, Santini L, Diniz AL, Munhoz CF (2016) Microsatellite markers: what they mean and why they are so useful. Genet Mol Biol 39(3):312–328
Vogel J, Hubschmann T, Borner T, Hess WR (1997) Splicing and intron-internal RNA editing of trnK-matK transcripts in barley plastids: support for matK as an essential splice factor. J Mol Biol 270:179–187

Wang Y, Zhan D, Jia X, Mei W, Dai H, Chen X, Peng S (2016) Complete chloroplast genome sequence of *Aquilaria sinensis* (Lour.) Gilg and evolution analysis within the Malvales Order. Front Plant Sci 7(280):1–13

Wang H, Park SY, Lee AR, Jang SG, Im DE, Jun TH, Lee J, Chung JW, Ham TH, Kwon SW (2018) Next-generation sequencing yields the complete chloroplast genome of *C. goeringii* acc. smg222 and phylogenetic analysis. Mitochondrial DNA Part B 3(1):215–216

Wilson C (2004) Phylogeny of Iris based on chloroplast *matK* gene and *trnK* intron sequence data. Mol Phylogenet Evol 33(2):402–412

Wyman SK, Jansen RK, Boore JL (2004) Automatic annotation of organellar genomes with DOGMA. Bioinformatics 20:3252–3255

Yamada KD, Tomii K, Katoh K (2016) Application of the MAFFT sequence alignment program to large data–reexamination of the usefulness of chained guide trees. Bioinformatics 32(21): 3246–3251

Yan L, Lai X, Li X, Wei C, Tan X, Zhang Y (2015) Analyses of the complete genome and gene expression of the chloroplast of sweet potato [*Ipomoea batata*]. PLoS One 10(4):e0124083

Yan M, Xiong Y, Liu R, Deng M, Song J (2018) The application and limitation of universal chloroplast markers in discriminating east Asian evergreen oaks. Front Plant Sci 9(569):1–15

Yang M, Zhuang X, Liu G, Yin Y, Chen K, Yun Q, Zhao D, Al-Mssallem IS, Yu J (2010) The complete chloroplast genome sequence of date palm (*Phoenix dactylifera* L.). PLoS One 5(9):e12762

Yao X, Tang P, Li Z, Li D, Liu Y, Huang H (2015) The first complete chloroplast genome sequences in Actinidiaceae: genome structure and comparative analysis. PLoS ONE 10(6):e0129347

Yi DK, Lee HL, Sun BY, Chung MY, Kim KJ (2012) The complete chloroplast DNA sequence of *Eleutherococcus senticosus* (Araliaceae); comparative evolutionary analyses with other three asterids. Mol Cell 33(5):497–508

Zang M, Su Q, Weng Y, Lu L, Zheng X, Ye D, Zheng R, Cheng T, Shi J, Chen J (2019) Complete chloroplast genome of *Fokienia hodginsii* (Dunn) Henry et Thomas: insights into repeat regions variation and phylogenetic relationships in Cupressophyta. Forests 10(7):1–15

Chapter 11
Genomics, Phenomics, and Next Breeding Tools for Genetic Improvement of Safflower (*Carthamus tinctorius* L.)

Abdurrahim Yılmaz, Mehmet Zahit Yeken, Fawad Ali, Muzaffer Barut, Muhammad Azhar Nadeem, Hilal Yılmaz, Muhammad Naeem, Burcu Tarıkahya Hacıoğlu, Yusuf Arslan, Cemal Kurt, Muhammad Aasim, and Faheem Shehzad Baloch

Contents

A. Yılmaz · M. Z. Yeken · Y. Arslan
Department of Field Crops, Faculty of Agriculture and Natural Sciences, Bolu Abant Izzet Baysal University, Bolu, Turkey

F. Ali
Department of Plant Sciences, Quaid-I-Azam University, Islamabad, Pakistan

M. Barut · C. Kurt
Department of Field Crops, Faculty of Agriculture, Çukurova University, Adana, Turkey

M. A. Nadeem · M. Aasim · F. S. Baloch (✉)
Faculty of Agricultural Sciences and Technologies, Sivas University of Science and Technology, Sivas, Turkey

H. Yılmaz
Department of Plant and Animal Production, Izmit Vocational School, Kocaeli University, Kocaeli, Turkey

M. Naeem
Department of Plant Breeding and Genetics, Faculty of Agriculture and Environmental Sciences, The Islamia University of Bahawalpur, Bahawalpur, Punjab, Pakistan

B. Tarıkahya Hacıoğlu
Department of Biology, Faculty of Science, Hacettepe University, Ankara, Turkey

H. Tombuloglu et al. (eds.), *Oil Crop Genomics*,
https://doi.org/10.1007/978-3-030-70420-9_11

11.1 Introduction

Oilseed plants are important products that are able to adapt to different agricultural conditions of the world from temperate to tropical regions (Singh and Nimbkar 2016). Safflower (*Carthamus tinctorius* L.), a perennial oilseed plant, is a member of the Compositae family (Weiss 2000), having $2n = 24$ chromosomes (Ashri 1957; Singh and Nimbkar 2007; Golkar 2014). Safflower is known as one of the oldest crop plants in the human history (Weiss 2000). It has been in existence for more than 4000 years (Pearl et al. 2014). Traditionally, this crop has been produced from Mediterranean region to China and along the Nile valley to Ethiopia (Weiss 1971). It has been cultivated in India for centuries, especially for the production of red-orange paint extracted from bright flowers (Knowles 1958; Singh and Nimbkar 2007; Weiss 1971). Safflower was anciently used as fat in meals and dye sector in Iran-Afghanistan area and China (Weiss 2000; Sehgal and Nath Raina 2011). On the other hand, the flowers of this crop were used in Mexico and have been grown for different purposes in many other regions of the world (Sehgal and Nath Raina 2011).

Besides the production of safflower for edible oil purposes, it also imparted a key role in the dye sector, and its presence in the medical and cosmetic sectors is incontrovertible. To an attractive orange color, the florets of safflower are mixed into bread, rice, and pickles (Sehgal and Nath Raina 2011). It is widely used in herbal products in China (Li and Mündel 1996). Its shoots and tender leaves are rich in calcium, phosphorus, iron, and vitamin A and are used in salads (Nimbkar 2002). Also, young plants are evaluated and marketed as vegetables in India and some other neighbor countries. Thanks to its monounsaturated and polyunsaturated fatty acids and its ability to cure many chronic diseases, it is rich in pharmaceutical dyes and nature dyes (Weiss 1971; Li and Mündel 1996). Furthermore, safflower oil plays a very important role in human nutrition due to its high α-tocopherol and linoleic acid content (Furuya et al. 1987; Velasco and Fernandez-Martinez 2001).

It has been shown that at the flowering stage, safflower is sensitive to water (Quiroga et al. 2001; Bassil and Kaffka 2002; Mirzahashemi et al. 2015) and well-adapted to growth in dry environments (Merrill et al. 2002; Kar et al. 2007; Pearl et al. 2014) Additionally, safflower is more resistant to salt stress compared to other oilseed crops and can be grown in arid soils where soil salinity is the most fundamental problem (Bayramin and Kaya 2009). Safflower sowing is a much more profitable product for agricultural laborer in some countries than lentil, chickpea, and barley and can be offered as an alternative to agricultural areas such as fallow, arid and semiarid, and salty areas (Weiss 1971; Dajue and Mundel 1996; Yau 2004).

In present days, it is cultivated in a wide range of countries including the USA, Argentina, Mexico, India, Kazakhstan, Pakistan, Uzbekistan, Iran, Russia, China, Spain, Turkey, Ethiopia, Tanzania, and Australia. The adaptation of this crop to such different climates since ancient times shows the higher sustainability level of safflower. In short, this crop is going to be one of the important plants of the future owing to its facilities and conveniences at the stage of breeding and the use and advantages of the product obtained after harvesting in many areas of the world (Singh and Nimbkar 2016).

11.2 Name

Safflower has been called with a higher variety of names by humans in different parts of the world: zaffer, zaffrone, zaffrole, zafaran-golu, usfur, usfar, ssuff, thistle saffron, suff, suban, su, snecus, safflor, saffiore, qurtum, qortom, ostur, osfor, onickus, muswar, maswarh, ma, laba torbak, kusumba, kusuma, kusum, kouchangule, kosheh, khoube'e khasq, khasdonah, khartum, kharkhool, khariah, khardam, kazireh, kazhirak, kavisheh, kasumbha, kariza, kardi, karar, kar, kamal lotarra, kajireh, kajena-goli, kahil, kafshe, kafsha, kafesheh, ihriz, hung hua, hubulkhortum, hong hua, hebu, golzardu, golrang, golbar aftab, ghurtom, flase, false saffron, dyer's saffron, dikken, cusumba, cnikos, cnicus, cnecus, carthamos, carthamo, carthami flos, cartham, cartamo, brarta, benihana, benibana, bastard saffron, azafrancillo, assfrole, assfore, aspir, asper, asfiore, and agnisikha (Chavan 1961; Weiss 1971; Salunkhe et al. 1992; Smith 1996).

11.3 Phenomics of Safflower

Safflower plant is a dwarf, herbaceous annual plant with several branches classified as primary, secondary, and tertiary with a terminal capitulum (Singh and Nimbkar 2016). The plant can grow up to 1 meter even if in poor and arid soil under full sun. There are two different forms of safflower: prickly spiny and spineless (Gautam et al. 2014). In those with spiny ones, stems and branches have numerous spiny leaves. The plant has smooth, bright, and white seeds (maybe with or without pappus) weighing between 0.01 and 0.10 g (Singh and Nimbkar 2016). Stems are shining white and glabrous. Cauline leaves are green, glabrous, sessile, ovate to linear ovoid, margins entire to finely spiny-serrate, and with 3–9 x 1–2 cm dimensions. Outer phyllaries are ovate and shorter than to slightly exceeding inner, inner arachnoid, acuminate. The capitula are in loose corymbs. All achenes are epappose. The

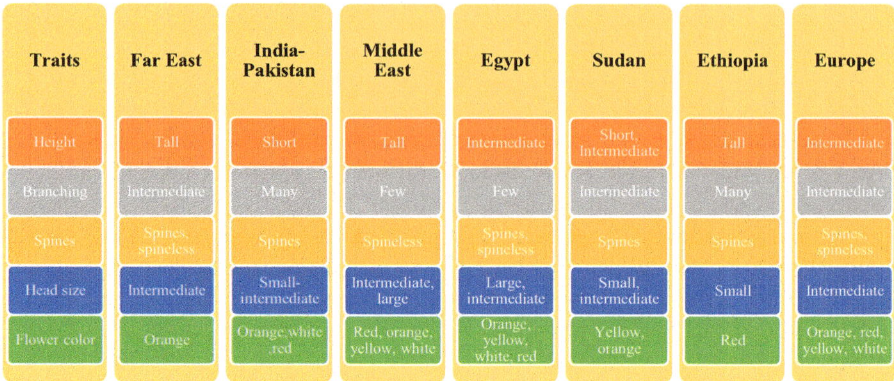

Fig. 11.1 Safflower characteristics of different geographical regions. (Adapted from Knowles 1969 and Smith 1996)

flowers are yellow-orange in color (Davis 1975). Knowles (1969) identified seven geographical regions to compare safflower characteristics. The plant appearance has some differences to seven geographical regions. These differences are separated in Fig. 11.1.

Growth is slow in the initial stage after safflower germination. This stage is called as rosette; several leaves are shaped on the stem base, lasting between 20 and 25 days. Rapid elongation of the root and abundant branching occurs after this stage. The globular flower capitulum occurred at each branch end, which is fully supported by the plant and surrounded by tightly secured supports. Safflower roots have a taproot system of stretching up to 2–3 m in soils with sufficient depth (Singh and Nimbkar 2016).

Safflower plant completes maturation between 110 and 150 days as a spring crop and in 200 or more days as a fall crop. The plant is waited to dry completely for harvest (Duke 1983). Capitulum diameter, capitula per plant, branches per plant, plant height, 1000-seed weight, and seeds per capitulum are the most important features to follow in order to increase seed yield (Hamadi et al. 2001; Rudra Naik et al. 2001). These traits are in indirect or direct correlation with seed yield (Mahasi et al. 2006; Camas and Esendal 2006). Some views of these traits were presented in Fig. 11.2.

Adalı and Öztürk (2017) tested Remzibey-05, Black Sun 2, KS-07, Balci, AC Stirling, Ole, V 50/63, Dinçer, Ayaz, BDYAS-4, Linas, Yenice, and TRE-Aso 12/08 safflower varieties under Konya ecological conditions of Turkey. They found that KS-07 variety was identified as the highest yielding as 3920 kg/ha. In another study conducted by Hatipoğlu et al. (2017), Dinçer and Remzibey-05 standard varieties were experimented for 3 years in order to determine the correct planting time, and as the result of the study, Remzibey-05 had the highest seed yield value as 4260 kg/ha sown on October 30. As mentioned above, the average seed yields of the countries are below 200 kg. However, it is seen that safflower seed yield can reach more

Fig. 11.2 Safflower capitulum, flowers, and seeds, respectively (first view, with permission of the Yaman 2014; second and third views were taken by Yılmaz A)

than 400 kg when proper management and correct ecotype selection is maintained with these results.

11.4 Chemical Compositions of Essential Oil in Safflower

The concentrated hydrophobic liquid containing volatile chemical compounds from different plant species is known as an essential oil. Essential oils may also be referred to as volatile oils, ethereal oils, aetherolea, or simply as the oil of the plant from which they were extracted. The term "essential" is applied to the essential oil as it contains the "essence of" of that specific plant characteristic fragrance from which it is derived. Ziarati et al. (2012) evaluated the essential oils composition from the dried flowers of safflower cultivated in Iran via GC-MS (gas chromatography-mass spectrophotometry) and GC (gas chromatography). They reported a total of 29 compounds as the essential oils composition. The major compounds comprised of 1-hydroxy-3-propyl-5-(4-methyl-penten)-2-methylbenzene (25.2%), 2,5,5-trimethyl-3-propyl,tetra-hydro1-naphtol (19.8%), and benzaldehyde (8.0%). Complete list of the 29 essential oils compounds are presented in Table 11.1.

11.5 Chemical Compositions of Fatty Acids in Safflower

Safflower is not popular only for its colorful petals but also utilized as edible oil in cooking. Sufficient variability for fatty acid composition is present in its oil (Camas and Esendal 2006). Safflower seeds contain oil (30–40%), protein (15–20%), and hull (35–45%) (Rahamatalla et al. 2001). Approximate compositions of fatty acids in safflower seeds are about 71–75% (linoleic acid), 16–20% (oleic acid), 6–8% (palmitic acid), and 2–3% (stearic acid) (Nagaraj 1993). Safflower cultivars containing high proportion of linoleic acid (87–89%) and oleic acid (over 85%) are also reported (Kumar et al. 2016). Recently, safflower is more popularized as oilseed

Table 11.1 Exploring the chemical composition from essential oils of *C. tinctorius* flower

Compound[a]	%	RI[b]	RI[c]
2,5,5-Trimethyl-3-n-propyl,tetra hydro1-naphtol	19.8	2034	–
1-Hydroxy-3-propyl-5-(4-methyl-panten)-2-methyl (benzene)	25.2	2016	–
Myristic acid	0.5	1943	–
β-Turmerone	1	1868	1664
Caryophylla-4(12),8(13)-diene-5-beta-ol	2.8	1859	1639
Caryophyllene oxide	6.5	1807	1581
Spathulenol	0.6	1797	1576
Lauric acid	5.1	1746	1568
Myristicin	0.5	1724	1520
β-Bisabolene	0.3	1707	1509
β-Lonone	0.6	1690	1485
Curcumene	0.3	1682	1483
α-Humulene	0.3	1672	1454
β-Caryophyllene (z)	4.3	1638	1404
α-Elemene	0.6	1620	1391
α-Copaene	0.3	1588	1376
Eugenol	0.7	1557	1356
α-Terpinyl acetate	3.2	1547	1340
Carvacrol	2.9	1493	1298
2,6,6-Trimethyl aldehyde, 1,3-cyclohexadiene-1-carboxaldehyde	0.9	1489	1293
Thymol	2.6	1482	1290
Benzaldehyde	8	1443	1257
α-Terpineol	0.4	1390	1189
Decanone <3>	0.2	1384	1186
Terpinen-4-ol	0.7	1379	1177
Linalool	0.5	1289	1098
γ-Terpinene	0.3	1251	1062
n-Decane	0.3	1180	999
n-Octane	0.3	974	800
Total	89.7		

Adapted from Ziarati et al. (2012)
[a]Compounds listed in order of elution; [b]*RI* retention index, [c]*RI* retention index

crop due to its good oil quality and fatty acid composition (Camas and Esendal 2006; Yeilaghi et al. 2012). It leads to the development of safflower breeding programs for the improvement of oil content and fatty acid profile by determining its genetic control. Safflower germplasm with oil content above 43%, oleic acid content above 75%, and linoleic acid content above 80% were recommended for future breeding programs by Kumar et al. (2016) and Li et al. (1993). With the knowledge of quantitative genetics study, it has been revealed that safflower oil content is controlled by nonadditive gene action (Golkar et al. 2011). Similarly, Yermanos et al. (1967) resulted in epistatic gene effects for the control of oil content. That is why both broad and narrow sense heritability estimates have been reported for the oil content of the safflower crop (Golkar et al. 2011). Additive gene actions are reported to be operated in linoleic acid (Hamdan et al. 2008a, b), palmitic acid and stearic acid (Hamdan et al. 2009a), and oleic acid (Hamdan et al. 2009b). Furthermore, Golkar et al. (2011) described that stearic acid and linoleic acid contents of the safflower are also influenced by the maternal effects. Fernandez-Martinez et al. (1993) proposed that recessive alleles are responsible for high oleic acid content, while Ladd and Knowles (1971) suggested that stearic acid is single gene inherited. Based on the above discussion, it is highly suggested to explore the safflower oil content and fatty acid profile conducting more research work through quantitative genetics study.

11.6 Origin and Diffusion

According to Tarıkahya Hacıoğlu et al. (2014), ancestral area reconstruction analysis suggests that the genus *Carthamus*, in which cultivated safflower belonged, originated in North Africa/West Asia. There are many claims on the origin and diffusion of safflower. For example, Vavilov (1951) and Kupsow (1932) thought that there are three main origins (India, Iran-Afghanistan region, and Ethiopia). Afterward, De Candolle (1890) thought that Arabia is the center of origin. Then, some studies are carried out as cytogenetic and divergence of evolution of this crop; Ashri and Knowles (1960, 1977) and Ashri (1973) have stated that the Euphrates basin is the true origin of safflower. However, archaeological sources have revealed that the earliest evidence of safflower was found in Egypt (Weiss 1971).

Additionally, domestication of safflower was initially started in the Fertile Crescent before 4000 years. Thus, it can be considered as one of the oldest plants in the world (Weiss 1971; Ashri et al. 1975). As reported by many scientists, this crop has been cultivated in Egypt, China, India, and Iran since prehistoric times (Ashri 1975; Abel and Discroll 1976; Dajue and Yuzhou 1993; Bergman and Flynn 2001; Popov and Kang 2011). In Europe (France, Italy, and Spain), the crop was firstly cultivated in the middle age (Duke 1983).

Knowles and Ashri (1958) and Conners (1943) claimed that the first detection of safflower coincided before the 1900s in North America. Besides this detection, Duke (1983) claimed that America first met this plant in 1925. However, it was brought to Mexico much earlier than probably from Spain (Wiesner 1927), which corresponds to the time immediately after the discovery of America. Duke (1983) reported that the Spanish first took the safflower to Mexico after the discovery of America. Then, it was transferred to Colombia and Venezuela. To clarify all this information, the origin and diffusion of the safflower plant have been combined as an extract on the basis of all these researches in Fig. 11.3.

Nowadays, there are some differences in the production pattern in the world according to the regions where safflower first diffuses. For example, until 1980, Mexico was the country with the largest safflower production in the world. Mexico, which had 528,000 ha cultivation area in 1980, has decreased its production rapidly

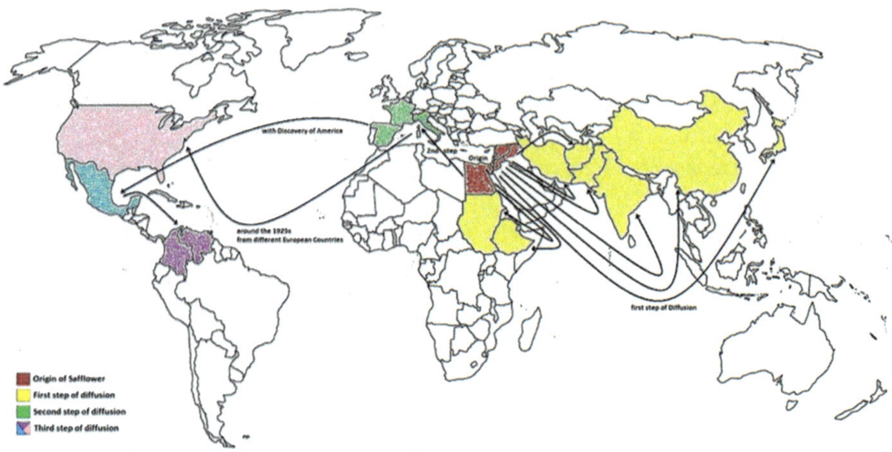

Fig. 11.3 Origin and diffusion stages of safflower (Painted by Yılmaz A)

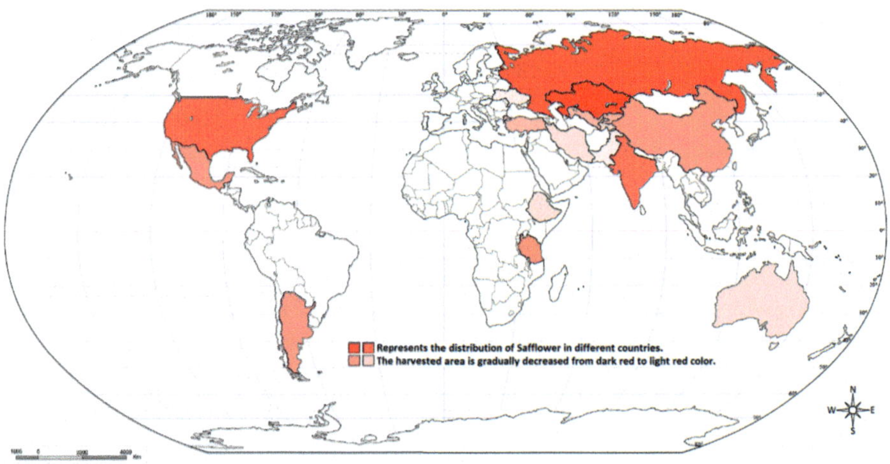

Fig. 11.4 Distribution of safflower all over the world presently (FAO 2021)

since then (Cervantes-Martinez 2001). The current cultivation area is even less than 10% of the Mexico's 1980 cultivation area. On the other hand, while America just started commercial safflower production in the 1950s (Esendal 2001), it ranks fifth in the world sowing area rankings in present days (FAO 2021). Accordingly, countries such as Kazakhstan, Russia, the USA, India, and Argentina have been drawn attention as countries showing growth in production so far (FAO 2021, Fig. 11.4). Kazakhstan (262,768 ha) is the leading country for harvested area. Subsequently, Russia (106,952 ha), the USA (61,800 ha), India (45,890 ha), and Argentina (28,646 ha) were identified as other countries that have completed the top 5 list, respectively. Other countries are as follows: Mexico (27,828 ha), Tanzania (27,823 ha), China (22,570 ha), Uzbekistan (16,218 ha), Turkey (15,860 ha), and Kyrgyzstan (12,414 ha). Countries producing safflower in less areas can be listed as follows: Ethiopia (7561 ha), Australia (6389 ha), Tajikistan (5061 ha), and Iran (4714 ha) (FAO 2021, Fig. 11.4).

In the safflower, yield per field varies significantly from region to region. The highest value of yield belongs to Mexico (1856 kg/ha) all over the world. This country is followed by Tajikistan (1825 kg/ha), China (1468 kg/ha), and the USA (1426 kg/ha), respectively. Kazakhstan (760 kg/ha), Russia (759 kg/ha), and India (537 kg/ha), which have high sowing areas, have very low yields compared to these countries (FAO 2021). This is due to the variability of varieties as being winter or summer version, as well as variability of climatic, environmental, and soil conditions.

There are also serious differences in the production quantity ranking due to the changes in yield. The country ranking in terms of yield is as follows: Kazakhstan (199.789 tonnes), the USA (88.130 tonnes), Russia (81.189 tonnes), Mexico (51.655 tonnes), China (33.128 tonnes), India (24.640 tonnes), Argentina (24.327 tonnes), and Turkey (21.883 tonnes) (FAO 2021).

11.7 Safflower Similarity Centers

In 1969, Knowles described a system of matching traits of safflower in different geographic regions and called as similarity centers. These centers have been proposed by various researchers using different parameters like plant height, plant branching patterns, spines, capitulum size, and flower color, and they revealed dissimilarities among these centers. Safflower accessions within the same similarity center observed relatedness to each other compared to other similarity centers. Knowles (1969) proposed seven similarity centers: Europe, Middle East, Egypt, India-Pakistan, Far East, Ethiopia, and Sudan. Then, Ashri (1975) expanded this list, and offered ten similarity centers Europe, Near East, Turkey, Iran-Afghanistan, India-Pakistan, Egypt, Sudan, Ethiopia, Kenya, and Far East for safflower based on various morphoagronomic traits. Chapman et al. (2010) suggested five similarity centers (Europe, the Far East-Pakistan-India, Israel-Syria-Jordan, Egypt-Ethiopia, and Turkey-Iraq-Iran-Afghanistan) for safflower using EST-SSR markers. But still, accessions similarity extent within the same center compared to other similarity centers was unclear at molecular level. Very recently, Ali et al. (2019, 2020a, b) comprehensively studied the genetic diversity and safflower similarity centers pattern conducting two field experiments at two diverse locations (Pakistan and Turkey) and utilizing iPBS-retrotransposon markers. They supported the presence of the seven similarity centers hypothesis throughout the world for safflower.

11.8 Weed and Wild Relatives of *Carthamus tinctorius* L. (*Carthamus* spp.)

The genus *Carthamus* includes 25 species, and among these species, only *Carthamus tinctorius* L. is cultivated on a global scale (Ashri and Knowles 1960; Arslan, 2018). Information about the taxonomic relationships serves as a basis for the active utilization of desirable traits in wild relatives of safflower in breeding activities (Dajue and Mundel 1996). In the previous studies, several classifications were proposed for the genus *Carthamus* because of morphologic and genetic differences by Ashri and Knowles (1960), Cassini (1819), De Candolle (1838), Estilai (1977), Hanelt (1961), Knowles (1958), Lopez-Gonzalez (1989), Vilatersana (2005), and Sehgal et al. (2009). According to Ashri and Knowles (1960), the genus was classified under four sections based on chromosome numbers and morphological traits. These four sections and additional species are presented as follows.

Section I (2n = 24)

***C. tinctorius* L.**: Cultivated species of safflower was cultured in India, Pakistan, Iran, Turkey, Jordan, Israel, and Syria (Ashri and Knowles 1960). Present days, this species is cultivated in the different parts of the world. The plant height of this crop is approximately 25–45 cm. Stems are glabrous, flowers vermilion to yellow, and all achenes epappose (Davis 1975; Arslan et al. 2010).

***C. palaestinus* Eig. (Syn. *C. persicus*)**: This species is the genetically and morphologically closest relative to the cultivated safflower (Chapman and Burke 2007; Sasanuma et al. 2008; Tarıkahya Hacıoğlu et al. 2014). Arslan (2018) reported that this species has been generally found near Ankara, Mersin, and Sanliurfa provinces in Turkey.

***C. oxyacantha* M. Bieb.**: This species, generally called as "Poli" or "Peeli kandiary," is a winter season plant growing up in the wheat fields (Ahmad Zadeh et al. 2011; Bukhsh et al. 2014). The dispersal of this species is in Western Iraq, North-West India, Iran, Kazakhstan, Turkmenistan, Uzbekistan, and Turkey (Ashri et al. 1974).

The abovementioned species were identified as glabrous or pubescent; external involucral bracts green, ovate to linear; internal bracts entire at apex; florets not saccate; corollas red, yellow, white, or orange; pollen grains yellow; and pappus none or chaffy (Singh and Nimbkar 2006). These species cross easily and generate fertile hybrids, and they are strictly linked to each other (Ashri and Knowles 1960).

Section II (2n = 20)

***C. syriacus* (Boiss.) Dinsm.**: The species naturally grow on the eastern side of the Mediterranean region (Singh and Nimbkar 2006). It has been considered as a weed in chickpea fields by Singh and Diwakar (1995).

***C. glaucus* Bieb.**: This species is a weedy plant. It is different from the other species, as it owns a bigger head and ovate bracts (Singh and Nimbkar 2006). The plant height is between 20 and 65 cm. Stem colors are changed from brownish to straw-colored, and flower colors are purplish-pink (Davis 1975; Arslan et al. 2010).

***C. tenuis* (Boiss & Blanche) Bornm.**: The distribution of this species is in western parts of the Middle East (Azab 2018). V. Kuete et al. (2012) indicated that the plant was traditionally used in Egypt to stop abortion and to enlarge fertility and acts as an aphrodisiac. The plant height is about 80 cm. Stem and flower colors of this species are pale brown and purplish (Davis 1975; Arslan et al. 2010).

***C. alexandrinus* (Boiss. & Heldr.) Bornm.**: Schank and Knowles (1964) stated that this species was confinement to Northern Egypt and has explicit characteristics. Some of these explicit characteristics are long external involucral bracts, dark-purple anthers, decreased number of rudimentary ovaries, and a great degree of self-fertility (Schank and Knowles 1964).

The species given above have generally blue or pink flowers (Singh and Nimbkar 2006). Further species in this group are *C. dentatus* Vahl, *C. boissieri* Halacsy, *C. leucocaulos* Sibth. and Sm., *C. glaucus* subsp. *anatolicus* (Boiss.) Hanelt and subsp. *glandulosus* Hanelt, *C. rechingeri* Davis, *C. ambiguus* Heldr., *C. ruber* Link, and *C. sartori* Held. (Singh and Nimbkar 2006). Ashri and Knowles (1960) stated that section I and II species are not closely linked to each other.

Section III (2n = 44)

***C. lanatus* L.**: This species is native to the Mediterranean region and known as saffron thistle (Morin and Sheppard 2012). It spreads naturally in Turkey, Spain, Greece, Morocco, and Portugal and has antitumor, interferon-inducing, and sedative activities (Bocheva et al. 2003; Singh and Nimbkar 2006). The plant height is about 15–75 cm, stems are brownish to straw-colored, and flowers are yellow (Davis 1975; Arslan et al. 2010).

Section IV (2n = 64)

***C. baeticus* (Boiss. & Reuter) Nyman**: This species is an allopolyploid plant holding three diverse genomes: two different nonhomologous genomes of 10 chromosomes and one 12-chromosome genome (Singh and Nimbkar 2006). This species naturally grows in the Mediterranean region, Spain, and North Africa (Ashri and Knowles 1960).

***C. turkestanicus* Popov.**: This species, which originated from *C. lanatus* and *C. glaucus* subsp. *glaucus*, is allopolyploid existing in West Asia and Ethiopia (Garnatje et al. 2006).

Additional Carthamus Species (Unclassified by Ashri and Knowles)

***C. divaricatus* (Beg. & Vaccari) Pamp. (2*n* = 22)**: This species, which is the only safflower species having 22 chromosomes, grows in Libya (Knowles 1988). It has purple, white, or yellow flowers with yellow pollen (Ashri and Knowles 1960). It is not capable of self-pollination, and it crosses easily with species having 20 chromosomes but generates partly fertile hybrids (Singh and Nimbkar 2006). Additionally, this species can be crossed with *C. tinctorius* and it generates sterile progeny (Ashri and Knowles 1960).

***C. caeruleus* L. (2*n* = 24):** It has been found in North Africa and Iberian Peninsula (Ashri and Knowles 1960).

***C. arborescens* L. (2*n* = 24):** A study revealed that this species is the most primitive species in the genus (Sasanuma et al. 2008). It has been found in Spain and some areas of North Africa.

C. arborescens and *C. caeruleus* do not cross with any other *Carthamus* species due to their explicit morphological characteristics (Ashri and Knowles 1960). Thus, *C. caeruleus* and *C. arborescens* were not grouped to the four sections.

***C. rhiphaeus* Font Quer & Pau:** The species grows in Morocco and appears to be morphologically strictly related to *C. arborescens* (Ashri and Knowles 1960).

***C. nitidus* Boiss.:** This species was detected as having 24 chromosomes by Lopez-Gonzalez (1990).

Section I, which includes cultivated safflower, formed a separate cluster in the ITS phylogeny inference and split from the other *Carthamus* lineages around 0.24 million years ago (Tarıkahya Hacıoğlu et al. 2014). According to Bülbül et al. (2013), *C. lanatus* was the most dissimilar species, and *C. tenuis* and *C. glaucus* had the most similar pollens in pollen analyses of five wild species naturally grown in Turkey. Very recently, Arslan (2018) indicated that significant variations were observed among some wild safflower species on several morphological traits such as rosette period, plant height, 1000-seed weight, branches per plant, and days to flowering and maturity. To understand the taxonomic relationship of the genus, it is needed to consider both morphologic and genetics studies. Parallel to morphological and chromosomal classifications of safflower given above, genetic diversity studies confirmed that there was great polymorphism between the wild relatives of safflower (Sabzalian et al. 2009; Barati and Arzani 2012; Derakhshan et al. 2014; Yaman et al. 2014). These studies revealed that some morphological traits of the wild species of safflower can be contributed to understand the taxonomy and distribution of *Carthamus* species. The views of achene and flower morphology and light and electron microscope micrographs of pollen grains of some wild *Carthamus* species are presented in Fig. 11.5.

11.9 Safflower Genetic Resources and the Idea of Core Collection

Germplasm is the gene pool for traits diversity and discloses a significant role to the crop improvement. Larger population size and heterogeneous structure of germplasm confine its easy availability and usage for different breeding programs (Noirot et al. 1996; van Hintum 2000). Worldwide different gene/seed banks conserved the collected safflower germplasm. The National Bureau of Plant Genetic Resources in New Delhi (India) and Project Coordinating Unit for Safflower in Solapur (India) contain 2393 and 7525 safflower accessions, respectively. The Western Regional Plant Introduction Station (WRPIS) (USA) contains more than 2400 accessions,

Fig. 11.5 The views of seed morphology, flower morphology, and light and electron microscope micrographs of pollen grains of some wild *Carthamus* species: **a**, *C. dentatus*; **b**, *C. glaucus*; **c**, *C. lanatus*; **d**, *C. persicus*; **e**, *C. tenuis*. (with permission of Bülbül et al. 2013 and Arslan 2018)

while Iran and Turkey possess 200 and 125 accessions. Countries like, Iraq, Syria, Kazakhstan, Uzbekistan, and Tajikistan also contain some safflower germplasm (Johnson et al. 2008) (Table 11.2). For the most effective organization and application of the crop germplasm resources, Frankel (1984) familiarized the idea of "core collection." A core collection is the minimum subset of that crop germplasm containing huge amount of variability widespread in the whole germplasm. So, it is an easy job to characterize and evaluate the core collection with respect to the whole germplasm collection. Initial attempts were made to characterize and evaluate the core collection using agro-morphological traits and geographical distribution (Huaman et al. 1999; Upadhyaya and Ortiz 2001; Li et al. 2005; Mahalakshmi et al. 2007; Bhattacharjee et al. 2007; Upadhyaya et al. 2009). With development of molecular markers, they were used to elucidate genetic variability with greater efficacy. Usage of molecular markers greatly facilitated the robust germplasm core collections development either alone (Zhang et al. 2009) or in combination with phenotypic data (Wang et al. 2006; Ebana et al. 2008; Shehzad et al. 2009; Belaj et al. 2012; Díez et al. 2012; Liu et al. 2015).

Johnson et al. (1993) developed the first safflower core collection that comprised of 210 accessions from the evaluation of 2042 accessions obtained from 50 countries of the world. Similarly, Dwivedi et al. (2005) also developed safflower core collection of 570 accessions from the evaluation of 5522 accessions collected from 38 countries. One thing must be kept in mind that most of the agronomic traits are quantitative and are considerably affected by genotype × environment (G × E) interactions. Therefore, the available data regarding agro-morphological and geographical distribution might be changing due to genotype × environment interactions. It is therefore highly recommended that strong efforts are required to explore genetic diversity utilizing different marker systems and to develop more effective and robust safflower core collections. Safflower germplasm resources conserved in different gene/seed banks are given in Table 11.2. List of safflower germplasm suggested for future breeding programs is presented in Table 11.3.

11.10 Trade in Safflower

Safflower plant is used in many different areas of the world for a variety of uses. Colorful flowers (petals) of safflower are evaluated as fabric and food coloring. Seeds with 30–45% oil content are used as edible oil and this oil has high linoleic acid content. Also, the oil is evaluated as dye, polish, varnish, and soap. Safflower, which can also be converted as biofuel, is also used as animal feed with its cossette having a 25% protein content. When all of these used areas are transformed into trade in the world, they are marketed as two basic elements: safflower oil trade and safflower seed trade.

Total value in the world of safflower seed trade is 38,9 million dollars. The highest rate in world of safflower seed exports belongs to Russia with 15,2 million

Table 11.2 Safflower germplasm resources conserved in different gene/seed banks

Germplasm conservation institute/country name	Accessions	References
India/National Bureau of Plant Genetic Resources in New Delhi	2393	
India/Project Coordinating Unit for Safflower in Solapur	7525	
USA/Western Regional Plant Introduction Station (WRPIS)	More than 2400	Johnson et al. (2008)
Iran	200	
Turkey	125	
Iraq, Syria, Kazakhstan, Uzbekistan, and Tajikistan	Few	

Table 11.3 List of various germplasm with its PI number concerning important traits of safflower

Germplasm having seed yield above 80 g/plant			
PI number	Grain yield (g/plant)	Origin	References
426.187	86	Afghanistan	Li et al. (1993)
401.477	87	Bangladesh	
279.054	82	India	
306.883	84	India	
307.014	89	India	
307.066	87	India	
307.085	87	India	
307.132	83	India	
340.072	82	Turkey	
Variety name	Grain yield (kg/ha)	Released in	Saxena et al. (2008)
JSI-97	1500	2004	
Germplasm producing more than 70 capitula per plant (Beijing evaluations)			
PI number	Capitula per plant	Country of origin	References
426.187	90	Afghanistan	Li et al. (1993)
198.844	82	France	
248.801	78	India	
306.873	80	India	
304.467	73	Iran	
305.530	70	Sudan	
Germplasm containing seed oil content above 43%			
Beijing accession number	Oil content (%)	Origin	Reference
430	44.8	China	Li et al. (1993)
1134	43.4	Turkey	
30	43.8	USA	
32	44	USA	
33	47.5	USA	
35	44.4	USA	
38	43	USA	
42	44	USA	
401	46	USA	
402	45.2	USA	
404	43.3	USA	
547	44.3	USA	
PI number	**Oil content (%)**	**Origin**	**Reference**

(continued)

Table 11.3 (continued)

Germplasm containing seed oil content above 43%			
Beijing accession number	Oil content (%)	Origin	Reference
537635	50	USA	Kumar et al. (2016)
537701	48	USA	
560169	47	USA	
537662	46	USA	
560175	45	USA	
560171	45	USA	
560168	43	USA	
537693	43	USA	
537110	43	USA	

Germplasm containing oleic content above 75%			
PI number	Origin	Oleic content (%)	References
613394	USA	82	Kumar et al. (2016)
560177	USA	81	
560165	USA	81	
560173	USA	81	
560166	USA	80	
560169	USA	79	
401474	Bangladesh	78	
560172	USA	77	
560168	USA	77	
401589	India	77	
537712	USA	77	
470942	Bangladesh	77	
401470	Bangladesh	76	
401477	Bangladesh	76	
401476	Bangladesh	76	
401479	Bangladesh	76	

(continued)

Table 11.3 (continued)

Germplasm containing linoleic content above 80%			
PI number	Origin	Linoleic content (%)	References
250081	Egypt	87	Kumar et al. (2016)
544025	China	82	
560188	USA	81	
305198	India	81	
543992	China	81	
514624	China	80	
613459	Portugal	80	
537654	USA	80	
560185	USA	80	
560176	USA	80	
537645	USA	80	
537653	USA	80	
251987	Turkey	80	
426186	Afghanistan	80	
306685	Israel	80	

Germplasm with high degree of resistance to selected biotic stresses (Phaltan evaluations)		
PI number	Origin	References
Fungal pathogen (leaf blight): *Alternaria carthami*		Anonymous (1985)
199.935C	India	
209.281A	Israel	
209.287	Romania	
240.409	Egypt	
248.362	India	
248.362B	India	
Fungal pathogen (leaf spot): *Ramularia carthami*		
181.866A	Syria	
199936A	India	
209.281	Israel	
240.409	Egypt	
248.362A	India	
248.383	India	
248.620A	Pakistan	
Fungal pathogen (leaf spot): *Cercospora carthami*		
173.883A	India	
173.885A	India	
175.624D	Turkey	
199.892A	India	
199.925	India	
Insect pest (safflower fly): *Acanthiophilus helianthi*		
199.935C	India	
248.806	India	

dollars. This amount corresponds to 39,1% of total world safflower seed exports. The other highest rates in export values are followed by Kazakhstan, the Netherlands, India, and Germany as shown in Fig. 11.6.

The highest rate in world safflower seed imports belongs to Turkey with 8,48 million dollars. This amount corresponds to 21,8% of total world safflower seed imports. In terms of import values, this country is followed by China, Belgium, the Philippines, the Netherlands, and Uzbekistan, respectively, as shown in Fig. 11.7.

Total value in the world safflower oil trade is nearly 7,08 billion dollars. The highest rate in world safflower oil exports belongs to Ukraine with 3,74 billion dollars. This amount corresponds to 52,8% of total world safflower oil exports. The other highest rates in export values are followed by Russia, Argentina, the Netherlands, and Bulgaria as shown in Fig. 11.8.

The highest share in world safflower oil imports belongs to India with 1,86 billion dollars. This amount corresponds to 26,2% of total world safflower oil imports. In terms of import values, this country is followed by China, the Netherlands, Italy, and Spain, respectively, as shown in Fig. 11.9.

11.11 Safflower Breeding Activities in the World

11.11.1 Biotic and Abiotic Factors

With the increasing world population , there are many serious concerns like biotic and abiotic stresses which render global food production from plants. Drought and salinity are known to be the most crucial plant abiotic stresses which threaten crop production worldwide (Guo et al. 2014). According to FAO, 1/3 of the world population lives in water-scarce areas (FAO 2003). Safflower irrigation at the interval of

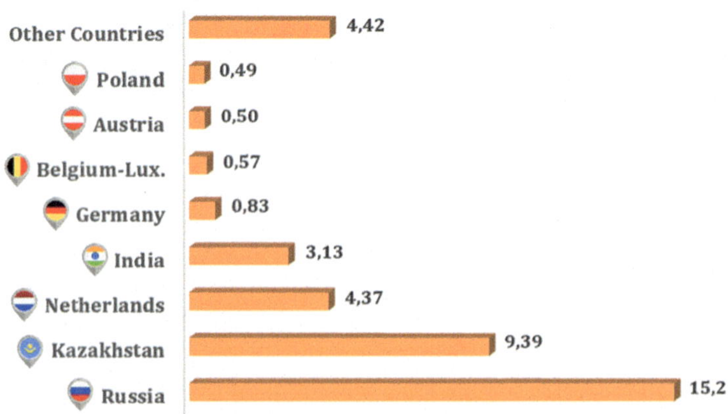

Fig. 11.6 The main exporter countries of safflower seeds in 2018 all over the world (million $). Source: https://oec.world/en/profile/hs92/120760/

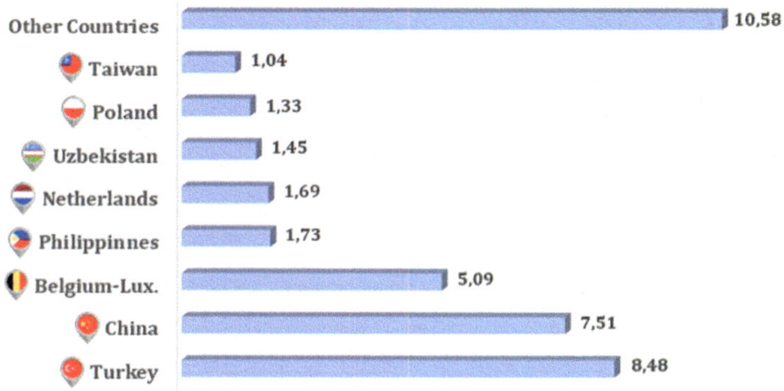

Fig. 11.7 The main importer countries of safflower seeds in 2018 all over the world (million $). Source: https://oec.world/en/profile/hs92/120760/

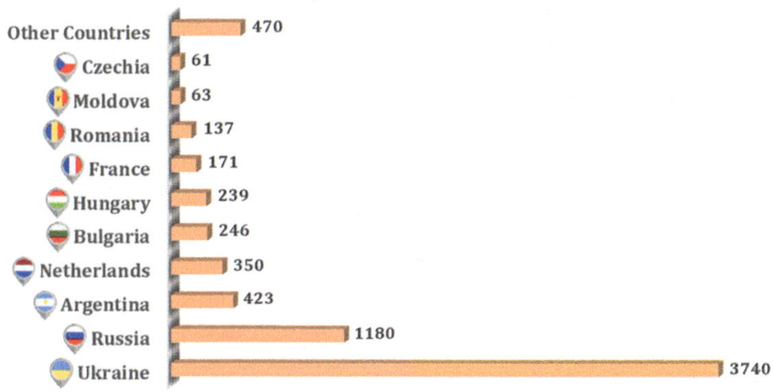

Fig. 11.8 Top exporter countries of safflower oil in 2018 all over the world (million $) Source: https://oec.world/en/profile/hs92/sunflower-seed-or-safflower-oil-crude

15 days does not affect its yield, but increasing the irrigation interval from 22 to 28 days can reduce the yield by 18 to 29.8% at vegetative growth stage (Jalali et al. 2012). Safflower seed oil content is badly affected with drought stress. Drought stress reduces stearic and palmitic acid contents up to 57%, while 8 and 14% reduction was observed in linoleic and oleic acid contents, respectively (Ashrafi and Razmjoo 2010). Rahmani et al. (2019) reported that the foliar application of zinc (1.2 kg ha^{-1}) alleviates drought stress effects at seed filling stage and remains an effective way to safflower in water-scare conditions. Chavoushi et al. (2019) documented that application of salicylic acid and sodium nitroprusside alleviates drought stress in safflower during vegetative growth. Salicylic acid (0 and 250μM) and sodium nitroprusside (0 and 25μM) were applied to safflower plants at 23rd day of germination having 25 and 100% field capacity. It was observed that salicylic acid promoted quick activation of nonenzymatic defense system coupled with high level of osmolytes. The nonenzymatic scavengers (like proline) were increased

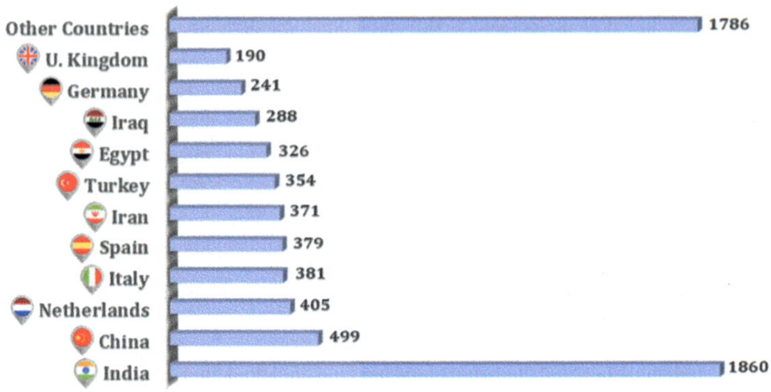

Fig. 11.9 Top importer countries of safflower oil in 2018 all over the world (million $) Source: https://oec.world/en/profile/hs92/sunflower-seed-or-safflower-oil-crude

simultaneously. Similarly, sodium nitroprusside utilization activated the enzymatic defense system and thus enhanced the superoxide dismutase and catalase activities and gene expression of two subunits of Fe- and Cu-superoxide dismutase. The mentioned mechanisms minimized free radicals and lipid peroxidation, thus helped in membrane stability, and also improved drought tolerance of the crop plants.

In the very similar way, salt stresses also affect the oil contents of safflower cultivars (Bassil and Kaffka 2002; Irving et al. 1988). Salt stresses in safflower mainly decrease capitula/plant, seeds/capitula, and oil content (Irving et al. 1988). Germination stage is the most critical stage for salt stress in comparison to the late developmental stages. Salt stresses reduce plant height and stem diameter, and plant becomes succulent with thick and darkened leaves (Weiss 1971; Beke and Volkmar 1995; Bassil and Kaffka 2002). High-oleate cultivars are disturbed in comparison to linoleate cultivars (Irving et al. 1988) due to salt stress. Shaki et al. (2018) revealed that application of penconazole (PEN) operates as endogenous signal molecule that significantly induces stress tolerance in plants. They studied safflower biochemical and molecular responses to the effects of PEN (15 mg l^{-1}) and sodium chloride (0, 100, and 200 mM NaCl). The exogenously applied PEN had a positive effect on anthocyanin, flavonoid, chlorophyll, carotenoid, soluble protein, and carbohydrate contents. Furthermore, RT-qPCR analysis exhibited the induced expression of SOS1 and NHX1 genes due to exogenous PEN utilization in both salt-treated and untreated plants (Table 11.4). They highly suggested the exogenous application of PEN to better cope with salt stress in safflower.

Safflower is susceptible to foliar diseases caused by different insects and organisms that lead to the root rot (Singh and Nimbkar 1993, 2006). Safflower fly is known as one of the most limiting factors in its cultivation and distribution and mostly found in Asia, Europe, and Africa. Fifty-seven pathogens including 40 fungi, 2 bacteria, 14 viruses, and 1 mycoplasma are known to infect safflower crop (Patil

Table 11.4 Sequences of SOS1 and NHX1 genes used for RT-qPCR

Gene name	Produced product	Forward sequence	Reverse sequence
SOS1	Sodium proton antiporter	CACGTCCCAAGAAGGCGCTGAT	CCAGCGCTTTGCCCATCATTTC
NHX1	Na+/H+ antiporter	TCGGGGAAGGCGTCGTGAAT	CCCAACAAAACCCCGCTCACTC

Adapted from Shaki et al. (2018)

et al. 1993). *Fusarium* wilt, disease caused by *Fusarium oxysporum* Schlecht f.sp. *carthami*, is the most infectious (Klisiewicz and Houston 1962) and can reduce the yield severely (Smith 1996). *Alternaria carthami* Chowdhury cause leaf spot disease and reduce the yield from 10% to 25% in India (Indi et al. 1988). Severe *Alternaria* attack can reduce the yield losses up to the level of 50% (Indi et al. 1986 and Singh and Prasad 2005). *Alternaria* blight disease also reduces the yield by 25–60% and can reach up to 90% when it occurs at early stage (Chowdhury 1944). *Ramularia* leaf spot usually occurs in irrigated conditions and reduces the yield up to 18–23%. Aphids also contribute to the reduction of safflower yield up to 74% (Bhardwaj et al. 1990). Safflower cannot compete strongly against weeds at early growth stages when branching is not initiated (Blackshaw 1993). Yield losses up to 75% have been reported in safflower due to weeds depending on the type of species and density (Agyman et al. 2002). Jalali et al. (2012) observed 29% yield losses due to weeds in the safflower crop. Therefore, it is mandatory to adapt proper selection criteria for different stress conditions. Safflower germplasm with high degree of resistance to selected biotic stresses like leaf blight, leaf spot, and safflower fly were proposed for safflower breeding against these stresses by Anonymous (1985) (Table 11.5).

From the above discussion, it is very clear that both biotic and abiotic stresses badly effect safflower cultivation and render its wide distribution. Some of the recent research studies significantly highlighted prominent research outcomes that help safflower sustainability. Still there is a need to devise prominent strategies regarding different biotic and abiotic stresses that will aid safflower cultivation in wider areas than now.

11.11.2 Classical Breeding

Historically, humans have used different conventional breeding (classical or traditional) techniques in self-pollinated species to change the characters of plants in order to produce desired features and develop new cultivars (Poehlman and Sleper 1995; Acquaah 2012). Conventional breeding contains different methods in

Table 11.5 List of various germplasm with its PI number concerning important traits of Safflower

Germplasm with high degree of resistance to selected biotic stresses (Phaltan evaluations)		
PI number	Origin	References
Fungal pathogen (leaf blight): ***Alternaria carthami***		Anonymous (1985)
199.935C	India	
209.281A	Israel	
209.287	Romania	
240.409	Egypt	
248.362	India	
248.362B	India	
Fungal pathogen (leaf spot): ***Ramularia carthami***		
181.866A	Syria	
199936A	India	
209.281	Israel	
240.409	Egypt	
248.362A	India	
248.383	India	
248.620A	Pakistan	
Fungal pathogen (leaf spot): ***Cercospora carthami***		
173.883A	India	
173.885A	India	
175.624D	Turkey	
199.892A	India	
199.925	India	
Insect pest (safflower fly): ***Acanthiophilus helianthi***		
199.935C	India	
248.806	India	

self-pollinated species. In the development of varieties, two main activities are extremely important for breeders: (I) to create or combine variability and (II) selection of desirable genotypes (Fehr 1987; Acquaah 2012). In this stage, the main purpose of the strategy is to develop a variety having genetic purity and productivity in the natural process (Acquaah 2012). Pedigree method has been used in safflower breeding (Knowles 1989; Dajue and Mundel 1996). Safflower cultivars are developed by using this method in different countries such as India, A-1, Nira, JSI-73, Girna, and NARI-6; the USA, Leed, Sidwill, Hartman, Rehbein, Oker, Girard, and Finch; Mexico, Sahuaripa 88, San Jose 89, and Ouiriego 88; and Canada, AC Sunset and AC Stirling (Hegde et al. 2002; Singh and Nimbkar 2007). Selection is one of the most important breeding processes for cultivar development. There are safflower cultivars that were developed via selection methods in different countries worldwide such as Turkey, Yenice and Olas; Canada, Saffire; the USA, Nebraska-10 and Nebraska-5; and India, HUS-305, N-62-8, Nagpur-7, A-300, N-630, Manjira, S-144, K-1, JSF-1, APRR-3, Type-65, CO-1, Bhima, Sharda, A-2, JSI-7, and PBNS-12 (Hegde et al. 2002; Singh and Nimbkar 2007). Köse (2016) determined agricultural performances of some safflower lines developed by a single plant selection method. Some lines determined that they might be candidate varieties in terms of different characteristics. Another study is conducted to determine yield, yield-related components, and some technological characteristics using some safflower genotypes from the World Safflower Gene Collection in Samsun province of Turkey (Şenel 2019). Based on results of this study, it was decided that it would be appropriate to continue adaptation and selection studies with 15 genotypes. Very recently, Ali et al. (2020a) investigated morphoagronomic performance evaluating 94 safflower accessions from 26 different countries for breeding perspective by carrying out two field trials in Turkey and Pakistan. A good level of genetic diversity was revealed among the accessions. According to hierarchical clustering, safflower panel is separated into three main groups (A, blue; B, green; C, red) (Fig. 11.10) and in line with the patterns of seven similarity centers. Additional, promising safflower accessions were selected for future breeding programs (Table 11.6). The views of the experimental area of Ali et al.'s (2020a) work are given in Fig. 11.11.

The hybridization breeding method was also applied successfully in the improvement of safflower (Golkar et al. 2011). There are some safflower cultivars such as Asol and Safir that were developed using hybridization breeding method in Turkey (https://arastirma.tarimorman.gov.tr/ttae/Sayfalar/EN/AnaSayfa.aspx). Pahlavani et al. (2004) studied the flower color and spininess inheritance while exploring some Iranian safflower genotypes. Köse (2016) determined seed setting ratio in safflower and reported that the mean of seed setting ratio varied between 89.5 and 39.7%. Erbaş and Haydar (2017) investigated the flower color and leaf spininess genetics in safflower and found that each of these characters was controlled by a single dominant gene. Backcross breeding technique is used effectively in the improvement of safflower varieties resistant to root rot caused by *Phytophthora drechsleri* in the USA (Thomas 1964; Singh and Nimbkar 2007). To the best of our knowledge, very less work has been done utilizing conventional safflower breeding approaches and needs to be focused in the future.

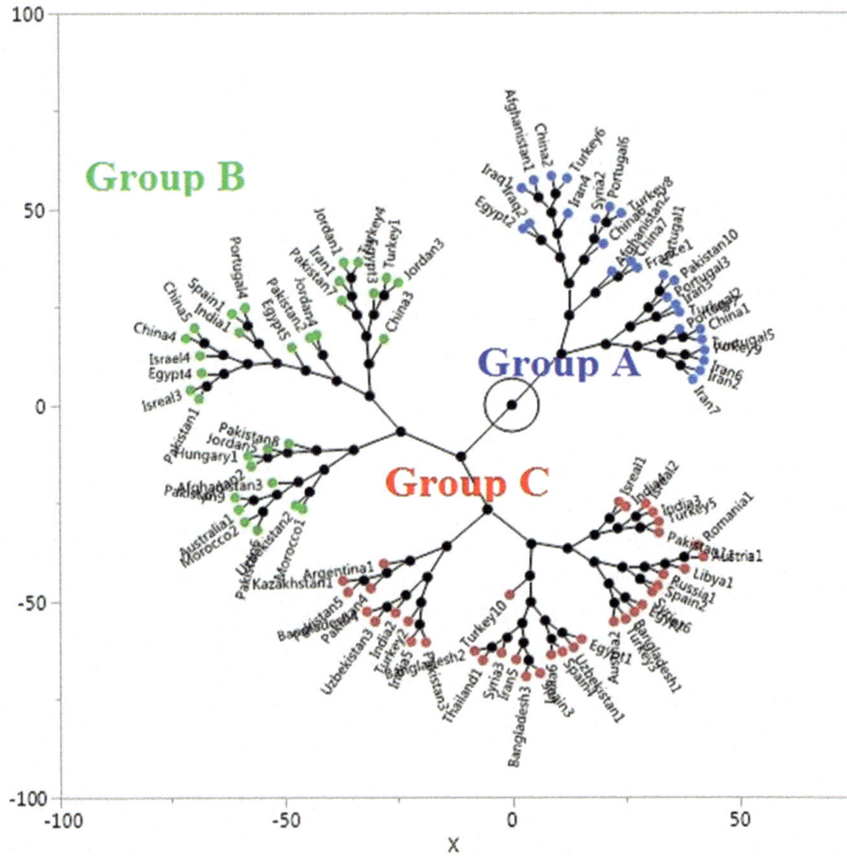

Fig. 11.10 Hierarchical clustering analysis separated the assessed 94 international safflower panel into three groups (with permission of Ali et al. 2020a)

11.11.3 Mutation Breeding

Traditional breeding techniques have always not been sufficient in improving the important characters like oil content and seed yield of safflower. Enhancement of agronomic traits (e.g., seed yield and oil content) in the safflower is a complex process. These features in safflower can be generated via mutagenesis, and this diversity can be used in breeding studies (Khadeer and Anwar 1991). Mutation is a valuable tool to create novel characters for crops (Verma and Shrivastava 2014). Another study stated that mutations lead to beneficial major changes in the genetic architecture of the plant properties (Bagawan and Ravikumar 2001). It is reported that induced mutation is commonly used to create new genetic diversity in crop plants by Maluszynski and Kasha (2002). Mozaffari and Asadi (2006) irradiate seeds of safflower cultivar Zarghan 279 by gamma rays (80, 100, 150, and 200 Gy)

Table 11.6 Selected promising safflower accessions for different crucial morphoagronomic features to improve production (with permission of Ali et al. 2020a)

Genotypes	DFI	DFF	DFC	DM	LL	LW	PH	BPP	CPP	CD	SPC	SY	100-SW
Pakistan-9	121.5	126.5	133.5	146.0	15.39	4.90	81.28	11.9	49.6	22.5	27.8	26.2	2.9
Egypt-5	119.5	124.0	130.5	144.5	20.14	6.11	104.6	11.0	26.0	26.7	19.6	32.8	5.3
Jordan-4	117.0	123.5	129.5	143.5	20.24	5.79	82.1	8.9	44.5	22.9	24.9	20.4	4.0
Portugal-4	121.5	128.5	133.5	150.0	16.53	5.80	96.6	8.6	14.7	26.6	26.0	20.8	3.7
Turkey-9	121.0	128.0	134.0	150.5	13.31	4.36	100.4	8.6	32.0	26.0	31.0	21.5	3.0
Israel-4	122.5	127.0	133.0	146.5	18.54	5.33	98.4	8.1	32.0	22.5	20.2	22.7	4.1
Jordan-3	119.0	125.0	132.0	145.5	12.83	4.70	90.5	9.5	37.6	25.4	24.9	23.3	4.0
China-1	121.5	128.0	133.0	152.5	15.15	5.15	98.3	9.2	36.2	24.6	32.4	24.5	3.5
Jordan-5	120.5	125.5	133.5	147.0	15.53	5.12	86.7	11.4	55.6	23.6	23.7	28.2	3.6
Iran-1	119.5	124.5	131.0	150.0	16.67	5.04	84.5	13.0	36.1	25.6	30.3	29.5	3.9
Turkey-4	120.0	123.5	130.0	145.5	15.02	4.75	87.2	10.0	30.0	25.4	28.4	30.5	4.3
Jordan-1	121.5	124.5	131.5	145.5	16.62	5.32	95.8	9.7	38.9	25.4	30.3	31.5	4.0
Hungary-1	120.5	126.5	133.5	150.0	18.99	5.46	93.3	13.1	44.8	24.1	21.3	31.6	3.5
China-4	120.0	127.0	138.5	150.5	16.90	4.67	95.1	10.4	24.4	23.6	22.0	33.1	3.7
Pakistan-8	121.0	125.5	129.5	145.5	14.96	5.23	76.8	13.9	80.4	21.8	26.6	33.6	2.9
Egypt-3	122.5	128.0	136.0	148.5	14.55	5.04	97.9	10.5	36.4	28.3	22.1	35.9	3.8
China-5	119.5	127.5	135.5	150.5	18.90	4.76	95.7	9.5	26.2	25.4	24.4	35.9	4.5
Jordan-2	119.0	122.5	130.5	146.5	17.02	5.23	90.7	12.3	46.1	24.0	18.5	39.2	3.4
Pakistan-7	120.0	124.0	131.0	149.5	17.26	5.14	88.1	9.2	39.9	25.2	33.3	43.3	3.3
China-3	117.0	123.0	131.5	145.5	10.94	3.65	85.3	10.8	21.4	27.1	35.1	51.0	4.5

DFI days to flower initiation, *DFF* days to 50% flowering, *DFC* days to flower completion, *DM* days to maturity, *LL* leaf length, *LW* leaf width, *PH* plant height, *BPP* branches per plant, *CPP* capitula per plant, *CD* capitulum diameter, *SPC* seeds per capitulum, *SYP* seed yield per plant, *100-SW* 100-seed weight

Fig. 11.11 Experimental area of Turkey in 2018. (with permission of Ali et al. 2020a)

and sown in field. The mutants of M5 generation are cultivated in field under different conditions (irrigated and drought stress), and it was observed that the studied mutants varied for some traits in both cases. But, no significant differences were found in terms of oil content and number of capitula in among the evaluated mutants. Kaya et al. (2009) aimed to determine the responses of different gamma doses (0, 100, 200, 300, 400, 500, 600, 700, and 800 Gy) applied to the seed during the early developmental stages of safflower line (Taek-Uslu) and cultivars (Dinçer, Remzibey-05, and Shifa). Based on the results of this study, it was determined that gamma doses between 200 and 400 Gy can be administered without a decrease in viability. Yaman (2014) applied different gamma rays (200, 300, 400, 500, and 600 Gy) on seeds of safflower cultivars (Remzibey, Dinçer, and Shifa) using Co-60 source and investigated in gamma rays affection to agricultural traits of M1 and M2 safflower cultivars. The result presented that the best genetic diversities were obtained from 300–400 Gy doses (Fig. 11.12 and Table 11.7).

Fig. 11.12 The views of table formation and blossoming in Remzibey safflower cultivars exposed to different gamma rays (300 and 400 Gy). (with permission of Yaman 2014)

11.11.4 Biotechnological Tools

11.11.4.1 Tissue Culture

In vitro regeneration via tissue culture discloses evident improvement of intermediate propagation compared to traditional breeding activities in safflower. Important plant growth regulators application was done for successful induction of callus (Lijiao and Meili 2013). Very few studies have been conducted regarding rhizogenesis improvement of safflower in vitro generated shoots. Further research studies of safflower regeneration via rhizogenesis are limited due to the lack of an efficient protocol. Safflower regeneration via tissue culture comprised of callus-mediated regeneration, direct organogenesis, and somatic embryogenesis. For brief description of the safflower regeneration through tissue culture, please refer to the Sujatha and Gupta (2013). Constraints related to safflower regeneration include low rate of shoot regeneration and its sensitivity to high humidity in culture vessel and diverse rooting response in different cultivars (Sujatha and Gupta 2013). Sufficient level of somaclonal variation regarding flower color, plant height, leaf shape, and seed oil is obtained (Seeta et al. 2000). Nowadays, in vitro research activities concern abiotic stresses, mainly salinity conducted to explore salinity tolerance in the safflower germplasm (Hamedi et al. 2016).

Tissue culture remains very important to improve salt tolerance in wheat, durum wheat, rice, sunflower, sugarcane, fennel, and potato (Barakat and Abdel-Latif 1996; Arzani and Mirodjagh 1999; Lutts et al. 1999; Basu et al. 2002; Alvarez et al. 2003; Gandonou et al. 2005; Golkar et al. 2007; Errabii et al. 2007; Khorami and Safarnejad 2011; Hasan and Sarker 2013) by revealing consistent results. Tissue culture techniques were applied to improve different abiotic stresses like cold

Table 11.7 Suggested dose (SD) for practical application in safflower

Organs	Suggested gamma radiation doses (Gy)	References
Seeds	400	Yatou (1985)
Seeds	200–400	Kaya et al. (2009)
Seeds	300–400	Yaman (2014)
Seedling	60	Yatou (1985)
Callus	120	Yatou (1985)

hardiness and salt and drought stress (Zair et al. 2003; Bajji et al. 2004; Gawande et al. 2005). Tissue culture observed promising results mainly against salt stress in safflower with the help of various important in vitro physiological traits (Chawla 2000; Lutts et al. 2004; Hasan and Sarker 2013). Genetic exploitation with tissue culture and selection of the potential variable strains are observed as supplementary techniques to conventional plant breeding to obtain plants containing sufficient amount of stress resistance (Borsani et al. 2003; Ashraf and Foolad 2013). Yaman (2014) observed that different doses of gamma ray stimulated callus formation and adventitious shoot regeneration in safflower cultivars, and higher doses negatively affected callus formation and adventitious shoot regeneration. Additionally, adventitious shoot regeneration was surged in MS medium supplemented with 4 mg/l TDZ and 0.2 mg/l NAA in Remzibey cultivar, 1 mg/l TDZ in Dinçer cultivar, and 2 mg/l NAA and 2 mg/l BAP in Shifa cultivar in the study. Some views of this study were presented in Fig. 11.13.

Recently, Vijayakumar et al. (2017) determined the meta-Topolin (mT) effects for development of an efficient protocol of safflower regeneration utilizing cv. NARI-H-15. They reported that safflower plants regenerated via micropropagation and organogenesis hardened in pots under greenhouse with the survival rate of about 67% and 42%, respectively. So far, the new optimized protocol is believed to be useful in the safflower regeneration through tissue culture approach and lead to the improvement of various important agronomic traits. It is also highly suggested to utilize tissue culture approach to other safflower abiotic stresses by formulating and developing efficient and feasible protocols.

11.11.4.2 Genomics of Safflower

Genetic resources are very important for crop genetic and breeding programs (Nadeem et al. 2018a; Yeken et al. 2018; Arystanbekkyzy et al. 2019; Yildiz et al. 2019; Ekincialp et al. 2019; Barut et al. 2020). Therefore, it is very important to collect genetic resources of crop species and to identify agronomic, quality traits and other important characteristics responsible for stress factors (biotic and abiotic) and to perform molecular characterization using large number of markers. Modern plant breeding has been exceptionally successful in increasing crop productivity parallel to the growing human population. International research has demonstrated the need to protect and manage local populations because these materials contain

Fig. 11.13 (**a**) Callus formation from shoot tip explants after 30 days in MS nutrient medium containing 4 mg/l TDZ and 0.2 mg/l NAA. (**b**) Callus formation from hypocotyl explants after 20 days in MS nutrient medium containing 2 mg/l TDZ. (**c**) Direct multiple shoot formation from shoot tip explants 21 days after culture in MS nutrient medium containing 4 mg/l TDZ and 0.2 mg/l NAA. (**d**) Induced shoot formation from shoot tip explants 20 days after culture in MS nutrient medium containing 4 mg/l TDZ and 0.2 mg/l NAA. (with permission of Yaman 2014)

valuable genes for future breeding studies. From this perspective, differences in phenotype and genotype have been used to characterize and evaluate genetic variety in the germplasm panel of numerous crop species (Baloch et al. 2017; Nadeem et al. 2018a; Arystanbekkyzy et al. 2019; Ekincialp et al. 2019; Karık et al. 2019; Yeken et al. 2019; Yildiz et al. 2019).

Safflower is a self-pollinated crop belonging to the Compositae family. It has a haploid genome size with nearly 1.4 GB/$2n$ = 24 chromosomes (Kumari et al. 2017). Productivity, seed quality, seed mineral contents, and resistance to stress factors and many other traits of safflower will be future breeding aims to meet the fast-increasing industrial evolution of the world (Ali et al. 2019). To enhance the selection efficiency, maximum utilization of natural genetic diversity is one of the most critical and significant objectives of the safflower improvement programs. Most of the traits of agricultural importance are complex and quantitative in nature and controlled by single and multiple genes. In order to define a desirable trait, it is important to exploit about the location of a gene or genes. Therefore, genomics-assisted crop improvement is crucial to identify important loci affecting targeted characteristics and to select safflower with desired allele combinations (Varshney and Tuberosa 2013; Zargar et al. 2016). Genetics and genomics of safflower are mostly unexplored, and the lack of convenient molecular markers is a key limitation for the improvement of valuable modern breeding programs (Garcia-Moreno et al. 2010; Hamdan et al. 2011; Kumari et al. 2017). Various markers such as RAPD (randomly amplified polymorphic DNA), ISSR (inter-simple sequence repeat), AFLP (amplified fragment length polymorphism), SSRs (simple sequence repeats), and SNP (single nucleotide polymorphism) have been used to investigate the genetic diversity (GD) and population structure (PS) of safflower (Sehgal and Raina 2005; Yang et al. 2007; Johnson et al. 2007; Amini et al. 2008; Sabzalian et al. 2009; Khan et al. 2009; Sehgal et al. 2009; Chapman et al. 2010; Hacioglu et al. 2013; Pearl and Burke 2014; Lee et al. 2014; Kumar et al. 2015; Ambreen et al. 2015; Kumari et al. 2017; Ali et al. 2019, 2020b; Hassani et al. 2020).

Sehgal and Raina (2005) carried out molecular characterization studies at the DNA level in order to determine the GD of safflower world germplasm sources. They used a combination of 22 RAPD primers, 18 SSR primers, and 10 AFLP primers to determine the GD of 85 accessories obtained from 24 countries. Yang et al. (2007) examined the GD and genetic relationships in safflower using ISSR markers and found polymorphic 355 bands out of 429 bands. Sabzalian et al. (2009) used ISSR primers to determine the relationship between *Carthamus oxyacanthus* and other safflower varieties and compared morphological features. As a result of the study, they found GD among wild safflower varieties at molecular level was higher than morphological characteristics. Hacioglu et al. (2013) performed GD using 8 RAPD markers in 32 safflower accessions from Afghanistan, China, Pakistan, India, Iran, Turkey, and Japan. They detected polymorphism in 53 out of 56 band arrays in accessions. Pearl and Burke (2014) genotyped a total of 190 safflower accessions using 133 SNP markers, and their results showed that most of the safflowers originated from a single pool of diversity in the Old World of safflower germplasm. Kumar et al. (2015) investigated the GD and PS of huge safflower panel (531 accessions) obtained from 43 countries of different regions of the world. While the Near East, Iran, and Afghanistan accessions had major diversity, American accessions revealed minor molecular variability as compared to previous reports. Kumari et al. (2017) investigated GD of 20 safflower genotypes using morphological traits and SSR markers. Their results showed that morphological characters (oil content, test weight, branches/plant, and days to maturity) strongly associated with SSR markers, and SES-129, SSR-5, SES-81, and SES-85 SSR markers can be further used to marker-assisted selection. Very recently, Ali et al. (2019) explored the GD and PS of 131 safflower accessions using 13 iPBS (inter-primer binding site) retrotransposon markers, and major genetic diversity was found between safflower accessions (Fig. 11.14).

In this study, they determined that the safflower accessions were divided into A, B, C, D, and unclassified populations (Fig. 11.15).

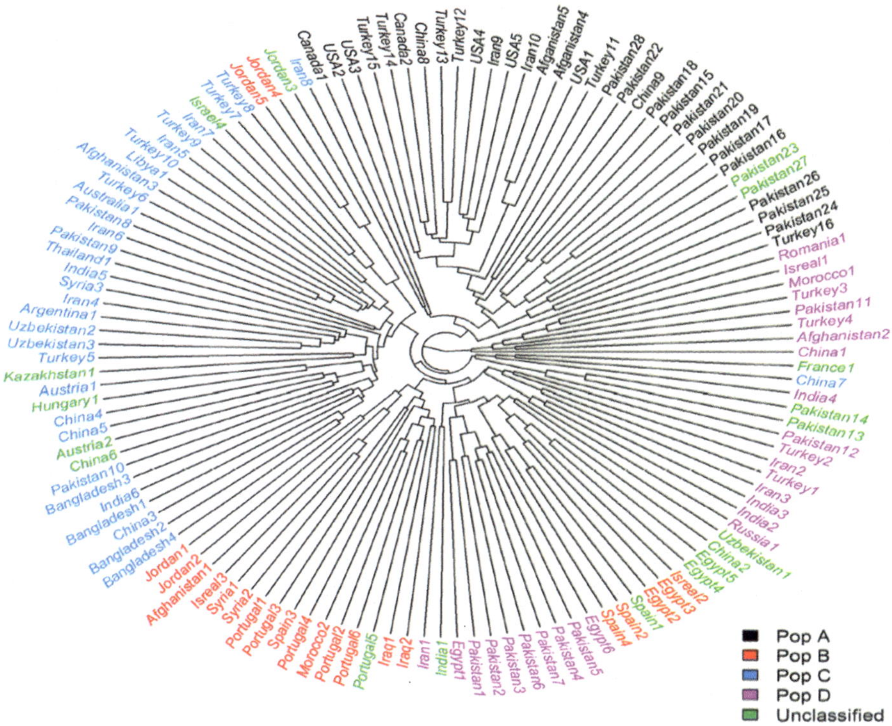

Fig. 11.14 UPGMA clustering analysis in safflower panel using iPBS markers. (with permission of Ali et al. 2019)

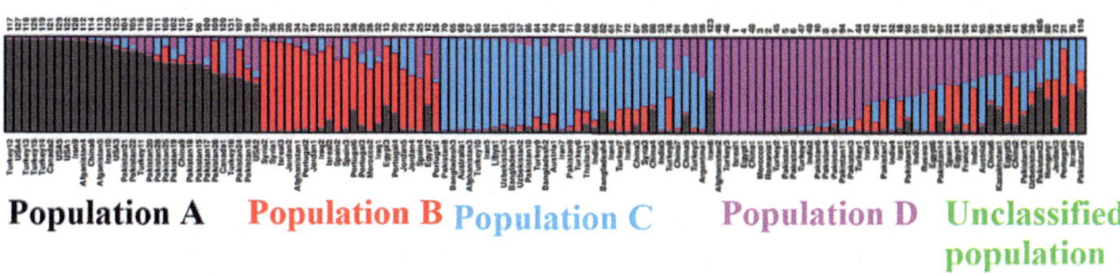

Fig. 11.15 Structure-based clustering analysis in safflower panel using iPBS markers. (with permission of Ali et al. 2019)

QTL Mapping

Traditional plant breeding techniques have made a significant contribution to the improvement of safflower, but marker-assisted selection has significantly transformed the science of plant breeding by minimizing the discrepancies facing conventional breeding. Marker-assisted selection is an advanced improvement technique in which linked molecular markers are used for the phenotypic selection. Marker-assisted selection is commonly used to rise the effectiveness of backcross breeding (Das et al. 2017). In order to increase the productivity of marker-assisted selection for quantitative traits, convenient field experimental designs and methodologies have to be employed. Additionally, in marker-assisted selection programs, we need molecular markers that can be obtained through the QTL mapping or GWAS (Nadeem et al. 2018b). Molecular marker techniques continue to evolve day by day. Among the recently developed molecular markers, diversity array technology (DArT) which is hybridization-based has gained considerable importance due to produced simultaneously hundreds of thousands of markers across the genome. DArT technique has only been used in many crops such as rice, wheat, barley, common bean, pigeon pea, *Eucalyptus*, etc. (Jaccoud et al. 2001; Wenzl et al. 2004; Lezar et al. 2004; Semagn et al. 2006; Akbari et al. 2006; Yang et al. 2006; Baloch et al. 2015; Nadeem et al. 2018a) for germplasm diversity, population structure, QTL mapping, and genome-wide association studies. However, this technique has not been used in safflower. In the near future, DArT technique will provide a promising alternative to meet the requirements of genome coverage, with higher reproducibility and transferability in safflower, and will help the plant breeders. QTL mapping or biparental mapping involved the selection of two diverse parents and crossed them to develop a mapping population for their improvement purpose (Collard et al. 2005; Nadeem et al. 2018b). In safflower, the first genetic linkage was developed by Mayerhofer et al. (2010). In addition, different studies in safflower have been conducted on the mapping of genes controlling male sterility and high oleic acid (Hamdan et al. 2008b; Hamdan et al. 2012). For the investigation of the genetic construction of safflower domestication, 61 QTLs underlying 24 domestication-related characters were mapped by Pearl et al. (2014). The first study to determine the QTL that controls seed yield and their components under drought stress at the reproductive stage in safflower was reported by Mirzahashemi et al. (2015). These researchers determined that four major QTLs and three link groups (2, 4, and 6) play a crucial role in the drought tolerance of safflower. Various locus and markers associated with important traits of safflower were listed in Table 11.8. Bowers et al. (2016) conducted whole-genome shotgun sequencing on 96 F6 recombinant inbred lines (RILs) of a cross between *Carthamus tinctorius* L. and *Carthamus palaestinus* Eig. The resulting map of this study included 2,008,196 genetically positioned SNPs in 1178 unique locations. These results will be useful in MAS breeding programs and identifying candidate genes for diverse features in safflower.

Association Mapping

Association mapping (AM) is referred to as an alternative approach to handle the limitations of biparental QTL mapping (Morton 2006). Diverse germplasm is used in AM when compared with conventional QTL mapping approaches. Markers

Table 11.8 List of various locus and markers associated with important traits of safflower

Marker	Markers name	Mapped traits	References
AFLP	M51/E41-6, M51/E41-4, M61/E40-6, and M62/E40-17	Seed and oil yield/drought-stress conditions	
AFLP	M62/E40-35 and M47/E37-13	Seed and oil yield/normal conditions	
AFLP	M62/E41-11	Oil yield/drought-stress conditions	Ebrahimi et al. (2017)
AFLP	M61/E40-6, M47/E378, and M51/E32-9	Oil content/under drought-stress conditions	
AFLP	M61/E2-2, M61/E40-6, and M51/E41-12	Oil content/under normal conditions	
SSR/ISSR	qPh6_1, qPh6_2	Plant height	
ESTs	qBpno4_1, qBpno 4_2, qBpno6	Branches/plant	
ESTs	qCpno2	Capsules/plant	
ESTs	qDw2, qDw4, qDw6	Dry weight/plant	Mirzahashemi et al. (2015)
ESTs	qSpno2, qSpno3, qSpno4, qSpno7, qSpno9, qSpno18	Seeds/plant	
ESTs	qSyp2, qSyp9	Seed yield/plant	
ESTs	qThsw5	1000-seed weight	
ESTs	B353, H113	Average leaf size	
ESTs	D378, G154, H40, L116	Average leaf roundness	
ESTs	E190, H327, L116	Spininess	
ESTs	D271, H113, I253	Days to flower	
ESTs	A69, A117, D129, D275, H312, I276, L333	Primary capitulum height	
ESTs	A199, H113, I253, L333	Primary disk diameter	
ESTs	H76	Number of heads	
ESTs	D234	Flower color	
ESTs	E201, E359, H130, I276	Stem height	
ESTs	A245, E354, L219, L339, L116	Internode length	
ESTs	G100	Lowest branch height	Pearl et al. (2014)
ESTs	C278, H255, I175	Number of self-seeds	
ESTs	C278, H231, I111, K35	Achene weight	
ESTs	C120, D213, I223, K35	Achene length	
ESTs	C278, I111, J232, K35	Achene width	
ESTs	E190	Seed dormancy	
ESTs	I203, I92, J232, L221	Seed oil	
ESTs	E140	Palmitic acid	
ESTs	C98, G26, H327	Oleic acid	
ESTs	G110, H327	Linoleic acid	
ESTs	C200, L116	Number of internodes	

(continued)

Table 11.8 (continued)

Marker	Markers name	Mapped traits	References
RAPD	Loci OPM1900, OPH2750, OPH8670, and OPA1-OPH7300	Coupling phase to the allele *li determining very high linoleic acid cont.*	Hamdan et al. (2008b)
SSR	ct365	*Ol* gene responsible for high oleic acid content	Hamdan et al. (2012)
SSR	NGSaf_15 and NGSaf_300 locus	Oil content	
SSR	NGSaf_67, NGSaf_210, and NGSaf_309 locus	Oleic acid content	
SSR	NGSaf_67 and NGSaf_309 locus	Linoleic acid content	
SSR	NGSaf_156 and NGSaf_296	Plant height	
SSR	NGSaf_92 locus	Days to 50% flowering	Ambreen et al. (2018)
SSR	NGSaf_279	Primary branches	
SSR	NGSaf_279 locus	Number of capitula/plant	
SSR	NGSaf_306 and NGSaf_309	100-seed weight	

associated with major QTL can be evaluated in MAS programs, if new QTLs are identified. The correlation among markers and traits is statistically accessed in unrelated genotypes in the AM approach. Natural genetic diversity and ancestral recombination in natural populations constitute the basis for the identification of nonrandom co-segregation of alleles between loci and traits. AM is more complex than QTL biparental mappings such as genetic drift, conscious and unconscious selection by farmers and breeders, and the admixture of the population which can bias the marker-trait association. AM needs a large number of polymorphic markers such single nucleotide polymorphisms (SNPs) which referred single DNA base differences between homologous DNA fragments and small nucleotide insertions and deletions (indels). Utilization of SNPs is preferred over other molecular marker systems as it discloses in-depth genetic analysis even among closely related germplasm (Souza et al. 2012). SNPs are biallelic and codominant molecular markers that aid to the safflower improvement including discovery of different genes and QTLs, plant traits association analysis, and marker-assisted selection. SNP markers are preferred over other marker systems as (1) they disclose the most abundant genetic variation within genomes due to their presence at single nucleotide base level (Zhu et al. 2003) and (2) the development of wide array of technologies for high-throughput SNP analysis (Fan et al. 2006). For this perspective, SNPs are useful for population genetics and phylogeny studies (Jin et al. 2003). Another important feature of SNP marker is their transfer between plant closely related species and also beneficial to microsynteny analysis. For all the reasons mentioned above, SNPs technique could be highly suggested for AM in safflower as a start. AM has been explored as a beneficial approach to classify marker-trait associations for various agronomic properties in various crop types by Yang et al. (2010), Li et al. (2011), Upadhyaya et al. (2013), Xu et al. (2013), and Zhang et al. (2014).

Ebrahimi et al. (2017) explored the marker-trait associations under two moisture conditions in worldwide safflower panel through the association mapping using AFLP markers. AM was carried out among 8 significant traits, and 341 polymorphic AFLP markers generated via 10 primer combinations in safflower genotypes. Ambreen et al. (2018) reported AM for significant agronomic features in a safflower panel comprised of 124 accessions using microsatellite markers and determined the first map QTLs for oil content in safflower. Additionally, they identified important marker-trait associations for seed oil content, oleic acid content, linoleic acid content, days to 50% flowering, primary branches, number of plant height, number of capitula per plant, and 100-seed weight. These results will help plant breeders to improve new safflower cultivars having desirable traits.

Genomic Selection

Genomic selection is a technique to estimate the genetic value of selection individuals based on the GEBV (genomic estimated breeding value) estimated from a large number of markers (SNP arrays) positioned to whole genome (Newell and Jannick 2014). Genomic selection, also recognized as a type of MAS, needs genome-wide molecular data for efficient selection to QTL (Goddard and Hayes 2007; Bernardo 2008; Massman et al. 2013; Würschum et al. 2013). In safflower, molecular markers have been efficiently used for MAS programs for diverse traits such as the controlling gene for nuclear male sterility and very high linoleic acid (Hamdan et al. 2008b), high gamma-tocopherol genes (Garcia-Moreno et al. 2010), high oleic acid content (Hamdan et al. 2012), related to drought-responsive genes (Thippeswamy et al. 2013), linkage between non-spiny and male sterility marker (Kammili 2013), fatty acid content, flower color and number of heads (Pearl et al. 2014), linoleic acid content, oleic acid content, seed oil content, primary branches, number of plant height, number of capitula per plant, flowering time, and 100-seed weight (Ambreen et al. 2018). Genomic selection has been provided in breeding of different field crops such as oat (Asoro et al., 2011), barley (Lorenz et al. 2012), sugar beet (Würschum et al. 2013), maize (Massman et al. 2013; Cantelmo et al. 2017; Bandeira e Sousa et al. 2017; Lyra et al. 2017), wheat (Bassi and Sanchez-Garcia 2017), soybean (de Azevedo Peixoto et al. 2017), and potato (Habyarimana et al. 2017). To our knowledge, no study has been found in literature about genomic selection of safflower. We envisage that this technique will help to accelerate safflower breeding programs and will be attended to notable developments of modern safflower cultivars in the near future. The broadest extent of genomic selection for breeding schemes given by Voss-Fels et al. (2019) can be assisted in the selection of developed genotypes in the various selection levels.

11.11.4.3 Functional Genomics

Transgenic Breeding

Transgenic is described as "an organism that has been converted with a foreign DNA sequence" (Acquaah 2012). In present days, transgenic plants are produced in different field crops. However, soybean (*Glycine max* (L.) Merr.), corn (*Zea mays* L.), cotton (*Gossypium hirsutum* L.), and canola (*Brassica napus* L. and *B. rapa* L.) constitute the biggest part (91%, Fig. 11.16) of commercial genetically modified plants worldwide (Visarada et al. 2009; James 2014). The USA, Brasil, Argentina, Canada, and India are major countries that produced genetically modified (GM) crops in 75, 51.3, 23.9, 12.7 and 11.6 million hectares' areas, respectively (ISAAA 2019). On the other hand, the production of GM safflower has just been started commercially. Numerous genetic transformation studies conducted about safflower were found in literature.

For example, transformation in safflower via callus-mediated regeneration was reported by Ying et al. (1992), Orlikowska et al. (1995), and Rao and Rohini (1999). Seedling explants were transformed into MS (Murashige and Skoog 1962) medium including BA (6-benzyladenine) and NAA (naphthaleneacetic acid) for regenerated shoots (Ying et al. 1992; Rao and Rohini 1999). Orlikowska et al. (1995) investigated the effects of co-cultivation situations on transformation effectiveness and direct shoot regeneration from seedling explants of safflower cv. 'Centennial'. Another study conducted by Rohini and Rao (2000) used zygotic embryo transformation via in planta in safflower (Table 11.9). "In planta" transformation is described as direct transformation technique without any tissue culture stages (Jan et al. 2016). There are lots of advantages of this technique such as (I) can constitute large number of uniform plants in short time, (II) fewer labor energies, and (III)

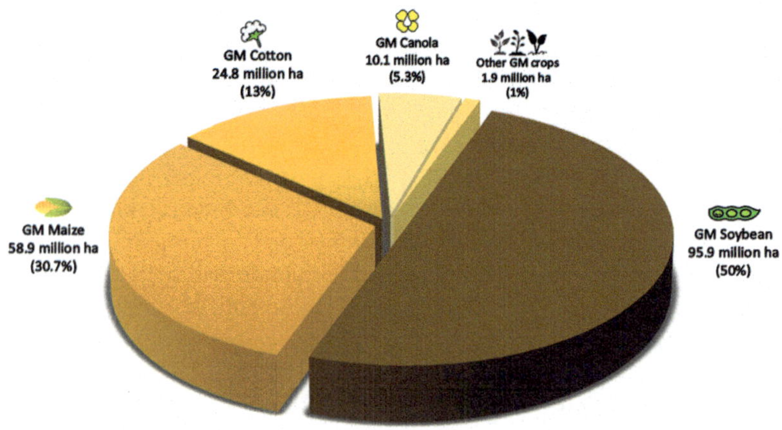

Fig. 11.16 The rates of GM crops in global area in 2018. (Data was adapted from ISAAA 2018 to construct the figure)

Table 11.9 Genetic transformation studies in safflower (adapted from Sujatha and Gupta 2013)

Explant	STC	Medium used (μM)	Reference
PSE	Ts	5.37 NAA + 4.44 BAP + MS	Ying et al. (1992) Sujatha and Gupta (2013)
PSE	Ts	0.53 NAA + B5 vitamins+2.5 AgNO3 + 0.053 TDZ + 500 carbenicillin+MS salts	Orlikowska et al. (1995) Sujatha and Gupta (2013)
PSE	Ts	0.53 NAA + 4.44 BA+MS	Rao and Rohini (1999) Sujatha and Gupta (2013)
Embryos	In planta Ts		Rohini and Rao (2000) Sujatha and Gupta (2013)
Cotyledons	T0 plants	1.07 NAA + 0.91 TDZ	Shilpa et al. (2010) Sujatha and Gupta (2013)
Cotyledons	T0 and T1 plants	0.53 NAA + 4.54 TDZ 0.53 NAA + 4.44 BA	Srinivas et al. (2011) Sujatha and Gupta (2013)

PSE primary seedling explants, *STC* stage of transformants characterized, *Ts* transformed shoots

minimal reagents requirements. According to Rohini and Rao (2000), the effectiveness of transformation may diverge between safflowers, but the procedure can be used in all of the safflower cultivars and genotypes having susceptibility to *Agrobacterium tumefaciens* infection. In the above studies, npt II was used as plant selection marker, and Uid A and GFP genes were used as reporter gene (Sujatha and Gupta 2013). Then, validation was done via PCR (polymerase chain reaction) for vector genes (npt II, virC, Uid A) and genes of interest, non-denaturing PAGE, southern blot, dot blot, and western blot analysis by Ying et al. (1992), Orlikowska et al. (1995), Rao and Rohini (1999), Rohini and Rao (2000), Shilpa et al. (2010), and Srinivas et al. (2011).

An effective and reproducible procedure for genetic transformation method was developed by Srinivas et al. (2011) for both high oleic acid content (S-137, transformation efficiency 4.8%) and high linoleic acid content (WT, transformation efficiency 3.1%) safflower genotypes (Fig. 11.17). In the study, cysteine (50 mg/l), iota-carrageenan (1.5 g/l), and ascorbic acid (1.5 mg/l) compounds were efficiently used for controlling hyperhydration and necrosis of *Agrobacterium*-infected cotyledons. A method was developed to overcome poor in vitro root regeneration for safflower, and approximately 50% of the transgenic shoots were transformed into mature plants containing viable transgenic T1 seed. Afterward, validation was done by PCR, western blot, and southern blot analysis for GFP and hygromycin genes.

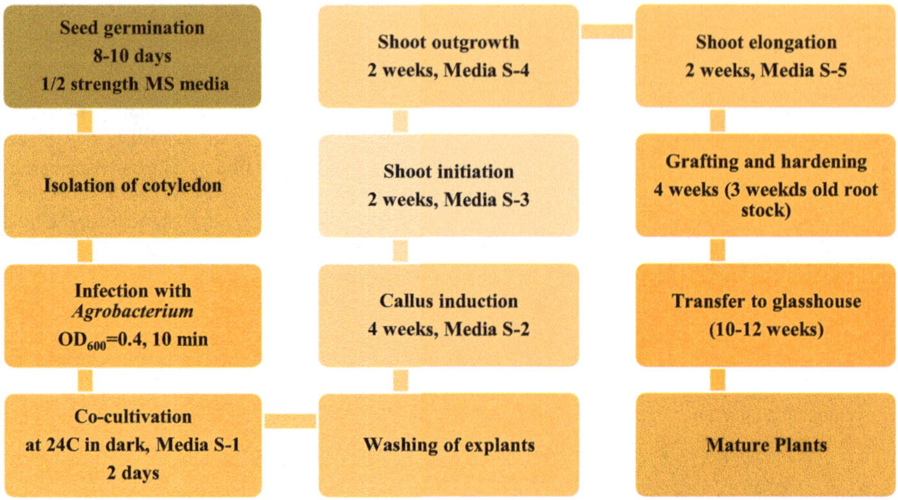

Fig. 11.17 Schematic view of the whole procedure and optimized parameter for the generation of GM safflower. (Adapted from Srinivas et al. 2011)

Result of the southern blot analysis revealed that there were one to seven transgenes in each line and Mendel inheritance was seen in T_1 progeny. This was the first evidence for a complete process to reliably produce GM safflower seed.

The first commercial GM safflower lines were released in Australia in 2018. GO Resources Pty Ltd. developed two GM safflowers (code: GOR-73226-6 and GOR-7324Ø-2) and took a license from the Office of the Gene Technology Regulator that authorizes the commercial cultivation of GM safflower in Australia (ISAAA 2019). Details of GM safflowers were given in Table 11.10. The GM safflowers were developed for industrial oil uses, not for human food (OGTR 2018). The seeds of GM Safflowers contain a high level of a particular oil and oleic acid (approximately 92%). High purity oleic acid can be used instead of petroleum-based precursors in the industry (OGTR 2018). The GM safflowers include a selectable marker gene that presents antibiotic resistance from a common soil bacterium. Antibiotic resistance gene was used in plant selection during laboratory development of GM safflower, and no function in field (OGTR 2018).

Genome Editing

Genome editing (GE) includes various techniques such as meganuclease, transcription activator-like effector nucleases (TALENs), zinc-finger nucleases (ZFNs), and clustered regularly interspaced short palindromic repeats (CRISPR) (Khan 2019). Especially, CRISPR/Cas9 which is the latest and most popular method of GE techniques has been applied effectively in numerous plant types, such as *Arabidopsis*, rice, tobacco, maize, wheat, sorghum, poplar, tomato, petunia, soybean, citrus,

Table 11.10 Summary of basic genetic modification in GM safflowers developed by GO Resources Pty Ltd

Carthamus tinctorius L. (safflower)	Gene introduced	Gene source	Product	Function	Developer	Method of trait introduction
GOR-73226-6 and GOR-7324Ø-2	**fatB**	*C. tinctorius*	No functional enzyme produced (production of FATB enzymes or acyl-acyl carrier protein thioesterases is suppressed by RNA interference)	Downregulates fatB gene	Go Resources Pty Ltd.	*A. tumefaciens*--mediated plant transformation
	fad2.2	*C. tinctorius*	No functional enzyme is produced (production of delta-12 desaturase enzyme is suppressed by RNA interference)	Downregulation of fad2.2 gene	Go Resources Pty Ltd.	*A. tumefaciens*--mediated plant transformation
	hph[a]	*Streptomyces* sp.	Hygromycin phosphotransferase	Allows selection for resistance to the antibiotic hygromycin B	Go Resources Pty Ltd.	*A. tumefaciens*--mediated plant transformation

Adapted from ISAAA (2019)
Source: http://www.isaaa.org/gmapprovaldatabase
[a]selection marker/reporter

grape, apple, camelina, and canola crops by Jia and Wang (2014), Fan et al. (2015), Woo et al. (2015), Song et al. (2016), Zhang et al. (2016), Nishitani et al. (2016), Ren et al. (2016), Malnoy et al. (2016), Osakabe et al. (2016), Jiang et al. (2017), Osakabe et al. (2018), and Okuzaki et al. (2018). However, there is no work about safflower's genome editing in literature. In regard to *Brassica napus*, a mutant of FAD2_Aa was edited with the CRISPR/Cas9 system. While oleic acid content was increased (73–80%), linoleic acid content (16–9%) was decreased in *B. napus* (Okuzaki et al. 2018). In another study, the oleic acid content of soybean was increased by Calyxt Inc., knocking out Fad2.1 gene. Thus, shelf life of soybean was prolonged (Cebrailoglu et al. 2019). Furthermore, CRISPR/Cas9 system was also used in *Camelina sativa* and determined variations in fatty acid content. As a result of the study, as matched to the wild type, T4 seeds of *Camelina sativa* included high levels of oleic acid and low contents of linoleic and linolenic acid (Jiang et al. 2017). Similar studies can be carried out to improve oleic acid content of safflower using CRISPR technique.

11.11.5 Speed Breeding

Speed breeding (SB) accelerates the transition between generations by shortening the vegetation times of plants. Especially, the production times of some long-day plants or neutral-day plants are reduced, and these plants are benefited positively by obtaining more products in less time. Speed breeding was used for numerous crops such as barley, bread/durum wheat, oat, quinoa, different *Brassica* species, pea, chickpea, grass pea, peanut, amaranth, and *Brachypodium distachyon* by O'Connor et al. (2013), Mobini and Warkentin (2016), Stetter et al. (2016), Hickey et al. (2017), Alahmad et al. (2018), Ghosh et al. (2018), and Watson et al. (2018). Current protocols were provided on how to conduct SB presented by Ghosh et al. (2018) and Watson et al. (2018). To our knowledge, no study has been conducted about speed breeding of safflower until now. We envisage that safflower improvement can be accelerated with speed breeding via combining different modern technologies such as high-throughput phenotyping/genotyping (cameras, sensors, robotics, and computers), genomic selection, and genome editing (CRISPR/Cas9). Thus, new safflower cultivars having high yield and resistant to stress factors can be developed by applying these technologies.

11.12 Conclusion

Safflower is an important crop having great potential of oil, which can be utilized to meet the oil requirements. Safflower shows diversity in its phenotype upon which similarity centers are proposed for this crop. Advancement in sequencing technology leads in the genome sequencing of nearly all important crops. However, genome sequencing of this crop is not performed and made this crop as neglected one. There is a need to perform genome sequencing which will provide deep insight in the genome of this plant and opens the ways to identify the genomic regions having association for traits of interest in the near future.

References

Abel GH, Discroll MF (1976) Sequential traits development and breeding for high yield. Crop Sci 16:213–216

Acquaah G (2012) Principles of plant genetics and breeding, 2nd edn. Wiley, Hoboken

Adalı M, Öztürk Ö (2017) Konya Koşullarında Bazı Aspir Çeşitlerinin Verim ve Verim Unsurlarının Belirlenmesi. Selçuk Tarım Bilimleri Dergisi 3(2):233–237

Agyman GA, Loiland L, Karow R et al. (2002) Safflower. Dry land cropping systems [internet]. Oregon State University. Available from: http://www.eesc.oregonstate.edu

Ahmad Zadeh AK, Almassi M, Meighani HM et al (2011) Suitability of *Carthamus oxyacantha* plant as biodiesel feedstock. Aust J Crop Sci 5(12):1639

Akbari M, Wenzl P, Caig V et al (2006) Diversity arrays technology (DArT) for high-throughput profiling of the hexaploid wheat genome. Theor Appl Genet 113:1409–1420

Alahmad S, Dinglasan E, Leung KM et al (2018) Speed breeding for multiple quantitative traits in durum wheat. Plant Methods 14(1):36

Ali F, Yılmaz A, Nadeem MA et al (2019) Mobile genomic element diversity in world collection of safflower (*Carthamus tinctorius* L.) panel using iPBS-retrotransposon markers. PLoS ONE 14(2):e0211985

Ali F, Yilmaz A, Chaudhary HJ et al (2020a) Investigation of Morpho-Agronomic Performance and Selection Indices in the International Safflower Panel for Breeding Perspectives 2. Turk J Agric For:43

Ali F, Nadeem MA, Barut M et al (2020b) Genetic diversity, population structure and marker-trait association for 100-seed weight in international safflower panel using SilicoDArT marker information. Plants 9(5):652

Alvarez I, Tomaro LM, Benavides PM (2003) Changes in polyamines, proline and ethylene in sunflower calluses treated with NaCl. Plant Cell Tiss Org Cult 74:51–59

Ambreen H, Kumar S, Variath MT et al (2015) Development of genomic microsatellite markers in *Carthamus tinctorius* L. (Safflower) using next generation sequencing and assessment of their crossspecies transferability and utility for diversity analysis. PLoS ONE. https://doi.org/10.1371/journal.pone.0135443

Ambreen H, Kumar S, Kumar A et al (2018) Association mapping for important agronomic traits in safflower (*Carthamus tinctorius* L.) core collection using microsatellite markers. Front Plant Sci 9:402

Amini F, Saeidi G, Arzani A (2008) Study of genetic diversity in safflower genotypes using agro-morphological traits and RAPD markers. Euphytica 163:21–30. https://doi.org/10.1007/s10681-007-9556-6

Anonymous (1985). Safflower improvement. Thirteen research report, Nimbkar Agricultural Research Institute, Phaltan, District Satara, Maharashtra, India.69 p

Arslan Y (2018) Agro-morphological characterization of wild safflower (*Carthamus* L., Asteraceae) species in Turkey. Pak. J Bot 50(2):685–692

Arslan Y, Katar D, Güneylioğlu H et al (2010) Türkiye Florasındaki Yabani *Carthamus* L. Türleri ve Aspir (C. tinctorius L.) Islahında Değerlendirme Olanakları. Tarla Bitkileri Merkez Araştırma Enstitüsü Dergisi 19(1-2):36–43

Arystanbekkyzy M, Nadeem MA, Aktas H et al (2019) Phylogenetic and taxonomic relationship of turkish wild and cultivated emmer (Triticum turgidum ssp. dicoccoides) revealed by iPB-Sretrotransposons markers. Int. J. Agric. Biol, 21, 155–163.
Arzani A, Mirodjagh SS (1999) Response of durum wheat cultivars to immature embryo culture, callus induction and in vitro salt stress. Plant Cell Tiss Org Cult 58:67–72
Ashraf M, Foolad MR (2013) Crop breeding for salt tolerance in the era of molecular markers and marker-assisted selection. Plant Breed 132:10–20
Ashrafi E, Razmjoo K (2010) Effect of irrigation regimes on oil content and composition of safflower (*Carthamus tinctorius* L.) cultivars. J Am Oil Chem Soc 87(5):499–506
Ashri A (1957) Cytogenetics and morphology of *Carthamus* L. species and hybrids. Ph. D. Thesis Univ. of Calif., Davis
Ashri A (1973) Divergence and Evolution in the Safflower Genus, *Carthamus*, Final Research Report for USDA PL 480 Project No. A-10-CR-18, Hebrew University, Rehovot, Israel
Ashri A (1975) Evaluation of the germ plasm collection of safflower, *Carthamus tinctorius* L. V distribution and regional divergence for morphological characters. Euphytica 24(3):651–659
Ashri A, Knowles PF (1960) Cytogenetics of safflower (*Carthamus* L.) species and their hybrids. Agron J 52:11–17
Ashri A, Knowles PF (1977) Abst. A. Mtg. Am. Soc. Agron. p. 50
Ashri A, Zimmer DE, Urie AL et al (1974) Evaluation of world collection of safflower *Carthamus tinctorius* L. yield and yield components and their relationships. Crop Sci 14:799–802
Ashri A, Knowles PF, Urie AL et al. (1975). Evaluation of the germ plasm collection of safflower, C. tinctorius L. III. Oil content and iodine value and their associations with other characters. Econ Bot
Asoro FG, Newell MA, Beavis WD et al (2011) Accuracy and training population design for genomic selection on quantitative traits in elite North American oats. Plant Genome 4(2):132–144
Azab A (2018) Total Phenolic Content, Antioxidant Capacity and Antifungal Activity of Extracts of *Carthamus tenuis* and *Cephalaria joppensis*. Eur Chem Bull 7(4–6):156–161
Bagawan II, Ravikumar RL (2001) Strong undesirable linkages between seed yield and oil components-a problem in safflower improvement. In: Proceedings of the 5th international Safflower conference, Williston, North Dakota and Sidney, Montana, USA, 23–27 July, 2001. Safflower: a multipurpose species with unexploited potential and world adaptability, pp. 103–107. Department of Plant Pathology, North Dakota State University
Bajji M, Lutts S, Kinet JM (2004) Physiological changes after exposure to and recovery from polyethylene glycolinduced water deficit in callus cultures issued from durum wheat (Triticum durum Desf.) cultivars differing in drought resistance. J Plant Physiol 156:75–83
Baloch FS, Alsaleh A, de Miera LES et al (2015) DNA based iPBS-retrotransposon markers for investigating the population structure of pea (Pisum sativum) germplasm from Turkey. Biochem Syst Ecol 61:244–252
Baloch FS, Alsaleh A, Shahid MQ et al (2017) A whole genome DArTseq and SNP analysis for genetic diversity assessment in durum wheat from central fertile crescent. PloS one 12(1):e0167821
Barakat MN, Abdel-Latif TH (1996) In vitro selection of wheat callus tolerant to high levels of salt and plant regeneration. Euphytica 91:127–140
Barati M, Arzani A (2012) Genetic diversity revealed by EST-SSR markers in cultivated and wild safflower. Biochem Syst Ecol 44:117–123
Barut M, Nadeem MA, Karaköy T, Baloch FS (2020) DNA fingerprinting and genetic diversity analysis of world quinoa germplasm using iPBS-retrotransposon marker system. Turk J Agric For 44
Bassi FM, Sanchez-Garcia M (2017) Adaptation and stability analysis of ICARDA durum wheat elites across 18 countries. Crop Sci 57(5):2419–2430
Bassil ES, Kaffka SR (2002) Response of safflower (*Carthamus tinctorius* L.) to saline soils and irrigation: II. Crop response to salinity. Agric Water Manag 54(1):81–92

Basu S, Gangopadhyay G, Mukherjee BB (2002) Salt tolerance in rice in vitro: Implication of accumulation of Na^+, K^+ and proline. Plant Cell Tiss Org Cult 96:55–64

Bayramin S, Kaya MD (2009) Advancement of safflower and rapeseed production of Turkey in recent years. Tarla Bitkileri Merkez Araştırma Enstitüsü Dergisi 18:43–47

Beke GJ, Volkmar KM (1995) Mineral composition of flax (Linum usitatissimum L.) and safflower (*Carthamus tinctorius* L.) on a saline soil high in sulfate salts. Can J Plant Sci 75:399–404

Belaj A, del Carmen Dominguez-García M, Atienza SG et al (2012) Developing a core collection of olive (Olea europaea L.) based on molecular markers (DArTs, SSRs, SNPs) and agronomic traits. Tree Genet Genomes 8:365–378. https://doi.org/10.1007/s11295-011-0447-6

Bergman JW, Flynn CR (2001) High oleic safflower as a diesel fuel extender-a potential new market for Montana safflower. In: Proceedings of the 5th international Safflower conference, Williston, North Dakota and Sidney, Montana, USA, 23–27 July, 2001. Safflower: a multipurpose species with unexploited potential and world adaptability, pp 289–293. Department of Plant Pathology, North Dakota State University.

Bernardo R (2008) Molecular markers and selection for complex traits in plants: learning from the last 20 years. Crop Sci 48(5):1649–1664

Bhardwaj SC, Kumar A, Bhargavand GK et al (1990) Assessment of losses caused by insect pests in safflower (*Carthamus tinctorius* L.) Indian. J Appl Ent 4:61–70

Bhattacharjee R, Khairwal IS, Bramel PJ et al (2007) Establishment of a pearl millet [Pennisetum glaucum (L.) R. Br.] core collection based on geographical distribution and quantitative traits. Euphytica 155:35–45. https://doi.org/10.1007/s10681-006-9298-x

Blackshaw RE (1993) Safflower (*Carthamus tinctorius*) density and row spacing effects on competition with green foxtail (Setaria viridis). Weed Sci. 41:403–408

Bocheva A, Mikhova B, Taskova R et al (2003) Antiinflammatory and analgesic effects of *Carthamus lanatus* aerial parts. Fitoterapia 74(6):559–563

Borsani O, Valpuesta V, Botella MA (2003) Developing salt tolerant plants in a new century: a molecular biology approach. Plant Cell Tiss Org Cult 73:101–115

Bowers JE, Pearl SA, Burke JM (2016) Genetic mapping of millions of SNPs in safflower (*Carthamus tinctorius* L.) via whole-genome resequencing. G3 6(7):2203–2211

Bukhsh E, Malik SA, Ahmad SS et al (2014) Hepatoprotective and hepatocurative properties of alcoholic extract of *Carthamus oxyacantha* seeds. Afr J Plant Sci 8(1):34–41

Bülbül AS, Tarıkahya-Hacıoğlu B, Arslan Y et al (2013) Pollen morphology of *Carthamus* L. species in Anatolian flora. Plant Syst Evol 299(3):683–689

Camas N, Esendal E (2006) Estimates of broad-sense heritability for seed yield and yield components of safflower (*Carthamus tinctorius* L.). Hereditas 143:55–57

Cantelmo NF, Von Pinho RG, Balestre M (2017) Genome-wide prediction for maize single-cross hybrids using the GBLUP model and validation in different crop seasons. Mol Breed 37(4):51

Cassini H (1819) Dictionnaire de Sciences Naturelles. Paris. Cited by King R, Dawson HW (eds), 1975. Cassini on Compositae. Oriole Editions, New York

Cebrailoglu N, Yildiz AB, Akkaya O et al (2019) CRISPR-Cas: Removing Boundaries of the Nature. Eur J Biol 78(2):157–164

Cervantes-Martinez JE (2001) Safflower production and research in Mexico: status and prospects. In Proceedings of the 5th international Safflower conference, Williston, ND, and Sidney, MT, July 23-27, 2001. Bergman, J.W. and H.H. Mundel, Eds., p. 282

Chapman MA, Burke JM (2007) DNA sequence diversity and the origin of cultivated safflower (*Carthamus tinctorius* L.; *Asteraceae*). BMC Plant Biology 7(1):60

Chapman MA, Hvala J, Strever J et al (2010) Population genetic analysis of safflower (*Carthamus tinctorius*; *Asteraceae*) reveals a Near Eastern origin and five centers of diversity. Am J Bot 97(5):831–840

Chavan VM (1961) Niger and Safflower. Indian Central Oilseeds Committee Publication, Hyderabad

Chavoushi M, Najafi F, Salimi A et al (2019) Improvement in drought stress tolerance of safflower during vegetative growth by exogenous application of salicylic acid and sodium nitroprusside. Ind Crops Prod 134:168–176

Chawla HS (2000) Introduction to plant biotechnology. Science Publisher, New Hampshire

Chowdhury S (1944) An Alternaria disease of safflower. J Indian Bot Sci 23:59–65

Collard BCY, Jahufer MZZ, Brouwer JB et al (2005) An introduction to markers, quantitative trait loci (QTL) mapping and marker-assisted selection for crop improvement. The basic concepts. Euphytica 142:169–196

Conners IL (1943) The rusts of safflower. Phytopathology 33:789–796

Dajue L, Mundel H (1996) Safflower. *Carthamus tinctorius* L. Promoting the conservation and use of underutilized and neglected crops. International Plant Genetic Resources Institute, Rome, Institute of Plant Genetics and Crop Plant Research, Gatersleben

Dajue L, Yunzhou H (1993) The development and exploitation of safflower tea. In; Proceedings of the 3rd international safflower conference, June 14–18, Beijing, China, pp: 837-843

Das G, Patra JK, Back KH (2017) Corrigendum: Insight into MAS: a molecular tool for development of stress resistant ad quality of rice trough gene stacking. Front plant Sci 8:1321

Davis PH (1975) Flora of Turkey and The East Aegeans İslands, vol 5. The University Press, Edinburg

De Azevedo Peixoto L, Moellers TC, Zhang J et al (2017) Leveraging genomic prediction to scan germplasm collection for crop improvement. PloS one 12(6):e0179191

De Candolle AP (1838) Prodromus systematis naturalis regni vegetabilis. Sumptibus Sociorum Treuttel et Würtz, Paris, p 6

De Candolle A (1890) Origin of cultivated plants. R.W. Hofner Co, New York. [1890] 1967

Derakhshan E, Majidi MM, Sharafi Y et al (2014) Discrimination and genetic diversity of cultivated and wild safflowers (*Carthamus* spp.) using EST-microsatellites markers. Biochemical systematics and ecology 54:130–136

Díez CM, Imperato A, Rallo L (2012) Worldwide core collection of olive cultivars based on simple sequence repeat and morphological markers. Crop Sci 52:211–221. https://doi.org/10.2135/cropsci2011.02.0110

Duke JA (1983) Handbook of energy crops

Dwivedi SL, Upadhyaya HD, Hegde DM (2005) Development of core collection using geographic information and morphological descriptors in safflower (*Carthamus tinctorius* L.) germplasm. Genet Resour Crop Evol 52(7):821–830

E Sousa MB, Cuevas J, de Oliveira Couto EG et al (2017) Genomic-enabled prediction in maize using kernel models with genotype× environment interaction. G3 7(6):1995–2014

Ebana K, Kojima Y, Fukuoka S et al (2008) Development of mini core collection of Japanese rice landrace. Breed Sci 58:281–291. https://doi.org/10.1270/jsbbs.58.281

Ebrahimi F, Majidi MM, Arzani A et al (2017) Association analysis of molecular markers with traits under drought stress in safflower. Crop Pasture Sci 68:167–175. https://doi.org/10.1071/cp16252

Ekincialp A, Erdinc C, Turan S et al (2019) Genetic Characterization of Rheum ribes (Wild Rhubarb) Genotypes in Lake Van Basin of Turkey through ISSR and SSR Markers. Int J Agric Biol 21(4):795–802

Erbaş S, Haydar H (2017) Aspir (*Carthamus tinctorius* L.)'de yaprak dikenliliği ve çiçek renginin genetiği Anadolu Tarım Bilim. Derg./Anadolu J Agric Sci 32:244–248

Errabii T, Gandonou CB, Essalmani H, Abrini J, Idaoma M, Senhaji NS (2007) Effects of NaCl and mannitol induced stress on sugarcane (Saccharum sp.) callus cultures. Acta Physiol Plant 29:95–102

Esendal E (2001) Global adaptability and future potential of safflower. In: Proceedings of the 5th international Safflower conference, Williston, ND, and Sidney, MT, July 23–27, 2001. Bergman, J.W. and H.H. Mundel, Eds., pp xi–xii.

Estilai A (1977) Genus *Carthamus* as an example of plant evolution. Acta Ecol Iran 2:70–76

Fan JB, Chee MS, Gunderson KL (2006) Highly parallel genomic assays. Nat Rev Genet 7(8):632

Fan D, Liu T, Li C et al (2015) Efficient CRISPR/Cas9-mediated targeted mutagenesis in Populus in the first generation. Sci Rep 5:12217

FAO (2003) State of the world's forests 2003. Rome. ftp.fao.org/docrep/fao/005/y7581e/

FAO (2021). FAOSTAT. http://www.fao.org/faostat/en/#data/QC/visualize [accessed 07 January 21]

Fehr WR (1987) Principles of cultivar development. McMillan, New York

Fernandez-Martinez JM, Rio M, Haro M (1993) Survey of safflower (*Carthamus tinctorius* L.) germplasm for variants in fatty acid composition and other seed characters. Euphytica 69:115–122

Frankel O (1984) Genetic perspectives of germplasm conservation. Genetic manipulation: impact on man and society. Cambridge University Press, Cambridge 61(3):161–170

Furuya T, Yoshikawa T, Kimura T et al (1987) Production of tocopherols by cell culture of safflower. Phytochemistry 26:2741–2747

Gandonou C, Abrini J, Idaomar M, Skali Senhaji N (2005) Response of sugarcane (Saccharum sp.) varieties to embryogenic callus induction and in vitro salt stress. Afr J Biotechnol 4:350–354

Garcia-Moreno MJ, Velasco L, Perez-Vich B (2010) Transferability of non-genic microsatellite and genebased sunflower markers to safflower. Euphytica 175(2):145–150

Garnatje T, Garcia S, Vilatersana R et al (2006) Genome size variation in the genus *Carthamus* (*Asteraceae*, *Cardueae*): systematic implications and additive changes during allopolyploidization. Ann Bot 97(3):461–467

Gautam S, Bhagyawant SS, Srivastava N (2014) Detailed study on therapeutic properties, uses and pharmacological applications of safflower (*Carthamus tinctorius* L.). Int J Ayur Pharma Res 2(3):1–12

Gawande ND, Mahurkar DG, Rathod TH (2005) In vitro screening of wheat genotypes for drought tolerance. Ann Plant Physiol 19:162–168

Ghosh S, Watson A, Gonzalez-Navarro OE et al (2018) Speed breeding in growth chambers and glasshouses for crop breeding and model plant research. Nat Protoc 13(12):2944

Goddard ME, Hayes BJ (2007) Genomic selection. J Anim Breed Genet 124(6):323–330

Golkar P (2014) Breeding improvements in safflower ('Carthamus tinctorius' L.): a review. Aust J Crop Sci 8(7):1079

Golkar P, Arzani A, Maibodi SAM (2007) Evaluation of bread wheat (Triticum aestivum L.) cultivars for in vitro salt tolerance. Agric Sci Technol J 20:191–200. (In Persian)

Golkar P, Arzani A, Rezai AM (2011) Genetic analysis of oil content and fatty acid composition in safflower (*Carthamus tinctorius* L.). J Am Oil Chem Soc 88:975–982

Guo J, Ling H, Wu Q et al (2014) The choice of reference genes for assessing gene expression in sugarcane under salinity and drought stresses. Sci Rep 4:7042

Habyarimana E, Parisi B, Mandolino G (2017) Genomic prediction for yields, processing and nutritional quality traits in cultivated potato (Solanum tuberosum L.). Plant Breed 136(2):245–252

Hacioglu BT, Yaman H, Arslan Y et al (2013) Investigation of molecular diversity of Asian safflower (*Carthamus tinctorius* L.) accessions by RAPD markers for using in hybridization programme. Res Crops 14(1):169–174

Hamadi BS, Hamrouni I, Marzouk B (2001) Comparison of yield components and oil content of selected Safflower (*Carthamus tinctorius* L.) accessions in Tunisia. In: International Safflower conference.

Hamdan YAS, Perez-Vich B, Fernandez-Martinez JM et al (2008a) Inheritance of veryhigh linoleic acid content and its relationship with nuclear male sterility in safflower. Plant Breed 127:507–509

Hamdan YAS, Velasco L, Perez-Vich B (2008b) Development of SCAR markers linked to male sterility and very high linoleic acid content in safflower. Mol Breed 22:385–393

Hamdan YAS, Perez-Vich B, Fernandez-Martinez JM et al (2009a) Novel safflower germplasm with increased saturated fatty acid content. Crop Sci 49:127–132

Hamdan YAS, Pérez-Vich B, Velasco L et al (2009b) Inheritance of high oleic acid content in safflower. Euphytica 168(1):61–69

Hamdan YAS, Garcia-Moreno MJ, Redondo-Nevado J (2011) Development and characterization of genomic microsatellite markers in safflower (*Carthamus tinctorius* L.). Plant Breed 130(2):237–241

Hamdan YAS, García-Moreno MJ, Fernández-Martínez JM et al (2012) Mapping of major and modifying genes for high oleic acid content in safflower. Mol Breed 30(3):1279–1293

Hamedi M, Golkar P, Arzani A (2016) In vitro salt tolerance of safflower (*Carthamus tinctorius* L.) genotypes using different explants. Plant Tiss Cult Biotech 26(2):231–242

Hanelt P (1961) Systematic study of the genus *Carthamus* L. (Compositae) a monographic review, Ph.D. Thesis (in German), Martin-Luther University, Halle-Wittenburg, Germany

Hasan M, Sarker RH (2013) In vitro selection for NaCl salt tolerance in aromatic rice (Oryza sativa) genotypes. Indian. J Agric Sci 83:1221–1226

Hassani SMR, Talebi R, Pourdad SS et al (2020) In-depth genome diversity, population structure and linkage disequilibrium analysis of worldwide diverse safflower (Carthamus tinctorius L.) accessions using NGS data generated by DArTseq technology. Mol Biol Rep 47(3):2123–2135

Hatipoğlu H, Nacar AS, Saraçoğlu M et al (2017) Safflower studies specially in Şanlıurfa. Selcuk J Agric Food Sci 31(2):44–53

Hegde DM, Singh V, Nimbkar N (2002) Safflower. In: Singh CB, Khare D (eds) Genetic improvement of field crops. Scientific Publishers, Jodhpur, pp 199–221

Hickey LT, Germán SE, Pereyra SA et al (2017) Speed breeding for multiple disease resistance in barley. Euphytica 213(3):64

Huaman Z, Aguilar C, Ortiz R (1999) Selecting a Peruvian sweetpotato core collection on the basis of morphological, eco-geographical, and disease and pest reaction data. Theor Appl Genet 98:840–844. https://doi.org/10.1007/s001220051142

Indi DV, Lukade GM, Patil PS (1986) Influence of Alternaria leaf spot (Alternaria carthami Chowdhary) on growth and yield of safflower. Curr Res Rep 2(1):137–139

Indi DV, Lukade GM, Patil PS et al (1988) Estimation of yield losses due to Alternaria leaf spot in safflower (c.o. Alternaria carthami Chowdhary) under dryland conditions. Pesticides 22(1):41–43

Irving DW, Shannon MC, Breda VA et al (1988) Salinity effects on yield and oil quality of high-linoleate and high-oleate cultivars of safflower. J Agric Food Chem 36(1):37–42

ISAAA's GM Approval Database (2019) Retrieved 30 December 2019, from http://www.isaaa.org/gmapprovaldatabase/

Jaccoud D, Peng K, Feinstein D et al (2001) Diversity arrays: a solid state technology for sequence information independent genotyping. Nucl Acids Res 29(4):25

Jalali A, Salehi F, Bahrani M (2012) Effects of different irrigation intervals and weed control on yield and yield components of safflower (*Carthamus tinctorius* L.). Arch Agron Soil Sci 58(11):1621–1269

James C (2014) Global status of transgenic crops in 2014. ISAAA Briefs No. 49

Jan SH, Shinwari ZK, Shah SH et al (2016) In-planta transformation: recent advances. Romanian Biotechnol Lett 21(1):11085–11091

Jia H, Wang N (2014) Targeted genome editing of sweet orange using Cas9/sgRNA. PloS One 9(4):e93806

Jiang WZ, Henry IM, Lynagh PG et al (2017) Significant enhancement of fatty acid composition in seeds of the allohexaploid, Camelina sativa, using CRISPR/Cas9 gene editing. Plant Biotechnol J 15(5):648–657

Jin Q, Waters D, Cordeiro GM et al (2003) A single nucleotide polymorphism (SNP) marker linked to the fragrance gene in rice (Oryza sativa L.). Plant Sci 165(2):359–364

Johnson R, Stout D, Bradley V (1993) The us collection: A rich source of safflower germplasm. In: Proceedings of the third international Safflower conference. China, Beijing, pp 9–13

Johnson RC, Kisha TJ, Evans MA (2007) Characterizing safflower germplasm with AFLP molecular markers. Crop Sci 47:1728–1736. https://doi.org/10.2135/cropsci2006.12.0757

Johnson R, Bradley V, Kisha T (2008) Safflower germplasm. Past, present, and future. In: safflower: unexploited potential and world adaptability. 7th International Safflower Conference,

Wagga Wagga, New South Wales, Australia, 3–6 November. Agri-MC Marketing and Communication. pp 1–7.
Kammili A (2013) Genetic linkage between male sterility and non-spiny trait in safflower (*Carthamus tinctorius* L.). Plant Breed 132(2):180–184
Kar G, Kumar A, Martha M (2007) Water use efficiency and crop coefficients of dry season oilseed crops. Agric Water Manag 87(1):73–82
Karık Ü, Nadeem MA, Habyarimana E et al (2019) Exploring the genetic diversity and population structure of Turkish laurel germplasm by the iPBS-Retrotransposon marker system. Agronomy 9(10):647
Kaya MD, Bayramin S, Kayaçetin F et al (2009) Determination of proper gamma radiation (60Co) dose to induce variation in safflower. Ziraat Fakültesi Dergisi-Süleyman Demirel Üniversitesi 4(2):28–33
Khadeer MA, Anwar SY (1991) Induced mutations in the improvement of safflower (*Carthamus tinctorius* L.). In: Plant mutation breeding for crop improvement, vol 1
Khan SH (2019) Genome-editing technologies: concept, pros, and cons of various genome-editing techniques and bioethical concerns for clinical application. Mol Ther. Nucleic Acids 16:326
Khan MA, von Witzke-Ehbrecht S, Maass BL et al (2009) Relationships among different geographical groups, agromorphology, fatty acid composition and RAPD marker diversity in safflower (*Carthamus tinctorius* L.). Genet Resour Crop Evol 56:19–30. https://doi.org/10.1007/s10722-008-9338-6
Khorami R, Safarnejad A (2011) In vitro selection of Foeniculum vulgare for salt tolerance. Not Sci Biol 3:90–97
Klisiewicz JM, Houston BR (1962) Fusarium wilt of safflower. Plant Dis Rep 46(10):748–749
Knowles PF (1958) Safflower. Adv Agron 10:289–323
Knowles P (1969) Centers of plant diversity and conservation of crop germplasm: Safflower. Econ Bot 23(4):324–329
Knowles PF (1988) Carthamus species relationships. Lecture presented at Beijing botanical garden. Institute of Botany, Chinese Academy of Sciences, Beijing
Knowles PF (1989) Importance And Distribution. Oil crops of the world: their breeding and utilization 363
Knowles P, Ashri A (1958) Wild safflower in California: Improvement of cultivated safflower through plant-breeding program to obtain desirable characteristics of wild species. Calif Agric *12*(4):4–5
Köse A (2016) A research on determining of seed setting rate in safflower (*Carthamus tinctorius* L.). Türkiye Tarımsal Araştırmalar Dergisi 3(2):152–158
Kuete V, Wiench B, Hegazy MEF et al (2012) Antibacterial activity and cytotoxicity of selected Egyptian medicinal plants. Planta Medica 78(02):193–199
Kumar S, Ambreen H, Murali TV et al (2015) Assessment of genetic diversity and population structure in a global reference collection of 531 accessions of *Carthamus tinctorius* L. (Safflower) using AFLP markers. Plant Mol Biol Rep 33:1299–1313. https://doi.org/10.1007/s11105-014-0828-8
Kumar S, Ambreen H, Variath MT et al (2016) Utilization of molecular, phenotypic, and geographical diversity to develop compact composite core collection in the oilseed crop, safflower (*Carthamus tinctorius* L.) through maximization strategy. Front Plant Sci 7:1554
Kumari S, Choudhary RC, Kumara Swamy RV et al (2017) Assessment of genetic diversity in safflower (*Carthamus tinctorius* L.) genotypes through morphological and SSR marker. J Pharmacogn Phytochem 6(5):2723–2731
Kupsow AI (1932) Bull. Appt. Bot. Genet. Pl. Breed. Ser. 9: 99
Ladd SL, Knowles PF (1971) Interactions of alleles at two loci regulating fatty acid composition of the seed oil of safflower (*Carthamus tinctorius* L.). Crop Sci 11:681–684. https://doi.org/10.2135/cropsci1971.0011183X001100050024x

Lee GA, Sung JS, Lee SY et al (2014) Genetic assessment of safflower (*Carthamus tinctorius* L.) collection with microsatellite markers acquired via pyrosequencing method. Mol Ecol Resour 14:69–78. https://doi.org/10.1111/1755-0998.12146

Lezar S, Myburg AA, Berger DK (2004) Development and assessment of microarray-based DNA fingerprinting in Eucalyptus grandis. Theor Appl Genet 109(7):1329–1336

Li D, Mündel HH (1996) Safflower: *Carthamus tinctorius* L. International Plant Genetic Resources Institute (IPGRI).

Li D, Zhou M, Ramanatha Rao V (1993) Characterization and evaluation of Safflower Germplasm. Geological Pub. House, Beijing, China. 260 text and 16 colour p. [Outline of origin, distribution, biology of safflower, collecting and conservation strategy, characterization resulting from evaluations of germplasm of safflower, including world collection grown in China; copies may be purchased by sending a certified cheque or money order for US$45, payable to the academy, to Beijing Botanical Garden, Institute of Botany, Chinese Academy of Sciences, Beijing 1000093, China]

Li Y, Shi Y, Cao Y et al (2005) Establishment of a core collection for maize germplasm preserved in Chinese National Genebank using geographic distribution and characterization data. Genet Resour Crop Evol 51:845–852. https://doi.org/10.1007/s10722-005-8313-8

Li ZM, Ding JQ, Wang RX et al (2011) A new QTL for resistance to Fusarium ear rot in maize. J Appl Genet 52:403–406. https://doi.org/10.1007/s13353-011-0054-0

Lijiao FAN, Meili GUO (2013) Progress of safflower (*Carthamus tinctorius* L.) regeneration through tissue culture. J Med Coll PLA 28(5):289–301

Liu W, Shahid MQ, Bai L et al (2015) Evaluation of genetic diversity and development of a core collection of wild rice (Oryza rufipogon Griff.) populations in China. PLoS ONE 10:e0145990. https://doi.org/10.1371/journal.pone.0145990

Lopez-Gonzalez G (1989) Anales del Jardin Botánico de Madrid 47:11–34

Lopez-Gonzalez G (1990) Acerca de la clasificacion natural del genero *Carthamus* L., s.1. Anales Jardin Bot Madrid 47:11–34

Lorenz AJ, Smith KP, Jannink JL (2012) Potential and optimization of genomic selection for Fusarium head blight resistance in six-row barley. Crop Sci 52(4):1609–1621

Lutts S, Kinet JM, Bouharmont J (1999) Improvement of rice callus regeneration in the presence of NaCl. Plant Cell Tiss Org Cult 57:3–11

Lutts S, Almansouri M, Kinet JM (2004) Salinity and water stress have contrasting effects on the relationship between growth and cell viability during and after stress exposure in durum wheat callus. Plant Sci 167:9–18

Lyra DH, de Freitas Mendonça L, Galli G et al (2017) Multi-trait genomic prediction for nitrogen response indices in tropical maize hybrids. Mol Breed 37(6):80

Mahalakshmi V, Ng Q, Lawson M (2007) Cowpea [Vigna unguiculata (L.) Walp.] core collection defined by geographical, agronomical and botanical descriptors. Plant Genet Resour 5:113–119. https://doi.org/10.1017/S1479262107837166

Mahasi MJ, Pathak RS, Wachira FN et al (2006) Correlations and path coefficient analysis in exotic safflower (*Carthamus tinctorious* L.) genotypes tested in the arid and semi arid lands (Asals) of Kenya. Asian J Plant Sci 5(6):1035–1038

Malnoy M, Viola R, Jung MH et al (2016) DNA-free genetically edited grapevine and apple protoplast using CRISPR/Cas9 ribonucleoproteins. Front Plant Sci 7:1904

Maluszynski M, Kasha KJ (2002) Mutations. Vitro and molecular techniques for environmentally sustainable crop improvement. Kluwer Academic Publishers, Dordrecht. 246p

Massman JM, Jung HJG, Bernardo R (2013) Genomewide selection versus marker-assisted recurrent selection to improve grain yield and stover-quality traits for cellulosic ethanol in maize. Crop Sci 53(1):58–66

Mayerhofer R, Archibald C, Bowles V et al (2010) Development of molecular markers and linkage maps for the *Carthamus* species C. tinctorius and C. oxyacanthus. Genome 53(4):266–276

Merrill SD, Tanaka DL, Hanson JD (2002) Root length growth of eight crop species in Haplustoll soils. Soil Sci Soc Am J 66(3):913–923

Mirzahashemi M, Mohammadi-Nejad G, Golkar P (2015) A QTL linkage map of safflower for yield under drought stress at reproductive stage. Iran J Genet Plant Breed 4(2):20–27
Mobini SH, Warkentin TD (2016) A simple and efficient method of in vivo rapid generation technology in pea (Pisum sativum L.). In Vitro Cell Dev Biol Plant 52(5):530–536
Morin L, Sheppard AW (2012) *Carthamus lanatus* L.-saffron thistle. Biological control of weeds in Australia. CSIRO Publishing, Collingwood, pp 139–145
Morton NE (2006) Linkage disequilibrium maps and association mapping. J Clin Investig 115(6):1425–1430. https://doi.org/10.1172/JCI25032
Mozaffari K, Asadi AA (2006) Relationships among traits using correlation, principal components and path analysis in safflower mutants sown in irrigated and drought stress condition. Asian J Plant Sci 5(6):977–983
Murashige T, Skoog F (1962) A revised medium for rapid growth and bio assays with tobacco tissue cultures. Physiol Plant 15(3):473–497
Nadeem MA, Habyarimana E, Ciftci V et al (2018a) Characterization of genetic diversity in Turkish common bean gene pool using phenotypic and whole-genome DArTseq-generated silicoDArT marker information. PlosONE. https://doi.org/10.1371/journal.pone
Nadeem MA, Aasim M, Kırıcı S (2018b) Laurel (Laurus nobilis L.): a less-known medicinal plant to the world with diffusion, genomics, phenomics, and metabolomics for genetic improvement. In: Biotechnological approaches for medicinal and aromatic plants. Springer, Singapore, pp 631–653
Nagaraj G (1993) June. Safflower seed composition and oil quality-a review. In: 3rd International Safflower conference, Beijing, pp 58–71.
Newell MA, Jannink JL (2014) Genomic selection in plant breeding. In: Crop breeding. Humana Press, New York, pp 117–130
Nimbkar N (2002) Safflower rediscovered. Times Agric J 2(1):32–36
Nishitani C, Hirai N, Komori S et al (2016) Efficient genome editing in apple using a CRISPR/Cas9 system. Sci Rep 6:31481
Noirot M, Hamon S, Anthony F (1996) The principal component scoring: a new method of constituting a core collection using quantitative data. Genet Resour Crop Evol 43(1):1–6
O'Connor DJ, Wright GC, Dieters MJ et al (2013) Development and application of speed breeding technologies in a commercial peanut breeding program. Peanut Sci 40(2):107–114
OGTR 2018. http://www.ogtr.gov.au/
Okuzaki A, Ogawa T, Koizuka C et al (2018) CRISPR/Cas9-mediated genome editing of the fatty acid desaturase 2 gene in Brassica napus. Plant Physiol Biochem 131:63–69
Orlikowska TK, Cranston HJ, Dyer WE (1995) Factors influencing *Agrobacterium tumefaciens* mediated transformation and regeneration of the safflower cultivar centennial. Plant Cell Tiss Org Cult 40:85–91
Osakabe Y, Watanabe T, Sugano SS et al (2016) Optimization of CRISPR/Cas9 genome editing to modify abiotic stress responses in plants. Sci Rep 6:26685
Osakabe Y, Liang Z, Ren C et al (2018) CRISPR-Cas9-mediated genome editing in apple and grapevine. Nat Protoc 13(12):2844
Pahlavani M, Mirlohi A, Saeidi G (2004) Inheritance of flower color and spininess in safflower (*Carthamus tinctorius* L.). J Hered 95(3):265–267
Patil MB, Shinde YM, Attarde KA (1993) Evaluation of safflower cultures for resistance to alternaria leaf spot (Alternariacarthami) and management strategies. In: Proceeding of the third international Safflower conference, June 14–18, Beijing, China. pp 269–278
Pearl SA, Burke JM (2014) Genetic diversity in Carthamus tinctorius (Asteraceae; safflower), an underutilized oilseed crop. Am J Bot 101(10):1640–1650
Pearl SA, Bowers JE, Chin-Wo SR et al (2014) Genetic analysis of safflower domestication. BMC Plant Biol 14(43)
Poehlman JM, Sleper DA (1995) Breeding field crops, 5th edn. Iowa State University, Ames

Popov AM, Kang D (2011) Analgesic and other medicinal properties of Safflower (*Carthamus tinctorius* L.) seeds. In: Nuts and seeds in health and disease prevention, pp 995–1002. Academic Press

Quiroga AR, Díaz-Zorita M, Buschiazzo DE (2001) Safflower productivity as related to soil water storage and management practices in semiarid regions. Commun Soil Sci Plant Anal 32(17–18):2851–2862

Rahamatalla AB, Babiker EE, Krishna AG et al (2001) Changes in fatty acids composition during seed growth and physicochemical characteristics of oil extracted from four safflower cultivars. Plant Foods Human Nutr 56(4):385–395

Rahmani F, Sayfzadeh S, Jabbari H et al (2019) Alleviation of drought stress effects on Safflower yield by foliar application of zinc. Int J Plant Prod:1–12

Rao SK, Rohini VK (1999) Gene transfer into Indian cultivars of safflower (*Carthamus tinctorius* L.) using *Agrobacterium tumefaciens*. Plant Biotechnol 16:201–206

Ren C, Liu X, Zhang Z et al (2016) CRISPR/Cas9-mediated efficient targeted mutagenesis in Chardonnay (Vitis vinifera L.). Sci Rep 6:32289

Rohini VK, Rao KS (2000) Embryo transformation, a practical approach for realizing transgenic plants of safflower (*Carthamus tinctorius* L.). Ann Bot 86:1043–1049

Rudra Naik V, Gulganji GG, Mallapur CP et al. (2001) Association analysis in safflower under rainfed conditions. In: 5th international safflower conference, Montana, July, pp 23–27

Sabzalian MR, Mirlohi A, Saeidi G et al (2009) Genetic variation among populations of wild safflower, *Carthamus oxyacanthus* analyzed by agro-morphological traits and ISSR markers. Genet Resour Crop Evol 56(8):1057–1064

Salunkhe DK, Charan JK, Adjule RN et al (1992) World oilseeds, Van Nostrand. Reinhold, New York, p 326

Sasanuma T, Sehgal D, Sasakuma T et al (2008) Phylogenetic analysis of *Carthamus* species based on the nucleotide sequence of the nuclear SACPD gene and chloroplast trn L-trn F IGS region. Genome 51(9):721–727

Saxena M, Singh J, Deshpande S (2008) Two decades of safflower in madhya pradesh from 1984-2004. In: Safflower: unexplored potential and world adaptability. Proceedings of the 7th International Safflower Conference, New South Wales, Australia, Wagga Wagga.

Schank SC, Knowles PF (1964) Cytogenetics of hybrids of *Carthamus* species (Compositae) with ten pairs of chromosomes. Am J Bot 51(10):1093-1102

Seeta P, Talat K, Anwar S (2000) Somaclonal variation – an alternative source of genetic variability in safflower. J Cytol Genet 1:127–135

Sehgal D, Raina SN (2005) Genotyping safflower (*Carthamus tinctorius*) cultivars by DNA fingerprints. Euphytica 146(1-2):67–76

Sehgal D, Raina SN (2011) *Carthamus*. In: Wild crop relatives: genomic and breeding resources. Springer, Berlin/Heidelberg, pp 63–95

Sehgal D, Rajpal VR, Raina SN et al (2009) Assaying polymorphism at DNA level for genetic diversity diagnostics of the safflower (*Carthamus tinctorius* L.) world germplasm resources. Genetica 135:457–470. https://doi.org/10.1007/s10709-008-9292-4

Semagn K, Bjornstad A, Skinnes H et al (2006) Distribution of DArT, AFLP, and SSR markers in a genetic linkage map of a doubled-haploid hexaploid wheat population. Genome 49:545–555

Şenel AA (2019) Dünya Aspir Gen Koleksiyonunda Yer Alan Bazi Aspir Hatlarinin Samsun Ekolojik Koşullarinda Verim ve Verim Unsurlari İle Bazi Teknolojik Özelliklerinin Belirlenmesi Üzerine Bir Araştirma. Ondokuz Mayis Üniversitesi, Fen Bilimleri Enstitüsü, Yüksek Lisans Tezi, Samsun

Shaki F, Ebrahimzadeh Maboud H, Niknam V (2018) Penconazole alleviates salt-induced damage in safflower (*Carthamus tinctorius* L.) plants. J Plant Interact 13(1):420–427

Shehzad T, Okuizumi H, Kawase M et al (2009) Development of SSR-based sorghum (Sorghum bicolor (L.) Moench) diversity research set of germplasm and its evaluation by morphological traits. Genet Resour Crop Evol 56:809–827. https://doi.org/10.1007/s10722-008-9403-1

Shilpa KS, Kumar VD, Sujatha M (2010) Agrobacterium-mediated genetic transformation of safflower (*Carthamus tinctorius* L.). Plant Cell Tiss Organ Cult 103(3):387–401
Singh F, Diwakar B (1995) Chickpea botany and production practices.
Singh V, Nimbkar N (1993) Genetics of aphid resistance in safflower (*Carthamus tinctorius* L.). Sesame Safflower. Newsletter 8:101–106
Singh RJ, Nimbkar N (2006) Chapter 6: Safflower (*Carthamus tinctorius* L.). In: Singh RJ (ed) Genetic resources, chromosome engineering, and crop improvement. CRC Press, New York, pp 167–194
Singh V, Nimbkar N (2007) Safflower (*Carthamus tinctorius* L.). In: Singh RJ (ed) Genetic resources, chromosome engineering, and crop improvement. CRC Press, Boca Raton, pp 167–194
Singh V, Nimbkar N (2016) Safflower. In: Breeding oilseed crops for sustainable production, pp 149–167. Academic Press
Singh V, Prasad RR (2005) Integrated management of pests and diseases in safflower. Directorate of Oilseeds Research, Hyderabad, India, p 49
Smith JR (1996) Safflower. American Oil Chemists' Society Press, Champaign. 606 p
Song G, Jia M, Chen K et al (2016) CRISPR/Cas9: a powerful tool for crop genome editing. Crop J 4(2):75–82
Souza TL, de Barros EG, Bellato CM (2012) Single nucleotide polymorphism discovery in common bean. Mol Breed 30(1):419–428
Srinivas B, Luch H, Green AG et al (2011) *Agrobacterium*-mediated transformation of safflower and the efficient recovery of transgenic plants via grafting. Plant Methods 7:12. https://doi.org/10.1186/1746-4811-7-12
Stetter MG, Zeitler L, Steinhaus A et al (2016) Crossing methods and cultivation conditions for rapid production of segregating populations in three grain amaranth species. Front Plant Sci 7:816
Sujatha M, Gupta SD (2013) Tissue culture and genetic transformation of safflower (*Carthamus tinctorius* L.). In: Biotechnology of neglected and underutilized crops. Springer, Dordrecht, pp 297–318
Tarıkahya Hacıoğlu B, Karacaoğlu Ç, Özüdoğru B (2014) The speciation history and systematics of *Carthamus* (*Asteraceae*) with special emphasis on Turkish species by integrating phylogenetic and ecological Niche modelling data. Plant Syst Evol 300(6):1349–1359
Thippeswamy M, Sivakumar M, Sudhakarbabu O et al (2013) Generation and analysis of drought stressed subtracted expressed sequence tags from safflower (*Carthamus tinctorius* L.). Plant Growth Regul 69(1):29–41
Thomas CA (1964) Registration of US 10 Saflower1 (Reg. No. 2). Crop Sci 4(4):446–447
Upadhyaya HD, Ortiz R (2001) A mini core subset for capturing diversity and promoting utilization of chickpea genetic resources in crop improvement. Theor Appl Genet 102:1292–1298. https://doi.org/10.1007/s00122-001-0556-y
Upadhyaya HD, Pundir RPS, Dwivedi SL (2009) Developing a mini core collection of sorghum for diversified utilization of germplasm. Crop Sci 49:1769–1780. https://doi.org/10.2135/cropsci2009.01.0014
Upadhyaya HD, Wang YH, Gowda CLL et al (2013) Association mapping of maturity and plant height using SNP markers with the sorghum mini core collection. Theor Appl Genet 126.:2003-2015. https://doi.org/10.1007/s00122-013-2113-x
Van Hintum TJ, Brown AHD, Spillane C (2000) Core collections of plant genetic resources. Biovers Int
Varshney RK, Tuberosa R (2013) Translational genomics in crop breeding for biotic stress resistance: an introduction. Transl Genom Crop Breed Biotic Stress 1:1–9
Vavilov NI (1951) The origin, variation, immunity, and breeding of cultural plants. Ronald Press Company, New York
Velasco L, Fernandez-Martinez J (2001) Breeding for oil quality in safflower. In: Proceeding 5th international safflower conference. Montana, USA 23–27 July. pp 133–137

Verma RC, Shrivastava P (2014) Radiation-induced reciprocal translocations in safflower (*Carthamus tinctorius* L.). Cytologia 79(4):541–545

Vijayakumar J, Ponmanickam P, Samuel P (2017) Influence of Meta-Topolin on Efficient Plant Regeneration via Micropropagation and Organogenesis of Safflower (*Carthamus tinctorius* L.) cv. NARI-H-15. Am J Plant Sci *8*(04):688

Vilatersana R, Garnatje T, Susanna A et al (2005) Bot J Linn Soc 147:375–383

Visarada KBRS, Meena K, Aruna C et al (2009) Transgenic breeding: perspectives and prospects. Crop Sci 49(5):1555–1563

Voss-Fels KP, Cooper M, Hayes BJ (2019) Accelerating crop genetic gains with genomic selection. Theor Appl Genet 132(3):669–686

Wang L, Guan Y, Guan R et al (2006) Establishment of Chinese soybean Glycine max core collections with agronomic traits and SSR markers. Euphytica 151:215–223. https://doi.org/10.1007/s10681-006-9142-3

Watson A, Ghosh S, Williams MJ (2018) Speed breeding is a powerful tool to accelerate crop research and breeding. Nat Plants 4(1):23

Weiss EA (1971) Castor, sesame, and Safflower. Barnes and Noble, Inc., New York, pp 529–554

Weiss EA, (2000) Safflower. In: Oilseed crops. pp 93–129. BlackwellScience Ltd., Victoria, Australia.

Wenzl P, Carling J, Kudrna D et al (2004) Diversity Arrays Technology (DArT) for whole-genome profiling of barley. Proc Natl Acad Sci U S A 101:9915–9920

Wiesner JV (1927) The raw materials of the plant kingdom [In German]. Vol. 1, 4th ed

Woo JW, Kim J, Kwon SI (2015) DNA-free genome editing in plants with preassembled CRISPR-Cas9 ribonucleoproteins. Nat Biotechnol 33(11):1162

Würschum T, Reif JC, Kraft T (2013) Genomic selection in sugar beet breeding populations. BMC Genet 14(1):85

Xu J, Ranc N, Muños S et al (2013) Phenotypic diversity and association mapping for fruit quality traits in cultivated tomato and related species. Theor Appl Genet 126:567–581. https://doi.org/10.1007/s00122-012-2002-8

Yaman H, (2014) The effects of different gamma radiation doses on the agricultural characters of m1 and m2 plants of safflower (*Carthamus tinctorius* L.) cultivars and on in vitro adventitious shoot regeneration. Phd thesis. Available from: https://tez.yok.gov.tr/UlusalTezMerkezi/giris.jsp

Yaman H, Tarıkahya-Hacıoğlu B, Arslan Y et al (2014) Molecular characterization of the wild relatives of safflower (*Carthamus tinctorius* L.) in Turkey as revealed by ISSRs. Genetic Resour Crop Evol 61(3):595–602

Yang S, Pang W, Ash G (2006) Low level of genetic diversity in cultivated Pigeonpea compared to its wild relatives is revealed by diversity arrays technology. Theor Appl Genet 113:585–595

Yang YX, Wu W, Zheng YL (2007) Genetic diversity and relationships among safflower (*Carthamus tinctorius* L.) analyzed by inter-simple sequence repeats (ISSRs). Genet Resour Crop Evol 54:1043. https://doi.org/10.1007/s10722-006-9192-3

Yang X, Yan J, Shah T et al (2010) Genetic analysis and characterization of a new maize association mapping panel for quantitative trait loci dissection. Theor Appl Genet 121:417–431. https://doi.org/10.1007/s00122-010-1320-y

Yatou O (1985) Radiosensitivity of callus of safflower, *Carthamus tinctorius* L. IRB Technical News 27:1–2

Yau S (2004) Yield, agronomic performance, and economics of safflower in comparison with other rainfed crops in a semi-arid, high-elevation Mediterranean environment. Exp Agric 40:453–462

Yeilaghi H, Arzani A, Ghaderian M et al (2012) Effect of salinity on seed oil content and fatty acid composition of safflower (*Carthamus tinctorius* L.) genotypes. Food Chem 130(3):618–625

Yeken MZ, Kantar F, Çancı H et al (2018) Breeding of dry bean cultivars using *Phaseolus vulgaris* landraces in Turkey. International Journal of Agricultural and Wildlife. Sciences 4:45–54

Yeken MZ, Nadeem MA, Karaköy T et al (2019) Determination of Turkish common bean germplasm for morpho-agronomic and mineral variations for breeding perspectives in Turkey. KSU. J Agric Nat 22(Suppl:1):38–50
Yermanos S, Hemstreet S, Garber MJ (1967) Inheritance of quality and quantity of seed-oil in safflower (*Carthamus tinctorius* L.). Crop Sci 7:417–422
Yildiz M, Koçak M, Nadeem MA et al (2019) Genetic diversity analysis in the Turkish pepper germplasm using iPBS retrotransposon-based markers. Turk J Agric For 43
Ying MC, Dyer WE, Bergman JW (1992) *Agrobacterium tumefaciens* mediated transformation of safflower (*Carthamus tinctorius* L.) cv centennial. Plant Cell Rep 11:581–585
Zair A, Chlyah A, Sabounji K (2003) Salt tolerance in some wheat cultivars after application of in vitro pressure. Plant Cell Tiss Org Cult 73:237–244
Zargar SM, Gupta N, Nazir M et al. (2016) Omics-A New Approach to Sustainable Production. In: Breeding oilseed crops for sustainable production, pp 317–344). Academic Press
Zhang C, Chen X, Zhang Y (2009) A method for constructing core collection of Malus sieversii using molecular markers. Sci Agric Sin 42(2):597–604
Zhang P, Liu X, Tong H et al (2014) Association mapping for important agronomic traits in core collection of rice (Oryza sativa L.) with SSR markers. PLoS ONE 9:e111508. https://doi.org/10.1371/journal.pone.0111508
Zhang Y, Liang Z, Zong Y (2016) Efficient and transgene-free genome editing in wheat through transient expression of CRISPR/Cas9 DNA or RNA. Nature Commun 7:12617
Zhu YL, Song QJ, Hyten DL et al (2003) Single-nucleotide polymorphisms in soybean. Genetics 163(3):1123–1134
Ziarati P, Asgarpanah J, Kianifard M (2012) The essential oil composition of *Carthamus tinctorius* L. flowers growing in Iran. Afr J Biotechnol 11(65):12921–12924

Chapter 12
Genomics of Mustard Crops

Umair Riaz, Wajiha Anum, Ghulam Murtaza, Moazzam Jamil, Tayyaba Samreen, Irfan Sohail, Qamar-uz-Zaman, Rashid Iqbal, and Muhammad Ameen

Contents

U. Riaz (✉)
Soil and Water Testing Laboratory for Research, Agriculture Department, Government of Punjab, Bahawalpur, Pakistan
e-mail: umair.riaz@uaf.edu.pk

W. Anum
Department of Agronomy, Regional Agricultural Research Institute, Bahawalpur, Pakistan

G. Murtaza · T. Samreen · I. Sohail · M. Ameen
Institute of Soil and Environmental Sciences, University of Agriculture, Faisalabad, Pakistan

M. Jamil
Department of Soil Science, Islamia University of Bahawalpur, Bahawalpur, Pakistan
e-mail: moazzam.jamil@iub.edu.pk

Qamar-uz-Zaman
Department of Environmental Sciences, University of Lahore, Lahore, Pakistan

R. Iqbal
Department of Agronomy, Islamia University of Bahawalpur, Bahawalpur, Pakistan

H. Tombuloglu et al. (eds.), *Oil Crop Genomics*,
https://doi.org/10.1007/978-3-030-70420-9_12

12.1 Introduction

Mustard (*Brassica juncea* (L.)) is an important crop from ancient times and still has important position in beneficial crops in modern world. It is utilized mainly as an oil crop but also used as vegetable and condiment and also has medicinal properties, thus benefiting the basic components of an ecosystem. As it is highly used in medicines, the world population is obtaining health benefits also (Szőllősi 2020). Brassicaceae family comprises a diverse group of plants that ranges from noxious weeds to economically important plants like mustard. A variation is observed related to the number of genus and species under this family. Heywood (1993) recorded 380 genera and 3000 species, while the figure changed to 365 genera and 3250 species by Mabberley (1997). After few years Judd et al. (1999) recorded about 419 genera having 4130 species in this family. However Warwick et al. (2006) stated 338 genera and 3709 species. The oilseed crop under brassica group occupies about 34 million hectares of cropped land in the world (Balalic et al. 2017). Among all the commodities moving in world trade, only petroleum has a higher value than vegetable oils. About 15% of the edible oil in the world comes from the brassica oilseeds and stands at the third position for the source of edible oil after soybean and palm. No doubt, brassica crops have been receiving attention of taxonomics from earlier times but still, because of the discrepancy in systemic ranking along with problems in nomenclature, the current situation of brassica crops is still confusing. *Brassica* genus comprises species like *B. napus* (oilseed rape), *B. carinata* (Abyssinian mustard), *B. nigra* (black mustard), *B. juncea* (Indian mustard) and *B. rapa* (turnip rape) that are grown for the oil purpose. Two species, i.e. *B. rapa* and *Brassica napus*, are called rapeseed, whereas three species, namely, *B. nigra, B. carinata* and *B. juncea*, are called mustard. *B. nigra* is grown as condiment crop. It also includes a biennial crop utilized as vegetable (cauliflower, kale and cabbage), and this comes under the species *B. oleracea* (Šamec et al. 2019) and is not grown for oil purposes. The oil consists of long-chain fatty acids and glucosinolates (defence-type secondary metabolites) (Booth and Gunstone 2004). Among the cultivated species, there are diploid (*B. rapa*, *B. nigra* and *B. oleracea*) and amphidiploids (*B. napus*, *B. juncea* and *B. carinata*). The amphidiploids contain combined chromosome sets of the diploid species. The relationship between the six species was first highlighted in 1935 by Nagaharu. *B. nigra* is widely spread mostly throughout the temperate climate regions while widely spread in southern and central Europe. Mediterranean centre is the basic centre identified for its origin, while secondary centre is Near East. Firstly its seeds were utilized for condiment purposes; however, now it has been replaced by the *B. juncea* species.

Mediterranean and Near East are also identified as the region for origin of white mustard (*Sinapis alba* L.) which is widely used in salads, as condiment, as green manure and as fodder crop. Its generic status is variously described in recent publications as *Brassica hirta* Moench or referred to as *Sinapis alba* L. *B. juncea* commonly known as brown or Indian mustard is originated in Asiatic while importantly diverse in China. However in Europe and North America, it is used as condiment while for seed oil in Indian subcontinent. In Far East it is a salad or vegetable plant. *B. carinata*, commonly called Abyssinian mustard, is found in Ethiopia and its neighbouring regions and used as vegetable as well as seed oil since ancient times (Vaughan 1977). Brassicas have ability to successfully survive in cold temperatures making it a crop which can be grown in subtropics as winter crops and also cultivated at higher elevations.

12.2 History and Distribution of Mustard Crops

Basin is already well identified. The distribution areas and the main ecological preferences of most of them are also well known. Comparison of this evidence with the archaeological findings reveals that with practically all early crops, the first signs of domestication appear in the same general areas where the wild ancestral stocks abound today (Zohary et al. 2012).

Mustard both as a weed and crop is extensively distributed in Indian subcontinent (easternmost regions) and also occupies areas from Iberian Peninsula to Asia Minor. It is also present in many of the near eastern countries like Israel (Yaniv et al. 1994). The presence of wild seeds or the cultivated ones of *Brassica* or *Sinapis* as revealed by archaeo-botanical studies revealed that they belonged to earlier times of tenth millennium BP. Willcox (2002) stated that at a kitchen site present in Syria (Jerf el Ahmar), the remains of the cake revealed that it was prepared by finely ground mustard seeds; however, at that time this discovery was rather rare. In Mediterranean and adjacent areas ranging from Neolithic and onwards, the presence of wild and cultivated mustard is somehow scarce as compared with other cereals. This fact is associated with the reason that mustard has higher degradable rates owing to high concentration of oil (about 40%) (Marjanovic-Jeromela et al. 2010), resembling, in a way, legume seeds, which are also highly degradable, due to their rich protein content (Smykal et al. 2014). Mustard was first introduced to Britain via the Roman conquest in 43 CE and, by the eighteenth century, had emerged as a common ingredient in the British kitchen (Alcock 2011). Vegetable *B. juncea* varieties are widely distributed in China and have been cultivated for a long period of 6000–7000 years (Yang et al. 2016). Oilseed *B. juncea*, among the major sources of edible oil in Brassica, is mainly distributed in the Indian subcontinent and Northwest China (Banga and Banga 2016; Yang et al. 2016). It is also used for canola crops in Canada and Australia (Burton et al. 2004) and condiment crops in Europe, China and other regions (Yang et al. 2016, 2018). Mustard species are also considered native to temperate regions of Europe, which were one of the first domesticated crops. This crop's economic value resulted in its wide dispersal, and it has been grown as an herb in Asia, North Africa and Europe for thousands of years. Ancient Greeks and Romans enjoyed mustard (*Sinapis*) seed as a paste and powder. In about 1300, the name "mustard" was given to the condiment made by mixing mustum, which is the Latin word for unfermented grape juice, with ground mustard seeds. Mustard has been a major specialty crop in North America since supplies from western Europe were interrupted by World War II. China has a long history of *B. juncea* cultivation. Its first mention is in literature from the Chou Dynasty (1122–247 BC). Its use as a flavouring agent is recorded during the West Han Dynasty (206 BC–24 AD). Dai, in his work Liji (The Book of Rites), mentioned it as "sliced jam of fish with mustard". Uses for its seeds and leaves are also mentioned in Chia-ssu-hsich's book Himin-yao-shu of the late fifth or early sixth century (Li 1969). Frequent references are found in Su's (10–61 AD) work Tu-Jin-Bin-Cao (Illustrated Book of Medicinal Herbs) indicating that mustard was a popular crop of the time. Wang (1576–1588) mentioned root forms in his work 'Gua Guo Shz ~ (Explanations of Cucurbits and Vegetable Crops). A fascinating account regarding the origin of variations in Chinese *B. juncea* has been presented by Li (1980). Description in ancient literature from the fifth century AD indicates that the primitive type was a small annual plant with poor leaf growth cultivated for its pungent seeds. Subsequent variations of leaf shape, size and colour, petiole width, heading type and spread of leaves evolved and were selected. Forms with luxurious broad leaves were developed in the Tang Dynasty (618–907 AD) and used as greens in temperate and humid South China. A form with deeply dissected leaves adapted to arid environments was developed in northern China. This, in turn, produced a tillering form that was more productive, was branched early during vegetative growth and was good for pickles. Forms with broad, thick mid-ribs and petioles were developed during the Chin Dynasty (1644–1911 AD). Later, headed forms with leaves with fleshy midribs and petioles evolved. Types with swollen stem were also bred. Fleshy root forms evolved independently from broad-leaved forms, presumably after the twelfth century. *B. juncea* was also a component of the agriculture of the Indus Valley civilization which flourished around 2300–1750 BC. The art of extracting oil was known to this civilization. Seeds of *B. juncea* have been excavated from Chanhudaro, a site of this civilization (Allchin 1969). When the Aryans came to India c. 1500 BC, they adopted *B. juncea* oil as a preservative. Its use was later extended to cooking and massage purposes. Around 1000 BC, it spread eastwards with migrating people. Reports of Chinese travellers Huen Tsang (c. 640 AD) and Itsing (c. 690 AD) reveal that it was established as an oil crop in the Indo-Gangetic Plains by 700 AD. Though conflicting views have been expressed regarding the route of entry of *B. juncea* into India, it seems that *B. juncea* reached northwest India from the Middle East, its place of origin, through Afghanistan between 5500 and 2300 BC (Hinata and Prakash 1984). New hybridizations later complemented intensive differentiation of many agro-ecotypes between the constituent parents in northwest India (Gómez-Campo and Prakash 1999). Carbonized mustard seeds stored in a gallipot were excavated from Banpo site in China in 1963, and 14C analysis indicated that the seeds belonged to New Stone Age (about 4800 BC; Institute of Archaeology of Chinese Academy of Sciences). The earliest Chinese literature record of brassicas being used as vegetable appeared in Xiaxiaozhen (ancient almanac) in Xia Dynasty (about 3000 BC) and Shijin-Gufeng (a collection of poems) in Zhou Dynasty (1122–247 BC) (Wu et al. 2009).

12.3 Origin of Mustard Crop

Two aspects are crucial for considering the origin of any crop species. The first aspect is the taxon origin which determines its evolutionary processes in wild while the second aspect is the origin where it was first cultivated. The origin of cultivation highlights the place of domestication and its uses and diversification. In this regard turnip rape (*B. rapa*) was the first domesticated species. Moreover black mustard (*B. nigra*) has also been in use from ancient times; however, it is believed that Indian mustard (*B. juncea*) originated as a result of the two mentioned species and is amphidiploid. Kale and cabbage (*B. oleracea*) seem to be domesticated later because its natural origin area (Atlantic coasts) was far from the domestication places. Therefore, amphidiploid species in which *B. oleracea* intervenes as a parent, namely, *B. napus* (rape) and *B. carinata* (Ethiopian mustard), should have been the last brassicas to be incorporated into agriculture (Gómez-Campo and Prakash 1999). The three diploid Brassica species *B. nigra*, *B. rapa* and *B. oleracea* were the first to be domesticated, and these species have been cultivated for a very long time. The amphidiploid species were domesticated later but probably still very early in mankind's agrarian evolution (Hedge 1976).

Many studies are present that reveal the history of cultivation. According to Prakash (1980), rapeseed (*B. rapa*) was cultivated in India in 4000 BC. However recently Gupta and Pratap (2007) and Snowden et al. (2007) stated it was cultivated in Greece, Japan and China about 2000–25,000 years ago while in Europe its cultivation dates back to 800 years (Mcvety and Duncan 2015), and 60 years back it was cultivated in North America (Prakash 1980). The original centres identified provide information about its possible areas of primary domestication. *Brassica rapa* is thought to have a primary centre of origin in the Indian subcontinent with secondary centres of origin in Europe, in the Mediterranean area and in Asia. *Brassica oleracea* and *B. napus* are presumed to have originated in the Mediterranean area (McVetty and Duncan 2015). *Brassica nigra* and *B. juncea* species are thought to have originated in the Middle East (Arias et al. 2014), while *B. carinata* is presumed to have originated in Ethiopia (Khedikar et al. 2020). *Brassica rapa* appears to be the most widely distributed oilseed *Brassica* species. At least 2000 years ago, it was distributed from the Atlantic islands in the west to the eastern shore of China and Korea and from northern Norway to the Sahara and northern India (From McVetty and Duncan 2015). About half a century ago, Vavilov (1951) proposed and still recent studies like that of Yang et al. (2018) claim that Afghanistan and its adjoining regions were the primary origin centre of *B. juncea* and that Asia Minor, central/western China and eastern India were secondary centres of diversity. Some subsequent studies proposed that *B. juncea* originated from several independent hybridizations based on cytogenetic and biochemical evidence (Yang et al. 2018). Recently, simple sequence repeat (SSR) marker analysis of oilseed *B. juncea* suggested a polyphyletic origin and secondary centres of genetic diversity in China and India (Chen et al. 2013). However, there is also another view that China is the primary origin or centre of *B. juncea*, which is supported by the discovery of wild *B. rapa*

(Chinese variety) and *B. nigra* in Northwest China, the abundance of *B. juncea* varieties in China and archaeological evidence of a long history of cultivation (Yang et al. 2016). According to genomic analysis based on the re-sequencing of A-subgenomes of *Brassica* species, we suggested monophyletic origin and evolution into vegetable and oilseed varieties for *B. juncea* (Yang et al. 2016). However, the phylogenetic evolutionary history of *B. juncea* crops has been debated for decades. In conclusion, we propose that China was the primary origin and diversity centre of *B. juncea* and that more than one migration event occurred from the primary to the secondary centres of domestication in both China and India (Yang et al. 2018).

12.4 An Overview of Genetics

Brassica is an ancient plant, and as it has diverse uses and types so it is an ideal plant for interpreting its origin and evolution scenario. The literature from ancient times, archaeological studies, phylogenetic relationships and genetic diversity of different morphotypes can provide interesting information about the origin and evolutionary process of a specie. There are diverse mustard cultivated species from Mediterranean centre. Mikić (2016) revealed chromosome number as black mustard (2n = 16), brown mustard (2n = 36), rapeseed (2n = 38), turnip rape (2n = 20) and white mustard (2n = 24). *B. juncea* allopolyploidy evolution is verified through its taxonomy, artificial synthesis, molecular analysis and chromosome mapping as revealed by Prakash et al. (2009). Chen et al. (2016) stated that about more than 1000 cultivars of mustard are present throughout China making the country with richest mustard resources. In China mustard is preferred as vegetable and also for making cooking oil. Also, mustard has wide ranges of agronomic characteristics (Christopher et al. 2005). The cytogenetic relationship of the six main economically important species of the *Brassica* genus was depicted in the U triangle (Demeke et al. (1992). *Brassica* can adapt to various habitats and under different environmental conditions (Hong et al. 2008). The genome structures are assembled, sequenced and analysed by methods like DNA sequencing, recombinant DNA and bioinformatics to sequence and to know genomic functions (NHGRI 2014). Sequence studies determine the bases' order in the respective DNA segment. Another technique called next-generation sequencing combines a set of high-throughput sequencing technologies. SOLiD, Ion Torrent Proton/PGM, Roche 454 and Illumina (Solexa) are examples of next-generation sequencing (NHGRI 2014); they are easier and quicker DNA and RNA sequencers as compared to technologies used previously, e.g. Sanger sequencing. DNA sequencing is also used to identify mutations that are responsible for causing diseases (NHGRI 2014). During mutation, a single pair of base or thousands of bases are inserted or deleted (INDELs) or else substituted (SNPs). SNP markers are amenable,,codominant (high-throughput automation) and abundant in genomic studies (Tsuchihashi and Dracopoli 2002). Many studies related to taxonomy, genetic variation estimation, phylogenetic analysis, genome-wide association, genetic linkage maps and population structure are recently studied in brassica and other plants by using SNP markers and whole-genome sequencing (Trick et al. 2009; Bancroft et al. 2011; Huang et al. 2013). In a study conducted by Lai et al. (2012), leaf transcriptomes were sequenced across a mapping population of *B. napus*. Transcript abundance and the sequenced variation analysis helped in construction of a single nucleotide polymorphism linkage map. They revealed that *B. napus* consists of 23.037 markers and also used such analysis in alignation of *B. napus* genome with genome of *Arabidopsis thaliana*, along with the *B. oleracea* and *B. rapa* genome sequenced assemblies. In a study Huang et al. (2013) stated that *B. napus* populations were identified with a number of bi-allelic SNPs as 892,536. AFLP markers revealed that *B. oleracea* single accession has average genetic diversity of 0.13 while total diversity (HT) was 0.24 (van Hintum et al. 2007). In another study seven AFLP primer pairs were used to study the genetic variation and relationship among different taxa of *B. rapa*. The results showed 84–97% of polymorphisms, similar in amount among the accessions (Warwick et al. 2008). In 17 *Brassica* accessions, the genetic structure and diversification were studied by application of AFLP markers (Christensen et al. 2011). The wild populations were less diverse as compared to the several land races. The overall variation among the accessions was about 62%. Similarly, in another study, the genetic diversity of *B. oleracea* accessions revealed 806 polymorphic fragments. AFLP markers were used, as the accessions were grouped into two main groups, special subgroups revealed the origin of the accessions (Faltusová et al. 2011).

12.5 Utilization and Oil Content

As the brassica crops are mainly grown as oilseed plant, its four species (*Brassica rapa Brassica juncea, Brassica carinata and Brassica napus*) contribute about 12% of the vegetable oil in the world (Labana and Gupta 1993). The brassica seed oil is consumed by humans, utilized in petrochemical industry or else used as biofuel or renewable source of energy. Brassicas are source of vitamins (A,C and E), phenolics, dietary fibres and potassium and contain anticancer compounds along with many health-enhancing compounds (Cartea et al. 2011). Studies have revealed that brassicas protect human against heart diseases and cancers and also limit the tumour development as they have isothiocyanates (produced as a result of breakdown of glucosinolates that are produced in brassicas). Several biotic and abiotic factors affect Brassicaceae growth (Consentino et al. 2015; Jourdan et al. 2015). The brassica vegetable species comprises phytochemicals and nutrients and has nutraceutical as well as antioxidant activities. They contain minerals, phenolics, soluble sugars, carotenoids, vitamins and dietary fibres (Wagner et al. 2013). They also produce glucosinolates and metabolites that have sulphur in them which acts as anticancer agents as they have capability of producing detoxifying enzymes in mammalian cells and help to decrease the tumour development rate. Isothiocyanates are modulators of phase 1 and 2 enzyme activity and neutralize cancer-causing chemicals that damage cells by interfering with tumour growth (King 2005). They also produce secondary products capable of modulating the steroid metabolism and have antibacterial, antiviral and antioxidant effects and help in inducing immune system (King 2005). The phytochemicals produced by brassica plants have ability to respond to various genetic pathways and have chemopreventive action (Wagner et al. 2013). Those phytochemicals also have anti-inflammatory properties (Juge et al. 2007). The stimulation of antioxidants along with prohibition of pro-inflammatory signalling pathways is a characteristic that is responsible for their beneficial activities. The stimulation and prohibition occur as different transcription factors are regulated and further controlled by epigenetic modifications and miRNAs (Wang et al. 2009; Wagner et al. 2013). Yanaka et al. (2009) also stated that phytochemicals derived from brassica have anti-infective and anti-viral properties (Yanaka et al. 2009).

Brassica napus contain about 35–45% of oil (Tariq et al. 2020). A higher genetic variation in fatty acids is observed in brassica oil as compared to oil obtained from other vegetable oils (Sovero 1993). *Brassica* consists of long-chained monosaturated fatty acids, especially erucic acid, which is otherwise absent in any other oil (Tsunoda 1980). However oil with high erucic acids is useful in industrial usage and not recommended for human consumption (Sharafi et al. 2015). In a recent study, it was revealed that six species of *Brassica* were analysed for their oil contents. The average oil contents for *B. napus*, *B. juncea*, *B. carinata*, *B. oleracea*, *B. nigra* and *B. rapa* were 38–56%, 20–31%, 25–29%, 26–29%, 17–24% and 22–41%, respectively (Sharafi et al. 2015). Wen et al. (2015) and Bauer et al. (2015) claimed that nutritional qualities of rapeseed oil are dependent on its fatty acid composition. The fatty acid consists of 2% stearic acid, 4% palmitic acid and 60% oleic acid.

12.6 History of Genetic Improvement in Mustard

The cultivated *Brassica* has six significant species, which are diploid or allotetraploids. This category includes the presence of AA genome, BB genome and CC genome in *B. rapa*, *B. nigra* and *B. oleracea,* respectively. The allotetraploids include AACC genome, AABB genome and BBCC genome in *B. napus*, *B. juncea* and *B. carinata*, respectively. The relationship between the diploid and allotetraploids was first discovered in 1934 by Morinaga and after 1 year visualized by Dr. Woo Jang Choon (1935). Conclusively it was revealed that allotetraploid occurs when the diploid species undergoes interspecific hybridization (Hayward et al. 2012).

Brassica mapping was first done by the use of practices involving the utilization of isozymes and protein markers. Further advancement in technology leads to the development of PCR-based markers like RFLPs (restriction fragment length polymorphisms), RAPD (randomly amplified polymorphic DNA) and AFLP (amplified fragment length polymorphisms). In *Brassica*, first linkage maps were identified with the help of these technologies (Snowdon and Friedt 2004). Batley et al. (2007) and Hopkins et al. (2007) elaborated that microsatellite markers (also called Single

Sequence Repeat Marker) were the next successful technique allowing the amplification of highly polymorphic microsatellite DNA repeat sequences.

B. rapa is the progenitor of A genome in amphidiploid *Brassica* (*B. napus* and *B. juncea*), the major oilseed species. The origin of this species is a Mediterranean region and Central Asia where its natural growing is still persistent (Gomez Campo 1999, Gomez Campo and Prakash 1999). *B. rapa* is believed to be developed into two distinctive types as European oleiferous (originated in Mediterranean region) and Asian (originated in Central Asia, Afghanistan and northwest India) type (Sinskaja 1928). In this regard, the conclusive evidence was found through molecular marker analysis, cytogenetics and morphological markers about the two separate regions of origin (Denford and Vaughan 1977; Song et al. 1988; Zhao et al. 2004). Morphological distinctive characters in a large number of recognizable subspecies of *B. rapa* are observed to be present as a result of natural selection or else through breeding practices for obtaining a particular trait or as a result of adaptation to different geographical regions (Ramchiary and Lim 2011).

12.7 Basic Genomics of Mustard Crop

Brassica is a genus of plants in the mustard family, and it comprises few of the highly utilized and economically important food crops. Among them, *Brassica* includes numerous horticultural and crops as compared to another plant genus (Taiyan et al. 2001). *Brassica* is the second-largest oilseed crop grown after soybean (Raymer 2002). It consists of 37 species. However there are four (*Brassica rapa*, *Brassica juncea*, *Brassica napus* and *Brassica carinata*) cultivated species utilized as oilseed crop and vegetables. *Brassica oleiferous* is derived from *B. napus* and *B. campestris*. Other names of competes are Toria, Sumer turnip, Sarson, Polish rape and Turnip rape (Gupta and Pratap 2007). The cultivated *Brassica* species are highly polymorphic in nature that comes in forms like oilseeds, Chinese cabbage, broccoli and few condiment crops (Gupta 2015).

The basic genome analysis categorized the six cultivated species into diploid and amphidiploid. It is stated that amphidiploids are originated from different origins. *B. rapa* consists of AA genome (2n = 36). *B. nigra* consists of BB genome with 2n = 16. *B oleracea* contains genome CC (2n = 18) (Rehman et al. 2018). As a result of genomic recombinations, three amphidiploids are formed which are widely cultivated, i.e. the *B. juncea* comprising AABB genome (2n = 36), *B. carinata* BBCC (2n = 34) and *B. napus* comprising AACC genome with 2n = 38 (Sharma et al. 2014). Axelsson et al. (2000) performed a molecular analysis of *B. juncea* and revealed that it consists of a conserved genome of the progenitor species.

The major cultivated species also include *Arabidopsis thaliana* (model plant). Hybridization and cytogenetic studies carried out have successfully proven that amphidiploid is the natural hybrids formed by recombination of diploids and the interlinked species of *Brassica* (Chalhoub et al. 2014). This interrelationship of six brassica species at the molecular level has also been confirmed through comparative sequence analysis and gene evolution studies (Schmidt and Bancroft 2011).

In *Brassica* crops, primary, secondary and tertiary generic pools pose significant importance for the identification of available genetic resources in breeding programs (Branca and Cartea 2011). The primary gene pool was identified as *Brassica oleracea,* and various studies suggested the importance and identification of other gene pools (Branca and Cartea 2011). The pachytene chromosome morphology helped in the identification of the secondary gene pool and suggested the necessary genomes of *B. rapa*, *B. nigra* and *B. oleracea* as (2n = 20), (2n = 16) and (2n = 18), respectively (Branca and Cartea 2011). Branca and Cartea (2011) stated that *Brassica* genus and few related genera are as a result of evolution from a common ancestor having n = 6; however, there is an increase in chromosome numbers and partial homology of A, B and C genomes. The tertiary gene pool comprises 36 cytokines. Few of them are *Trachystoma*, *Sinapodendron*, *Sinapis*, *Phynchosinapis*, *Hirschfeldia*, *Erucastrum*, *Enarthrocarpus* and *Diplotaxis* genera (6; Branca and Cartea 2011). All of the identified tools help develop new traits and alleles of favourable nature by the use of novel methodology (El-Esawi 2017).

12.8 Genome Identification and Variation-Causing Tools

The identification of the genome sequence is the door to success in plant breeding and genetics. Without knowing the actual hereditary set-up of a crop, it is impossible to assess its phenotypic capability under various conditions further. With the changing environmental conditions and the need to fulfil the requirements of ever-increasing populations, it has been necessitated to improve consumable products. Advanced genetic tools have continuously improved the wild types of *Brassica*. In ten species of *Brassica*, partial or complete sequencing has been done so far as depicted in Table 12.1.

The tools or techniques for genetic mapping are genotyping through sequencing (Shea et al. 2018; Lee et al. 2016), genomic selection (Rai et al. 2019) and studies related to genetic diversity (Gao et al. 2020). These not only reveal the genomic set-up but also reveal links to the ancestral genomes.

Thirty *Brassica* genotypes were selected and analysed by using 24 SSR markers. A variation of 1–8 (BRMS 14) occurred in a total of 84 alleles while depicting 72% polymorphism. Out of the total 24 SSRs, 9 was able to produce 100% polymorphism. The size of amplicon was 99 bp and 383 bp for BRMS-26 and BRMS-31, respectively. The SSR marker (BRMS-17) gave specific bands for *B. carinata.* The genotype profiles of few species as depicted by PCR by using SSR primers in depicted in Table 12.2 (Prajapat et al. 2014).

Linkage disequilibrium (LD) takes place when genomic regions are inherited at a higher frequency than usual based on recombination. It indicates a relationship between the linked regions in the genome. In a species when there is high or moderate linkage disequilibrium, then it is found as SNP haplotypes. They contain several

Table 12.1 Sequenced *Brassica* species

Species	Number of predicted genes	% of genes orthologous to *A. thaliana*	References
Arabidopsis thaliana	28,710	100	Kaul et al. (2000)
Schrenkiella parvula	28,901	80.2	Dassanayake et al. (2011)
Arabidopsis lyrata	27,379	92	Hu et al. (2011)
Eutrema salsugineum	26,521	82.7	Yang et al. (2013)
Brassica rapa	41,174	78.2	Wang et al. (2011)
Capsella rubella	26,521	88	Slotte et al. (2013)
Sisymbrium irio	28,917	82.9	Haudry et al. (2013)
Leavenworthia alabamica	30,343	67.7	Haudry et al. (2013)
Brassica oleracea	45,758	–	Yu et al. (2013)
Aethionema arabicum	23,167	72.4	Haudry et al. (2013)

Table 12.2 Species as revealed by PCR profiling by using SSR markers

Species	Genotypes	Species	Genotypes
B. carinata	Pusa Swarnim	*B. rapa*	IC365684
B. carinata	Kiran	*B. rapa*	IC346013
B. napus	GSL-1	*B. rapa*	NRCYS
B. napus	H.N.S-0004	*B. rapa*	JT-1(toria)
B. napus	Hyola-401	*B. rapa*	IC363713
B. napus	Neelam	*B. rapa*	IC363710
B. napus	IC560699	*B. rapa*	IC363714
B. napus	IC399819	*B. juncea*	Kranti
B. napus	IC399790	*B. juncea*	Bio-34192
B. rapa	IC398101	*B. juncea*	Varuna

linked SNP alleles present in particular allelic combinations. According to Edwards et al. (2007), they may extend to the length of the gene or gene clusters. Such linked genomic regions can provide a subset of SNPs that can be utilized for screening out the appropriate breeding method. It is a fast-genomic selection aiding in the development of breeding programs. This procedure also helps in defining haplotypes (Cowling and Balázs 2010). A robust data set is provided through high-density LD in SNP maps in *Arabidopsis* (Atwell et al. 2010).

12.9 Different Studies Used for the Improvement of Sequencing and Gene Structure

Next-generation sequencing is a cost-effective technique which is successful in studying single nucleotide polymorphism (SNP), genotyping and gene expression. SNP markers and linkage maps are identified in *Brassica* by NGS. Biotic and abiotic stress also has the capability of altering the gene structure to some extent. In this regard, the transcriptome analysis technique is used to identify the gene expression file; it also helps to understand the gene regulatory mechanisms (Sharma et al. 2014).

Another technique for sequence analysis for further implication in gene structure is a comparative mapping approach used in *Brassica*. This process is used to identify the candidate lines and regions of functional genomics (Schmidt and Bancroft 2011). Many studies are reported in which the candidate genes have been identified through this process: for instance, Schranz et al. (2002) reported the cloning of FLC genes (cloning of flowering time); Suwabe et al. (2006) reported clubroot resistance genes in *Arabidopsis* chromosome, while Li et al. (2009) reported QTLs for flowering time. Similarly, a comparative analysis was performed between *B. rapa* and *B. oleracea* concerning resistance genes *(*Nagaoka et al. 2010). In *B. juncea*, comparative mapping was performed, and the candidate genes regulating the biosynthetic pathway for aliphatic glucosinolate were identified (Bisht et al. 2009).

The oil contents in canola were increased through the development of transgenic canola (Tan et al. 2011). The researchers revealed that no negative impact was found on other dominant agronomic traits, while the seed oil contents were significantly increased. The process was done by placing the seed-specific overexpression genes (*BnLEC1* and *BnL1L*) under one of the canola storage proteins (2S-1 promoter) at an appropriate level. This increased the oil content of the transgenic oilseed plant without affecting the other traits, specially the agronomic or yield-determining traits (Villanueva-Mejia and Alvarez 2017).

Association genetics is another research tool developed for mapping and character identification. Association mapping is especially crucial for the identification of genes for complex traits. The genetic mapping was previously done by observing the segregating populations which were produced as a result of hybridization between two different parents. That is why it is related to QTL mapping. And it is therefore related to methods such as QTL mapping. Nevertheless, association genetics make use of plants from different resources (wild populations, breeding lines and germplasm collection) (Murphy 2014). It makes use of novel methods to discover large-scale SNPs, hasten the crop genetic profiling and high throughput and increase availability of genomic sequencing.

The significant association that occurs as a result of a change in DNA and how it is expressed in its phenotypes in non-related genetic lines in species are successfully studied through association genetics. Previously it was used to study the underlying genetics in humans and animals influencing the complex traits; however, recently, it is being utilized in plants also. Initially, this tool was used to analyse single traits, but with the passage of time, quantitative trait analysis is also done. It is different from traditional linkage mapping in populations as a result of the crossing of two parents because it is capable of exploring mutations and recombination events in diverse populations. Due to higher mapping resolution, the dominant genes responsible for complex traits are identified more accurately, thus creating more feasibility and reliability (Murphy 2014).

Brassica marker-assisted genomic research began in the 1980s, giving rise to the development of first RFLP linkage maps for *B. oleracea*, *B. napus* and *B. rapa* (Slocum et al. 1990: Landry et al. 1991: Song et al. 1991). RFLPs and RAPDs have been extensively used in genetic mapping of brassica crops along with their phylogenetic studies (Dos Santos et al. 1994). With the PCR discovery, there is more potential to increase the marker density and its variations in already discovered genetic maps through microsatellites, AFLP and ISSR (Grist et al., 1993). Another simple and relatively inexpensive analyses includes SSRs. They are polymorphic in nature and help in distinguishing map-based alignment among distinct crosses. With the advent in research, the number of available microsatellite primers (SSRs) in *Brassica* sp. is increasing (Kumar et al. 2015).

12.10 Which Genes Cope with Environmental Stresses

Plant stress is defined as the unfavourable environmental condition or any agent which can cause stress in healthy plant development and produce a specific action. As a result, the plant can show a stress-specific response or general non-specific response (Haag 2013). Brassica crops undergo both biotic and abiotic stresses like insect pests, viral diseases and fungal attacks and temperature extremes, salinity and water shortage or waterlogging, respectively. The plants can adapt themselves under the abiotic streasses and undergo long terms changes like enhancing stress metabolites and photo protecting enzymes, change in leaf size and shape, stomatal distribution and density (Obidiegwu et al. 2015).

Due to the changing climate, the abiotic stresses have reached to a peak dangerous condition in brassica crop production (Kayum et al. 2016). The different genes respond differently in *B. rapa* and *B. oleracea* when exposed to stressed conditions. In *B. rapa*, the abiotic stresses are linked to transcriptor family-like and WRKY transcriptions, and they are well documented (Kayum et al. 2015a, b). The genes during WRKY transcriptions in chiffu showed higher expressions during the cold stress, the expression of BrWRKY 22 was 175-fold higher than BrWRKY44, BrWRKY70, and BrWRKY72 were 32-, 54-, and 42-fold respectively. Hence, we can say the introgression or overexpression of BrWRKY22 and BrWRKY44 genes can help in producing cold-tolerant lines. BrAL2, BrAL3, BrAL7, BrAL9, BrAL12, BeAL13, BrAL14 and BrAL15 showed higher expression under abiotic stresses like low temperature, less water and salt (Kayum et al. 2015b).

12.11 Genomics and Radiation

With the advancement in crop evolution, a reduction has been observed in genetic diversity because breeders are focused on the development of elite cultivars (Sikora et al. 2011). This genetic diversity deterioration gave rise to the advancement of genetic tools required in artificial variation caused by inducing mutation (Smartt and Simmonds 1995). In the start, X-ray radiation was utilized as mutagen because of its availability to scientists. The physiological and biochemical procedures in plants are altered through a fast and reliable technique of gamma radiation. It successfully improves various plant characteristics and productivity of plants (Hanafy and Akladious 2018). It stands at a significant position in genetic studies targeting increased productivity of crops under normal and stressed conditions. The studies reveal that mustard seeds exposed to Gy (25–50 Gy) increase in their dry weights (Hamideldin and Eliwa 2015). It is because an enhancement in RNA activation occurs at the early stages of germination, leading to enhanced protein synthesis (Aly 2010).

12.12 Conclusion

Mustard being the essential and economically significant crop is gaining huge research attentions, particularly for improving its oil content both in quality and quantity. Also the oil qualities like health-enhancing factors, vitamins, dietary fibres and antibacterial, antioxidant, antiviral and anticancer compounds make the crop ideal for consumption. The progress in genetic improvement is being carried out since ancient times; however, due to current facilities, the scientists and breeders are successful in identifying the dominant and major genes in mustard. Brassica oil has higher genetic variation as compared to other edible oils. Vast research had been carried out to study genetics, however in order to study and identify the best genes affecting phenotypic expression for the complete physiological processes still needs to be studied by implanting recent technique. This will result in deeper insights of brassica genomic structure and functions.

References

Alcock JP (2011) A brief history of Roman Britain. Constable & Robinson, New York

Allchin FR (1969) Early cultivated plants in India and Pakistan. Domestication Exploit Plants Anim S:323–329

Aly AA (2010) Biosynthesis of phenolic compounds and water soluble vitamins in Culantro (Eryngium foetidum L.) plantlets as affected by low doses of gamma irradiation. Analele Univ din Oradea Fasc Biol 17:356–361

Arias T, Beilstein MA, Tang M, McKain MR, Pires JC (2014) Diversification times among Brassica (Brassicaceae) crops suggest hybrid formation after 20 million years of divergence. Am J Bot 101(1):86–91
Atwell S, Huang YS, Vilhjálmsson BJ, Willems G, Horton M, Li Y et al (2010) Genome-wide association study of 107 phenotypes in Arabidopsis thaliana inbred lines. Nature 465(7298):627–631
Axelsson T, Bowman CM, Sharpe AG, Lydiate DJ, Lagercrantz U (2000) Amphidiploid Brassica juncea contains conserved progenitor genomes. Genome 43(4):679–688
Balalic I, Marjanovic-Jeromela A, Crnobarac J, Terzic S, Radic V, Miklic V, Jovicic D (2017) Variability of oil and protein content in rapeseed cultivars affected by seeding date. Emir J Food Agric:404–410
Bancroft I, Morgan C, Fraser F, Higgins J, Wells R, Clissold L et al (2011) Dissecting the genome of the polyploid crop oilseed rape by transcriptome sequencing. Nat Biotechnol 29(8):762
Banga SS, Banga S (2016) Genetic diversity and germplasm patterns in Brassica juncea. In: Gene pool diversity and crop improvement. Springer, Cham, pp 163–186
Batley J, Hopkins CJ, Cogan NO, Hand M, Jewell E, Kaur J et al (2007) Identification and characterization of simple sequence repeat markers from Brassica napus expressed sequences. Mol Ecol Notes 7(5):886–889
Bauer B, Kostik V, Gjorgjeska B (2015) Fatty acid composition of seed oil obtained from different canola varieties. Farm Glas 71(1):1–7
Bisht NC, Gupta V, Ramchiary N, Sodhi YS, Mukhopadhyay A, Arumugam N et al (2009) Fine mapping of loci involved with glucosinolate biosynthesis in oilseed mustard (Brassica juncea) using genomic information from allied species. Theor Appl Genet 118(3):413–421
Booth EJ, Gunstone FD (2004) Rapeseed and rapeseed oil: agronomy, production and trade. In: Gunstone FD (ed) Rapeseed and canola oil: production, processing properties and uses. Blackwell Publishing/CRC Press, New York, pp 1–16
Branca F, Cartea E (2011) Brassica. In: Wild crop relatives: genomic and breeding resources. Springer, Berlin/Heidelberg, pp 17–36
Burton WA, Ripley VL, Potts DA, Salisbury PA (2004) Assessment of genetic diversity in selected breeding lines and cultivars of canola quality Brassica juncea and their implications for canola breeding. Euphytica 136(2):181–192
Cartea ME, Francisco M, Soengas P, Velasco P (2011) Phenolic compounds in Brassica vegetables. Molecules 16(1):251–280
Chalhoub B, Denoeud F, Liu S, Parkin IA, Tang H, Wang X et al (2014) Early allopolyploid evolution in the post-Neolithic Brassica napus oilseed genome. Science 345(6199):950–953
Chen FB, Liu HF, Yao QL, Fang P (2016) Evolution of mustard (Brassica juncea Coss) subspecies in China: evidence from the chalcone synthase gene. Genet Mol Res 15(2)
Chen S, Wan Z, Nelson MN, Chauhan JS, Redden R, Burton WA et al (2013) Evidence from genome-wide simple sequence repeat markers for a polyphyletic origin and secondary centers of genetic diversity of Brassica juncea in China and India. J Hered 104(3):416–427
Christensen S, Bothmer R, Poulsen G, Maggioni L, Phillip M, Andersen BA, Jørgensen RB (2011) AFLP analysis of cultivars of canola quality Brassica juncea and their implications for canola breeding. Euphytica 136:181–192
Christopher GL, Andrew JR, Geraldine ACL, Clare JH, Jacqueline B, Gary B, German CS, David E (2005) Brassica ASTRA: an integrated database for Brassica genomic research. Nucleic Acids Res 33:D656–D659
Consentino L, Lambert S, Martino C, Jourdan N, Bouchet PE, Witczak J et al (2015) Blue-light dependent reactive oxygen species formation by Arabidopsis cryptochrome may define a novel evolutionarily conserved signaling mechanism. New Phytol 206(4):1450–1462
Cowling WA, Balázs E (2010) Prospects and challenges for genome-wide association and genomic selection in oilseed Brassica species. Genome 53(11):1024–1028
Dassanayake M, Oh DH, Haas JS, Hernandez A, Hong H, Ali S et al (2011) The genome of the extremophile crucifer Thellungiella parvula. Nat Genet 43(9):913–918
Denford KE, Vaughan JG (1977) A comparative study of certain seed isoenzymes in the ten chromosome complex of Brassica campestris and its allies. Ann Bot 41(2):411–418

Demeke T, Adams RP, Chibbar R (1992) Potential taxonomic use of random amplified polymorphic DNA (RAPD): a case study in Brassica. Theor Appl Genet 84(7–8):990–994

Dos Santos JB, Nienhuis J, Skroch P, Tivang J, Slocum MK (1994) Comparison of RAPD and RFLP genetic markers in determining genetic similarity among Brassica oleracea L. genotypes. Theor Appl Genet 87(8): 909–915

Edwards D, Batley J, Cogan NOI, Forster JW, Chagné D (2007) Single nucleotide polymorphism discovery. In: Oraguzie NC et al (eds) Association mapping in plants. Springer, New York

El-Esawi MA (2017) Genetic diversity and evolution of Brassica genetic resources: from morphology to novel genomic technologies-a review. Plant Genet Resour 15(5):388

Evolution of mustard (Brassica juncea Coss) subspecies in China: evidence from the chalcone synthase gene

Faltusová Z, Kuera L, Ovesná J (2011) Genetic diversity of Brassica oleracea var. capitata gene bank accessions assessed by AFLP. Electron J Biotechnol 14(3):11–11

Gao Y, Gong W, Li R, Zhang L, Zhang Y, Gao Y et al (2020) Genetic diversity analysis of Tibetan turnip (Brassica rapa L. ssp. rapifera Matzg) revealed by morphological, physiological, and molecular marker. Genet Resour Crop Evol 67(1):209–223

Gómez-Campo C (ed) (1999) Biology of Brassica coenospecies. Elsevier, Amsterdam

Gómez-Campo C, Prakash S (1999) Origin and domestication. Dev Plant Genet Breed Elsevier 4:33–58

Grist SA, Firgaira FA, Morley AA (1993) Dinucleotide repeat polymorphisms isolated by the polymerase chain reaction. BioTechniques 15(2):304–309

Gupta SK (ed) (2015) Breeding oilseed crops for sustainable production: opportunities and constraints. Academic, New York

Gupta SK, Pratap A (2007) History, origin and evolution. In: Gupta SK (ed) Rapeseed breeding, Advances in botanical research. Elsevier, New York, pp 1–21

Haag J (2013) Improving photosynthetic efficiency in sports turf. Xlibris Corporation

Hamideldin N, Eliwa NE (2015) Gamma irradiation effect on growth, physiological and molecular aspects of mustard plant. Am J Agric Sci 2(4):164–170

Hanafy RS, Akladious SA (2018) Physiological and molecular studies on the effect of gamma radiation in fenugreek (Trigonella foenum-graecum L.) plants. J Genet Eng Biotechnol 16(2):683–692

Haudry A, Platts AE, Vello E, Hoen DR, Leclercq M, Williamson RJ et al (2013) An atlas of over 90,000 conserved noncoding sequences provides insight into crucifer regulatory regions. Nat Genet 45(8):891–898

Hayward A, Morgan JD, Edwards D (2012) Reviews; SNP discovery and applications in Brassica napus. J Plant Biotechnol 39(1):49

Heywood VH, Moore DM, Richardson IBK, Stearn WT (1993) Flowering plants of the world (No. 582.13 F644). Oxford University Press, Oxford

Hinata K, Prakash S (1984) Ethnobotany and evolutionary origin of Indian oleiferous Brassicae. Indian. J Genet 44:102–112

Hong CP, Kwon SJ, Kim JS, Yang TJ, Park BS, Lim YP (2008) Progress in understanding and sequencing the genome of Brassica rapa. Int J Plant Genomics:1–9

Hopkins CJ, Cogan NO, Hand M, Jewell E, Kaur J, Li XI et al (2007) Sixteen new simple sequence repeat markers from Brassica juncea expressed sequences and their cross-species amplification. Mol Ecol Notes 7(4):697–700

Hu TT, Pattyn P, Bakker EG, Cao J, Cheng JF, Clark RM et al (2011) The Arabidopsis lyrata genome sequence and the basis of rapid genome size change. Nat Genet 43(5):476–481

Huang S, Deng L, Guan M, Li J, Lu K, Wang H et al (2013) Identification of genome-wide single nucleotide polymorphisms in allopolyploid crop Brassica napus. BMC Genet 14(1):1–10

Jourdan N, Martino FC, El-Esawi M, Witczak J, Bouchet PE, d'Harlingue A, Ahmad M (2015) Blue-light dependent ROS formation by Arabidopsis cryptochrome-2 may contribute toward its signaling role. Plant Signal Behav 10(8):e1042647

Judd WS, Campbell CS, Kellogg EA, Stevens PF, Donoghue MJ (1999) Plant systematics: a phylogenetic approach. Ecol Mediterr 25(2):215
Juge N, Mithen RF, Traka M (2007) Molecular basis for chemoprevention by sulforaphane: a comprehensive review. Cell Mol Life Sci 64(9):1105
Kaul S, Koo HL, Jenkins J, Rizzo M, Rooney T, Tallon LJ et al (2000) Analysis of the genome sequence of the flowering plant Arabidopsis thaliana. Nature 408(6814):796–815
Kayum MA, Jung HJ, Park JI, Ahmed NU, Saha G, Yang TJ, Nou IS (2015a) Identification and expression analysis of WRKY family genes under biotic and abiotic stresses in Brassica rapa. Mol Gen Genomics 290(1):79–95
Kayum MA, Park JI, Ahmed NU, Jung HJ, Saha G, Kang JG, Nou IS (2015b) Characterization and stress-induced expression analysis of Alfin-like transcription factors in Brassica rapa. Mol Gen Genomics 290(4):1299–1311
Kayum MA, Kim HT, Nath UK, Park JI, Kho KH, Cho YG, Nou IS (2016) Research on biotic and abiotic stress related genes exploration and prediction in Brassica rapa and B. oleracea: a review. Plant Breed Biotechnol 4(2):135–144
Khedikar Y, Clarke WE, Chen L, Higgins EE, Kagale S, Koh CS et al (2020) Narrow genetic base shapes population structure and linkage disequilibrium in an industrial oilseed crop, Brassica carinata A. Braun Sci Rep 10(1):1–11
King GJ (2005) A white paper for the multinational Brassica genome project. Published online http://www.BrasicaInfowhite_paper/wp_draft.htm
Kumar M, Choi JY, Kumari N, Pareek A, Kim SR (2015) Molecular breeding in Brassica for salt tolerance: importance of microsatellite (SSR) markers for molecular breeding in Brassica. Front Plant Sci 6:688
Labana KS, Gupta ML (1993) Importance and origin. In: Breeding oilseed Brassicas. Springer, Berlin/Heidelberg, pp 1–7
Landry BS, Hubert N, Etoh T, Harada JJ, Lincoln SE (1991) A genetic map for Brassica napus based on restriction fragment length polymorphisms detected with expressed DNA sequences. Genome 34(4):543–552
Lee J, Izzah NK, Choi BS, Joh HJ, Lee SC, Perumal S et al (2016) Genotyping-by-sequencing map permits identification of clubroot resistance QTLs and revision of the reference genome assembly in cabbage (Brassica oleracea L.). DNA Res 23(1):29–41
Li HL (1969) The vegetables of ancient China. Econ Bot 23(3):253–260
Li CW (1980) Classification and evolution of mustard crops (Brassica juncea) in China. Cruciferae Newsl 5:33–35
Li F, Kitashiba H, Inaba K, Nishio T (2009) A Brassica rapa linkage map of EST-based SNP markers for identification of candidate genes controlling flowering time and leaf morphological traits. DNA Res 16(6):311–323
Mabberley DJ (1997) The plant-book: a portable dictionary of the vascular plants. Cambridge University Press, Cambridge
Marjanovic-Jeromela A, Marinkovic R, Miladinovic D, Miladinovic F, Jestrovic Z, Stojsin V, Miklic V (2010) Field & vegetable crops research. Ratarstvo i povrtarstvo 47(1):173 178
McVetty PB, Duncan RW (2015) Canola, rapeseed, and mustard: for biofuels and bioproducts. In: Industrial crops. Springer, New York, pp 133–156
Mikić, A (2016) Reminiscences of the cultivated plants early days as treasured by ancient religious traditions: the mustard crop (Brassica spp. and Sinapis spp.) in earliest Christian and Islamic texts. Genet Resour Crop Evol 63(1):1–6
Murphy DJ (2014) Using modern plant breeding to improve the nutritional and technological qualities of oil crops. OCL 21(6):D607
Nagaharu U (1935) Genome analysis in Brassica with special reference to the experimental formation of B. napus and peculiar mode of fertilization. Jpn J Bot 7(7):389–452
Nagaoka T, Doullah MAU, Matsumoto S, Kawasaki S, Ishikawa T, Hori H, Okazaki K (2010) Identification of QTLs that control clubroot resistance in Brassica oleracea and comparative

analysis of clubroot resistance genes between B. rapa and B. oleracea. Theor Appl Genet 120(7):1335–1346

NHGRI (National Human Genome Research Institute) (2014) A brief guide to genomics. Genome.gov. Available at http://www.genome.gov/18016863

Obidiegwu JE, Bryan GJ, Jones HG, Prashar A (2015) Coping with drought: stress and adaptive responses in potato and perspectives for improvement. Front Plant Sci 6:542

Prajapat P, Sasidharan N, Kumar M, Prajapati V (2014) Molecular characterization and genetic diversity analysis in four Brassica species using microsatellite markers. Bioscan 9(4):1521–1527

Prakash S, Bhat SR, Quiros CF, Kirti PB, Chopra VL (2009) 2 Brassica and its close allies: cytogenetics and evolution. Plant Breed Rev 31(23):21–187

Rai SK, Bawa V, Dar ZA, Sofi NR, Mahdi SS, Qureshi AMI (2019) Use of modern molecular biology and biotechnology tools to improve the quality value of oilseed brassicas: in quality breeding in field crops. Springer, Cham, pp 255–266

Ramchiary N, Lim YP (2011) Genetics of Brassica rapa L. In: Genetics and genomics of the Brassicaceae. Springer, New York, pp 215–260

Raymer PL (2002) Canola: an emerging oilseed crop. In: Janick J, Whipkey A (eds) Trends in new crops and new uses. ASHS Press, Alexandria, pp 122–126

Rehman HM, Nawaz MA, Shah ZH, Ludwig-Müller J, Chung G, Ahmad MQ et al (2018) Comparative genomic and transcriptomic analyses of Family-1 UDP glycosyltransferase in three Brassica species and Arabidopsis indicates stress-responsive regulation. Sci Rep 8(1):1–18

Šamec D, Urlić B, Salopek-Sondi B (2019) Kale (Brassica oleracea var. acephala) as a superfood: review of the scientific evidence behind the statement. Crit Rev Food Sci Nutr 59(15):2411–2422

Schranz ME, Quijada P, Sung SB, Lukens L, Amasino R, Osborn TC (2002) Characterization and effects of the replicated flowering time gene FLC in Brassica rapa. Genetics 162(3):1457–1468

Sharafi Y, Majidi MM, Goli SAH, Rashidi F (2015) Oil content and fatty acids composition in Brassica species. Int J Food Prop 18(10):2145–2154

Sharma A, Li X, Lim YP (2014) Comparative genomics of Brassicaceae crops. Breed Sci 64(1):3–13

Shea DJ, Shimizu M, Itabashi E, Miyaji N, Miyazaki J, Osabe K, … Fujimoto R (2018) Genome re-sequencing, SNP analysis, and genetic mapping of the parental lines of a commercial F1 hybrid cultivar of Chinese cabbage. Breed Sci 68:375–380

Sikora P, Chawade A, Larsson M, Olsson J, Olsson O (2011) Mutagenesis as a tool in plant genetics, functional genomics, and breeding. Int J Plant genomics 2011:1–13

Sinskaja EN (1928) The oleiferous plants and root crops of the family Cruciferae. Instituta Irikladnoi Botaniki i Novich Kultur, Leningrad

Slocum MK, Figdore SS, Kennard WC, Suzuki JY, Osborn TC (1990) Linkage arrangement of restriction fragment length polymorphism loci in Brassica oleracea. Theor Appl Genet 80(1):57–64

Slotte T, Hazzouri KM, Ågren JA, Koenig D, Maumus F, Guo YL et al (2013) The Capsella rubella genome and the genomic consequences of rapid mating system evolution. Nat Genet 45(7):831–835

Smartt J, Simmonds NW (1995) Evolution of crop plants (No. Sirsi) i9780470233726)

Smýkal P, Jovanović Ž, Stanisavljević N, Zlatković B, Ćupina B, Đorđević V et al (2014) A comparative study of ancient DNA isolated from charred pea (Pisum sativum L.) seeds from an early Iron Age settlement in Southeast Serbia: inference for pea domestication. Genet Resour Crop Evol 61(8):1533–1544

Snowdon RJ, Friedt W (2004) Molecular markers in Brassica oilseed breeding: current status and future possibilities. Plant Breed 123(1):1–8

Song KM, Osborn TC, Williams PH (1988) Brassica taxonomy based on nuclear restriction fragment length polymorphisms (RFLPs). Theor Appl Genet 75(5):784–794

Song KM, Suzuki JY, Slocum MK, Williams PM, Osborn TC (1991) A linkage map of Brassica rapa (syn. campestris) based on restriction fragment length polymorphism loci. Theor Appl Genet 82(3):296–304
Sovero M (1993) Rapeseed, a new oilseed crop for the United States. New Crops:302–307
Suwabe K, Tsukazaki H, Iketani H, Hatakeyama K, Kondo M, Fujimura M et al (2006) Simple sequence repeat-based comparative genomics between Brassica rapa and Arabidopsis thaliana: the genetic origin of clubroot resistance. Genetics 173(1):309–319
Szőllősi R (2020) Indian mustard (Brassica juncea L.) seeds in health in: nuts and seeds in health and disease prevention. Academic, London, pp 357–364
Taiyan Z, Lianli L, Guang Y, Al-Shehbaz IA (2001) Brassicaceae (Cruciferae). Flora China 8:1–193
Tan H, Yang X, Zhang F, Zheng X, Qu C, Mu J et al (2011) Enhanced seed oil production in canola by conditional expression of Brassica napus LEAFY COTYLEDON1 and LEC1-LIKE in developing seeds. Plant Physiol 156(3):1577–1588
Tariq H, Khan FA, Firdous H, Ullah Z, Javaid RA, Vaseer SG, Zulfiqar M (2020) Cluster analysis of morphological and yield attributing trait of Brassica napus genotypes. Life Sci J 17(8)
Trick M, Long Y, Meng J, Bancroft I (2009) Single nucleotide polymorphism (SNP) discovery in the polyploid Brassica napus using Solexa transcriptome sequencing. Plant Biotechnol J 7(4):334–346
Tsuchihashi Z, Dracopoli NC (2002) Progress in high throughput SNP genotyping methods. Pharmacogenomics J 2(2):103–110
Tsunoda S (1980) Biosynthesis of seed oil and breeding for improved oil quality of rapeseed. Brassica crops and wild allies: biology and breeding. Tokyo, pp 253–283
van Hintum TJ, van De Wiel CCM, Visser DL, Van Treuren R, Vosman B (2007) The distribution of genetic diversity in a Brassica oleracea gene bank collection related to the effects on diversity of regeneration, as measured with AFLPs. Theor Appl Genet 114(5):777–786
Vaughan JG (1977) A multidisciplinary study of the taxonomy and origin of Brassica crops. Bioscience 27(1):35–40
Vavilov NI (1951) The origin, variation, immunity and breeding of cultivated plants, vol 72, no 6. LWW, p 482
Villanueva-Mejia D, Alvarez JC (2017) Genetic improvement of oilseed crops using modern biotechnology. Adv Seed Biol 295
Wagner AE, Terschluesen AM, Rimbach G (2013) Health promoting effects of brassica-derived phytochemicals: from chemopreventive and anti-inflammatory activities to epigenetic regulation. Oxidative Med Cell Longev 2013:964539
Wang J, Kaur S, Cogan NOI, Dobrowolski MP, Salisbury PA, Burton WA et al (2009) Assessment of genetic diversity in Australian canola (Brassica napus L.) cultivars using SSR markers. Crop Pasture Sci 60(12):1193–1201
Wang X, Wang H, Wang J, Sun R, Wu J, Liu S et al (2011) The genome of the mesopolyploid crop species Brassica rapa. Nat Genet 43(10):1035–1039
Warwick SI, Francis A, Al-Shehbaz IA (2006) Brassicaceae: species checklist and database on CD-Rom. Plant Syst Evol 259(2–4):249–258
Warwick SI, James T, Falk KC (2008) AFLP-based molecular characterization of Brassica rapa and diversity in Canadian spring turnip rape cultivars. Plant Genet Resour 6(1):11
Wen J, Xu J, Long Y, Xu H, Wu J, Meng J, Shi C (2015) Mapping QTLs controlling beneficial fatty acids based on the embryo and maternal plant genomes in Brassica napus L. J Am Oil Chem Soc 92(4):541–552
Willcox G (2002) Charred plant remains from a 10th millennium BP kitchen at Jerf el Ahmar (Syria). Veg Hist Archaeobot 11(1–2):55–60
Wu XM, Chen BY, Lu G, Wang HZ, Xu K, Guizhan G, Song Y (2009) Genetic diversity in oil and vegetable mustard (Brassica juncea) landraces revealed by SRAP markers. Genet Resour Crop Evol 56(7):1011

Yanaka A, Fahey JW, Fukumoto A, Nakayama M, Inoue S, Zhang S et al (2009) Dietary sulforaphane-rich broccoli sprouts reduce colonization and attenuate gastritis in Helicobacter pylori–infected mice and humans. Cancer Prev Res 2(4):353–360

Yang R, Jarvis DJ, Chen H, Beilstein M, Grimwood J, Jenkins J et al (2013) The reference genome of the halophytic plant Eutrema salsugineum. Front Plant Sci 4:46

Yang J, Liu D, Wang X, Ji C, Cheng F, Liu B et al (2016) The genome sequence of allopolyploid Brassica juncea and analysis of differential homoeolog gene expression influencing selection. Nat Genet 48(10):1225–1232

Yang J, Zhang C, Zhao N, Zhang L, Hu Z, Chen S, Zhang M (2018) Chinese root-type mustard provides phylogenomic insights into the evolution of the multi-use diversified allopolyploid Brassica juncea. Mol Plant 11(3):512–514

Yaniv Z, Schafferman D, Elber Y, Ben-Moshe E, Zur M (1994) Evaluation of Sinapis alba, native to Israel, as a rich source of erucic acid in seed oil. Ind Crop Prod 2(2):137–142

Yu J, Zhao M, Wang X, Tong C, Huang S, Tehrim S et al (2013) Bolbase: a comprehensive genomics database for Brassica oleracea. BMC Genet 14(1):664

Zhao JJ, Wang XW, Deng B, Lou P, Wu J, Sun RF, … Bonnema G (2004) Phylogenetic relationships within Brassica rapa inferred from AFLP fingerprints. In: Joint meeting of the 14th Crucifer Genetics Workshop and 4th ISHS Symposium on Brassicas

Zohary D, Hopf M, Weiss E (2012) Domestication of plants in the old world: the origin and spread of domesticated plants in Southwest Asia, Europe, and the Mediterranean Basin. Oxford University Press on Demand

Chapter 13
Integrated Omics Analysis of Benzylisoquinoline Alkaloid (BIA) Metabolism in Opium Poppy (*Papaver somniferum* L.)

Kuaybe Yucebilgili Kurtoglu and Turgay Unver

Contents

K. Yucebilgili Kurtoglu (✉)
Faculty of Science, Department of Molecular Biology and Genetics, Istanbul Medeniyet University, Istanbul, Turkey

T. Unver
Ficus Biotechnology, Ostim Teknopark, Ankara, Turkey

H. Tombuloglu et al. (eds.), *Oil Crop Genomics*,
https://doi.org/10.1007/978-3-030-70420-9_13

Abbreviations

2D	two dimensional
4-HPAA	4-hydroxyphenylacetaldehyde.
7OMT	7-O-methyltransferase
AKR	aldo-keto reductase
BBE	berberine bridge enzyme
BIA	Benzylisoquinoline alkaloid.
CAS	canadine synthase
CFS or CheSyn	cheilanthifoline synthase
CNMT	coclaurine *N*-methyltransferase
CNV	copy number variation
CODM	codeine-*O*-demethylase
CoOMT	columbamine *O*-methyltransferase
COR	codeinone reductase
DBOX	dihydrobenzo-phenanthridine oxidase
DRR	1,2-dehydroreticuline reductase
DRS	1,2-dehydroreticuline synthase
EMS	ethyl methane sulfonate
EST	expressed sequence tag
FT-ICR-MS	Fourier transform ion cyclotron resonance mass spectrometry
HPLC	high-pressure liquid chromatography
JA	Jasmonic Acid
MALDI	matrix-assisted laser desorption/ionization
MeJA	Methyl jasmonate
miRNAs	microRNAs
MP	morphine poor
MR	morphine rich
MS	mass spectrometer
MSH	*N*-methylstylopine 14-hydroxylase
NCS	norcoclaurine synthase
NGS	next generation sequencing
NMCH	*N*-methylcoclaurine-3′-hydroxylase
NOS	noscapine synthase
P6H	protopine 6-hydroxylase
PDS	*phytoene desaturase*
qRT-PCR	quantitative real time PCR
REPI	reticuline epimerase
Text	RNAseq
RNAi	RNA interference
ROS	reactive oxygen species
SalAT	salutaridinol 7-*O*-acetyltransferase
SalR	salutardine reductase
SalSyn	salutardine synthase

SanR	sanguinarine reductase
SOMT1	scoulerine 9-*O*-methyltransferase
SPS or StySyn	stylopine synthase
STORR	(S)- to (R)-reticuline
STOX	*S*-tetrahydroprotobererine oxidase
T6ODM	thebaine 6-*O*-demethylase
TF	transcription factor
TNMT	tetrahydroprotoberberine *cis*-*N*-methyltransferase
TRV	tobacco rattle virus
TYDC	Tyrosine/DOPA decarboxylase.
TyrAT	Tyrosine aminotransferase.
VIGS	virus-induced gene silencing

13.1 General Characteristics of *Papaver somniferum* L.

Opium poppy (*Papaver somniferum* L., $2n = 22$) systematically belongs to the *Papaveraceae* family of Ranunculales order (Fig. 13.1). The species name was inspired from a Latin word meaning "sleep-inducing," and the genus was named *papaver* which means poppy in Greek.

The plant is grown in varying climate conditions such as tropical, subtropical, and warm regions of the world. It is an annual herbaceous plant, a source of opium that has been cultivated and utilized as a food and drug since the beginning of civilizations (Brownstein 1993; Huang and Kutchan 2000). So, its medicinal properties have been known to mankind for thousands of years. The opium poppy remains the one and only commercial source to manufacture special metabolites including narcotic analgesics morphine and codeine, cough-suppressant noscapine, semi-synthetic precursor thebaine, antimicrobial agent sanguinarine, and vasodilator papaverine (Ye et al. 1998; Desgagné-Penix et al. 2010, 2012; Guo et al. 2018). As one of the most important crops, it affects the socioeconomic and political relations of the countries (Wijekoon and Facchini 2012). In addition to its medicinal use, opium poppy is rich in oil content in seeds which makes it a valuable source of food. Also, it is known as "common garden poppy" and widely grown for ornamental purposes in Europe. On the dark side, illegal cultivation of the plant for the heroin production is a serious problem with negative consequences worldwide.

13.2 Benzylisoquinoline Alkaloids (BIA) and Opium Poppy

Plants accumulate a bunch of secondary metabolites in response to environmental conditions such as biotic/abiotic stresses, various elicitors or signal molecules. They are also unique sources for human use as pharmaceuticals, food supplement,

Fig. 13.1 *Papaver somniferum* L. (by the courtesy of Özlem Korkmaz, Science Illustrator, Yucebilgili Kurtoglu 2016)

flavors, and fragrances. It is important to understand biochemical pathways to optimize their commercial production (Samanani et al. 2005; Zhao et al. 2005). Benzylisoquinoline alkaloids (BIA) are a large group of secondary metabolites, with more than 2500 products, some of which is of great pharmacological importance such as narcotic analgesics morphine and codeine, antimicrobial sanguinarine and berberine, cough-suppressant, and anticancer drug noscapine and vasodilator papaverine (Facchini and Park 2003; Samanani et al. 2005; Hagel and Facchini 2013; Onoyovwe et al. 2013) (Fig. 13.2).

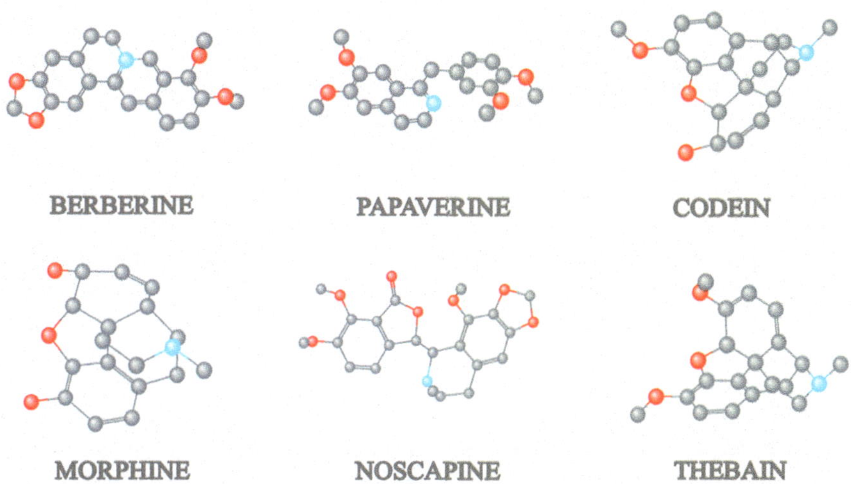

Fig. 13.2 Major alkaloids in opium poppy

Opium poppy produces vast amount of BIA alkaloids belonging to morphinan, aporphine, protopine, phthalideisoquinoline, benzophenanthridine, protoberberine, pavine, and rhoeadine subgroups (Hagel and Facchini 2013). The morphinan alkaloids (morphine as the most abundant and relatively low levels of thebaine and codeine), benzylisoquinoline papaverine, and phthalideisoquinoline noscapine are accumulated largely in opium latex, the cytoplasm of laticifers, which are in contact with phloem channels. Conversely, roots are rich in benzophenanthridine sanguinarine (Frick et al. 2005).

13.3 Biosynthesis of the Major Alkaloids in Opium Poppy

BIA metabolism in terms of biochemistry and physiology has long been studied. In this respect, radiotracer technique was adapted in 1960s, enzyme extraction and characterization methods developed around 1980s, and recombinant DNA era started in 1990s (Beaudoin and Facchini 2014). Along with recent applications of these methods in opium poppy, the knowledge of BIA metabolism is rapidly increased.

Biosynthesis of BIAs begins with the conversion of two L-tyrosine molecules to dopamine and 4-hydroxyphenylacetaldehyde (4-HPAA), via tyrosine/DOPA decarboxylase (TYDC) (Facchini and De Luca 1994) and tyrosine aminotransferase (TyrAT) enzymes respectively (Liscombe et al. 2005; Lee and Facchini 2010). The Pictet–Spengler condensation reaction of these tyrosine derivatives coupled by norcoclaurine synthase (NCS) (Samanani et al. 2005) to yield a central BIA precursor: (*S*)-norcoclaurine (Lee and Facchini 2010).

13.3.1 (S)-Norcoclaurine to (S)-Reticuline

(*S*)-norcoclaurine is further converted to the last common intermediate (*S*)-reticuline in the biosynthesis of most BIAs such as morphinans (e.g., morphine, thebaine, codeine, oripavine,), benzophenanthridines (e.g., sanguinarine), phthalideisoquinolines (e.g., noscapine), simple benzylisoquinolines (e.g., laudanine, papaverine), and protoberberines (e.g., berberine). This reaction involves four sequential steps including *O*-methylation at sixth position by norcoclaurine 6-*O*-methyltransferase (6-OMT) and *N*-methylation of norcoclaurine by coclaurine *N*-methyltransferase (CNMT) respectively to produce (*S*)-*N*-methylcoclaurine which will be further hydroxylated at 3′ by a P450-dependent monooxygenase enzyme (*S*)-*N*-methylcoclaurine-3′-hydroxylase (NMCH) and lastly 4′-*O*-methylation step of (*S*)-3′-hydroxy-*N*-methylcoclaurine with 3-hydroxy-*N*-methyl coclaurine 4′ methyltransferase (4-OMT) produces (*S*)-reticuline (Pauli and Kutchan 1998; Morishige et al. 2000; Huang and Kutchan 2000; Choi et al. 2002; Facchini and Park 2003; Ounaroon et al. 2003; Ziegler et al. 2005).

Four pathways diverge from (*S*)-reticuline, leading to benzophenanthridine, protoberberine, phthalideisoquinolines, as well as morphinan alkaloids. Papaverine biosynthesis partially involves (*S*)-reticuline (Beaudoin and Facchini 2014).

13.3.2 Papaverine Biosynthesis

Papaverine biosynthesis remain controversial which two different metabolic routes have been proposed. The first proposed pathway (NH pathway) involves demethylation of (S)-norreticuline, while the second (NH3 pathway) proceeds through *N*-methylation of (*S*)-reticuline and unspecified intermediates (Mishra et al. 2013). In the first one, (*S*)-coclaurine is converted to (*S*)-norreticuline utilizing 3′-hydroxylase and 7-O-methyltransferase (N7OMT) (Pienkny et al. 2009). On the other side, NH3 pathway is proposed to commence with the activity of 7-O-methyltransferase (7OMT) yielding (*S*)-laudanine from (*S*)-reticuline (Ounaroon et al. 2003). Both routes have successive methylation, demethylation, and dehydrogenation steps to form papaverine at the end. (Han et al. 2010).

In a recent transcriptome sequencing and virus-induced gene silencing (VIGS) study, intermediates of both pathways were supported to take place in papaverine biosynthesis (Dang and Facchini 2012). However another study of the same group later reported that not (*S*)-reticuline but (*S*)-norreticuline is involved in the formation of papaverine (Desgagné-Penix et al. 2012). In another recent article, comparative transcriptomics were performed on high papaverine mutants (*pap1*) of opium poppy, suggesting (*S*)-coclaurine as the primary route for papaverine biosynthesis (Pathak et al. 2013).

13.3.3 Protoberberine, Protopine, and Benzophenanthridine Biosynthesis

(*S*)-scoulerine is a branch-point intermediate in the biosynthesis of several protoberberines, protopines, and benzophenanthridine alkaloids. The formation of (*S*)-scoulerine from (*S*)-reticuline is catalyzed by berberine bridge enzyme (BBE) through a cyclization reaction involving oxidation of (*S*)-reticuline to form a C–C bond between the phenolic ring and *N*-methyl group. BBE encoding cDNAs have been isolated from opium poppy and California poppy (Dittrich and Kutchan 1991; Facchini 2006). The enzyme has also been well characterized (Winkler et al. 2008).

Through methylenedioxy bridging and O-methylation, (*S*)-scoulerine can lead to the formation of several protoberberines such as (*S*)-canadine and (*S*)-stylopine. In berberine pathway, (*S*)-scoulerine is first converted to (*S*)-tetrahydrocolumbamine by a FAD-linked enzyme scoulerine 9-*O*-methyltransferase (SOMT1) (Takeshita et al. 1995). Successively (*S*)-canadine is formed by a methylenedioxy bridging enzyme canadine synthase (CAS) belonging to CYP subfamily. Ultimately, aromatization by (*S*)-tetrahydroprotobererine oxidase (STOX) enzyme gives rise to *berberine*. Alternatively, (*S*)-tetrahydrocolumbamine can be methylated to yield (*S*)-tetrahydropalmatine with a protoberberine-specific enzyme columbamine *O*-methyltransferase (CoOMT) (Morishige et al. 2002).

The initial steps in the biosynthesis of benzophenanthridine-type alkaloids such as sanguinarine were elucidated by tracer experiments and advanced at molecular and biochemical levels (Battersby et al. 1975; Takao et al. 1983; Hagel and Facchini 2013). In *sanguinarine* pathway, conversion of (*S*)-scoulerine to (*S*)-stylopine is the first committed step. Cytochrome P450–dependent monooxygenases cheilanthifoline synthase (CFS or CheSyn) and stylopine synthase (SPS or StySyn) are the two enzymes responsible for the formation of methylenedioxy bridges that yield (*S*)-stylopine (Bauer and Zenk 1989, 1991). cDNAs that encode SPS have been isolated in *E. californica* (Ikezawa et al. 2007). (*S*)-stylopine is then converted to (*S*) *cis-N*-methylstylopine by the enzyme tetrahydroprotoberberine *cis-N*-methyltransferase (TNMT) (Liscombe and Facchini 2007) and further hydroxylated by another P450 dependent *N*-methylstylopine 14-hydroxylase (MSH) to yield protopine (Ruffer and Zenk 1987). TNMT encoding cDNAs were isolated in opium poppy and characterized (Liscombe and Facchini 2007). Benzophenanthridine alkaloid dihydrosanguinarine (which is known to be root specific) synthesis from protopine is catalyzed by a CYP82 member protopine 6-hydroxylase (P6H) (Tanahashi and Zenk 1990; Takemura et al. 2013). The resulting 6-hydroxyprotopine is spontaneously rearranged to dihydrosanguinarine and which is successively oxidized to sanguinarine by oxygen-dependent oxidoreductase dihydrobenzo-phenanthridine oxidase (DBOX) (Hagel et al. 2012). Furthermore, a sanguinarine reductase (SanR) enzyme is also involved in the benzophenanthridine alkaloid metabolism which catalyzes the reverse conversion to dihydrosanguinarine (Weiss et al. 2006; Vogel et al. 2010).

13.3.4 Noscapine Biosynthesis

As mentioned above, phthalideisoquinoline alkaloids such as noscapine are known for their cough-suppressing and anticancer activity (Ye et al. 1998; Facchini et al. 2007; Barken et al. 2008; Dang and Facchini 2012). Although noscapine is one of the first isolated alkaloids, its biosynthesis has not been investigated thoroughly. However, a metabolic scheme was proposed to include (*S*)-*N*-methylcanadine in the pathway. Distinguishing from berberine route, TNMT converts a common intermediate (*S*)-canadine to (*S*)-*N*-methylcanadine (Liscombe and Facchini 2007). In a recent research, a developed scheme has been revealed with all proposed enzymes identified in opium poppy. Accumulation of compounds as intermediates in response to suppression of a specific gene by VIGS suggested the functions of the gene clusters (Dang and Facchini 2012; Winzer et al. 2012). Additionally, three more enzymes have been isolated and characterized in opium poppy: N-methylcanadine 1-hydroxylase (CYP82Y1), CAS, and noscapine synthase (NOS) (Chen and Facchini 2014; Dang and Facchini 2014). NOS was shown to be a NAD^+-dependent, dehydrogenase/reductase yielding noscapine through irreversible conversion of narcotine hemiacetal (Beaudoin and Facchini 2014).

13.3.5 Morphine Biosynthesis

The epimerization of (*S*)-reticuline to (*R*)-reticuline is required to initiate the biosynthesis of morphinan branch alkaloids. This is a two-step process that starts with the dehydrogenation of (*S*)-reticuline to 1,2-dehydroreticuline followed by the reduction to (*R*)-reticuline employing 1,2-dehydroreticuline synthase (DRS) and NADPH-dependent 1,2-dehydroreticuline reductase (DRR), respectively. These two enzymes have been isolated and characterized in poppy seeds (De-Eknamkul and Zenk 1992; Hirata et al. 2004). In a recent study, reticuline epimerase (REPI) has been detected in opium poppy and genes encoding REPI, DRS, and DRR have been isolated and characterized in the plant (Farrow et al. 2015). The first promorphinan alkaloid, salutardine is formed by intramolecular C–C phenol coupling of (*R*)-reticuline driven by a P450 monooxygenase salutardine synthase (SalSyn) (Gesell et al. 2009) which is subsequently reduced by NADPH-dependent salutardine reductase (SalR) to yield salutardinol (Ziegler et al. 2006). The following step in the pathway is the *O*-acetylation of salutardinol by acetyl-CoA-dependent enzyme salutardinol 7-*O*-acetyltransferase (SalAT) (Grothe et al. 2001). The conversion of salutardinol 7-*O*-acetate to the first pentacyclic morphinan alkaloid thebaine is proposed to happen either spontaneously (Lenz and Zenk 1995) or by a deacetylation enzyme isolated from protein extracts of opium poppy (Fisinger et al. 2007). Following to thebaine formation, the morphine pathway is divided into two routes. In the major route, thebaine is first converted to neopinone by thebaine 6-*O*-demethylase (T6ODM), and then spontaneously rearranged to yield codeinone (Hagel and Facchini 2010a, b). Subsequent to this step, codeinone is reduced to codeine by a NADPH-dependent codeinone reductase (COR) (Unterlinner et al. 1999). As a final step, morphine is produced by demethylation of codeine by codeine-*O*-demethylase (CODM) (Hagel and Facchini 2010a, b). As an alternative route in morphine synthesis, the CODM catalyzes 3-*O*-demethylation of thebaine to oripavine and further 6-*O*- demethylation to morphinone by T6ODM prior to be reduced to morphine by COR (Brochmann-Hanssen 1984). Genes for SalSyn (Cyp719B1), SalR, SalAT, and COR enzymes have been isolated in opium poppy by different studies (Unterlinner et al. 1999; Grothe et al. 2001; Wijekoon and Facchini 2012).

In opium poppy, although the biosynthesis of morphine has been almost completely revealed, the regulation remains largely unknown (Kutchan et al. 2008; Kempe et al. 2009).

13.4 Methyl Jasmonate Treatment of Opium Poppy

Methyl jasmonate (MeJA) is a volatile derivative of Jasmonic Acid (JA) and first identified from *Jasminum grandiflorum*. MeJA with its free acid is accepted as a plant hormone that plays regulatory roles in a diverse set of developmental processes such as germination and growth, as well as plant defense responses against wounding, pathogen attack, and abiotic stresses (Reymond and Farmer 1998; Wasternack and Hause 2002). Several gene expression studies have been conducted upon MeJA treatment to identify jasmonate-responsive genes in plants. Upregulated genes are found to encode defensive proteins and those involved in JA and secondary metabolite biosynthesis. On the contrary, downregulated genes included *Rubisco* and *chlorophyll a/b* binding proteins which are active in photosynthesis (Lorenzo et al. 2003).

The effect of MeJA on signal transduction begins with the perception of the elicitor signals. Once the receptors are stimulated, several ion channels and protein/protein kinases are also activated (Blume et al. 2000). Subsequently, Ca^{+2} concentration is increased, and NADPH oxidase is activated to produce reactive oxygen species (ROS). Afterwards, following responses are induced: ethylene and jasmonate production, expression of response genes, and accumulation of secondary metabolites (Zhao et al. 2005). For instance, analysis of Arabidopsis mutants that lack of a proper jasmonate biosynthesis analysis revealed that *coi1* functions in JA signaling and pathogen response. The gene is found to encode an LRR/F-box motif which calls regulatory proteins for ubiquitination. So that the protein has an important role in JA-signaling pathway activating JA-responsive gene expression (Xie et al. 1998; Devoto et al. 2002).

Additionally, exogenous MeJA treatment causes to enhance secondary metabolism by inducing secondary metabolite accumulation through signal transduction in stress response. Elicitors are also known to trigger endogenous JA and MeJA accumulation which causes an increase in secondary metabolite biosynthesis. (Memelink et al. 2001).

Previous studies indicated that MeJA elicitor treatment effects a set of reactions in opium poppy and *Catharanthus roseus* that cause alkaloid accumulation (Huang and Kutchan 2000; Ruiz-May et al. 2009; Holkova et al. 2010). Secondary metabolite accumulation is also triggered by elicitor treatment through altering the gene expression responsible for their biosynthesis (Memelink et al. 2001; Zhang et al. 2011).

Moreover, it has been shown that exogenous MeJA treatment increases the release of signal molecules such as flavonoids and indole as a result of activation of JA signaling pathway (Badri et al. 2008). Another study indicates that plants have different jasmonate-signaling cascade in different organs, thus responses may vary based on the wounded site (Tytgat et al. 2013).

13.5 Approaches to Study Specialized Metabolisms in Opium Poppy

In plants, metabolic networks are highly complex. They are tightly regulated and known to be linked with other pathways in the metabolism. Any kind of biotic/abiotic stresses alter the metabolic state of the plant which in turn start a process to restore the homeostasis by activating particular components of the metabolic networks (Reuben et al. 2013). High-throughput analysis of these components such as transcripts, proteins, and metabolites provide researchers a comprehensive understanding of the metabolic interactions (Rai et al. 2016).

The discovery of new genes active in alkaloid biosynthesis following forward genetics approach involves purification and sequencing of the protein-of-interest, primer design for the cloning of related gene, and further functional analysis in an appropriate host. In opium poppy, for gene discovery expressed sequence tags (ESTs) were first generated by Sanger sequencing. The EST database was derived from stem, seedlings, roots, laticifers, and cell cultures of opium poppy (Pilatzke-Wunderlich and Nessler 2001; Ziegler et al. 2005, 2006; Zulak et al. 2007). Thus, several homology-based enzymes are identified.

With the advances in technology, four fundamental omics fields, namely genomics, transcriptomics, proteomics, and metabolomics, have been emerged which we describe briefly in this section.

13.5.1 Genomics

As next-generation sequencing (NGS) technologies emerged and developed, genomics research based on genome sequencing have been revolutionized (Türktaş et al. 2014). Standard whole-genome sequencing applications of NGS have extensively contributed to comprehensive DNA analysis (Morozova and Marra 2008). Recently, a 2.72 Gb draft genome of the opium poppy has been reported (Guo et al. 2018). Combining different state of the art sequencing technologies (Illumina Paired-End/Mate-Pair (214X), PacBio (66.8X), 10X Genomics–linked reads (40X)), they assembled the draft sequence with a 94.8x coverage and annotated 51,213 protein-coding genes and 9494 noncoding RNAs. The team has achieved to map all of the BIA metabolism–related genes, that are functionally characterized, to chromosomes or unplaced scaffold positions. In the same study, noscapine gene cluster was found to occur in a 584 kb region on chromosome 11 along with the (S)- to (R)-reticuline (*STORR*) gene fusion and the remaining four genes in the thebaine biosynthetic pathway. All are co-expressed in stem cells and referred as "BIA gene cluster" in the article. Any other gene known to be related to the BIA metabolism (such as *BBE, TNMT, CODM, T6ODM,* and *COR*) are found to be in a biosynthetic gene cluster (Guo et al. 2018). This study provided a foundation for the further improvement of opium poppy.

It has been inferred from the biosynthetic pathway of noscapine in *opium poppy*, or biosynthesis of terpene in *Solanum lycopersicum* and *Lotus japonicus* that metabolic gene clusters seem to be common in the biosynthesis of specialized metabolites in medicinal plants, aiding coregulation and coinheritance. All of these studies put forward data mining in genomic sequences for metabolic gene clusters of specialized metabolic pathways (Rai et al. 2017). In a very recent study, existing genome assembly of opium poppy has been improved and patterns of clustering, copy number variation (CNV), and gene expression in the BIA pathway genes have been explored (Li et al. 2020). New chromosome-scale scaffolds that include 35 previously unanchored BIA genes were produced. They found that co-expression of BIA genes increases within clusters, and by using these clusters they also identified new candidates with unknown functions. CNV in some important BIA genes was found to be correlated with differences in the production of alkaloids, linking noscapine production with an 11-gene deletion, and decreased morphine/increased thebaine production with deletion of a T6ODM cluster. These findings support the idea that opium poppy genome evolves dynamically in a way contributing to medically and industrially important plants.

13.5.2 Transcriptomics

Transcriptomics or transcriptome profiling exhibits entire set of RNA transcripts present in a given cell population to investigate their biological state. Differential expression analysis under different conditions enables researchers to figure out the regulation of different biological processes in the plant response mechanism. High throughput transcriptomics is used to determine candidate genes producing BIA enzymes and/or measure relative transcript levels in cDNA libraries derived from different organs or cells with varying alkaloid contents. Hybridization-based DNA macro and microarray studies, in this sense, has been extensively used to understand gene expression profiles in different organs/tissues/conditions of the plant. EST libraries of relevant tissues should be utilized to construct the cDNA microarray. (Zulak et al. 2007; Hagel and Facchini 2010a, b; Gurkok et al. 2015).

Along with the breakthrough in NGS technologies, RNA sequencing (RNAseq) has emerged as a valuable tool for accurate transcriptome profiling with powerful advantages including high level of sensitivity and high throughput. This method is also known as whole transcriptome shotgun sequencing and mainly used for characterization of RNA content and composition of a given sample. Transcription start sites, exon–intron boundaries, splice variants, regulatory elements can be determined. Taking the advantage of being derived directly from mRNA constituent, mostly the coding elements, RNAseq provides a good source for functional genomics studies. RNAseq analysis is extensively used for gene expression profiling. It enables researchers to determine the differentially expressed genes under different stress treatments or different developmental stages. Small RNA sequencing (miRNA, tRNA, rRNA) is also implicated by RNAseq (Morozova and Marra 2008).

RNA-seq-based transcriptome analysis for the discovery of specialized metabolite biosynthetic pathways employs de novo transcriptome assemblies and allows the characterization of associated genes. Transcriptome studies in conjunction with other omics studies on BIA metabolism in opium poppy will be discussed in the next section.

13.5.3 Proteomics

Protein synthesis and modifications, interactions, and their regulation are crucial for the functioning of all biological processes. This functionality and regulation cannot be fully explained without understanding these parameters. Proteomics has played an important role in understanding plant response mechanism.

In recent years, gene discovery studies are followed by with the isolation of the protein and its further computational analysis, using sensitive and fast mass spectrometers (MS) which allowed researchers to identify and quantify almost any expressed proteins (Desgagné-Penix et al. 2010; Hu et al. 2015). Protein separation can be performed either by one- or two-dimensional (2D) SDS-PAGE. Subsequent to proteolysis, fragmentation of peptides or collision-induced dissociation type or mass spectroscopy (MS/MS) is utilized to determine the molecular weight and amino acid (aa) sequence. Furthermore, peptide-mass fingerprint of the protein is generated by using matrix-assisted laser desorption/ionization (MALDI) or electro spray ionization (ESI).

Proteome analysis could be used to improve plant genome annotations by verifying correct ORF-gene assignments and splicing patterns. Until today, proteomics has been used to identify several candidate proteins and enzymes that are involved in biosynthesis of specialized metabolites as will be discussed below (Rai et al. 2017; Desgagné-Penix et al. 2010).

13.5.4 Metabolomics

Lastly, metabolomics has become an important approach for functional genomics with improved technologies, high-throughput analysis, computational tools, and databases. It is also necessary in understanding the biosynthesis of BIA pathway metabolites. Profiling targeted metabolites with the use of analytical and chromatographical methods such as high-pressure liquid chromatography (HPLC) have been conducted in opium poppy (Frick et al. 2005; Hagel and Facchini 2008; Hagel et al. 2008; Zulak et al. 2009; Gurkok et al. 2015). This approach begins by utilizing MeJA or elicitor-induced production of specialized metabolites, in this case BIA alkaloids. Without metabolomics, other omics studies would not be able to provide structural information of the target metabolites which makes it a key approach to explore biosynthesis of specialized metabolites (Prosser et al. 2014; Rai and Saito 2016).

13.6 Integrative Omics–Based Studies to Unravel Complex Biological Interactions in Opium Poppy

Along with the recent advances in "omics" technologies, accumulating data become more valuable for the discovery of genes and related enzymes in opium poppy BIA metabolism. More specifically, correlating targeted metabolite profiling data with gene transcripts and corresponding proteins can result in selection of a reasonable number of potential genes for functional characterization. For example, virus-inducing gene silencing (VIGS) and metabolomics approaches have been successfully applied together in order to understand the function of genes in biosynthesis of BIA enzymes in opium poppy BIA metabolic pathway (Dang and Facchini 2012; Dang et al. 2012; Desgagné-Penix et al. 2012; Hileman et al. 2005; Hagel and Facchini 2010a, b; Lee and Facchini 2010; Hagel et al. 2012; Wijekoon and Facchini 2012; Farrow and Facchini 2013; Gurkok et al. 2016).

Functional genomics in opium poppy has been first performed by Decker and colleagues in which latex proteins have been isolated using 2D SDS-PAGE method and their sequences identified (Decker et al. 2000). Furthermore, a group of researchers have isolated gene transcripts of three enzymes (BBE, COR, and NMCH(CYP80B1)) in different tissues of the plant utilizing RNA gel blot analysis upon elicitor treatment (Huang and Kutchan 2000).

In another study, molecular characterization of salutardinol 7-O-acetyltransferase (SalAT) has been achieved through cloning of the cDNA using amino acid sequence of the purified enzyme which catalyzes the previous step of thebaine. Opium poppy genome has shown to contain only a single copy of the gene by genomic DNA blotting experiment (Grothe et al. 2001).

Identification of novel genes in these alkaloid biosynthesis pathways is also possible with mutant analysis. There are two main studies using this approach that characterize mutants and further perform transcriptome screening in opium poppy. The first one describes "*top1*" poppy mutant which was established by ethyl methane sulfonate (EMS) mutagen treatment and screened for alkaloid content of latex. The mutant was found to accumulate thebaine and oripavine as major alkaloids instead of codeine and morphine. This study provides an insight to grow morphine-free but thebaine/oripavine-producing crops for opioid addiction treatment. Global gene expression was explored using DNA microarrays (Millgate et al. 2004). In the second study, mutant poppy "*pap1*," which accumulates high papaverine in the latex compared to wild-type opium poppy (BR086), has been characterized and 454 pyrosequencing of the transcriptome enabled researchers to identify candidate genes associated with papaverine biosynthesis (Pathak et al. 2013).

Comparative gene expression analysis has been performed in nine Papaver species but only one with morphine-producing capability. AFLP-based DNA macroarray was used to reveal differential gene expression profile among inspected species. In morphine-containing opium poppy, *4OMT* gene transcripts were shown to have high expression levels indicating a role in BIA biosynthesis (Ziegler et al. 2005). Subsequently the same group has identified *SalR* gene using the same approach (Ziegler et al. 2006). More recently gene transcripts of *N7OMT* (Pienkny et al. 2009) and *SalSyn* were also identified using array-based transcriptomics (Gesell et al. 2009). Zulak et al. (2007) have conducted a DNA microarray study for a large-scale transcriptome analysis of sanguinarine-accumulation in opium poppy upon fungal elicitor treatment (Zulak et al. 2007). At first, an EST database was generated from random clones derived from cDNA library. ESTs related to sanguinarine pathway enzymes were identified and further hybridized with RNA samples belonging to different time points. In comparison to control RNAs, BIA enzyme and defense protein transcripts demonstrated a fast and significant upregulation in response to elicitor treatment. Besides, many other transcripts belonging to plant primary metabolism were also stimulated. Metabolite profiling was screened by using "Fourier transform ion cyclotron resonance mass spectrometry" (FT-ICR-MS). Recently, another microarray study has been conducted with MeJa-treated opium poppy capsules. Expression levels of BIA metabolism specific genes such as CNMT,

NCS, StySyn, and COR were found to be altered (Gurkok et al. 2015). Taking together the deep transcriptomics, proteomics, and targeted metabolomics, two comprehensive studies have been published that reveal the components of BIA metabolism in opium poppy cell cultures: In the first one, upon MeJA treatment 454 sequencing was used to generate deep transcriptome data and SDS-PAGE coupled with advanced LC-MS/MS-based protein profiling has enabled the identification of approximately a thousand peptides. In order to correlate abundance of the alkaloids with specific genes and polypeptides, the amount of reticuline, sanguinarine, and protopine at different time intervals has been measured. Elicitor treatment resulted in a significant increase in sanguinarine as much as 40-folds after 100 h of application. Several genes and enzymes potentially involved in BIA metabolism were identified as genetic and biochemical output of the study (Desgagné-Penix et al. 2010). Similarly, following study of the same group reports deep transcriptome sequences for eight different opium poppy cultivars (Desgagné-Penix et al. 2012). The alkaloid levels of each cultivar were again profiled by using LC–MS/MS. cDNA libraries were sequenced, and EST libraries were established. A wide range metabolite and transcript representation of differential presence of specific BIA has been demonstrated. Significant correlations between the temporal presence of alkaloid and transcripts suggests specific genes such as cytochromes p450 may be involved in the formation of papaverine and noscapine.

Lately, in a PhD thesis study, transcriptome profiles of elicitor (MeJA) induced opium poppy capsule and stem tissues from morphine-rich (MR) and morphine-poor (MP) cultivars are compared using an integrative metabolite and transcriptome approach. In response to elicitor treatment, a defensive mechanism has been induced and led to the accumulation of alkaloids in both MR and MP capsule tissues. Besides, a completely opposite pattern has been determined in stem tissues of both MR and MP, supporting the hypothesis of that the morphine synthesized in stem immediately was transported into the capsule to be accumulated there. According to GO and KEGG pathway analysis, biologic processes related to seconder metabolite biosynthesis, signal transduction, and stress response are found to be prominent. 2ODD, bHLH113, (S)-tetrahydroprotoberberine N-methyltransferase, (+)-neomenthol dehydrogenase, major latex protein 146, aquaporin TIP, calmodulin, NRT1/PTR family, F-box/LRR-repeat protein coding transcripts and are observed to be the most differentially expressed transcripts. Besides, morphine biosynthesis specific transcripts corresponding to SalAT, SalSyn, T6ODM, SalR, CODM, and COR enzymes are observed to be differentially expressed in different cultivars and tissues. This study has generated an important source of data for further functional genomic studies to elaborate BIA biosynthesis mechanism in the poppy plant (Yucebilgili Kurtoglu 2016).

In recent years, the data for transcriptome profiles of 33 opium poppy samples in different cultivars, growth phases, and tissues have been described in the literature (Zhao et al. 2019). Another data article provides the transcriptome data from the leaves of four different Papaver species captured at the plant's three developmental life cycles (Subramaniyam et al. 2019).

In conclusion, these studies show that the combination of transcriptome data obtained by NGS sequencing and metabolite profiling through LC–MS/MS analysis could be useful to elucidate the genetic background in specialized metabolisms and their regulation in plants.

13.7 Transcriptional Regulation in Opium Poppy

Following integrated approach of transcriptomics and metabolite analysis, another group has identified novel ESTs with one encoding WRKY under wounding stress. Gene expression analysis of *PsWRKY* was done utilizing quantitative real time PCR (qRT-PCR) and gel retardation assay showed specific binding of the protein to W-box element indicating a regulatory role (Mishra et al. 2013). Only a limited number of regulatory elements were found in opium poppy (Kawano et al. 2012; Mishra et al. 2013). In a recent study, transcriptome datasets from previous studies have been collected to identify transcription factor (TF) gene families. Hence, transcriptional regulation at specific BIA pathways could be investigated. WRKY was found to be the most abundant TF family in all varieties. According to the comparative transcriptomics analysis, one C3H-type and two WRKY-type TFs were found to be candidate regulators in papaverine and thebaine pathways in opium poppy (Agarwal et al. 2015).

Unver and his group identified a number of conserved microRNAs (miRNAs) for the first time in opium poppy (Unver et al. 2010). Bioinformatics approach has been used to elucidate miRNA-mediated regulation of gene expression. The expression of opium poppy miRNAs and their targets have been experimentally validated through qRT-PCR. In addition to that, the same group has studied miRNA regulation of BIA metabolism in a tissue-specific manner. Eleven novel miRNAs have been identified and their targets were determined in opium poppy. In conclusion, pso-miR13, pso-miR2161, and pso-miR408 were predicted to be involved in BIA biosynthesis regulation (Boke et al. 2015).

Almost a decade ago, nearly 90 opium poppy specific genic markers have been developed by using EST database and some of them were validated for forensic use and genetic diversity analysis (Lee et al. 2011; Şelale et al. 2013). In 2014, Celik and coworkers reported a comprehensive study on genomic SSR marker development utilizing DNA pyrosequencing (Celik et al. 2014). They found 53 genomic SSR markers to be useful. In a very recent study, 17 new EST-SSR markers have been added to the repertoire for individual genotyping of opium poppy cultivars (Vašek et al. 2019). It is the first study reporting marker reproducibility within or between Papaver species.

13.8 Metabolic Engineering in Opium Poppy

Morphinan alkaloids including morphine, thebaine, codeine, and oripavine are medicinally important metabolites used in production of narcotic analgesics and medicine for treatment of opiate addiction. Opium poppy remains one and only source producing these metabolites. Today, there are a limited number of countries which can legally produce opiates. In this sense, higher alkaloid content becomes an important issue in terms of financial efficiencies: land needed for cultivation, efforts in harvesting, and transportation. Hence, plant breeders desire to grow high alkaloid–producing opium poppy.

In this context, genetic engineering offers the opportunity to manipulate specific metabolic activity. More specifically, plant metabolic engineering can be used to alter the level of intermediates either by overexpressing the gene-of-interest for increased production or by silencing it through antisense and RNA interference (RNAi) in order to explore the nearly exact function in morphinan alkaloid biosynthesis (Hughes and Shanks 2002).

There are several metabolic engineering studies conducted in opium poppy using these methods. Frick et al. (2004) was the first to alter the alkaloid content in latex. In this study, an industrial elite line has been transformed by cDNA encoding BBE enzyme which is isolated from opium poppy, PCR, northern and southern hybridization methods have been employed to evaluate transformation of the plant while the alkaloid type and abundance in latex have been measured by HPLC and LC-MS. The concentration of several pathway intermediates such as laudanine, reticuline, salutaridine, and (S)-scoulerine have been shown to have an increasing pattern. Transformation caused an altered ratio of morphinan and tetrahydrobenzylisoquinoline alkaloids in latex but did not affect benzophenanthridine alkaloids in roots. This study is known to be the first metabolic engineering in opium poppy which succeeded in the transformation of a BIA pathway gene and further analyzed the changes caused by transgene in the off-springs (Frick et al. 2004).

In another research, COR enzyme which is responsible for the reduction of codeinone to codeine was reported to be silenced through RNAi. Following gene silencing, the transgenic opium poppy has shown to accumulate a 7-step upstream codeinone precursor (*S*)-reticuline. The authors evaluated this result as a feedback mechanism which inhibits morphinan alkaloid biosynthesis but does not affect the other BIA branches. Besides, this is the first gene silencing experience in transgenic opium poppy (Allen et al. 2004).

Moreover, opium poppy was genetically transformed with NMCH encoding sense and antisense copies of *cyp80b3*. In the latex, overexpression of the cDNA increased the alkaloid content up to 450% and antisense-suppression decreased the levels to 84%. *Cyp80b3* was found to have a regulatory role in morphine biosynthesis (Frick et al. 2007).

Concurrently, Larkin and his colleagues reported the first transgenic opium poppy with higher morphinan alkaloid synthesis. Upon transformation with constitutively expressed cDNA of codeinone reductase (*cor1*), the over-expressor plant showed up to 30% (dry weight) increase in capsule alkaloid content in both glasshouse and field while other organs remained unaffected. This increase was concordant with the increase in morphine, thebaine, and codeine content. The thebaine level was not expected to increase, since COR is a downstream enzyme. However, the researchers explained this situation in two ways: one was the feedback mechanism and the other was the inhibition of thebaine transport into vesicles. Compared to non-transgenic control parent, a high-morphine genotype has been obtained in the off-springs (Larkin et al. 2007).

In another metabolic engineering study, a morphinan branch enzyme SalAT cDNA was both overexpressed and suppressed by RNAi in opium poppy. SalAT over-expressor plant demonstrated an increased morphine, codeine, and thebaine levels, while the suppressed plant resulted in unexpected accumulation of an upstream intermediate salutaridine suggesting a channeling between thebaine in reversible manner in which both SalAT and SalR enzymes involved (Allen et al. 2008). In the meanwhile, RNAi-mediated suppression of SalAT was performed by another research group and similarly salutaridine accumulation was observed too. The interaction of two enzymes (SalAT and SalR) was determined by yeast two-hybrid and co-immunoprecipitation analyses supporting the previous explanation. Furthermore, not morphine but codeine and thebaine levels were found to be reduced in the latex (Kempe et al. 2009).

Recently, VIGS method has become a widely used functional genomics tool to investigate BIA metabolism in opium poppy (Hileman et al. 2005; Hagel and Facchini 2010a, b; Lee and Facchini 2011; Dang and Facchini 2012; Hagel et al. 2012; Wijekoon and Facchini 2012; Winzer et al. 2012; Farrow and Facchini 2013). Hileman and coworkers described the method in opium poppy for the first time and silenced an endogenous gene *phytoene desaturase* (*PDS*) with a tobacco rattle virus (TRV)–based vector. Consequently, *PDS* expression was reduced and photobleached appearance was observed. This easy loss-of-function method suggests a rapid and alternative strategy for functional genomics studies on BIA pathway genes and enzymes (Hileman et al. 2005). Hagel and Facchini used functional genomics to isolate T6ODM and CODM involved in morphine biosynthesis and explained their physiological roles by knocking the genes down using VIGS. Silencing of T6ODM and CODM caused the inhibition of thebaine metabolism and codeinone accumulation, respectively. No oripavine was detected in CODM-specific vector-treated plants, since the conversion of thebaine to oripavine was blocked (Hagel and Facchini 2010a, b). Recently a further VIGS study has been conducted in opium poppy in which T6ODM and CODM genes are silenced revealing extensive roles in protopine, sanguinarine, and rhoeadine production in addition to morphine biosynthesis (Farrow and Facchini 2013).

Isolation and characterization of TyrAT, an enzyme in upstream precursor production of BIAs have been achieved using 454 pyrosequencing. VIGS-based functional genomics are employed to investigate the role of TyrAT in alkaloid metabolism. Gene silencing reduced the level of *TyrAT* transcripts up to 80% and showed a decrease in total alkaloid content suggesting a role in the formation of BIA precursors (Lee and Facchini 2011).

In another study, the cDNA libraries from eight cultivars have been constructed for sequencing and further metabolite measurements have been undertaken. *SOMT1*, *SOMT2*, and *SOMT3* gene transcripts are found to be correlated with noscapine accumulation. Their metabolic functions have been explored using VIGS. *SOMT* silencing resulted in a significant reduction in noscapine and papaverine production (Dang and Facchini 2012).

Wijekoon and coworkers performed systematic and individual knockdown of six enzymes involved in morphine biosynthesis in order to investigate the regulation of the metabolism (Wijekoon and Facchini 2012). After gene silencing by VIGS, *SalSyn*, *SalR*, *T6ODM*, and *CODM* knockout plants showed reduced morphine levels and increased s-reticuline, salutaridine, thebaine, and codeine levels, respectively. On the other hand, *COR* and *SalAT* knockouts accumulated reticuline and salutaridine, respectively.

In another metabolic engineering study, a ten gene cluster encoding enzymes of five different classes expressed in high noscapine opium poppy cultivar were identified with RNA-seq-based transcriptome profiling and co-expression analysis. Furthermore, VIGS-based silencing of individual genes was performed, and subsequent accumulation of intermediates was evaluated. Consequently, six out of ten genes were found to be involved in noscapine pathway (Winzer et al. 2012).

Hagel and his colleagues reported the identification, characterization, and silencing of a cDNA encoding DBOX which catalyzes the last reaction in sanguinarine formation. VIGS of the gene caused reduced sanguinarine levels but papaverine accumulation in knockout plants suggesting the enzyme a multifunctional role in BIA metabolism (Hagel et al. 2012).

In another study, a cytochrome P450/aldo-keto reductase (AKR) fusion which catalyzes (*S*)-reticuline to (*R*)-reticuline isomerization is proposed. This reticuline epimerase fusion (REPI) has been detected in opium poppy. VIGS-based suppression of REPI mRNAs caused a significant decrease in (*R*)-reticuline level in accordance with the decrease in morphinan alkaloids. In contrast, (*S*)-reticuline levels increased, indicating a role in stereochemical epimerization from S to R with the activity of REPI (Farrow et al. 2015).

Concordantly, Winzer and coworkers identified a genetic locus (s to r: STORR) responsible for S to R epimerization that express P450 and aldo/keto reductase together. *Storr* mutants were subjected to metabolite analysis. Results confirmed that P450 converts (S)-reticuline to 1,2-dehydroreticuline and aldo/keto reductase family enzyme oxidoreductase catalyzes the further step to (R)-reticuline. Consequently, following a proteomics approach, a fusion protein is found to be responsible for two-step isomerization (Winzer et al. 2015).

13.9 Conclusion

It has been almost a century that BIA alkaloids in opium poppy have attracted people's attention. First, researchers have focused on the plant to elucidate the chemistry of the pathway, then along with the advances in sequencing technologies, isolation of novel genes have been expedited and a diverse set of enzymes and genes have been identified to reveal the biosynthetic pathway of BIA. The availability of the reference genome and a growing repository of transcriptome profiles involving different tissues and growing conditions of the plant will pave the pay in synthetic biology to generate alternative systems in microbes to produce BIAs as alternative commercial sources. Although a considerable amount of information on the pathway at genetic and biochemical levels, there are still important challenges remains to be studied including enzyme interactions as well as transcriptional and metabolic regulation of the pathway. Research on this area will not only provide a deeper understanding of BIA biosynthesis but also will aid developing new papaver cultivars rich in alkaloid-of-interest.

References

Agarwal P, Pathak S, Lakhwani D, Gupta P, Asif MH, Trivedi PK (2015) Comparative analysis of transcription factor gene families from Papaver somniferum: identification of regulatory factors involved in benzylisoquinoline alkaloid biosynthesis. Protoplasma 253(3):857–871

Allen RS, Millgate AG, Chitty JA, Thisleton J, Miller JA, Fist AJ, Gerlach WL, Larkin PJ (2004) RNAi-mediated replacement of morphine with the nonnarcotic alkaloid reticuline in opium poppy. Nat Biotechnol 22(12):1559–1566

Allen RS, Miller JA, Chitty JA, Fist AJ, Gerlach WL, Larkin PJ (2008) Metabolic engineering of morphinan alkaloids by over-expression and RNAi suppression of salutaridinol 7-O-acetyltransferase in opium poppy. Plant Biotechnol J 6(1):22–30

Badri DV, Loyola-Vargas VM, Du J, Stermitz FR, Broeckling CD, Iglesias-Andreu L, Vivanco JM (2008) Transcriptome analysis of Arabidopsis roots treated with signaling compounds: a focus on signal transduction, metabolic regulation and secretion. New Phytol 179(1):209–223

Barken I, Geller J, Rogosnitzky M (2008) Noscapine inhibits human prostate cancer progression and metastasis in a mouse model. Anticancer Res 28(6A):3701–3704

Battersby AR, Staunton J, Wiltshire HR, Francis RJ, Southgate R (1975) Biosynthesis. Part XXII. The origin of chelidonine and of other alkaloids derived from the tetrahydroprotoberberine skeleton. J Chem Soc [Perkin 1] 1(12):1147–1156

Bauer W, Zenk M (1989) Formation of both methylenedioxy groups in the alkaloid (S)-stylopine is catalyzed by cytochrome P-450 enzymes. Tetrahedron Lett 30(39):5257–5260

Bauer W, Zenk MH (1991) Two methylenedioxy bridge forming cytochrome P-450 dependent enzymes are involved in (S)-stylopine biosynthesis. Phytochemistry 30(9):2953–2961

Beaudoin GA, Facchini PJ (2014) Benzylisoquinoline alkaloid biosynthesis in opium poppy. Planta 240(1):19–32

Blume B, Nurnberger T, Nass N, Scheel D (2000) Receptor-mediated increase in cytoplasmic free calcium required for activation of pathogen defense in parsley. Plant Cell 12(8):1425–1440

Boke H, Ozhuner E, Turktas M, Parmaksiz I, Ozcan S, Unver T (2015) Regulation of the alkaloid biosynthesis by miRNA in opium poppy. Plant Biotechnol J 13(3):409–420

Brochmann-Hanssen E (1984) A second pathway for the terminal steps in the biosynthesis of morphine. Planta Med 50(4):343–345
Brownstein MJ (1993) A brief history of opiates, opioid peptides, and opioid receptors. PNAS 90(12):5391–5393
Celik I, Gultekin V, Allmer J, Doganlar S, Frary A (2014) Development of genomic simple sequence repeat markers in opium poppy by next-generation sequencing. Mol Breed 34(2):323–334
Chen X, Facchini PJ (2014) Short-chain dehydrogenase/reductase catalyzing the final step of noscapine biosynthesis is localized to laticifers in opium poppy. Plant J 77(2):173–184
Choi KB, Morishige T, Shitan N, Yazaki K, Sato F (2002) Molecular cloning and characterization of coclaurinen-methyltransferase from cultured cells of Coptis japonica. J Biol Chem 277(1):830–835
Dang TTT, Facchini PJ (2012) Characterization of three O-methyltransferases involved in noscapine biosynthesis in opium poppy. Plant Physiol 159(2):618–631
Dang TTT, Facchini PJ (2014) Cloning and characterization of canadine synthase involved in noscapine biosynthesis in opium poppy. FEBS Lett 588(1):198–204
Dang TTT, Onoyovwi A, Farrow SC, Facchini PJ (2012) Biochemical genomics for gene discovery in benzylisoquinoline alkaloid biosynthesis in opium poppy and related species. Methods Enzymol 515:231–266
Decker G, Wanner G, Zenk MH, Lottspeich F (2000) Characterization of proteins in latex of the opium poppy (Papaver somniferum) using two-dimensional gel electrophoresis and microsequencing. Electrophoresis 21(16):3500–3516
De-Eknamkul W, Zenk MH (1992) Purification and properties of 1, 2-dehydroreticuline reductase from Papaver somniferum seedlings. Phytochemistry 31(3):813–821
Desgagné-Penix I, Khan MF, Schriemer DC, Cram D, Nowak J, Facchini PJ (2010) Integration of deep transcriptome and proteome analyses reveals the components of alkaloid metabolism in opium poppy cell cultures. BMC Plant Biol 10(1):252
Desgagné-Penix I, Farrow SC, Cram D, Nowak J, Facchini PJ (2012) Integration of deep transcript and targeted metabolite profiles for eight cultivars of opium poppy. Plant Mol Biol 79(3):295–313
Devoto A, Nieto-Rostro M, Xie D, Ellis C, Harmston R, Patrick E, Davis J, Sherratt L, Coleman M, Turner JG (2002) COI1 links jasmonate signalling and fertility to the SCF ubiquitin-ligase complex in Arabidopsis. Plant J 32(4):457–466
Dittrich H, Kutchan TM (1991) Molecular cloning, expression, and induction of berberine bridge enzyme, an enzyme essential to the formation of benzophenanthridine alkaloids in the response of plants to pathogenic attack. PNAS 88(22):9969–9973
Facchini PJ (2006) Regulation of alkaloid biosynthesis in plants. Alkaloids Chem Biol 63:1–44
Facchini PJ, De Luca V (1994) Differential and tissue-specific expression of a gene family for tyrosine/dopa decarboxylase in opium poppy. J Biol Chem 269(43):26684–26690
Facchini PJ, Park S-U (2003) Developmental and inducible accumulation of gene transcripts involved in alkaloid biosynthesis in opium poppy. Phytochemistry 64(1):177–186
Facchini PJ, Hagel JM, Liscombe DK, Loukanina N, MacLeod BP, Samanani N, Zulak KG (2007) Opium poppy: blueprint for an alkaloid factory. Phytochem Rev 6(1):97–124
Farrow SC, Facchini PJ (2013) Dioxygenases catalyze O-demethylation and O, O-demethylenation with widespread roles in benzylisoquinoline alkaloid metabolism in opium poppy. J Biol Chem 288(40):28997–29012
Farrow SC, Hagel JM, Beaudoin GA, Burns DC, Facchini PJ (2015) Stereochemical inversion of (S)-reticuline by a cytochrome P450 fusion in opium poppy. Nat Chem Biol 11(9):728–732
Fisinger U, Grobe N, Zenk MH (2007) Thebaine synthase: a new enzyme in the morphine pathway in Papaver somniferum. Nat Prod Commun 2(3):249–253
Frick S, Chitty JA, Kramell R, Schmidt J, Allen RS, Larkin PJ, Kutchan TM (2004) Transformation of opium poppy (Papaver somniferum L.) with antisense berberine bridge enzyme gene (anti-bbe) via somatic embryogenesis results in an altered ratio of alkaloids in latex but not in roots. Transgenic Res 13(6):607–613

Frick S, Kramell R, Schmidt J, Fist AJ, Kutchan TM (2005) Comparative qualitative and quantitative determination of alkaloids in narcotic and condiment papaver s omniferum cultivars. J Nat Prod 68(5):666–673
Frick S, Kramell R, Kutchan TM (2007) Metabolic engineering with a morphine biosynthetic P450 in opium poppy surpasses breeding. Metab Eng 9(2):169–176
Gesell A, Rolf M, Ziegler J, Chávez MLD, Huang F-C, Kutchan TM (2009) CYP719B1 is salutaridine synthase, the CC phenol-coupling enzyme of morphine biosynthesis in opium poppy. J Biol Chem 284(36):24432–24442
Grothe T, Lenz R, Kutchan TM (2001) Molecular characterization of the salutaridinol 7-O-acetyltransferase involved in morphine biosynthesis in opium poppy Papaver somniferum. J Biol Chem 276(33):30717–30723
Guo L, Winzer T, Yang X, Li Y, Ning Z, He Z, Teodor R, Lu Y, Bowser TA, Graham IA, Ye K (2018) The opium poppy genome and morphinan production. Science 362(6412):343–347
Gurkok T, Turktas M, Parmaksiz I, Unver T (2015) Transcriptome profiling of alkaloid biosynthesis in elicitor induced opium poppy. Plant Mol Biol Report 33(3):673–688
Gurkok T, Ozhuner E, Parmaksiz I, Ozcan S, Turktas M, Ipek A, Demirtas I, Okay S, Unver T (2016) Functional characterization of 4′OMT and 7OMT genes in BIA biosynthesis. Front Plant Sci 7:98
Hagel JM, Facchini PJ (2008) Plant metabolomics: analytical platforms and integration with functional genomics. Phytochem Rev 7(3):479–497
Hagel JM, Facchini PJ (2010a) Biochemistry and occurrence of o-demethylation in plant metabolism. Front Physiol 1:14
Hagel JM, Facchini PJ (2010b) Dioxygenases catalyze the O-demethylation steps of morphine biosynthesis in opium poppy. Nat Chem Biol 6(4):273–275
Hagel JM, Facchini PJ (2013) Benzylisoquinoline alkaloid metabolism–a century of discovery and a brave new world. Plant Cell Physiol 54(5):647–672
Hagel JM, Weljie AM, Vogel HJ, Facchini PJ (2008) Quantitative 1H nuclear magnetic resonance metabolite profiling as a functional genomics platform to investigate alkaloid biosynthesis in opium poppy. Plant Physiol 147(4):1805–1821
Hagel JM, Beaudoin GA, Fossati E, Ekins A, Martin VJ, Facchini PJ (2012) Characterization of a flavoprotein oxidase from opium poppy catalyzing the final steps in sanguinarine and papaverine biosynthesis. J Biol Chem 287(51):42972–42983
Han X, Lamshöft M, Grobe N, Ren X, Fist AJ, Kutchan TM, Spiteller M, Zenk MH (2010) The biosynthesis of papaverine proceeds via (S)-reticuline. Phytochemistry 71(11):1305–1312
Hileman LC, Drea S, Martino G, Litt A, Irish VF (2005) Virus-induced gene silencing is an effective tool for assaying gene function in the basal eudicot species Papaver somniferum (opium poppy). Plant J 44(2):334–341
Hirata K, Poeaknapo C, Schmidt J, Zenk MH (2004) 1, 2-Dehydroreticuline synthase, the branch point enzyme opening the morphinan biosynthetic pathway. Phytochemistry 65(8):1039–1046
Holkova I, Bezakova L, Bilka F, Balazova A, Vanko M, Blanarikova V (2010) Involvement of lipoxygenase in elicitor-stimulated sanguinarine accumulation in Papaver somniferum suspension cultures. Plant Physiol Biochem 48(10–11):887–892
Hu J, Rampitsch C, Bykova NV (2015) Advances in plant proteomics toward improvement of crop productivity and stress resistance. Front Plant Sci 6:209
Huang F-C, Kutchan TM (2000) Distribution of morphinan and benzo [c] phenanthridine alkaloid gene transcript accumulation in Papaver somniferum. Phytochemistry 53(5):555–564
Hughes EH, Shanks JV (2002) Metabolic engineering of plants for alkaloid production. Metab Eng 4(1):41–48
Ikezawa N, Iwasa K, Sato F (2007) Molecular cloning and characterization of methylenedioxy bridge-forming enzymes involved in stylopine biosynthesis in Eschscholzia californica. FEBS J 274(4):1019–1035
Kawano N, Kiuchi F, Kawahara N, Yoshimatsu K (2012) Genetic and phenotypic analyses of a Papaver somniferum T-DNA insertional mutant with altered alkaloid composition. Pharmaceuticals 5(2):133–154

Kempe K, Higashi Y, Frick S, Sabarna K, Kutchan TM (2009) RNAi suppression of the morphine biosynthetic gene salAT and evidence of association of pathway enzymes. Phytochemistry 70(5):579–589

Kutchan TM, Frick S, Weid M (2008) Engineering plant alkaloid biosynthetic pathways: progress and prospects. Adv Plant Biochem Mol Biol 1:283–310

Larkin PJ, Miller JA, Allen RS, Chitty JA, Gerlach WL, Frick S, Kutchan TM, Fist AJ (2007) Increasing morphinan alkaloid production by over-expressing codeinone reductase in transgenic Papaver somniferum. Plant Biotechnol J 5(1):26–37

Lee E-J, Facchini P (2010) Norcoclaurine synthase is a member of the pathogenesis-related 10/Bet v1 protein family. Plant Cell 22(10):3489–3503

Lee E-J, Facchini PJ (2011) Tyrosine aminotransferase contributes to benzylisoquinoline alkaloid biosynthesis in opium poppy. Plant Physiol 157(3):1067–1078

Lee S, Park Y, Han E, Choi H, Chung H, Oh SM, Chung KH (2011) Thebaine in hair as a marker for chronic use of illegal opium poppy substances. Forensic Sci Int 204(1):115–118

Lenz R, Zenk MH (1995) Acetyl coenzyme A: salutaridinol-7-O-acetyltransferase from papaver somniferum plant cell cultures. The enzyme catalyzing the formation of thebaine in morphine biosynthesis. J Biol Chem 270(52):31091–31096

Li Q, Ramasamy S, Singh P, Hagel JM, Dunemann SM, Chen X, Chen R, Yu L, Tucker J, Facchini PJ, Yeaman S (2020) Gene clustering and copy number variation in alkaloid metabolic pathways of opium poppy. Nat Commun 11:1190

Liscombe DK, Facchini PJ (2007) Molecular cloning and characterization of tetrahydroprotoberberine cis-N-methyltransferase, an enzyme involved in alkaloid biosynthesis in opium poppy. J Biol Chem 282(20):14741–14751

Liscombe DK, MacLeod BP, Loukanina N, Nandi OI, Facchini PJ (2005) Evidence for the monophyletic evolution of benzylisoquinoline alkaloid biosynthesis in angiosperms. Phytochemistry 66(11):1374–1393

Lorenzo O, Piqueras R, Sanchez-Serrano JJ, Solano R (2003) ETHYLENE RESPONSE FACTOR1 integrates signals from ethylene and jasmonate pathways in plant defense. Plant Cell 15(1):165–178

Memelink J, Kijne JW, van der Heijden R, Verpoorte R (2001) Genetic modification of plant secondary metabolite pathways using transcriptional regulators. Adv Biochem Eng Biotechnol 72:103–125

Millgate AG, Pogson BJ, Wilson IW, Kutchan TM, Zenk MH, Gerlach WL, Fist AJ, Larkin PJ (2004) Analgesia: morphine-pathway block in top1 poppies. Nature 431(7007):413–414

Mishra S, Tripathi V, Singh S, Phukan UJ, Gupta M, Shanker K, Shukla RK (2013) Wound induced transcriptional regulation of benzylisoquinoline pathway and characterization of wound inducible PsWRKY transcription factor from Papaver somniferum. PLoS One 8(1):e52784

Morishige T, Tsujita T, Yamada Y, Sato F (2000) Molecular characterization of thes-adenosyl-l-methionine: 3′-hydroxy-n-methylcoclaurine 4′-o-methyltransferase involved in isoquinoline alkaloid biosynthesis in Coptis japonica. J Biol Chem 275(30):23398–23405

Morishige T, Dubouzet E, Choi KB, Yazaki K, Sato F (2002) Molecular cloning of columbamine O-methyltransferase from cultured Coptis japonica cells. Eur J Biochem 269(22):5659–5667

Morozova O, Marra MA (2008) Applications of next-generation sequencing technologies in functional genomics. Genomics 92(5):255–264

Onoyovwe A, Hagel JM, Chen X, Khan MF, Schriemer DC, Facchini PJ (2013) Morphine biosynthesis in opium poppy involves two cell types: sieve elements and laticifers. Plant Cell 25(10):4110–4122

Ounaroon A, Decker G, Schmidt J, Lottspeich F, Kutchan TM (2003) (R, S)-Reticuline 7-O-methyltransferase and (R, S)-norcoclaurine 6-O-methyltransferase of Papaver somniferum–cDNA cloning and characterization of methyl transfer enzymes of alkaloid biosynthesis in opium poppy. Plant J 36(6):808–819

Pathak S, Lakhwani D, Gupta P, Mishra BK, Shukla S, Asif MH, Trivedi PK (2013) Comparative transcriptome analysis using high papaverine mutant of Papaver somniferum reveals pathway and uncharacterized steps of papaverine biosynthesis. PLoS One 8(5):e65622

Pauli HH, Kutchan TM (1998) Molecular cloning and functional heterologous expression of two alleles encoding (S)-N-methylcoclaurine 3′-hydroxylase (CYP80B1): a new methyl jasmonate-inducible cytochrome P-450-dependent mono-oxygenase of benzylisoquinoline alkaloid biosynthesis. Plant J 13(6):793–801

Pienkny S, Brandt W, Schmidt J, Kramell R, Ziegler J (2009) Functional characterization of a novel benzylisoquinoline O-methyltransferase suggests its involvement in papaverine biosynthesis in opium poppy (Papaver somniferum L). Plant J 60(1):56–67

Pilatzke-Wunderlich I, Nessler CL (2001) Expression and activity of cell-wall-degrading enzymes in the latex of opium poppy, Papaver somniferum L. Plant Mol Biol 45(5):567–576

Prosser GA, Larrouy-Maumus G, de Carvalho LP (2014) Metabolomic strategies for the identification of new enzyme functions and metabolic pathways. EMBO Rep 15(6):657–669

Rai A, Saito K (2016) Omics data input for metabolic modeling. Curr Opin Biotechnol 37:127–134

Rai A, Umashankar S, Rai M, Kiat LB, Bing JA, Swarup S (2016) Coordinate regulation of metabolite glycosylation and stress hormone biosynthesis by TT8 in Arabidopsis. Plant Physiol 171:2499–2515

Rai A, Saito K, Yamazaki M (2017) Integrated omics analysis of specialized metabolism in medicinal plants. Plant J 90(4):764–787

Reuben S, Rai A, Pillai BV, Rodrigues A, Swarup S (2013) A bacterial quercetin oxidoreductase QuoA-mediated perturbation in the phenylpropanoid metabolic network increases lignification with a concomitant decrease in phenolamides in Arabidopsis. J Exp Bot 64:5183–5194

Reymond P, Farmer EE (1998) Jasmonate and salicylate as global signals for defense gene expression. Curr Opin Plant Biol 1(5):404–411

Ruffer M, Zenk MH (1987) Enzymatic formation of protopines by a microsomal cytochrome P-450 system of Corydalisvaginans. Tetrahedron Lett 28(44):5307–5310

Ruiz-May E, Galaz-Avalos RM, Loyola-Vargas VM (2009) Differential secretion and accumulation of terpene indole alkaloids in hairy roots of Catharanthus roseus treated with methyl jasmonate. Mol Biotechnol 41(3):278–285

Samanani N, Park S-U, Facchini PJ (2005) Cell type–specific localization of transcripts encoding nine consecutive enzymes involved in protoberberine alkaloid biosynthesis. Plant Cell 17(3):915–926

Şelale H, Çelik I, Gültekin V, Allmer J, Doğanlar S, Frary A (2013) Development of EST-SSR markers for diversity and breeding studies in opium poppy. Plant Breed 132(3):344–351

Subramaniyam S, Bae S, Jung M, Shin Y, Oh JH (2019) The transcriptome data from the leaves of four Papaver species captured at the plant's three developmental life cycles. Data Brief 28:104955

Takao N, Kamigauchi M, Okada M (1983) Biosynthesis of benzo [c] phenanthridine alkaloids sanguinarine, chelirubine and macarpine. Helv Chim Acta 66(2):473–484

Takemura T, Ikezawa N, Iwasa K, Sato F (2013) Molecular cloning and characterization of a cytochrome P450 in sanguinarine biosynthesis from Eschscholzia californica cells. Phytochemistry 91:100–108

Takeshita N, Fujiwara H, Mimura H, Fitchen JH, Yamada Y, Sato F (1995) Molecular cloning and characterization of S-adenosyl-L-methionine: scoulerine-9-O-methyltransferase from cultured cells of Coptis japonica. Plant Cell Physiol 36(1):29–36

Tanahashi T, Zenk MH (1990) Elicitor induction and characterization of microsomal protopine-6-hydroxylase, the central enzyme in benzophenanthridine alkaloid biosynthesis. Phytochemistry 29(4):1113–1122

Türktaş M, Kurtoğlu KY, Dorado G, Zhang B, Hernandez P, Unver T (2014) Sequencing of plant genomes–a review. Turk J Agric For 39(3):361–376

Tytgat TO, Verhoeven KJ, Jansen JJ, Raaijmakers CE, Bakx-Schotman T, McIntyre LM, van der Putten WH, Biere A, van Dam NM (2013) Plants know where it hurts: root and shoot Jasmonic acid induction elicit differential responses in Brassica oleracea. PLoS One 8(6):e65502

Unterlinner B, Lenz R, Kutchan TM (1999) Molecular cloning and functional expression of codeinone reductase: the penultimate enzyme in morphine biosynthesis in the opium poppy Papaver somniferum. Plant J 18(5):465–475

Unver T, Parmaksız I, Dündar E (2010) Identification of conserved micro-RNAs and their target transcripts in opium poppy (Papaver somniferum L.). Plant Cell Rep 29(7):757–769
Vašek J, Čílová D, Melounová M, Svoboda P, Vejl P, Štikarová R, Vostrý L, Kuchtová P, Ovesná J (2019) New EST-SSR markers for individual genotyping of opium poppy cultivars (*Papaver somniferum* L.). Plants (Basel) 9(1):10
Vogel C, de Sousa AR, Ko D, Le SY, Shapiro BA, Burns SC, Sandhu D, Boutz DR, Marcotte EM, Penalva LO (2010) Sequence signatures and mRNA concentration can explain two-thirds of protein abundance variation in a human cell line. Mol Syst Biol 6(1):400
Wasternack C, Hause B (2002) Jasmonates and octadecanoids: signals in plant stress responses and development. Prog Nucleic Acid Res Mol Biol 72:165–221
Weiss D, Baumert A, Vogel M, Roos W (2006) Sanguinarine reductase, a key enzyme of benzophenanthridine detoxification. Plant Cell Environ 29(2):291–302
Wijekoon CP, Facchini PJ (2012) Systematic knockdown of morphine pathway enzymes in opium poppy using virus-induced gene silencing. Plant J 69(6):1052–1063
Winkler A, Łyskowski A, Riedl S, Puhl M, Kutchan TM, Macheroux P, Gruber K (2008) A concerted mechanism for berberine bridge enzyme. Nat Chem Biol 4(12):739–741
Winzer T, Gazda V, He Z, Kaminski F, Kern M, Larson TR, Li Y, Meade F, Teodor R, Vaistij FE, Walker C, Bowser TA, Graham IA (2012) A Papaver somniferum 10-gene cluster for synthesis of the anticancer alkaloid noscapine. Science 336(6089):1704–1708
Winzer T, Kern M, King AJ, Larson TR, Teodor RI, Donninger SL, Li Y, Dowle AA, Cartwright J, Bates R (2015) Morphinan biosynthesis in opium poppy requires a P450-oxidoreductase fusion protein. Science 349(6245):309–312
Xie DX, Feys BF, James S, Nieto-Rostro M, Turner JG (1998) COI1: an Arabidopsis gene required for jasmonate-regulated defense and fertility. Science 280(5366):1091–1094
Ye K, Ke Y, Keshava N, Shanks J, Kapp JA, Tekmal RR, Petros J, Joshi HC (1998) Opium alkaloid noscapine is an antitumor agent that arrests metaphase and induces apoptosis in dividing cells. PNAS 95(4):1601–1606
Yucebilgili Kurtoglu K (2016) Transcriptome analysis of morphine biosynthesis in opium poppy (Papaver somniferum L.). Dissertation, Marmara university
Zhang H, Hedhili S, Montiel G, Zhang Y, Chatel G, Pre M, Gantet P, Memelink J (2011) The basic helix-loop-helix transcription factor CrMYC2 controls the jasmonate-responsive expression of the ORCA genes that regulate alkaloid biosynthesis in Catharanthus roseus. Plant J 67(1):61–71
Zhao J, Davis LC, Verpoorte R (2005) Elicitor signal transduction leading to production of plant secondary metabolites. Biotechnol Adv 23(4):283–333
Zhao Y, Zhang Z, Li M, Luo J, Chen F, Gong Y, Li Y, Wei Y, Su Y, Kong L (2019) Transcriptomic profiles of 33 opium poppy samples in different tissues, growth phases, and cultivars. Sci Data 6(1):1–10
Ziegler J, Diaz-Chávez ML, Kramell R, Ammer C, Kutchan TM (2005) Comparative macroarray analysis of morphine containing Papaver somniferum and eight morphine free Papaver species identifies an O-methyltransferase involved in benzylisoquinoline biosynthesis. Planta 222(3):458–471
Ziegler J, Voigtländer S, Schmidt J, Kramell R, Miersch O, Ammer C, Gesell A, Kutchan TM (2006) Comparative transcript and alkaloid profiling in Papaver species identifies a short chain dehydrogenase/reductase involved in morphine biosynthesis. Plant J 48(2):177–192
Zulak KG, Cornish A, Daskalchuk TE, Deyholos MK, Goodenowe DB, Gordon PM, Klassen D, Pelcher LE, Sensen CW, Facchini PJ (2007) Gene transcript and metabolite profiling of elicitor-induced opium poppy cell cultures reveals the coordinate regulation of primary and secondary metabolism. Planta 225(5):1085–1106
Zulak KG, Khan MF, Alcantara J, Schriemer DC, Facchini PJ (2009) Plant defense responses in opium poppy cell cultures revealed by liquid chromatography-tandem mass spectrometry proteomics. Mol Cell Proteomics 8(1):86–98x

Chapter 14
Transcriptome Analysis in Jatropha During Abiotic Stress Response

Joyce A. Cartagena and Gian Powell B. Marquez

Contents

14.1 Introduction

Sustainable development requires circular economic growth, sustainable utilization of natural resources, and inclusive carbon neutral society. However, reliable, affordable, and renewable energies and materials are necessary to support a sustainable growth and development while decoupling from fossil fuels and mitigating climate change. Among the available renewable energies such as solar heat and photovoltaic, wind, hydroelectric, tide and ocean wave, geothermal, green hydrogen and fuel cells, salinity gradient, and biomass, only biomass can be used as both bioenergy and biomaterials. While a balance mix of renewable energies is needed to make it more reliable and avoid overburdening one system over the other, biomass should be prioritized due to its more economical and efficient process of converting solar to marketable fuels (Carpita and Sage 2015) and its future role as major source of industrial chemicals and advanced structural materials (Allais et al. 2020).

J. A. Cartagena (✉)
Graduate School of Bioagricultural Sciences, Nagoya University, Nagoya, Japan
e-mail: joyce@nuagr1.agr.nagoya-u.ac.jp

G. P. B. Marquez
College of Global Liberal Arts, Ritsumeikan University, Osaka, Japan

H. Tombuloglu et al. (eds.), *Oil Crop Genomics*,
https://doi.org/10.1007/978-3-030-70420-9_14

Many crops can be utilized as biomass feedstock for bioenergy and biomaterial production, but most of them directly or indirectly compete against food crops. Hence, general criteria for the selection of viable crops were identified as (1) being nonedible or non-staple crops, (2) being grown easily in marginal land or marine environment, (3) having scalability in production to meet future demand, (4) being farmed without adverse impact on the environment, (5) having low-energy-input requirement, and (6) being resilient to uncertainties of climate variability. Among these viable crops, physic nut or *Jatropha curcas* L. came into limelight as a wonder crop in 2008 due to its supposed drought tolerance and high productivity even with minimal care. While this may be true in some cases of small plantations, Jatropha lost its spotlight after 2 years due to the failure of many plantations in Asia and Africa to deliver target outputs (Cartagena 2017). The failure was mainly attributed to the lack of basic research on crop improvement which limited the effectiveness of management practices for this emerging biofuel crop. Coupled with the worsening global warming, which drives the occurrences of extreme weather and climate events (Otto 2017) and consequently affects the global agricultural landscapes (Arora 2019), it is difficult for crops to naturally adapt to the rapid and erratic changes in environmental conditions. Hence, it is important to further develop new lines of Jatropha which can fit into various types of environmental degradation.

The impact of climate extremes can change land conditions which affects productivity and viability of crop plantations around the world (Jia et al. 2019). Feng et al. (2019) suggested that low- ($\leq$1 mm/day), middle- (1–30 mm/day), and high- ($\geq$30 mm/day) intensity precipitations will increase at the area of high latitudes (70°–90°), while occurrence of middle-intensity precipitation from the Tropics (0°–20°) to midlatitude (40°–70°) will accordingly decrease and shift to more low- and high-intensity precipitations. This means that dry season and arid regions will become drier while wet season and humid regions will become wetter. The change from middle- to low-intensity precipitation during hotter dry season can cause insufficient natural replenishment of groundwater which leads to prolonged drought and desertification in some region. Desertification consequently causes nutrient depletion by inducing the release of organic carbon and nitrogen to the atmosphere as greenhouse gas (An et al. 2019). On the other hand, more frequent high-intensity precipitation during wet season can cause flooding at the Tropics and midlatitude, and snowstorms and extremely cold winter at high altitude (Eccles et al. 2019). Flooding leads to waterlogging of land and may cause anaerobic stress on crops. The change of landscape is sometimes permanent, making the farmland unfit for its current crops. Nonetheless, direct anthropogenic activities such as overextraction of underground water and land clearing, along with overfertilization and use of pesticide create abiotic stresses in farmland such as soil salinization and heavy metal contamination, respectively, which are further exacerbated by climate extremes. The combination of climate variability and anthropogenic activities can change the global agricultural landscape, opening new regions for agricultural activities while degrading others into marginal lands. Reutilization of marginal lands for bioenergy crops to lessen further degradation can be an effective and beneficial mitigation response.

The ability of Jatropha to grow in any kind of soil quality and conditions made it an ideal candidate in reclaiming marginal lands as long as its harvest output is economically viable. Since the whole-genome sequence of Jatropha was reported by Sato et al. in 2011, research on molecular breeding of Jatropha to improve harvest yield and resistance against abiotic stress have been accelerating. Discovery of complex transcriptional processes and expression mechanisms of new genes in response against different abiotic stresses were increasingly available for Jatropha due to the advancement of high-throughput sequencing platforms. Jatropha has a relatively small genome ($C = 416$ Mb) and 22 (2n) diminutive chromosomes (1.24–1.71 μm) which make it a convenient organism for genetic transformation and analysis (Carvalho et al. 2008). Its short generation time (6 months) is also advantageous for functional genomic studies. The better resolution, specificity, and sensitivity of RNA-Seq transcriptome analysis powered by the new sequencing platforms rapidly advance the genetic understanding on how Jatropha cope with abiotic stresses. This chapter will present the progress in transcriptome analysis of gene families that contribute to stress response mechanisms in Jatropha and their potential contribution to the development of genetically superior Jatropha lines which are both high-yielding and adapted to future climatic conditions.

14.1.1 Abiotic Stresses Affecting Jatropha

Among the abiotic stresses that are worsened by climate change, drought, salinity, cold, waterlogging, and nutrient deficiency will be the focus on this chapter. The morphological and physiological effects of these abiotic stresses on Jatropha will be reviewed to develop a better understanding on Jatropha's mechanism of response. In conjunction with genetic analysis and transcriptome profiling of Jatropha, the regulation of gene expression involved in promoting high productivity and maintaining tolerance against abiotic stresses will be better elucidated. This will help develop new Jatropha lines which can be used as an economically viable method in mitigating climate change.

14.1.1.1 Drought

Drought is a prolonged shortage of water in an area which can be exhibited by dry soil or cracked land. Although cyclical drought can be experienced in various places around the world, its intensity has been magnified due to higher temperature and reduced precipitation brought by climate change. Jatropha was known to be a drought-tolerant crop which can grow in arid regions. However, basic understanding on its drought response mechanisms needs further elucidation.

Aside from growth suspension, the morphophysiological strategies of Jatropha against drought focus on water conservation (Sapeta et al. 2013). To reduce water loss through transpiration, stomatal closure is commonly observed on the leaves of

Jatropha. While this mechanism can conserve water, the leaves are also being deprived with CO_2 for photosynthetic carbon assimilation which alternatively results to photorespiration. Photorespiration process produces high amount of reactive oxygen species (ROS) which can damage cells. To scavenge ROS, higher activities of antioxidant enzymes such as superoxide dismutase, catalase, ascorbate peroxidase, and glutamine synthetase were observed by Pompelli et al. (2010) in Jatropha. This can be accompanied by delayed emergence of new leaf, increased leaf drop, and reorientation of leaf surface away from the light to reduce photooxidative damage (Sapeta et al. 2016). On the other hand, increase in root development was observed in Jatropha under drought stress which was an adaptation mechanism that enhances the ability of plants to absorb water (de Santana et al. 2017). However, under prolonged drought, membrane lipid peroxidation still occurred despite the presence of antioxidant enzymes (Pompelli et al. 2010) while complete fall off of mature leaves were observed by Cartagena et al. (2015), leading to plant dormancy and unproductive state.

14.1.1.2 Salinity

Soil salinization is the accumulation of inorganic salts in the soil, such as Cl^-, Na^+, Ca^{2+}, K^+, and Mg^{2+}, which is generated by either mineral weathering, irrigation, saline underground water, or increased inland reach of tide due to sea level rise. High salinity in soil can negatively affect plant growth by reducing the ability of plants to absorb water due to osmotic stress and consequently increasing the ionic salt concentration in tissues (Silva et al. 2015). Jatropha is a salt includer plant which means ionic salts such as Na^+ and Cl^- are absorbed and transported to their apical leaves and stems. As a consequence, efflux of K^+ among leaves and roots occurs, causing K^+ deficiency. Under salinity stress, it is crucial to maintain the cytosolic balance of K^+ and Na^+ concentration to tolerate osmotic stress. This means that K^+/Na^+ ratio in plant tissues should be above 1.0 value. However, it was found that Jatropha can only maintain K^+/Na^+ ratio of 1.0 at 25 mM NaCl treatment and exhibits lower ratios in both leaves and roots at higher NaCl concentration treatment, which suggest that Jatropha can only tolerate low-salinity stress (Silva et al. 2015). To adapt in a high-salinity stress condition, Jatropha sequester Na^+ into its vacuole (Jha et al. 2013) and retain more water in their leaves as a mechanism to osmoregulate the accumulation of NaCl (Silva et al. 2015). The mechanism to conserve water in the leaves is associated with stomatal closure which is the same response seen under drought stress (Sapeta et al. 2013). This leads to the accumulation of ROS, leading to the cellular production of low molecular weight antioxidant organic solutes such as ascorbic acid, glutathione, and phenols, and antioxidant enzymes such as catalase, peroxidase, superoxide dismutase, and ascorbate–glutathione cycle enzymes in order to scavenge ROS (Diaz-Vivancos et al. 2013; Chen et al. 2015). Leaf drooping and wilting were also observed at 150 mM NaCl concentration. (Eswaran et al. 2012). Hence, targeting the regulation of the osmoprotectant may help develop both salt- and drought-tolerant Jatropha.

14.1.1.3 Cold

Most economically important crops have difficulty acclimatizing in cold regions, limiting their cultivation in the tropics and subtropical regions. The need to expand agricultural output reignited the attention toward cold regions as potential area for agricultural expansion. Hence, the effect of cold stress on Jatropha has been assessed here to better understand its viability in cold climate.

Cold stress response can be classified as chilling-sensitive, chilling-tolerant but freezing-sensitive, and freezing-tolerant (Ao et al. 2013). Under chilling stress (0–15 °C), Jatropha arrests its growth and metabolism. It was observed that electrolyte leakage increased from 10% to 60% after subjecting Jatropha seedlings at 1 °C for 7 days, while malondialdehyde content in the leaves peaked (~45 $\mu mol/g_{DW}$) on the fifth day of chilling stress (1 °C) which means chilling stress damages the membrane systems and destabilizes cellular homeostasis, and initiate lipid peroxidation, respectively (Ao et al. 2013). As a response, osmoprotectant and cryoprotective proteins are synthesized to maintain homeostasis and prevent ice nuclei formation (Ding et al. 2019). Jatropha also employ reversible protein phosphorylation to regulate ion transport and maintain homeostasis (Liu et al. 2019). Moreover, instability of metabolism due to chilling stress causes ROS accumulation which consequently increases the activities of low molecular weight antioxidant organic solutes and antioxidant enzymes in order to scavenge ROS (Ao et al. 2013; Ding et al. 2019). This accumulation and scavenging of ROS are the same response mechanism exhibited under drought and salinity stresses. While employing chill hardening was shown to increase chilling tolerance of Jatropha, prolonged exposure to chilling stress at 1 °C still resulted to plant death (Ao et al. 2013).

14.1.1.4 Waterlogging

Waterlogging is the condition of soil characterized by the presence of amount of water beyond the soil's holding capacity. This is becoming a common occurrence due to the increasing frequency of heavier rainfall released in a narrow period of time. Waterlogging suffocates the soil by limiting gas exchange and creating hypoxia condition in the root areas of plants. With limited O_2 comes constrained oxidative phosphorylation which ultimately causes ATP shortage and leads to metabolic shifting from oxidative phosphorylation to anaerobic respiration in order to supply the lacking ATP (Bailey-Serres et al. 2012). This metabolic shifting requires activation of diverse physiological processes with particular enhancement in alternative oxidase expression in roots to prevent ROS production. NO_3^- utilization is also triggered to allow the reuse of NAD^+ in glycolysis and synthesis of nitrogen oxide (NO). Accumulation of NO further enhances alternative oxidase expression while inducing the production of citrate and assimilation of nitrogen. Together with the lowering of carbohydrate content in roots, it was suggested by Juntawong et al. (2014) that roots can adapt to waterlogging stress by altering its energy production and consumption pathways. However, in the plantation sites in Laos, Papua, and

Mali where land consists of high percentage of clay, the occurrence of abrupt heavy rainfall generated waterlogging condition. The Jatropha shrubs in these plantation sites, including the old Jatropha trees, failed to adapt in the waterlogging condition which resulted to crop failure (Degail and Chantry 2013). In the study of de Santana et al. (2017), 16 days of waterlogging stress caused leaf senescence and root decay in Jatropha by 62% and 38%, respectively. Clearly, Jatropha is sensitive to waterlogging stress and needs further genetic improvement if its future plantation sites will be exposed to cyclical flooding.

14.1.1.5 Nutrient Deficiency

Soil nutrients are essential for plant growth and development. They can be classified as macronutrient and micronutrient. Macronutrients such as nitrogen (N), phosphorus (P), magnesium (Mg), and potassium (K) are needed in large amount while micronutrients such as iron (Fe), zinc (Zn), manganese (Mn), and copper (Cu) are required in small quantities. Among the macronutrients, N limitation is commonly observed in the field due to leaching, erosion, gaseous loss, microbial uptake, or plant assimilation. In Jatropha, 16 days of N deprivation elicited increased activity of N transporters in roots for better N uptake. Amino acid synthesis was also reduced in roots. In leaf, synthesis of precursor chlorophyll and Rubisco was restrained while proteinaceous compounds were degraded to reallocate N to other plant parts that requires it. Amino acid reallocation from leaves to roots was also observed. With limited N, plant growth and development were impeded which may cause plant dwarfism (Kuang et al. 2017). This indicated that Jatropha has low tolerance against N deficiency. In contrast, P deficiency only caused small reduction of growth due to the efficient ability of Jatropha to redistribute P through dephosphorylation of its organic compounds. While K deficiency was only tolerated by Jatropha during its early development, longer exposure still resulted to reduced growth (Santos et al. 2017). Without proper irrigation and fertilization, Jatropha plantation has been reported to have low yield (Soto et al. 2018). Hence, abiotic stress-tolerant varieties of Jatropha should be developed to successfully demonstrate the economic viability of Jatropha plantation.

14.2 Transcriptome Analysis Approaches in Jatropha

Gene discovery studies in Jatropha are mostly based on microarray analysis and RNA sequencing. Many of the recent works on Jatropha are focused on understanding the molecular and physiological mechanisms of biotic and abiotic stress response in Jatropha as well as identification of key genes that function in response to stress. Aside from these, transcriptome analysis studies in Jatropha aim to address one of the biggest drawbacks in growing Jatropha in the field: low and inconsistent yields.

For example, target traits related to flowering and seed development are also widely studied in this biofuel crop.

The very first reports of transcriptome analysis in Jatropha involved the generation of cDNA expression libraries and EST libraries that were used for functional screening (Eswaran et al. 2010, 2012; Costa et al. 2010). Thereafter, high-throughput transcriptomics using microarray and next-generation sequencing were applied to Jatropha. A 44 K custom oligomicroarray was developed to evaluate gene expression profiles of Jatropha plants subjected to drought stress (Cartagena et al. 2015). Recent works also reported the use of custom 8×60K oligonucleotide microarray developed from Jatropha seeds (Maghuly et al. 2020). While microarray analysis is a practical approach for investigating gene expression in Jatropha, many researchers prefer RNA sequencing as it has the ability to discover novel genes and transcripts. RNA sequencing has been used to identify genes that are involved in Jatropha's response to different abiotic stresses such as drought, salinity, cold, waterlogging, nutrient deficiency, heavy metals, and even high CO_2 conditions. In many cases, transcriptome analysis is used in combination with morphological and physiological examination of stress-treated plants in order to get more detailed information about the adaptation strategies implemented by Jatropha to survive under different levels of stress conditions.

14.2.1 Transcriptome Profiling

Studies that investigate the whole genome transcriptome profile of Jatropha under stressed conditions would usually have objectives that include (1) understanding the molecular basis of stress response; (2) identifying key genes involved in stress response and adaptation; and (3) getting some clues as to how Jatropha plants regulate gene expression in response to stress. Transcriptome profiling works have been reported in Jatropha using different abiotic stressors such as drought, salinity, cold, waterlogging, and nutrient deficiency.

14.2.1.1 Drought

Jatropha is a biofuel crop with inherent drought tolerance (Kheira and Atta 2009; Sapeta et al. 2013), making it a good candidate to study drought response in plants. However, the level of drought tolerance in Jatropha requires further enhancement. For this reason, drought has become the most widely investigated of the abiotic stresses in this oil plant. In terms of transcriptome analysis, both microarray and RNA sequencing has been applied to Jatropha. An oligomicroarray was first developed and used to analyze leaf tissues of Jatropha plants subjected to 3 weeks of non-watering and plants that were allowed to recover (Cartagena et al. 2015). From this study, a total of 2214 differentially expressed genes (DEGs; 875 upregulated and 1339 downregulated) were found to be involved in drought stress response, and 3497 DEGs related to recovery. Several genes identified to be upregulated by drought stress include transcription factors such as *JcNAC050* and *JcDEAR2*, ROS biosynthesis-related *JcRbohD*, and an osmoprotectant *JcGolS2*. On the other hand, the senescence-related gene *JcSRG3*, aquaporins *JcPIP2D*, *JcPIP3*, *JcPIP1C*; and photosynthesis-related genes *JcLHB1B2* and *JcPSBX* showed downregulated expression (Cartagena et al. 2015). Gene ontology analysis of DEGs as illustrated in Fig. 14.1 revealed that gene expression was highly affected in Jatropha when subjected to drought stress and that priority was given to the response to reactive oxygen species (ROS) and oxidative stress. Meanwhile, plant growth was inhibited during stress response as indicated by the downregulation of genes involved in cell wall organization or biogenesis and cell division (Fig. 14.1).

Drought stress in Jatropha has been evaluated by RNA sequencing as reported by Zhang et al. (2015) and Sapeta et al. (2016), using short-term (1, 4, 7 days) and long-term (49 days) drought treatments, respectively. In both studies, comparisons were made between DEGs in roots and leaf tissues, showing that in short-term drought exposure, the numbers, and types of DEGs were different between the two types of tissues. There were higher numbers of DEGs (2900) which are involved in different biological and molecular processes in the leaves, as compared to those in the roots (1533) (Zhang et al. 2015). Aside from genes involved in abscisic acid (ABA) biosynthesis, ABA signal transduction, transcription factors, and osmotic adjustment identified in roots, DEGs related to ER stress responses, ethylene biosynthesis and signal transduction, chlorophyll degradation, photosynthesis, glycolysis and TCA cycle, wax biosynthesis, and fatty acid composition were also detected in leaves (Zhang et al. 2015). For the long-term drought treatment, the number of DEGs observed in roots (1896) and leaf tissues (1710) were comparable, but it was possible to distinguish between organ- and non-organ-specific transcripts (Sapeta et al. 2016). Furthermore, the RNA sequence data were combined with morphological and physiological analysis of Jatropha plants subjected to both moderate and maximum drought stress, thereby recognizing how chlorophyll metabolism was affected by drought stress in Jatropha.

From these three studies, it was shown that Jatropha copes with drought stress by activating ABA signaling pathways in both roots and leaves, coping with the osmotic stress, as well as limiting metabolic processes such as photosynthesis and growth. The gene expression patterns observed in drought-stressed Jatropha is consistent with the proposed drought avoidance mechanism that is morphologically exhibited by Jatropha (Krishnamurthy et al. 2012; Niu et al. 2012; Cartagena et al. 2015). Jatropha is able to survive in dry soil conditions by staying dormant and keeping very minimal metabolic activities.

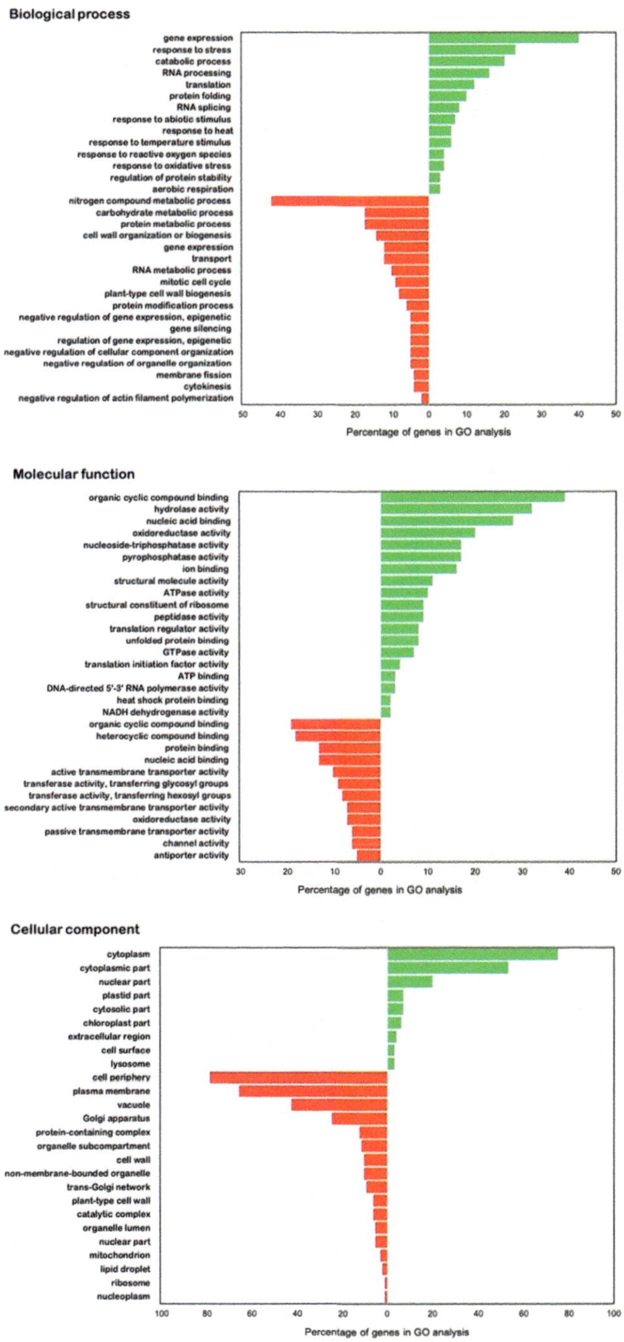

Fig. 14.1 Gene ontology classification of DEGs in Jatropha subjected to 3 weeks of cumulative drought stress

14.2.1.2 Salinity

The growth and productivity of Jatropha plants are significantly affected by salinity, which is comparable to drought condition that results in osmotic stress to plant cells. One of the first studies on transcriptomic analysis in Jatropha involved the generation of expressed sequence tag (EST) libraries as carried out by Eswaran et al. (2012). A total of 1240 ESTs were identified using root tissues of Jatropha seedlings subjected to 2 h salt treatment. Downregulated sequences included those related to metabolic processes, metal binding, structural proteins, and membrane-associated processes. Among the upregulated genes, a significant number of those involved in stress response, storage, and nuclear processes were identified. Subsequently, next-generation sequencing was applied by Zhang et al. (2014) and de Souza et al. (2020) to investigate the effect of salt stress to the gene expression patterns in Jatropha. Zhang and coworkers (2014) used both roots and leaf tissues of Jatropha plants subjected to 2 hours, 2 days, or 7 days NaCl treatment, and found that more DEGs were detected in roots than in leaves, especially in 2-hours and 7-days salt-treated plants. A total of 1504 genes were upregulated and 1115 genes were found to be downregulated, wherein among the DEGs are related to ABA and ethylene signaling, osmotic regulation, ROS scavenging, and cell structure (Zhang et al. 2014).

A more recent transcriptomics study by de Souza et al. (2020) utilized root tissues from two Jatropha accessions, one salt tolerant and the other salt-sensitive, and found 57 and 4646 DEGs, respectively. Genes involved in the metabolism of phytohormones, carbohydrates, amino acids, lipids were the major components of the DEGs. This study showed that the salt-tolerant accession used a protective mechanism to cope with salt stress by regulating phytohormone signaling, ROS detoxification and using compounds that serve as osmoprotectant (de Souza et al. 2020). Interestingly, the same group came up with a TFome (DEGs encoding transcription factors) using the same accessions subjected to salt stress (de Lima Cabral et al. 2020). In general, using root tissues to analyze the effect of salinity stress on Jatropha made more sense as they are the organs that first detect the osmotic stress and this was highlighted in the study by Zhang et al. (2014). It was established by these transcriptomics studies that the key players in salinity stress tolerance in Jatropha involves phytohormone signaling, osmotic regulation, and cell structure adjustment. Furthermore, the next generation sequencing proved to be effective in elucidating a more detailed molecular basis for salinity tolerance in Jatropha.

14.2.1.3 Cold

Jatropha as a tropical shrub is naturally cold-sensitive. To increase the potential of Jatropha as a biofuel crop, the area for suitable plantation should go beyond tropical and subtropical regions. For this reason, Wang et al. (2013) analyzed the gene expression profile of Jatropha plants exposed to 12 °C for 12, 24, or 48 h using RNA sequencing. This study reported that there are 4185 genes including novel genes related to cold resistance in Jatropha. The identified DEGs related to Jatropha cold response include fatty acid unsaturated genes, antioxidative enzymes and antioxidants regenerating related genes, transcripts coding for osmoprotectants and cold-responsive hydrophilic proteins, and those involved in signal transduction (Wang et al. 2013). Overall, a total of 3178 genes were upregulated and 1244 genes were downregulated in response to cold treatment in Jatropha (Wang et al. 2013). This study provided a clear understanding of the molecular mechanism of cold response related to the morphological and physiological data that were previously described in Jatropha.

14.2.1.4 Waterlogging

Jatropha is known to be very sensitive to waterlogging, as reported by several works evaluating how flooding affects the growth and productivity of this biofuel plant (Gimeno et al. 2012; Verma et al. 2014), which is more severe as compared to the effect of drought (Santana et al. 2017). However, limited information is available as to how gene expression patterns are regulated in Jatropha in response to waterlogging. Employing RNA sequencing, the transcriptome profile of Jatropha roots was evaluated after 24 h of growth in waterlogged condition (Juntawong et al. 2014). As a result, a total of 1968 DEGs were identified with the upregulated genes mostly related to hypoxia and anaerobic respiration, while downregulated genes were found to be associated with carbohydrate synthesis, cell wall biogenesis and growth. This study also reported 85 transcription factors that were induced by waterlogging (Juntawong et al. 2014). While Jatropha's response to waterlogging is similar to other crops, it was also possible to distinguish certain Jatropha-specific response by comparing the transcriptome profiles of plants such as Arabidopsis, gray poplar, and rice with Jatropha. Some of the processes specifically found in Jatropha include the non-mevalonate pathway biosynthesis of isoprenoids, response to stress such as heat and touch/wounding, calcium signaling, lipid metabolism, cold stress response, and cell organization (Juntawong et al. 2014).

14.2.1.5 Nutrient Deficiency

Soil nutrients such as nitrogen, phosphorus, and potassium are essential for plant growth and to carry out basic metabolic processes. However, soil quality is easily affected by the ever-changing climate making it more difficult to cultivate crops especially food crops. One of the advantages of using Jatropha as a biofuel crop is that it can be grown in marginal lands such as those with low fertility soil and contaminated areas. While this is true, studies have shown that sufficient amounts of soil nutrients such as nitrogen is necessary to support the growth and productivity of Jatropha (Agusta and Nisya 2015; Santos et al. 2017). Moreover, it was shown that the amount of available nitrogen affects the quality of Jatropha oil (Montenegro et al. 2014). Needless to say, it is essential to gain a full understanding of how

Jatropha regulates gene expression during conditions of nutrient deficiency. An RNA sequence analysis on Jatropha plants subjected to nitrogen starvation for 2 and 16 days reported that gene expression was highly regulated in both roots and leaves (Kuang et al. 2017). Overall, four major classes of genes were shown to be involved in the organ-specific response to nitrogen starvation, specifically those implicated in N uptake, N reutilization, C/N ratio balance, and cell structure and synthesis (Kuang et al. 2017). The intensive transcriptome analysis shed light on the strategies used by Jatropha plants in order to cope with a limited supply of nitrogen. Although nutrient deficiency is one of the biggest challenges faced by Jatropha, studies on its impact on Jatropha growth and productivity remain limited.

14.2.2 Genome-Wide Identification and Functional Analysis of Gene Families

In transcriptome analyses described in the previous section, there are several gene families that have been shown to be consistently associated with abiotic stress response in Jatropha. Another approach for transcriptomics is to carry out genome-wide identification of genes that comprise these families and provide a basis for the roles they play in stress response and/or tolerance in Jatropha. This approach is more straightforward since known gene families are evaluated by analyzing gene expression patterns in response to a particular abiotic stress or a combination of stress conditions. A survey of genome-wide studies to identify stress-responsive gene families showed that the most common categories were transcription factors and those involved in signal transduction (Table 14.1). Most of these gene families are involved in the response to multiple abiotic stresses especially drought and salinity.

In genome-wide studies of gene families, one of the first steps is to perform database searches in order to identify gene sequences and compare them with those from other plants. Based on the sequences, the gene structure and motifs are analyzed, and then chromosome mapping and phylogenetic analysis are carried out. The organ-specific expression profiles of some or all genes are analyzed by qRT-PCR and/or RNA sequencing. One common component of such studies is the evaluation of gene expression levels during exposure to one or multiple abiotic stressors. In some cases, one or a few genes will be analyzed in more detail which can include characterization and functional analysis in Jatropha or in other model plants. Overall, these studies provide foundations for further research on the functions of stress-inducible genes and their application in Jatropha breeding.

Table 14.1 Genome-wide studies to identify gene families that contribute to stress response mechanisms in Jatropha

Gene family	Number of genes identified	Abiotic stress	Reference
Transcription factors			
MYB	128	Salinity, freezing	Peng et al. (2016)
	125	Drought, salinity, nutrient deficiency	Zhou et al. (2015)
HD-ZIP	32	Drought, salinity	Tang et al. (2019a)
ARF	17	Drought, salinity	Tang et al. (2018)
NAC	100	Drought, salinity, nutrient deficiency	Wu et al. (2015)
WRKY	58	Drought, salinity, nutrient deficiency	Xiong et al. (2013)
AP2/ERF	119	Drought, salinity, nutrient deficiency	Tang et al. (2016)
MADS-box	63	Drought, salinity	Tang et al. (2020)
Hsf	17	Drought, salinity	Zhang et al. (2020)
Signal transduction			
HXK	9	Cold	Wang et al. (2019)
MAPK MAPKK MAPKKK	12 5 65	Cold	Wang et al. (2018)

14.3 Application of Jatropha Transcriptomics

14.3.1 Functional Analysis of Stress-Responsive Genes

The increasing number of transcriptome-profiling studies has provided a tremendous amount of information to Jatropha researchers to be utilized for functional analysis of stress-responsive genes. Table 14.2 presents a list of Jatropha genes that have been shown to play significant roles during abiotic stress response and tolerance using a variety of species, including plants, bacteria, and yeast. Majority of these genes include those that function in transcription regulation such as *JcERF2*, *JcWRKY2*, *JcNAC1*, *JcMYB2*, *JcCBF2*, *JcARF*, *JcDREB*, *JcHDZ16*, and *JcR1MYB1*. These transcription factors were shown to play important roles in many biological processes in Jatropha related to growth and development, stress response, and regulation of gene expression.

Both salinity and drought can cause osmotic stress to plant cells making it necessary for transporters and osmoprotectants to actively function in order for our crops to tolerate such stressors and to recover when the environmental conditions improve. Some of the genes that have been analyzed in Jatropha include *JcPIP1*, *JcPIP2*,

Table 14.2 List of Jatropha stress-responsive genes with characterized function in stress tolerance

Gene name	Function in stress response	Organism	Reference
JcBD1	Increased salt tolerance	*E. coli*	Zhang et al. (2008)
JcDREB2	Increased freezing and salt tolerance	Arabidopsis	Tang et al. (2011)
	Increased salt tolerance	Rice	Tang et al. (2017)
JcPIP1 *JcPIP2*	Recovery from drought Early response to drought	Jatropha	Jang et al. (2013)
JcLEA	Increased drought and salt tolerance	Arabidopsis	Liang et al. (2013)
JcMT2a	Heavy metal tolerance	*E. coli*	Mudalkar et al. (2014)
JcAPX	Increased salt tolerance	Tobacco Arabidopsis	Liu et al. (2013) Chen et al. (2015)
JcR1MYB1	Increased salt tolerance	Tobacco	Li et al. (2014)
JcERF1	Increased salt tolerance	Tobacco	Yang et al. (2014)
JcWRKY2	Improved antioxidative status during salinity stress	*E. coli* Tobacco	Agarwal et al. (2014) Dabi et al. (2019)
JcNAC1	Increased drought tolerance	Arabidopsis Jatropha	Qin et al. (2014)
JcSnRK2	Increased drought and salt tolerance	Arabidopsis	Chun et al. (2014)
JcPIP2;7 *JcTIP1;3*	Increased drought and salt tolerance	Arabidopsis	Khan et al. (2015)
JcERF2	Increased drought and salt tolerance	Tobacco	Wang et al. (2015a)
JcCu-ZnSOD	Increased salt tolerance	Arabidopsis	Liu et al. (2015)
JcUEP	Promoter activity under cold, salt and drought	Jatropha Arabidopsis	Tao et al. (2015)
JcCBF2	Increased freezing tolerance Increased drought tolerance	Arabidopsis *N. benthamiana*	Wang et al. (2015b) Wang et al. (2020)
JcAKR	Increased drought and salt tolerance	*E. coli* *S. cerevisiae*	Mudalkar et al. (2016)
JcVP1	Increased salt tolerance	*E. coli* *S. cerevisiae*	Yang et al. (2016)
JcMYB2	Increased freezing and salt tolerance	Arabidopsis	Peng et al. (2016)
JcHDZ16 *JcHDZ07*	Negative regulator of salt stress response	Rice Arabidopsis	Tang et al. (2019a) Tang et al. (2019b)

JcPIP2;7 and *JcTIP1;3*, which code for aquaporins. Membrane intrinsic proteins (PIPs) and tonoplast intrinsic proteins (TIPs) are examples of aquaporins, which are water channels that regulate the flow of water molecules in and out of plant cells ensuring that water balance is maintained (Chaumont and Tyerman 2014). On the other hand, Jatropha genes that are involved in providing osmoprotection to plant cells include *JcBD1* and *JcLEA* (Zhang et al. 2008; Liang et al. 2013). *JcBD1* codes for betaine-aldehyde dehydrogenase which functions in the biosynthesis of the osmoprotectant, glycine betaine (GB). The accumulation of GB was shown in Jatropha plants that were subjected to drought, salt, and heat stress while GB activity was also demonstrated in *E. coli* (Zhang et al. 2008). This is one of the earliest studies of GB in Jatropha that presents valuable information about osmotic stress response in Jatropha. Moreover, a plant-specific protein called LEA, which has been studied in other crops also provide protection to plant cells under stressed conditions. The Jatropha gene *JcLEA* was used to improve drought and salinity tolerance in Arabidopsis by preventing water loss, limiting electrolyte leakage, and maintaining the integrity of the plasma membrane (Liang et al. 2013).

During exposure to abiotic stress, one of the challenges faced by plant cells is to minimize the effect of toxic molecules; thus, genes involved in detoxification or tolerance such as *JcAPX*, *JcAKR*, *JcCu-ZnSOD*, *JcMT2a*, and *JcCBF2* are important players for plant survival. For example, *JcAPX* codes for ascorbate peroxidase, which is responsible for conversion of H_2O_2 to H_2O and O_2 with ascorbate serving as the specific electron donor (Chen et al. 2015). In transgenic Arabidopsis overexpressing *JcAPX*, it showed better survival in salinity conditions and displayed higher APX activity. Furthermore, the endogenous H_2O_2 content was found to be more than 30% lower in transgenic plants when compared to the wild type, giving further evidence that *JcAPX* functions in H_2O_2 scavenging (Chen et al. 2015).

Other important Jatropha genes so far studied in abiotic stress tolerance include those that function in signal transduction, namely *JcSnRK2* and *JcMYB2*, and in ATP production such as *JcVP1*. While *JcSnRK2* enhanced salinity and drought tolerance in Arabidopsis, it was shown that *JcSnRK2* is involved in the immediate response (within 1 hour) to salt and drought stress in Jatropha plants based on its expression pattern (Chun et al. 2014). *JcMYB2*, on the other hand, is involved in regulating the stress response signaling networks by interacting with methyl jasmonate and abscisic acid signaling pathway (Peng et al. 2016).

14.3.2 Generation of Transgenic Plants

One of the ultimate goals of gene discovery by transcriptome analysis in any plant species is to be able to identify candidate genes that can be used for crop improvement. With Jatropha being an undomesticated plant species, the development of abiotic stress tolerant lines will contribute greatly to the immediate and widespread adoption of this biofuel crop for oil production. Creation of transgenic Jatropha will first require the establishment of transformation method, and our group has contributed some work on this aspect (Khemkladngoen et al. 2011; Tsuchimoto et al. 2012) together with other researchers from all over the world. Although several reports on the creation of transgenic Jatropha plants are available, still only a few are aiming for the improvement of abiotic stress tolerance. Many of the target traits in Jatropha include lipid biosynthesis, seed/flower development, inactivation of curcin, virus and insect resistance, among others.

For improvement of abiotic stress tolerance in Jatropha using genetic engineering, the reports so far available utilized genes from other species that have been shown to function in stress tolerance. In an attempt to enhance drought tolerance in

Jatropha, our group introduced two Arabidopsis genes namely *AtNF-YB* and *AtPPAT*, as well as *GSMT* and *DMT* from the cyanobacterium *Synechococcus* (Tsuchimoto et al. 2012). The corresponding genes in Jatropha, *JcNF-YB* and *JcPPAT* genes were introduced into Jatropha and the transgenic plants are currently being evaluated (unpublished). In terms of salt tolerance, transgenic Jatropha plants were developed using the *SbNHX1* gene from an extreme halophyte, *Salicornia brachiata* (Joshi et al. 2013). It is expected that more efforts are being carried out by Jatropha researchers to take the challenge to genetically engineer abiotic stress tolerance in Jatropha.

The application of Jatropha stress-responsive genes for genetic engineering also extends to the development of abiotic stress tolerance in other crops. One such example is the development of transgenic rice using the dehydrin gene, *JcDHN-2* (Samar et al. 2018). Transgenic rice plants exhibited enhanced drought tolerance to rice plant which can be attributed to improved membrane stability and higher activity of osmoregulators, which resulted to better growth than wild type rice. Furthermore, a research group developed transgenic rice lines using Jatropha genes that are involved in salt stress response, namely *JcERF011* (Tang et al. 2016), *JcDREB2* (Tang et al. 2017), and *JcHDZ16* (Tang et al. 2019a). Although the main objective of their work was to characterize and perform functional analysis of Jatropha stress-responsive genes in transgenic rice plants, these steps will contribute to further the application of useful genes from Jatropha for the genetic improvement of important agricultural crops such as rice.

14.4 Conclusion

Jatropha is one of the popular oil crops that has gained attention for its ability to grow in marginal areas. However, its ability to survive comes with a tradeoff of low productivity. Several studies have shown that Jatropha is susceptible to damage caused by abiotic stresses present in marginal land and will benefit from crop improvement. Analyzing the transcriptome dynamics during exposure to stress is one way to understand how Jatropha responds and copes with abiotic stress. The results on gene expression and regulatory pathways from recent literature can contribute to developing improved Jatropha lines as well as the possible application to improving other crops. With climate change being a major challenge for farmers in many parts of the world, it is necessary to prepare crops which are productive and adapted to future climatic conditions. New Jatropha varieties that can tolerate abiotic stresses and produce good-quality oil in high amounts will be highly desirable as they can help alleviate various environmental problems and support a sustainable future society.

References

Agarwal P, Dabi M, Agarwal PK (2014) Molecular cloning and characterization of a group II WRKY transcription factor from *Jatropha curcas*, an important biofuel crop. DNA Cell Biol 33(8):503–513

Agusta H, Nisya FN (2015) *Jatropha curcas* Linn. response on nitrogen deficiency. KnE Energy 2(2):101–105

Allais F, Coqueret X, Farmer T, Raverty W, Rémond C, Paës G (2020) Editorial: from biomass to advanced bio-based chemicals and materials: a multidisciplinary perspective. Front Chem 8:131

An H, Tang Z, Keesstra S, Shangguan Z (2019) Impact of desertification on soil and plant nutrient stoichiometry in a desert grassland. Sci Rep 9:9422

Ao PX, Li ZG, Fan DM, Gong M (2013) Involvement of antioxidant defense system in chill hardening-induced chilling tolerance in *Jatropha curcas* seedlings. Acta Physiol Plant 35:153–160

Arora NK (2019) Impact of climate change on agriculture production and its sustainable solutions. Environ Sustain 2:95–96

Bailey-Serres J, Fukao T, Gibbs DJ, Holdsworth MJ, Lee SC, Licausi F, Perata P, Voesenek LACJ, van Dongen JT (2012) Making sense of low oxygen sensing. Trends Plant Sci 17(3):129–138

Carpita NC, Sage RF (2015) Plants and bioenergy. J Exp Bot 66(14):4093–4095

Cartagena JA (2017) Towards varietal improvement of Jatropha by genetic transformation. In: Tsuchimoto S (ed) The Jatropha Genome. Compendium of plant genomes. Springer, Cham, pp 177–190

Cartagena JA, Seki M, Tanaka M, Yamauchi T, Sato S, Hirakawa H, Tsuge T (2015) Gene expression profiles in Jatropha under drought stress and during recovery. Plant Mol Biol Rep 33:1075–1087

Carvalho CR, Clarindo WR, Praça MM, Araújo FS, Carels N (2008) Genome size, base composition and karyotype of *Jatropha curcas* L., an important biofuel plant. Plant Sci 174(6):613–617

Chaumont F, Tyerman SD (2014) Aquaporins: highly regulated channels controlling plant water relations. Plant Physiol 164(4):1600–1618

Chen Y, Cai J, Yang FX, Zhou B, Zhou LR (2015) Ascorbate peroxidase from *Jatropha curcas* enhances salt tolerance in transgenic Arabidopsis. Genet Mol Res 14(2):4879–4889

Chun J, Li FS, Ma Y, Wang SH, Chen F (2014) Cloning and characterization of a *SnRK2* gene from *Jatropha curcas* L. Genet Mol Res 13(4):10958–10975

Costa GG, Cardoso KC, Del Bem LE, Lima AC, Cunha MA, de Campos-Leite L, Vicentini R, Papes F, Moreira RC, Yunes JA, Campos FA, Da Silva MJ (2010) Transcriptome analysis of the oil-rich seed of the bioenergy crop *Jatropha curcas* L. BMC Genomics 11:462

Dabi M, Agarwal P, Agarwal PK (2019) Functional validation of JcWRKY2, a group III transcription factor toward mitigating salinity stress in transgenic tobacco. DNA Cell Biol 38(11):1278–1291

de Lima Cabral GA, Binneck E, de Souza MCP et al (2020) First expressed TFome of physic nut (*Jatropha curcas* L.) after salt stimulus. Plant Mol Biol Rep. 38:189–208

de Santana TA, da Silva LD, de Oliveira PS, Benjamin CS, Ramos EP, de Souza Jr JO, Gomes FP (2017) Leaf gas exchange and biomass partitioning in *Jatropha curcas* L. young plants subjected to flooding and drought stresses. Aust J Crop Sci 11(7):792–798

de Souza MCP, da Silva MD, Binneck E, Cabral GAL, Iseppon AMB, Pompelli MF, Endres L, Kido EA (2020) RNA-Seq transcriptome analysis of *Jatropha curcas* L. accessions after salt stimulus and unigene-derived microsatellite mining. Ind Crop Prod 147:112168

Degail AC, Chantry J (2013) Developing jatropha projects with smallholder farmers conditions for a sustainable win-win situation for farmers and the project developer. Field Actions Sci Rep 7:1–12

Diaz-Vivancos P, Faize M, Barba-Espin G, Faize L, Petri C, Hernández JA, Burgos L (2013) Ectopic expression of cytosolic superoxide dismutase and ascorbate peroxidase leads to salt stress tolerance in transgenic plums. Plant Biotechnol J 11(8):976–985

Ding Y, Shi Y, Yang S (2019) Advances and challenges in uncovering cold tolerance regulatory mechanisms in plants. New Phytol 222(4):1690–1704

Eccles R, Zhang H, Hamilton D (2019) A review of the effects of climate change on riverine flooding in subtropical and tropical regions. J Water Clim Change 10(4):687–707

Eswaran N, Parameswaran S, Anantharaman B, Kumar GRK, Sathram B, Johnson TS (2012) Generation of an expressed sequence tag (EST) library from salt-stressed roots of *Jatropha curcas* for identification of abiotic stress-responsive genes. Plant Biol 14(3):428–437

Feng X, Liu C, Xie F, Lu J, Chiu LS, Tintera G, Chen B (2019) Precipitation characteristic changes due to global warming in a high-resolution (16 km) ECMWF simulation. Q J R Meteorol Soc 145(718):303–317

Gimeno V, Syvertsen J, Simón I, Nieves M, Díaz-López L, Martínez V, García-Sánchez F (2012) Physiological and morphological responses to flooding with fresh or saline water in *Jatropha curcas*. Environ Exp Bot 78:47–55

Jang HY, Yang SW, Carlson JE, Ku YG, Ahn SJ (2013) Two aquaporins of Jatropha are regulated differentially during drought stress and subsequent recovery. J Plant Physiol 170(11):1028–1038

Jha B, Mishra A, Jha A, Joshi M (2013) Developing transgenic *Jatropha* using the *SbNHX1* gene from an extreme halophyte for cultivation in saline wasteland. PLoS One 8(8):e71136

Jia G, Shevliakova E, Artaxo P, De Noblet-Ducoudré N, Houghton R, House J, Kitajima K, Lennard C, Popp A, Sirin A, Sukumar R, Verchot L (2019) Land–climate interactions. In: Shukla PR, Skea J, Calvo Buendia E, Masson-Delmotte V, Pörtner H-O, Roberts DC, Zhai P, Slade R, Connors S, van Diemen R, Ferrat M, Haughey E, Luz S, Neogi S, Pathak M, Petzold J, Portugal Pereira J, Vyas P, Huntley E, Kissick K, Belkacemi M, Malley J (eds) Climate Change and Land: an IPCC special report on climate change, desertification, land degradation, sustainable land management, food security, and greenhouse gas fluxes in terrestrial ecosystems. In press

Joshi M, Jha A, Mishra A, Jha B (2013) Developing transgenic Jatropha using the SbNHX1 gene from an extreme halophyte for cultivation in saline wasteland. PLoS One 8(8):e71136

Juntawong P, Sirikhachornkit A, Pimjan R, Sonthirod C, Sangsrakru D, Yoocha T, Tangphatsornruang S, Srinives P (2014) Elucidation of the molecular responses to waterlogging in Jatropha roots by transcriptome profiling. Front Plant Sci 5:658

Khan K, Agarwal P, Shanware A, Sane VA (2015) Heterologous expression of two Jatropha Aquaporins imparts drought and salt tolerance and improves seed viability in transgenic *Arabidopsis thaliana*. PLoS One 10(6):e0128866

Kheira AAA, Atta NMM (2009) Response of *Jatropha curcas* L. to water deficit: yield, water use efficiency and oilseed characteristics. Biomass Bioenergy 33:1343–1350

Khemkladngoen N, Cartagena J, Fukui K (2011) Physical wounding-assisted agrobacterium-mediated transformation of juvenile cotyledons of a biodiesel-producing plant, *Jatropha curcas* L. Plant Biotechnol Rep 5:235–243

Krishnamurthy L, Zaman-Allah M, Marimuthu S, Wani SP, Rao AVRK (2012) Root growth in Jatropha and its implications for drought adaptation. Biomass Bioenergy 39:247–252

Kuang Q, Zhang S, Wu P, Chen Y, Li M, Jiang H, Wu G (2017) Global gene expression analysis of the response of physic nut (*Jatropha curcas* L.) to medium- and long-term nitrogen deficiency. PLoS One 12(8):e0182700

Li HL, Guo D, Peng SQ (2014) Molecular characterization of the *Jatropha curcas* JcR1MYB1 gene encoding a putative R1-MYB transcription factor. Genet Mol Biol 37(3):549–555

Liang J, Zhou M, Zhou X, Jin Y, Xu M, Lin J (2013) JcLEA, a novel LEA-like protein from *Jatropha curcas*, confers a high level of tolerance to dehydration and salinity in *Arabidopsis thaliana*. PLoS One 8(12):e83056

Liu Z, Bao H, Cai J, Han J, Zhou L (2013) A novel thylakoid ascorbate peroxidase from Jatrophacurcas enhances salt tolerance in transgenic tobacco. Int J Mol Sci 15(1):171–185

Liu ZB, Zhang WJ, Gong XD, Zhang Q, Zhou LR (2015) A Cu/Zn superoxide dismutase from Jatropha curcas enhances salt tolerance of *Arabidopsis thaliana*. Genet Mol Res 14(1):2086–2098

Liu H, Wang FF, Peng XJ, Huang JH, Shen SH (2019) Global phosphoproteomic analysis reveals the defense and response mechanisms of Jatropha curcas seedling under chilling stress. Int J Mol Sci 20(1):208

Maghuly F, Deák T, Vierlinger K, Pabinger S, Tafer H, Laimer M (2020) Gene expression profiling identifies pathways involved in seed maturation of *Jatropha curcas*. BMC Genomics 21:290

Montenegro RO, Magnitskiy S, Henao TMC (2014) Effect of nitrogen and potassium fertilization on the production and quality of oil in *Jatropha curcas* L. under the dry and warm climate conditions of Colombia. Agronomía Colombiana 32(2):255–265

Mudalkar S, Golla R, Sengupta D, Ghatty S, Reddy AR (2014) Molecular cloning and characterisation of metallothionein type 2a gene from *Jatropha curcas* L., a promising biofuel plant. Mol Biol Rep 41(1):113–124

Mudalkar S, Sreeharsha RV, Reddy AR (2016) A novel aldo-keto reductase from *Jatropha curcas* L. (JcAKR) plays a crucial role in the detoxification of methylglyoxal, a potent electrophile. J Plant Physiol 195:39–49

Niu G, Rodriguez D, Mendoza M, Jifon J, Ganjegunte G (2012) Responses of Jatropha curcas to salt and drought stresses. Int J Agron. https://doi.org/10.1155/2012/632026

Otto FEL (2017) Attribution of weather and climate events. Ann Rev Environ Res 42:8.1–8.20

Peng X, Liu H, Wang D, Shen S (2016) Genome-wide identification of the *Jatropha curcas* MYB family and functional analysis of the abiotic stress responsive gene *JcMYB2*. BMC Genomics 17:251

Pompelli MF, Barata-Luís R, Vitorino HS, Gonçalves ER, Rolim EV, Santos MG, Almeida-Cortez JS, Ferreira VM, Lemos EE, Endres L (2010) Photosynthesis, photoprotection and antioxidant activity of purging nut under drought deficit and recovery. Biomass Bioenergy 34:1207–1215

Qin X, Zheng X, Huang X, Lii Y, Shao C, Xu Y, Chen F (2014) A novel transcription factor JcNAC1 response to stress in new model woody plant *Jatropha curcas*. Planta 239(2):511–520

Samar AO, Nabil IE, Abdelnaser AE, Wacław S, Hazem MK (2018) Over expression of Jatropha's Dehydrin *Jcdhn-2* enhances tolerance to water stress in Rice plants. Environ Anal Eco stud 3(2):53–65

Santana TA, Silva LD, Oliveira PD, Benjamin CS, Ramos EP, Júnior JO, Gomes FP (2017) Leaf gas exchange and biomass partitioning in *Jatropha curcas* L. young plants subjected to flooding and drought stresses. Aust J Crop Sci 11(7):792–798

Santos EF, Macedo FG, Zanchim BJ, Lima GPP, Lavres J (2017) Prognosis of physiological disorders in physic nut to N, P, and K deficiency during initial growth. Plant Physiol Biochem 115:249–258

Sapeta H, Costa JM, Lourenço T, Maroco J, van der Linde P, Oliveira MM (2013) Drought stress response in *Jatropha curcas*: growth and physiology. Environ Exp Bot 85:76–84

Sapeta H, Lourenço T, Lorenz S, Grumaz C, Kirstahler P, Barros PM, Costa JM, Sohn K, Oliveira MM (2016) Transcriptomics and physiological analyses reveal co-ordinated alteration of metabolic pathways in *Jatropha curcas* drought tolerance. J Exp Bot 67(3):845–860

Sato S, Hirakawa H, Isobe S, Fukai E, Watanabe A, Kato M, Kawashima K, Minami C, Muraki A, Nakazaki N, Takahashi C, Nakayama S, Kishida Y, Kohara M, Yamada M, Tsuruoka H, Sasamoto S, Tabata S, Aizu T, Toyoda A, Shin-i T, Minakuchi Y, Kohara Y, Fujiyama A, Tsuchimoto S, Kajiyama S, Makigano E, Ohmido N, Shibagaki N, Cartagena JA, Wada N, Kohinata T, Atefeh A, Yuasa S, Matsunaga S, Fukui K (2011) Sequence analysis of the genome of an oil-bearing tree, *Jatropha curcas* L. DNA Res 18:65–76

Silva EN, Silveira JAG, Rodrigues CRF, Viégas RA (2015) Physiological adjustment to salt stress in *Jatropha curcas* is associated with accumulation of salt ions, transport and selectivity of K^+, osmotic adjustment and K^+/Na^+ homeostasis. Plant Biol 17(5):1023–1029

Soto I, Ellison C, Kenis M, Diaz B, Muys B, Mathijs E (2018) Why do farmers abandon jatropha cultivation? The case of Chiapas, Mexico. Energy Sustain Dev 42:77–86

Tang M, Liu X, Deng H, Shen S (2011) Over-expression of *JcDREB*, a putative AP2/EREBP domain-containing transcription factor gene in woody biodiesel plant *Jatropha curcas*, enhances salt and freezing tolerance in transgenic *Arabidopsis thaliana*. Plant Sci 181(6):623–631

Tang Y, Qin S, Guo Y, Chen Y, Wu P, Chen Y, Li M, Jiang H, Wu G (2016) Genome-wide analysis of the AP2/ERF gene family in physic nut and overexpression of the *JcERF011* gene in Rice increased its sensitivity to salinity stress. PLoS One 11(3):e0150879

Tang Y, Liu K, Zhang J, Li X, Xu K, Zhang Y, Qi J, Yu D, Wang J, Li C (2017) *JcDREB2*, a physic nut AP2/ERF gene, Alters plant growth and salinity stress responses in transgenic rice. Front Plant Sci 8:306

Tang Y, Bao X, Liu K, Wang J, Zhang J, Feng Y, Wang Y, Lin L, Feng J, Li C (2018) Genome-wide identification and expression profiling of the auxin response factor (ARF) gene family in physic nut. PLoS One 13(8):e0201024

Tang Y, Wang J, Bao X, Liang M, Lou H, Zhao J, Sun M, Liang J, Jin L, Li G, Qiu Y, Liu K (2019a) Genome-wide identification and expression profile of HD-ZIP genes in physic nut and functional analysis of the *JcHDZ16* gene in transgenic rice. BMC Plant Biol 19(1):298

Tang Y, Bao X, Wang S, Liu Y, Tan J, Yang M, Zhang M, Dai R, Yu X (2019b) A physic nut stress-responsive HD-zip transcription factor, JcHDZ07, confers enhanced sensitivity to salinity stress in transgenic Arabidopsis. Front Plant Sci 10:942

Tang Y, Wang J, Bao X, Wu Q, Yang T, Li H, Wang W, Zhang Y, Bai N, Guan Y, Dai J, Xie Y, Li S, Huo R, Cheng W (2020) Genome-wide analysis of *Jatropha curcas* MADS-box gene family and functional characterization of the JcMADS40 gene in transgenic rice. BMC Genomics 21(1):325

Tao YB, He LL, Niu LJ, Xu ZF (2015) Isolation and characterization of a ubiquitin extension protein gene (JcUEP) promoter from *Jatropha curcas*. Planta 241(4):823–836

Tsuchimoto S, Cartagena J, Khemkladngoen N, Singkaravanit S, Kohinata T, Wada N, Sakai H, Morishita Y, Suzuki H, Shibata D, Fukui K (2012) Development of transgenic plants in Jatropha with drought tolerance. Plant Biotechnology 29(2):137–143

Verma KK, Singh M, Gupta R, Verma C (2014) Photosynthetic gas exchange, chlorophyll fluorescence, antioxidant enzymes, and growth responses of *Jatropha curcas* during soil flooding. Turk J Bot 38:130–140

Wang H, Zou Z, Wang S, Gong M (2013) Global analysis of transcriptome responses and gene expression profiles to cold stress of *Jatropha curcas* L. PLoS One 8(12):e82817

Wang X, Han H, Yan J, Chen F, Wei W (2015a) A new AP2/ERF transcription factor from the oil plant *Jatropha curcas* confers salt and drought tolerance to transgenic tobacco. Appl Biochem Biotechnol 176(2):582–597

Wang L, Gao J, Qin X, Shi X, Luo L, Zhang G, Yu H, Li C, Hu M, Liu Q, Xu Y, Chen F (2015b) JcCBF2 gene from *Jatropha curcas* improves freezing tolerance of *Arabidopsis thaliana* during the early stage of stress. Mol Biol Rep 42(5):937–945

Wang H, Gong M, Guo J, Xin H, Gao Y, Liu C, Dai D, Tang L (2018) Genome-wide identification of *Jatropha curcas* MAPK, MAPKK, and MAPKKK gene families and their expression profile under cold stress. Sci Rep 8(1):16163

Wang H, Xin H, Guo J, Gao Y, Liu C, Dai D, Tang L (2019) Genome-wide screening of hexokinase gene family and functional elucidation of HXK2 response to cold stress in *Jatropha curcas*. Mol Biol Rep 46(2):1649–1660

Wang L, Wu Y, Tian Y, Dai T, Xie G, Xu Y, Chen F (2020) Overexpressing Jatropha curcas CBF2 in *Nicotiana benthamiana* improved plant tolerance to drought stress. Gene 742:144588

Wu Z, Xu X, Xiong W, Wu P, Chen Y, Li M, Wu G, Jiang H (2015) Genome-wide analysis of the NAC gene family in Physic Nut (*Jatropha curcas* L.). PLoS One 10(6):e0131890

Xiong W, Xu X, Zhang L, Wu P, Chen Y, Li M, Jiang H, Wu G (2013) Genome-wide analysis of the WRKY gene family in physic nut (*Jatropha curcas* L.). Gene 524(2):124–132

Yang H, Yu C, Yan J, Wang X, Chen F, Zhao Y, Wei W (2014) Overexpression of the *Jatropha curcas JcERF1* gene coding an AP2/ERF-type transcription factor increases tolerance to salt in transgenic tobacco. Biochemistry (Mosc) 79(11):1226–1236

Yang Y, Luo Z, Zhang M, Liu C, Gong M, Zou Z (2016) Molecular cloning, expression analysis, and functional characterization of the H^+-pyrophosphatase from *Jatropha curcas*. Appl Biochem Biotechnol 178(7):1273–1285

Zhang FL, Niu B, Wang YC, Chen F, Wang SH, Xu Y, Jiang LD, Gao S, Wu J, Tang L (2008) Jia YJA A novel betaine aldehyde dehydrogenase gene from *Jatropha curcas*, encoding an enzyme implicated in adaptation to environmental stress. Plant Sci 174:510–518

Zhang L, Zhang C, Wu P, Chen Y, Li M, Jiang H, Wu G (2014) Global analysis of gene expression profiles in physic nut (*Jatropha curcas* L.) seedlings exposed to salt stress. PLoS One 9(5):e97878

Zhang C, Zhang L, Zhang S, Zhu S, Wu P, Chen Y, Li M, Jiang H, Wu G (2015) Global analysis of gene expression profiles in physic nut (*Jatropha curcas* L.) seedlings exposed to drought stress. BMC Plant Biol 15:17

Zhang L, Chen W, Shi B (2020) Genome-wide analysis and expression profiling of the heat shock Transcription factor gene family in Physic Nut (*Jatropha curcas* L.). PeerJ 8:e8467

Zhou C, Chen Y, Wu Z, Lu W, Han J, Wu P, Chen Y, Li M, Jiang H, Wu G (2015) Genome-wide analysis of the MYB gene family in physic nut (*Jatropha curcas* L.). Gene 572(1):63–71

Part III
Oil Crop Biotechnology

Chapter 15
Oilseed Crops as the Alternate Source of Omega Fatty Acids: A Paradigm Shift

Sadaf Nazir and Insha Zahoor

Contents

15.1 Introduction

Omega fatty acids are long chain mono/polyunsaturated fatty acids (MUFA/PUFA) and mainly consist of omega-3 fatty acids and omega-6 fatty acids. The bioactive omega-3 fatty acids are eicosapentaenoic acid (EPA), docosahexaenoic acid (DHA), alpha-linolenic acid (ALA), docosapentaenoic acid (DPA), and oleic acid (OA) (Li et al. 2015). And bioactive omega-6 fatty acids include linoleic acid (LA), gamma linolenic acid (GLA), arachidonic acid (AA), and conjugated linoleic acid (CLA) (Ruiz-López et al. 2012). The designations of omega-3 (ω-3) and omega-6 (ω-6) are stated according to the position of double bond from methyl end in carbon chain. The number of double bonds relates to the bioactive properties of omega fatty acids (Gocen et al. 2018). Omega fatty acids are "essential" fatty acids as human physiologic processes are unable to synthesize them, except the fact that they convert dietary alpha-linolenic acid (ALA) into small quantities of eicosapentaenoic (EPA),

S. Nazir (✉)
Department of Food Technology, Institute of Engineering & Technology, Bundelkhand University, Jhansi, Uttar Pradesh, India

I. Zahoor (✉)
Department of Neurology, Henry Ford Hospital, Detroit, MI, USA

H. Tombuloglu et al. (eds.), *Oil Crop Genomics*,
https://doi.org/10.1007/978-3-030-70420-9_15

docosapentaenoic, and docosahexaenoic (DHA) acids. Therefore, omega fatty acids must be supplemented through diet (Miller et al. 2020). Primarily, the dietary source of omega fatty acids is marine sources, especially in the form of fish oils (Azizian et al. 2010). However, recently plant sources have emerged as a sustainable and alternative resource for omega fatty acids. Predominantly, various oil seeds such as flaxseed, chia seed, mustard, walnut, linseed, cottonseed, safflower, corn, soybean, grapeseed, and sunflower seeds have been widely used as sources of omega fatty acids. (Jain 2020; List 2016). A state of shift from marine to plant source is due to the enormous demand for omega fatty acids in due course of time. This demand has eventually led wild marine fish to the brink of extinction. A alter in consumption from marine sources to oilseeds can sufficiently reduce the burden on marine sources. Oilseeds are replenishable and the therapeutic and nutritional value of oilseeds in terms of omega fatty acids is equivalent to marine sources. They contain ALA and LA as precursor for omega-3 and omega-6 fatty acids, respectively (Sabikhi and Sathish Kumar 2012). Oilseeds can ameliorate the sustainability and ensure an ample supply of omega fatty acids (Wilson and Hildebrand 2010).

Omega fatty acids are well recognized for their therapeutic effects. They are known to confer significant nutritional benefits and have diverse effects on physiological processes in human body. In recent years, the consumption of omega fatty acids has appeared as a therapeutic asset to improve the nutritional status and well-being of individuals. The dietary supplementation of omega fatty acid can boost the health status of an individual (Watanabe and Tatsuno 2017). Omega fatty acids are considered to play a fundamental role in neurogenesis and neuroinflammation (Lange 2020). Literature studies report omega fatty acids to affect the development, functioning, and aging of brain. ALA which is the precursor for EPA/DHA is known to play a significant role in the formation of leukotrienes and thromboxanes that affect various physiological functions in human body (Moreau and Kamal-Eldin 2009; Webb 2019). Omega-3 fatty acids are also known to acts as calcium and sodium channel blockers and alleviate the cause of hypertension (Alexander 2014). Interestingly, they also have anti-inflammatory and cardioprotective properties by promoting inflammation resolution due to which they are emerging therapeutic candidates for several inflammatory conditions (Zahoor and Giri 2020). Considering the enormous potential of omega fatty acids, this chapter focuses on oilseeds as a novel source of omega fatty acids and their potential as health food supplement. The chapter discusses oilseeds from molecular and nutritional point of view so as to use oilseeds as a novel approach for the omega fatty acid production.

15.2 Oilseeds: Novel Sources of Omega Fatty Acids

In recent years, numerous oil seed crops have emerged as a novel and alternative source of omega fatty acids. The underutilized oil crops have attracted researchers across the globe for the production of omega fatty acids. The rise in global population has put an enormous pressure on wild marine resources. Therefore, there is an utmost need to switch toward novel sources of omega fatty acids. Overtime, researchers have shifted their focus toward various conventional and unconventional oilseeds for the production of omega fatty acids (Sabikhi and Sathish Kumar 2012).

Based on the intrinsic composition, oil seeds are categorized according to the fatty acid composition as omega-3 and omega-6. Oil crops containing monounsaturated fatty acids (MUFAs) composition include olive, peanut, almond, mustard, apricot kernel, avocado, sesame, rapeseed, canola, and macadamia. And oil crops with polyunsaturated fatty acid (PUFAs) composition include soybean, cottonseed, grapeseed, linseed, safflower, sunflower, corn, and walnut. The unsaturation level of fatty acids sets forth specific functional properties and oxidative stabilities to the oil crops. The perceived functionality, health benefits, sustainability of oil crops have led to their substantial use as a substitute for marine fatty acids (Hui 1996).

Flax is reported to be widely cultivated in Iran and Middle East countries. The oil extracted from flaxseed contains about 60% of ALA and 19% of LA (Chen et al. 2017). Safflower is cultivated in India, Mexico, and Middle East. It is reported to contain about 81% of LA and 0.40 ALA (Lee et al. 2004). Sesame is grown in tropical countries and reports 42% OA and 47% LA (Huang 2005). Corn is reported to contain 58% of LA and 1% ALA (Black et al. 1994). The major fatty acids reported in olive oil are 83% OA and 9% LA (Mailer 2006). However, investigations also report various unconventional seeds as source of omega fatty acids. Examples of such oilseeds are muskmelon seeds, watermelon seeds, pumpkin seeds, kernels of cucumber, sweet chilly seeds (*Capsicum frutescens*), drumstick seeds (*Moringa oleifera*), apple seeds, and Spanish sweet pomegranate seeds (*Punica granatum*) (Datta 1976; Nzikou et al. 2006; Anwar et al. 2007; Tian et al. 2010; Melgarejo and Artes 2000).

15.3 Omega Fatty Acid Composition of Oilseeds

The oilseeds consist of a glycerol molecule bonded to fatty acid chains. The unsaturated fatty acid chains are known as omega fatty acids. Discrete oils extracted from oilseeds consist of different fatty acid chains. The chains vary in the number of carbon atoms and the number of double bonds (Fig. 15.1). The chain length and extent of unsaturation affects the overall quality and chemical behavior of oils. Most fatty acids derived from oilseeds are 18-carbon long such as in canola, soyabean, sunflower, safflower (Hui 1996; McKevith 2005; Sabikhi and Sathish Kumar 2012; List 2016). Longer than 18-carbon fatty acid chains are found in mustard, camelina, and meadowfoam. The predominant omega fatty acids found in oil seeds are oleic acid (18:1; 9, octadecanoic acid), erucic acid (22:1; docosenoic acid), linoleic acid (18:2; 9, 12 octadecanoic acid), linolenic acid (18:3; 9, 12, 15 octadecenoic acid). However, arachidonic acid is present in minor quantities. Canola oil consists of 62% omega-3 fatty acids and 31% omega-6 fatty acids. Sesame 42% omega-3 fatty acids and 44% omega-6 fatty acids. Mustard 68% omega-3 fatty acids and 24% omega-6

Fig. 15.1 Omega fatty acids

fatty acids. Olive contains 77% omega-3 fatty acids and 9% omega-6 fatty acids. Almond contains 73% omega-3 fatty acids and 18% omega-6 fatty acids. Avocado contains 74% omega-3 fatty acids and 14% omega-6 fatty acids. Apricot kernel contains 63% omega-3 fatty acids and 31% omega-6 fatty acids. Peanut contains 49% omega-3 fatty acids and 34% omega-6 fatty acids. Corn contains 25% omega-3 fatty acids and 61% omega-6 fatty acids. Soyabean contains 24% omega-3 fatty acids and 61% omega-6 fatty acids. Walnut contains 24% omega-3 fatty acids and 67% omega-6 fatty acids. Sunflower contains 21% omega-3 fatty acids and 69% omega-6 fatty acids. Safflower contains 13% omega-3 fatty acids and 78% omega-6 fatty acids. Grapeseed contains 17% omega-3 fatty acids and 73% omega-6 fatty acids (Hui 1996; McKevith 2005; Sabikhi and Sathish Kumar 2012; List 2016).

15.4 Synthesis of Series of Omega Fatty Acids

Fatty acid biosynthesis takes place in the plastids of plant cells. Acyl chains are synthesized which majorly lead to the formation of ALA and LA. ALA and LA are 18-carbon chains which are converted to arachidonic acid. The omega fatty acids ALA and LA act as precursors for the synthesis of EPA and DHA inside the human body. The oil crops are rich in ALA (up to 70%) and LA (up to 20%) which leads to subsequent production for the same upon consumption. The EPA and DHA synthesis involve endoplasmic reticulum fatty acid elongation enzyme systems. Figure 15.2 presents the metabolic pathway for the synthesis of EPA and DHA (Webb 2019). The first metabolic step in the synthesis is slow. The enzyme Δ-6 desaturase corresponds for the rate limiting step. The desaturation reactions involves Δ-6 desaturases (Δ-6 -oleoyl(linolenoyl)-CoA desaturase), Δ-5 desaturases (Δ-5-eicosatrienoyl-CoA desaturase), and Δ-4 desaturases The Δ-6 and Δ-5 desaturases are encoded by FADS1 and FADS2 genes. The Δ-6, Δ-5, and Δ-4 desaturases

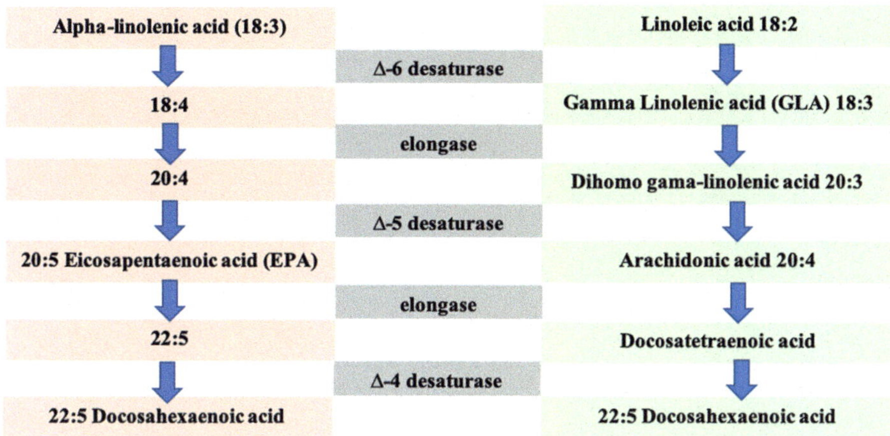

Fig. 15.2 Synthesis of omega fatty acids

catalyze the addition of double bond at carbon 6, 5, and 4 position in the carbon chain, respectively. The final step is a peroxisomal β-oxidation step which leads to synthesis of EPA/DHA (Lee et al. 2016). The end products produced in the pathway inhibit the activity of the enzyme (Δ-6 desaturase). Also, ALA and LA compete for the enzyme Δ-6 desaturase which slows down the metabolism. Therefore, it is inferred that low levels of EPA are produced from oilseed supplementation than EPA levels of fish oils consumed. However, DHA is synthesized from LA and there is no interconversion of EPA and DHA (Webb 2019).

15.5 Genetic Regulation of Omega Fatty Acid Concentration in Oilseeds

The genetic regulation of omega fatty acids in oilseeds is modulated by the lipid metabolism. Lipid metabolism is governed by the glycerol and fatty acid formation. The metabolism involves enzymatic pathway for the synthesis of triacylglycerol (Fig. 15.3). The enzyme Δ-6 desaturase catalyzes the fatty acid synthesis at position Δ9, Δ12, and Δ15. The genes that encode the Δ9, Δ12, and Δ15 desaturase are regulated to mediate the biosynthesis of omega fatty acids (Ohlrogge and Jaworski 1997). FADS genes code for desaturase enzymes involved in the synthesis of PUFA. The role of FADS have been assessed through genomic approaches (Lee et al. 2016). Biotechnologists have engineered the expression of high oleic acid concentration in canola, safflower, palm, peanut, cottonseed, sunflower, and soybean (Wilson and Hildebrand 2010). In addition, oilseed breeding traits are also enhanced by producing hybrid varieties. Such traits are reported to be improved in canola and soybean. Recently, DNA sequencing technologies have been used to

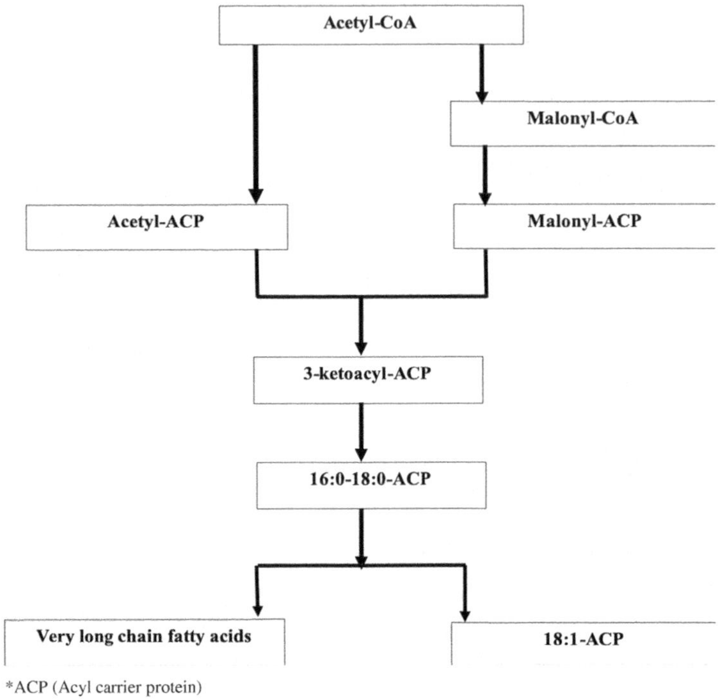

Fig. 15.3 Biosynthesis of fatty acids

reveal genome structure of oilseed crops so as to improve their omega fatty acid concentration (Wilson 2012).

15.6 Extraction of Omega Fatty Acids from Oil Seeds

Various methods are used for the extraction of oil/omega fatty acids from oilseeds. The most widely used method is Soxhlet extraction using solvents such as n-hexane, petroleum ether, and ethanol. A Soxhlet apparatus is used to extract oil from crushed seeds using a solvent by diffusion. The solvent is circulated throughout the process for about 8–10 h. The oil gets dissolved into the solvent and the solvent is later recovered by vacuum evaporation to leave behind a mixture of omega fatty acids. However, Soxhlet extraction is preferred less for its use of chemical solvents as it may pose a health risk. The oil can also be extracted by passing the seeds through a mechanical screw press. The yield of oil obtained from a screw press is lower than that of Soxhlet extraction. In addition, both the above-mentioned processes compromise the quality of oils extracted. This is due to the low oxidative stability of omega fatty acids (Gunstone 2004). Recently, supercritical fluid extraction (SFE) and pressure liquid extraction (PLE) are a better alternative to yield better purity and quality of omega fatty acids (Moreau and Kamal-Eldin 2009; Knowles and Watkinson 2014). SFE and PLE are better alternatives to prevent the thermal degradation of heat labile omega fatty acids. The oilseeds are pretreated, conditioned, and followed by coarse grinding. The bed of oilseeds acts as a solid matrix and carbon dioxide (CO_2) acts as a supercritical fluid. A temperature of 31 °C and a pressure of 74 bar is used to extract the mixture of omega fatty acids from the oil. The CO_2 diffuses into the matrix and dissolves away the oil. The CO_2 and oil are then separated by lowering the pressure to settle out the extracted oil. SFE is preferred for the use of non-toxicity of carbon dioxide (Shahidi and Wanasundara 1998; Melgosa et al. 2019). PLE is also a recent emerging technology that uses temperature and pressure with liquid solvents to efficiently extract omega fatty acids. It is a form of subcritical fluid extraction. It uses less solvent than conventional Soxhlet extraction. Enhanced solvation is used by raising the pressure of the extraction solvent to its critical point (Ruttarattanamongkol et al. 2014).

The extracted oil is later segregated into separate fatty acids. The omega fatty acids may be partially separated by the conventional distillation. Distillation may be fractional or molecular. It separates fatty acids on the basis of boiling points and molecular weights under reduced pressure. Recently, SFE, low temperature crystallization, enzymatic splitting, and urea complexation have been practiced. SFE separates the fatty acids on the basis of degree unsaturation of fatty acids. Low temperature crystallization separates fatty acids on the basis of molecular weights (Shahidi and Wanasundara 1998). The principle of separation involves solubility of fatty acids in organic solvents. The solubility decreases with increasing mean molecular weight and increases with increasing unsaturation. The enzyme lipase is used for the enzymatic splitting of fatty acids. Lipases are also used for the esterification and hydrolysis of fatty acids. And urea complexation involves the splitting of fatty acids in oil and formation of urea–fatty acid adducts. The widely practiced method for the separation of fatty acids is urea complexation. However, enzymatic methods have recently gained momentum (Shahidi and Wanasundara 1998; Knowles and Watkinson 2014).

15.7 Encapsulation of Omega Fatty Acids from Oil Seeds

Omega fatty acids are prone to oxidative deterioration which reduces their bioactive properties. Encapsulation can prevent the oxidation of omega fatty acids and extend their shelf life. It is a process by which omega fatty acids are entrapped within a coating (Martínez Rivas et al. 2017). The entrapped omega fatty acid is known as the core material or capsule and the coating is referred as wall material or encapsulation matrix. The wall materials prevent the core material from environmental effects such as oxygen, light, and humidity. The wall materials used include bio-based polymers such as cellulose, starch, gelatin, alginates, and chitosan (Comunian and Favaro-Trindade 2016). An emulsion of omega fatty acid and wall material is prepared which is further dried by spray drying or freeze drying, or by extrusion

(Kaushik et al. 2015; Geranpour et al. 2020). Sometimes coacervates of omega fatty acids are also prepared by complexing them with proteins (gelatin). These encapsulates are either micro or nano in size. The microencapsulation and nanoencapsulation of these omega fatty acids enhances their bioavailability. Omega fatty acids from flaxseed and sunflower have been successfully encapsulated for the target delivery of omega fatty acids (Chen et al. 2017; Komaiko et al. 2016).

15.8 Commercial Applications of Omega Fatty Acids from Oil Seeds

Commercial applications of omega fatty acids include in nutraceuticals and pharmaceuticals. Various encapsulated omega fatty acids are encapsulated in the form of nutraceuticals to preserve their attributes. Micro and nanoencapsulated omega fatty acids have enhanced bioavailability and target delivery inside human body. Storage and oxidative stability of such encapsulated omega fatty acids is high. These encapsulated nutraceuticals of omega fatty acids are pure bioactive compounds which are often used to fortify foods such as meat, milk, and cereal products. Fortification of food with omega fatty acids is used to improve the health status of individuals (Kaushik et al. 2015; Jamshidi et al. 2020). Omega fatty acids are also used in the production of various pharmaceuticals to treat neurodegenerative disorders, cardiovascular problems, skin-related disorders (photoaging and dermatitis) (Brainard et al. 2020; Huang et al. 2018). Certain cancers are also reported to be treated with the help of omega fatty acids (Farah and Begum 2003). Apart from improving health status certain omega fatty acids find applications in biodiesels (Moser and Vaughn 2010).

15.9 Therapeutic Effects of Omega Fatty Acids from Oilseeds

Omega fatty acids play an important role in various physiological processes in human body. They are recognized as essential in human diet and are therefore known as essential fatty acids. Essential fatty acids lead to eicosanoid production which regulate brain development, anti-inflammatory, cardioprotective, neurogenerative, and neurotransmission properties (Zahoor and Giri 2020). ALA, LA, EPA, DHA, and arachidonic acid act as precursors for the production of eicosanoids. Eicosanoids are regulatory molecules of short duration which exert effects close to their site of duration. They get rapidly inactivated after exerting their effects. Eicosanoids affect regulate secretory processes, immune/inflammatory responses, respiratory/cardiovascular functions (Hwang 2000, Levi et al. 2005, Petrukbina and Makarov 1998). Leukotrienes, prostaglandins, and thromboxane are the eicosanoids

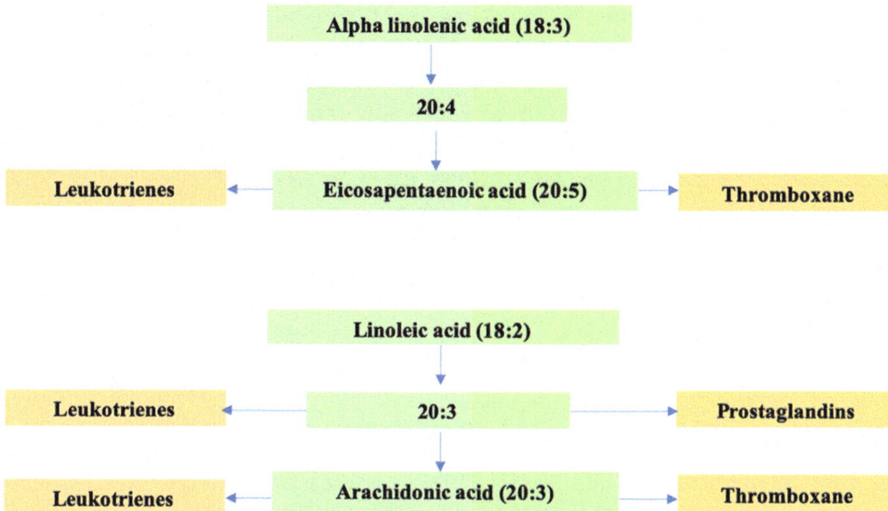

Fig. 15.4 Production of eicosanoids

produced from essential fatty acids. Figure 15.4 shows the production of eicosanoids from essential fatty acids. The eicosanoids are known to promote anti-inflammatory properties, platelet anti-aggregation (Webb 2019). Omega-3 fatty acids play a vital role in prevention of diabetes, certain cancers, hypertension, depression, thrombosis, and atherosclerosis (Verma et al. 2020). These fatty acids are known to regulate hypertension by blocking calcium and sodium channels (Alexander 2014). Electrical activity of heart (arrhythmia) and preconditioning of brain against stroke stresses is also regulated by omega fatty acids. Omega fatty acids are also reported to have beneficial and therapeutic effects in sickle cell disease (Daak et al. 2020).

15.10 Conclusions

Omega fatty acids are the essential fatty acids with numerous health implications, primarily obtained from marine sources especially fish oils. They mainly include oleic acid (OA), erucic acid (EA), linoleic acid (LA), alpha-linolenic acid (ALA), arachidonic acid (AA), eicosapentaenoic acid (EPA), docosapentaenoic acid (DPA), and docosahexaenoic acid (DHA). They significantly affect various physiological processes in human body by producing various mediators. Omega fatty acids produce metabolic by-product precursors such as leukotrienes and thromboxanes which are anti-inflammatory. The protective response has been observed against various neuro and cardio disorders. Numerous oilseeds are widely known for their omega fatty acid content such as canola, flaxseed, and chia. And so oilseeds can serve as a sustainable source of omega fatty acids from oilseed crops. But they predominantly prove to be a rich source of ALA. However, oilseed crops need to be engineered for serving as untapped reservoir of essential fatty acids. Recently, metabolic pathway of oil crops have been genetically modulated to enhance traits for omega fatty acid production (LA, EPA, DHA). The genes that regulate the enzyme desaturase are modulated so as to mediate the biosynthesis of omega fatty acids. Understanding the mechanisms behind genomics of oilseeds can further employ oilseeds for the supplementation of omega fatty acids.

References

Alexander W (2014) Hypertension: is it time to replace drugs with nutrition and nutraceuticals. P & T 39(4):291–295

Anwar F, Hussain AI, Iqbal S, Bhanger MI (2007) Enhancement of the oxidative stability of some vegetable oils by blending with Moringa oleifera oil. Food Chem 103:1181–1191

Azizian H, Kramer JKG, Ehler S, Curtis JM (2010) Rapid quantitation of fish oil fatty acids and their ethyl esters by FT-NIR models. Eur J Lipid Sci Technol 112(4):452–462

Black JM, Nesheim MC, Kinsella JE (1994) Dietary level of maize oil affects growth and lipid composition of Walker 256 carcinosarcoma. Br J Nutr 71:283–294

Brainard JS, Jimoh OF, Deane KHO, Biswas P, Donaldson D, Maas K, Abdelhamid AS, Hooper L, Ajabnoor S, Alabdulghafoor F, Alkhudairy L, Bridges C, Hanson S, Martin N, O'Brien A, Rees K, Song F, Thorpe G, Wang X, Winstanley L (2020) Omega-3, Omega-6, and polyunsaturated fat for cognition: systematic review and meta-analysis of randomized trials. J Am Med Dir Assoc. https://doi.org/10.1016/j.jamda.2020.02.022

Chen F, Fan GQ, Zhang Z, Zhang R, Deng ZY, McClements DJ (2017) Encapsulation of omega-3 fatty acids in nanoemulsions and microgels: impact of delivery system type and protein addition on gastrointestinal fate. Food Res Int 100:387–395

Comunian TA, Favaro-Trindade CS (2016) Microencapsulation using biopolymers as an alternative to produce food enhanced with phytosterols and omega-3 fatty acids: a review. Food Hydrocoll 61:442–457

Daak AA, Lopez-Toledano MA, Heeney MM (2020) Biochemical and therapeutic effects of omega-3 fatty acids in sickle cell disease. Complement Ther Med 52:102482

Datta N (1976) Vegetable oils from some nonconventional sources. Econ Bot 30:298

Farah IO, Begum RA (2003) Effect of Nigella sativa and oxidative stress on the survival pattern of MCF-7 breast cancer cells. Biomed Sci Instrum 39:359–364

Geranpour M, Assadpour E, Jafari SM (2020) Recent advances in the spray drying encapsulation of essential fatty acids and functional oils. Trends in Food Sci Tech 102:71–90

Gocen T, Bayarı SH, Guven MH (2018) Effects of chemical structures of omega-6 fatty acids on the molecular parameters and quantum chemical descriptors. J Mol Struct 1174:142–150

Gunstone FD (2004) The chemistry of oils and fats. Publishing Ltd, Blackwell, ISBN 1-4051-1626-9

Huang LS (2005) Sesame oil. In: Shahidi F (ed) Bailey's industrial oil and fat products, 6th edn. Wiley, New York, pp 537–576

Huang TH, Wang PW, Yang SC, Chou WL, Fang JY (2018) Cosmetic and therapeutic applications of fish oil's fatty acids on the skin. Mar Drugs 16(8):256

Hui YH (1996) Edible oils and fat products: oils and oilseeds. In: Bailey's industrial oil and fat products, vol 2, 5th edn. Wiley, New York

Hwang D (2000) Fatty acids and immune responses – a new perspective in searching for clues to mechanism. Annu Rev Nutr 20:431–456

Jain T (2020) Fatty acid composition of oilseed crops: a review. In: Thakur M, Modi V (eds) Emerging Technologies in Food Science. Springer, Singapore. https://doi.org/10.1007/978-981-15-2556-8_13
Jamshidi A, Cao H, Xiao J, Simal-Gandara J (2020) Advantages of techniques to fortify food products with the benefits of fish oil. Food Res Int 137:109353
Kaushik P, Dowling K, Barrow CJ, Adhikari B (2015) Microencapsulation of omega-3 fatty acids: a review of microencapsulation and characterization methods. J Funct Foods 19:868–881
Knowles J, Watkinson C (2014) Extraction of omega-6 fatty acids from speciality seeds. Lipid Tech 26:107–110
Komaiko J, Sastrosubroto A, McClements DJ (2016) Encapsulation of ω-3 fatty acids in nanoemulsion-based delivery systems fabricated from natural emulsifiers: sunflower phospholipids. Food Chem 203:331–339
Lange KW (2020) Omega-3 fatty acids and mental health. Global Health J 4(1):18–30
Lee YC, Oh SW, Chang J, Kim IH (2004) Chemical composition and oxidative stability of safflower oil prepared from safflower seed roasted with different temperatures. Food Chem 84:1–6
Lee JM, Lee H, Kang S, Park WJ (2016) Fatty acid desaturases, polyunsaturated fatty acid regulation, and biotechnological advances. Nutrients 8(1):23
Levy BD, Bonnans C, Silverman ES et al (2005) Diminished lipoxin biosynthesis in severe asthma. Am J Respir Crit Care Med 172(7):824–830
Li Y, Hruby A, Bernstein AM, Ley SH, Wang DD, Chiuve SE et al (2015) Saturated fat as compared with unsaturated fats and sources of carbohydrates in relation to risk of coronary heart disease: a prospective cohort study. J Am Coll Cardiol 66(14):1538–1548
List GR (2016) Oilseed composition and modification for health and nutrition. In: T. A. B. B. T.-F. D. L. Sanders (ed) Woodhead publishing series in food science, technology and nutrition. Woodhead publishing, Oxford, pp 23–46
Mailer R (2006) Chemistry and quality of olive oil. Primefacts. Profitable and Sustainable Primary Industries. www.dpi.nsw.gov.au/primefacts. ISSN 1832-6668
Martínez Rivas CJ, Tarhini M, Badri W et al (2017) Nanoprecipitation process: from encapsulation to drug delivery. Int J Pharm 532:66–81
McKevith B (2005) Nutritional aspects of oilseeds. Br Nutr Found Nutr Bull 30:13–26
Melgarejo P, Artes F (2000) Total lipid content and fatty acid composition of oilseed from lesser-known sweet pomegranate clones. J Sci Food Agric 80:1452–1454
Melgosa R, Sanz MT, Benito-Román Ó, Illera AE, Beltrán S (2019) Supercritical CO2 assisted synthesis and concentration of monoacylglycerides rich in omega-3 polyunsaturated fatty acids. J CO2 Util 31:65–74
Miller TC, Gutierre RC, Ingham SJM et al (2020) The effect of high doses of ω-3 fatty acid on the structure of the gastrocnemius muscle and on the lipidic profile of Wistar rats submitted to swimming. Nutrition 78:110832
Moreau RA, Kamal-Eldin A (2009) Gourmet and health-promoting speciality oils. Ed. AOCS Press, ISBN 978-189399797-4
Moser BR, Vaughn SF (2010) Coriander seed oil methyl esters as biodiesel fuel: unique fatty acid composition and excellent oxidative stability. Biomass Bioenergy 34:550–558
Nzikou JM, Mvoula-Tsieri M, Matouba E, Ouamba JM, Kapseu C, Parmentier M et al (2006) A study on gumbo seed grown in Congo Brazzaville for its food and industrial applications. Afr J Biotechnol 5:2469–2475
Ohlrogge JB, Jaworski JG (1997) Regulation of fatty acid synthesis. Annu Rev Plant Physiol Plant Mol Biol 48:109–136
Petrukbina GN, Makarov VA (1998) Natural eicosanoids in regulation of blood coagulation. Biochemist 63(1):93–101
Ruiz-López N, Sayanova O, Napier JA, Haslam RP (2012) Metabolic engineering of the omega-3 long chain polyunsaturated fatty acid biosynthetic pathway into transgenic plants. J Exp Bot 63(7):2397–2410

Ruttarattanamongkol K, Siebenhandl-Ehn S, Schreiner M, Petrasch AM (2014) Pilot-scale supercritical carbon dioxide extraction, physico-chemical properties and profile characterization of Moringa oleifera seed oil in comparison with conventional extraction methods. Ind Crop Prod 58:68–67

Sabikhi L, Sathish Kumar, MH (2012) Fatty acid profile of unconventional oilseeds. JB.-Ain Fand NRHenry (ed.); Vol. 67, Academic Press, pp 141–184

Shahidi F, Wanasundara UN (1998) Omega-3 fatty acid concentrates: nutritional aspects and production technologies. Trends in Food Sci Tech 9(6):230–240

Tian HL, Zhan P, Li KX (2010) Analysis of components and study on antioxidant and antimicrobial activities of oil in apple seeds. Int J Food Sci Nutr 61:395–403

Verma ML, Kishor K, Sharma D, Kumar S, Sharma KD (2020) Microbial production of omega-3 polyunsaturated fatty acids. ML Verma & A. K. B. T.-B. P. of B. C. Chandel (eds) pp 293–326

Watanabe Y, Tatsuno I (2017) Omega-3 polyunsaturated fatty acids for cardiovascular diseases: present, past and future. Expert Rev Clin Pharmacol 10:865–873

Webb GP (2019) Food and nutritional analysis | dietary supplements: a classification and overview of uses and efficacy. P Worsfold, C Poole, A Townshend, MBT-E of AS (Third E. Miró (eds)); pp 406–418 Academic Press

Wilson RF (2012) The role of genomics and biotechnology in achieving global food security for high-oleic vegetable oil. J Oleo Sci 61(7):357–367

Wilson RF, Hildebrand DF (2010) Engineering status, challenges and advantages of oil crops. In: Mascia PN, Scheffran J, Widholm JM (eds) In plant biotechnology for sustainable production of energy and co-products, Biotechnology in agriculture and forestry 66. Springer, Berlin, pp 209–259

Zahoor I, Giri S (2020) Specialized pro-resolving lipid mediators: emerging therapeutic candidates for multiple sclerosis. Clinic Rev Allerg Immunol. https://doi.org/10.1007/s12016-020-08796-4

Chapter 16
Genetic Manipulation for Developing Desired Engineered Oil Crops

Insha Nahvi, Thamer AlShammari, Touseef Amna, and Suriya Rehman

Contents

I. Nahvi
Department of Basic Sciences, Preparatory Year Deanship, King Faisal University, Hofuf, Alahsa, Saudi Arabia

T. AlShammari
Department of Genetic Research, Institute for Research & Medical Consultations (IRMC), Imam Abdulrahman Bin Faisal University, Dammam, Saudi Arabia

T. Amna
Department of Biology & Biotechnology, Faculty of Science, Albaha University, Albaha, Saudi Arabia

S. Rehman (✉)
Department of Epidemic Diseases Research, Institute for Research & Medical Consultations (IRMC), Imam Abdulrahman Bin Faisal University, Dammam, Saudi Arabia
e-mail: surrehman@iau.edu.sa

H. Tombuloglu et al. (eds.), *Oil Crop Genomics*,
https://doi.org/10.1007/978-3-030-70420-9_16

16.1 Introduction

Oil crops are considered as one of the most important agricultural items. At present, most of the vegetable oils are utilized for different edible items like margarine spreads, cooking oils, and other canned foods. Oil crops and other food crops are widely being modified which are known as genetically engineered plants and are defined as plants whose DNA is manipulated using different genetic engineering techniques like tissue culture methods, gene recombinant technology, and other gene transfer techniques (Fig. 16.1). Application of such techniques has resulted in the production of vast diversity of genetically modified plants including the oil crops (Murphy 1998). Genetically modified oil crops add a new attribute to the plant

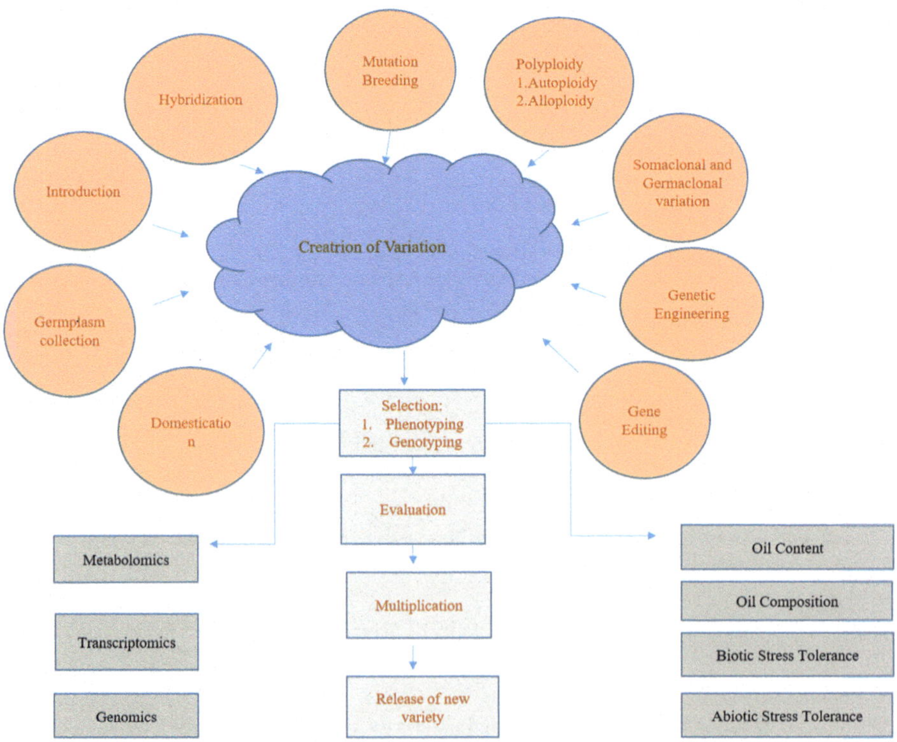

Fig. 16.1 Schematic diagram showing the criteria of genetic engineering for achieving the desired and sustainable oil crops

which is not naturally present in that species. Different traits may be introduced to the plant including resistance to certain bugs or pests and diseases, adaptability to external conditions, resistance to chemicals like tolerance toward herbicides, improvisation of the crop nutrient, etc. Engineered oil crops are designed to have more nutrients or fewer toxins (Diego Villanueva-Mejia and Javier Correa Alvarez 2017; Gasser and Fraley 1989).

Transgenic manipulation is an emerging tool to carry out transgenic breeding of oil plants. This perspective of transgenesis retrieves to a broader gene pool, as the genes may originate from different organisms like insects, viruses, fungi, bacteria, animals, and human beings and even in the laboratory via chemosynthesis (Ma et al. 2003). New approaches are being made in genetic manipulation of plants using biotechnological methods, paying more attention on oil crops like *Arachis hypogaea* commonly called as peanut, *Brassica napus* commonly called as canula, *Glycine max* commonly called as soyabean, *Elaeis guineensis* commonly called as palm, *Ricinus communis* commonly called as castor bean, *Helianthus annuus* commonly called as sunflower, *Gossypium* spp. commonly called as cotton, and *Olea europaea* commonly called as olive (Liu 1997). Agriculture sector has been facing big challenges since decades due to population growth, loss of biodiversity, increased life expectancy, climate change, etc. (Altman 2003; Kumar et al. 2020).

The challenges continue to exist in oil-producing crops with major obstruction being forced by biotic and abiotic stresses. Growing diverse transgenic oil crops which are resistant to abiotic stresses, like dry spell, temperature variations, salinity, and heavy metals, needs lot of efforts. It is also difficult to produce transgenic oil crops having better nutrient competence and good nutritional and processing standards (Margaret et al. 2010; Babar et al. 2020).

16.2 Methods to Obtain Transgenic Oil Crops

Oils produced by oil crops are made up of a variety of fatty acids like stearic acid, palmitic acid, linoleic acid, and lauric acid as well as many more other fatty acids. These oils have food as well as industrial use due to their structure and composition and even in product applications like cosmetics, paints, solvents, ink, soaps, etc. That is why there is a focus on the genetic improvement of these oil crops. Nowadays, transgenesis including genetic engineering techniques and new breeding techniques (NBTs) has gained the utmost importance (Maheshwari and Kovalchuk 2014).

The first step for transgenesis includes carrying the gene of interest, selective marker gene, and promoter and terminator sequences and introduces them to the plants using standard processes and the appropriate gene transfer technique, and the resulting plants are identified by following the phenotypic procedures and different molecular approaches. Various methods for gene transfer includes particle bombardment of soybean (*Glycine max* L.) using meristems as the target tissue (Sato et al. 1993) and microprojectile bombardment of half-shoot apices of sunflower (*Helianthus annuus L.*) that is co-cultured along with *Agrobacterium tumefaciens* for obtaining transgenic shoots (Knittel et al. 1994). From canola seeds some genes like BnLEC1 and BnL1L are overexpressed and then are kept under the control of protein 2S-1 promoter (also called as the napA promoter), at a standard level. This helps in increase of the seed oil content of the transgenic oilseed crop (Tan et al. 2011), PEG-mediated transfection for oil palm genetic engineering (Masani et al. 2014), etc. Other gene transfer methods like microinjection of DNA (Ledoux 1965), liposome encapsulation (Fraley et al. 1980), electroporation (Fromm et al. 1985), pollen-mediated transformation (Sanford et al. 1985), Whiskers method (Frame et al. 1994), *Agrobacterium rhizogenes* (Tepfer 1984), microprojectiles (Klein et al. 1987), and laser microbeam (Weber et al. 1989) are also used for developing transgenic oil crops.

In the recent times, some of the RNA molecules that hamper the gene expression (known as RNA interference, RNAi) or translation by counterbalancing targeted mRNA molecules (Kim and Rossi 2008; Gupta et al. 2013; Younis et al. 2014) and CRISPR/Cas9 system (Wang et al. 2016; Arora and Narula 2017) have widened the scope of genetic engineering (Zhou et al. 2020).

Once the transgenes are introduced, different screenable markers are used to screen the transgenics. Then transformation is assessed using GUS (β-glucuronidase) expression (Rakosy-Tican et al. 2007; Shimada et al. 2010, 2011), which is one among the reporter genes used in the gene transfer method like transformation. Later the transformed cells are placed in X-gluc solution for 1–12 h at 37 °C (in dark). The transgenic nature is indicated by appearance of blue spots. Transgenic tissues are grown in a suitable medium having selective agents (antibiotics) at suitable concentrations for a minimum of two cycles of selection of 2 weeks each. Therefore, the entire plant is regenerated by selected tissues grown on suitable medium. Then these regenerated plants are exposed to phenotypic assessments and molecular analysis (Deom et al. 1990; Guttikonda et al. 2016) using the methods already discussed above. Various other reporter genes like octopine synthase and nopaline synthase identified by electrophoresis and luciferase identified by bioluminescence are also used.

16.3 Techniques to Analyze or Characterize Putative Transgenic Oil Crops

16.3.1 Phenotypic Assays

Many selectable markers, like antibiotics, antimetabolites, hormone biosynthetic genes, and herbicide-resistant genes, are now available (Perl et al. 1993). Untransformed cells should be fully inhibited by the selection agent. Generally, selection agent with lowest concentration is used that suppresses the growth of untransformed cell. Tissue culture conditions, the plant genotype, the developmental stage, and the nature of explants determine the sensitivity of the plant cells to the selection agent. Eventually, the intensity of resistance depends on the control signals of transcription and translation to which the resistance gene is merged.

16.3.2 Polymerase Chain Reaction

PCR can be used to analyze transgenic crops by using (a) primers specific to transgene of interest, (b) plasmid (used as a positive control), and (c) DNA of a non-transformed plant (used as a negative control). PCR actually amplifies a part of the gene only, and the product can only indicate the existence or nonexistence of the transgene. But the problem with PCR is that it does not give the correct information of the transgene copy number.

16.3.3 Southern and Western Blot Hybridization

Southern blotting provides the confirmation of transgene integration and reveals the number of sites at which the transgene is integrated without any amplification. The basic steps involved in the technique are genomic DNA isolation, restriction enzyme digestion of genomic DNA, its treatment with NaOH to make ssDNA, electrophoresis, blotting, hybridization with DNA probe, and autoradiography. Western blotting is used to measure gene expression at the protein level by transgenic plant.

16.3.4 Next-Generation Sequencing Technologies

Nowadays, NGS is a powerful approach for characterization of the transgenic oil crops (Hall et al. 2008). Not only does it help in molecular characterization, but it also helps to calculate the number of copies, to understand how stable the transgene is; characterize the integration site within a host genome; and verify if the vector DNA is absent.

16.3.5 Progeny Analysis/Backcross Breeding

1. Backcross breeding is highly desirable in transgenic plant research to ensure the steady integration and inheritance of transgene. Varieties of basic techniques are being used to determine the zygosity of transgenic oil crops in T_1 generation (Nishat Passricha et al. 2016). This can be done by backcrossing of the putative transgenics.

16.3.6 Bioassay

The last step is to perform bioassay using transgenic plants. The gene that has been introduced is checked. For example, while genetic modification, insect-resistant gene is introduced in the crop, the larvae of target insect are permitted to nourish on these transgenic plants, and the death rate is documented. The transgenic crops having more transgenic expression and causing higher death rate of larvae are examined and then declared for commercial farming.

16.4 Modification of Oil Crops for Agricultural Traits

Adoption of genetically manipulated oil plants is flourishing remarkably since decades not only due to the production of oil but also because of the possibility of their approach in polyhydroxyalkanoates related to waste issues and hazardous effects on our environment. Also there is an application of these oilseed crops in the use of their metabolites in biofuel development (Durrett et al. 2008).

Transgenics in many oil-producing crops carrying useful agricultural traits have emerged successfully. Some of such genetically modified oilseed crops that have proven to be very useful in agronomics have been discussed below.

*16.4.1 Soybean (*Glycine max *L.)*

Oil produced by soybean crop is a mixture of five different fatty acids comprising of stearic acid, oleic acid, palmitic acid, linoleic acid, and linolenic acid. This soybean oil is present in a variety of foods like cooking oils, margarine spread, and industrial residues like plastics and biofuels. Lecithin, a natural emulsifier and lubricant that is extracted from soybean oil, has wide applications in pharmaceutical industry (Durrett et al. 2008). The crop's valuable traits have been improved drastically by genetic manipulation. The first transgenic soybean crop was introduced for commercial production in 1995 where herbicide-tolerant soybean which was resistant to glyphosate was produced (American Soybean Association 2017). An increase in the amount of soybean oil for a particular fatty acid or class of fatty acids was also put in by genetic manipulation. Some researchers even converted the monounsaturated oleic acid to the polyunsaturated linoleic acid by downregulating the expression of *FAD2* genes. Such genetically modified soybean seeds had a high content of oleic acid (around 80%) of the total oil, compared to the typical soybean oil that consists of oleic acid (around 25%) of the total oil (Adgette et al. 1995).

16.4.2 *Palm (*Elaeis guineensis*)*

Palm oil is one among the cheaper cooking oils in many parts of the world due to its extensive applications in food industry. The first ever genetic engineering of oil palm was done in the late 1980s at Malaysian Palm Oil Board (Kinney 1997). Insect pest resistance (*Metisa plana* being the major insect pest) *in* palm crops was achieved by introduction of a synthetic *cryIA(b)* gene into oil palm while biobal-listic that produced proteins which upon crystallization were malignant to Lepidoptera. A very sensitive process was developed where transgenic expressions among putative transformed tissues were combined with RT-PCR and southern blotting techniques. This way one can get rid of the delaying time phase for trans-genic plants to reach maturity (Masani et al. 2014).

16.4.3 *Peanuts (*Arachis hypogaea*)*

Fungal resistance has gained the utmost importance in the research about transgenic peanut oil crops (Lee et al. 2006). Chenault et al. reported that when glucanase from alfalfa and a chitinase from rice were introduced into somatic embryos biolistically, it helped in the development of transgenic peanuts. These researchers concluded that the resulting transgenes could be favorable because of high transgene expres-sion that would exhibit tolerance to a variety of fungal pathogens.

16.5 Genetic Engineering for Development of Insect Resistance in Oil Crops

Bacillus thuringiensis, a bacterium, gave the name Bt crops to the insect-resistant plants as originally the introduced genes were identified in this bacterium. The two ways of developing insect-resistant transgenic plant are either by introducing bacte-rial gene into the crop or by introduction of plant genes for insecticidal proteins. In the former, when the insect larvae consume the crystal protein produced by *Bacillus thuringiensis*, it gets solubilized in the midgut of insect due to alkaline conditions, and then proteases in the midgut produce a protease-resistant polypeptide which is poisonous to the insect. In the latter, the insecticidal proteins of plant origin, like lectins, amylase, and protease inhibitors, can slow down the growth of insect when consumed at increased doses. Researchers have been using genes like CpTi, PIN-1, PIN 11, and GNA in the modification techniques focusing on insect resistance (Chenault et al. 2002).

16.6 Genetic Engineering for Development of Disease Resistance in Oil Crops

16.6.1 Virus Resistance

Infection caused by some of the viruses affects many oil crops around the globe. Pathogen-derived resistance has been exploited nowadays due to genetic modification where the resistance is developed by the viruses' own genetic material. Many attempts have been employed to pathogen-derived resistance over viruses that are serologically different to attain in vivo resistance in many oil-producing crops. Some researchers (Xu et al. 2005; Yang et al. 1998) reported that particle bombardment of genetically modified groundnut consisting of TPWVNP gene resulted in the revival of genetically transformed plants containing single copy of the *TPWVNP* gene exhibiting its proteomic expression. During ex vitro assessments, they (Chenault et al. 2002) analyzed that genetically engineered crops are more tolerant to viral pathogens if compared with control lines. Introduction of coat resistance-mediated protein, satellite RNA-mediated resistance, and antisense-mediated protection has been very effective to transgenic resistance toward different viruses. Among these protections, coat resistance-mediated protein is the most effective.

16.6.2 Fungal Resistance

Fungal diseases are a threat to oil crops and can affect severely to the oil production. Antifungal genes are introduced in genetically modified oil crops to offer resistance against fungal pathogens. Enzymes like chitinase and glucanase that function in degradation of cell wall are used to produce fungal-resistant transgenic oil crops. Although genetic engineering for fungal resistance is limited, still some advances in this field are introduced. One of such advances is antifungal protein-mediated resistance. *Aspergillus giganteus* produces an antifungal protein. This AFP gene confers resistance against fungal pathogens in transgenic oil crops like olive plants. In transgenic olive plants, resistance against fungal pathogens like *Verticillium dahliae* and *Rosellinia necatrix* was evaluated by producing AFP gene under CaMV35S promoter. It was observed that this gene can be helpful in tolerating some fungi like *Rosellinia necatrix* (Yang et al. 2004).

16.6.3 Bacterial Resistance

Although bacterial resistance has not gained much importance in genetic engineering field, its development still has the potential to resist pathogens. Cationic antimicrobial peptides (CAPs) show resistance against bacterial diseases in plants. Both polar and nonpolar qualities give these CAPs the ability to interconnect with the phospholipid membrane that aids in the opening of the phospholipid bilayer followed by collapsing of the transmembrane electrochemical gradients that causes the cell to die (Yang et al. 2004).

16.7 Development of Herbicide-Resistant Oil Crops

Transgenic oil crops have grown interest over the past few decades in the development of resistance against selective and nonselective herbicides (Fast et al. 2020). There are two approaches in the engineering for herbicide resistance:

(a) Modification of plant enzyme or other sensitive biochemical target of herbicide action so that the plant becomes insensitive to that herbicide and thus the plant can be saved
(b) Introduction of an enzyme or enzyme system that detoxifies (detoxification) the herbicide in the plant before it reaches the site of action

2. One of the examples of herbicide resistance is by glufosinate whose tolerance is conferred by the bacterial gene bar, which processes the herbicide into a non-toxic compound (Table 16.1). By introducing agrobacterium gene CP4, the resistance of glyphosate may be achieved (Narwez et al. 2018).

16.8 Development of Plants Resistant to Various Abiotic Stresses

Abiotic stresses like drought, extreme temperature, frost, flood, and salt stress are the alarming environmental situations that affect the yields of almost all agricultural plants.

Table 16.1 Some important herbicide-resistant oil crops, their resistant traits, and trait genes

Crop	Resistant trait	Trait gene	References
Canola	Glyphosate	Cp4 esps and goxv 247	Alan and Earle (2002)
Soybean	Glyphosate	Cp4 esps	Goyal and Mattoo (2014)
Corn	Glyphosate	Three modified Cp4 esps	Maroti et al. (2011)

16.8.1 Drought Tolerance

It has been proven that transgenic oil crops are tolerant to abiotic stresses like drought and salinity when compared with non-transformed crops via different tests (Rana and Rana 2016). One such experiment was done with *Crambe abyssinica*, an oil-producing crop which is used as a moisturizer in cosmetic industry. *C. abyssinica* was subjected to water deficit medium to identify the strategies used by it to tolerate the effects of water stress. The results indicate the ability of this species to control water loss, and consequently photosynthesis and growth were maintained even under drought conditions (Saini and Rana 2018).

16.8.2 Heat Resistance

High temperature has a drastic effect on cellular proteins due to which they lose their structure and function, hence losing their biological activity. In high temperatures, the proteins that are involved with protein degradation processes get activated which in turn reduces the damage to other proteins. Heat shock proteins help in cell protection due to increased temperature. So upregulation of heat shock proteins is a benchmark of heat resistance response.

16.8.3 Salinity Tolerance

Salinity is one of the most widespread abiotic stresses that affect different crops including the oil-producing crops. Growth of transgenic plants in such saline soils and salt stress is really a challenge without compromising the quality of the seed oil and the seed yields. One among such oil crops is transgenic canola plant that is cultivated worldwide. By overexpressing a vacuolar Na^+/H^+ antiport, they were able to grow, flower, and produce seeds in the presence of 200 mM NaCl (Gangadhar et al. 2016).

16.9 Improvement in Nutritional Quality and Oil Production

Desired amount of oil production by oil crops may be achieved by various biotechnological methods with an improvement in the nutritional quality as well. Earlier this was restricted because of the lack of understanding of plant metabolism which involves various metabolic pathways. Metabolic modification of plants is described as the alteration of reactions involving enzymes to build new compounds, upgrade the yield of existing compounds, or degrade undesirable compounds (de Freitas

Table 16.2 List of some transgenic oil crops with nutritionally upgraded features

	Crop (feature detail)	Feature	References
1.	Soybean (amino acid balance)	Amino acid	Zhang et al. (2001)
2.	Canola (Lys up)	Amino acid	Hall et al. (2008)
3.	Palm oil (palmitic acid down, oleic acid up, stearic acid up)	Oils + fatty acids	Rapp (2002) and Falco et al. (1995)
4.	Rapeseed (lauric acid up, GLA up; +ω-3 fatty acids; 8:0 and 10:0 fatty acids up; lauric and myristic acids up; oleic acid up)	Oils + fatty acids	Jalani et al. (1997), Parveez (2003), Roesler et al. (1997), Dehesh et al. (1996), and Del Vecchio (1996)
5.	Flaxseed (+ω-3 and ω-6 fatty acids)	Oils + fatty acids	James et al. (2003)
6.	Rapeseed (vitamin E up)	Vitamins	Agbios (2008)
7.	Soybean	Fructans	Abbadi et al. (2004)

Moura et al. 2018). There are many genetically engineered oil crops that have nutritionally improved traits and improved oil production. A few examples are given in Table 16.2.

As far as health is concerned, monounsaturated fatty acids are chosen over polyunsaturated fatty acids. So genetically modified oil crops produce oils that have ample amount of monounsaturated fatty acids which makes it nutritious for human consumption and also boosts oil stability. Oleic acid, a monounsaturated fatty acid, can provide more stability than the polyunsaturated fatty acids like linoleic acid and linolenic acid. Genetically engineered soybean produced more than 80% oleic acid due to antisense inhibition of oleate desaturase expression (Shintani and DellaPenna 1998).

16.10 Conclusion

Transgenic manipulation is an emerging tool to carry out transgenic breeding of oil crops. Undoubtedly, transgenic oil crops have proven to be a boon for both edible and industrial use by increasing the amount and diversity of required oils. Advantages of genetic engineering in agricultural world are increased oil production, increased crop production, tolerance to pesticides, upgraded nutrients and better food quality, substantial food security, and therapeutic benefits to the world.

Acknowledgments Prof. Ebtesam A. Al-Suhaimi, Dean, Institute for Research and Medical Consultation, IAU, is highly acknowledged for encouraging to write the book chapters.

Competing Interest None.

References

Abbadi A, Domergue F, Bauer J, Napier JA, Welti R, Zahringer U, Cirpus P, Heinz E (2004) Biosynthesis of very-long-chain polyunsaturated fatty acids in transgenic oilseeds: constraints on their accumulation. Plant Cell 16:2734–2748

Adgette SR, Kolacz KH, Delannay X et al (1995) Development, identification, and characterization of a glyphosate-tolerant soybean line. Crop Sci 35:1451

Agbios (2008) GM crop database, March 11. http://www.agbios.com/dbase.php?action=ShowForm

Alan AR, Earle ED (2002) Sensitivity of bacterial and fungal plant pathogens to the lytic peptides, MSI-99, magainin II, and cecropin B. MPMI 15:701–708

Altman A (2003) From plant tissue culture to biotechnology: scientific revolutions, abiotic stress tolerance and forestry. In Vitro Cell Dev Biol Plant 39:75–84

American Soybean Association (2017) SoyStats: a reference guide to important soybean facts and figures. http://soystats.com/wp-content/uploads/17ASA-006_Soy-Stats-2017_1F-web.pdf

Arora L, Narula A (2017) Gene editing and crop improvement using CRISPR-Cas9 system. Front Plant Sci 2017(8). https://doi.org/10.3389/fpls.2017.01932

Babar U, Nawaz MA, Arshad U, Azhar MT, Atif RM, Golokhvast KS, Tsatsakis AM, Shcerbakova K, Chung G, Rana IA (2020) Transgenic crops for the agricultural improvement in Pakistan: a perspective of environmental stresses and the current status of genetically modified crops. GM Crops Food 11(1):1–29. https://doi.org/10.1080/21645698.2019.1680078. Epub 2019 Nov 3. PMID: 31679447; PMCID: PMC7158921

Chenault KD, Burns JA, Melouk HA et al (2002) Hydrolase activity in transgenic peanut. Peanut Sci 29:89–95

de Freitas Moura LM et al (2018) Drought tolerance in potential oilseed plants for biofuel production. Aust J Crop Sci 12(02):289–298

Dehesh K, Jones A, Knutzon DS, Voelker TA (1996) Production of high levels of 8:0 and 10:0 fatty acids in transgenic canola by overexpression of Ch FatB2, a thioesterase cDNA from Cuphea hookeriana. Plant J 9:167–172

Del Vecchio AJ (1996) High-laurate canola: how Calgene's program began, where it's headed. INFORM Int News Fats Oils Relat Mater 7:230–243

Deom CM, Schubert KR, Wolfs S, Holt CA, Lucas WJ, Beachy RN (1990) Molecular characterization and biological function of the movement protein of tobacco mosaic virus in transgenic plants. Proc Natl Acad Sci USA 87:3284–3288

Durrett TP, Benning C, Ohlrogge J (2008) Plant triacylglycerols as feedstocks for the production of biofuels. Plant J 54:593–607

Falco SC, Guida T, Locke M, Mauvais J, Sanders C, Ward RT, Webber P (1995) Transgenic canola and soybean seeds with increased lysine. Bio/Technology 13:577–582. [PubMed]

Fast BJ, Shan G, Gampala SS, Herman RA (2020) Transgene expression in sprayed and non-sprayed herbicide-tolerant genetically engineered crops is equivalent. Regul Toxicol Pharmacol 111:104572. https://doi.org/10.1016/j.yrtph.2019.104572. Epub 2019 Dec 26. PMID: 31884154

Fraley R, Wilschut J, Düzgüneş N, Smith C, Papahadjopoulos D (1980) Studies on the mechanism of membrane fusion: role of phosphate in promoting calcium ion induced fusion of phospholipid vesicles. Biochemistry 19(26):6021–6029

Frame BR, Drayton PR, Bagnall V, Lewnau CJ, Bullock WP, Wilson HM, Dunwell JM, Thompson JA, Wang K (1994) Production of fertile transgenic maize plants by silicon carbide whisker-mediated transformation. Plant J 6(6):941–948

Fromm M, Taylor LP, Walbot V (1985) Expression of genes transferred to monocot and dicot plant cells by electroporation. Proc Natl Acad Sci U S A 82:5824–5828

Gangadhar BH, Sajeesh K, Venkatesh J, Baskar V, Abhinandan K, Yu JW, Prasad R, Mishra RK (2016) Enhanced tolerance of transgenic potato plants over-expressing non-specific lipid transfer protein-1 (*StnsLTP1*) against multiple abiotic stresses. Front Plant Sci 7:1228. https://doi.org/10.3389/fpls.2016.01228

Gasser CS, Fraley RT (1989) Genetically engineering plants for crop improvement. Sci New Ser 244(4910):1293–1299

Goyal RK, Mattoo AK (2014) Multitasking antimicrobial peptides in plant development and host defense against biotic/abiotic stress. Plant Sci Int J Exp Plant Biol 228:135–149. https://doi.org/10.1016/j.plantsci.2014.05.012

Gupta B, Saha J, Sengupta A, Gupta K (2013) Recent advances on virus induced gene silencing (VIGS): plant functional genomics. J Plant Biochem Physiol 1:e116. https://doi.org/10.4172/2329-9029.1000e116

Guttikonda SK, Marri P, Mammadov J, Ye L, Soe K, Richey K et al (2016) Molecular characterization of transgenic events using next generation sequencing approach. PLoS One 11(2):e0149515

Hall RD, Brouwer ID, Fitzgerald MA (2008) Plant metabolomics and its potential application for human nutrition. Plant Physiol 132:162–175

Jalani BS, Cheah SC, Rajanaidu N, Darus A (1997) Improvement of palm oil through breeding and biotechnology. J Am Oil Chem Soc 74:1451–1455

James MJ, Ursin VM, Cleland LG (2003) Metabolism of stearidonic acid in human subjects: comparison with the metabolism of other n-3 fatty acids. Am J Clin Nutr 77:1140–1145

Jewell MC, Campbell BC, Godwin ID (2010) Chapter 2: transgenic plants for abiotic stress resistance. In: Kole C et al (eds) Transgenic crop plants. Springer, Berlin/Heidelberg

Kim DH, Rossi JJ (2008) RNAi mechanisms and applications. BioTechniques 44:613–616

Kinney AJ (1997) Genetic engineering of oilseeds for desired traits. In: Genetic engineering. Springer US, Boston, pp 149–166

Klein TM, Wolf ED, Wu R, Sanford JC (1987) High velocity microprojectiles for delivering nucleic acids into living cells. Nature 327:70–73

Knittel N, Gruber V, Hahne G et al (1994) Transformation of sunflower (*Helianthus annuus* L.): a reliable protocol. Plant Cell Rep 14–14:81–86

Kumar K, Gambhir G, Dass A, Tripathi AK, Singh A, Jha AK, Yadava P, Choudhary M, Rakshit S (2020) Genetically modified crops: current status and future prospects. Planta 251(4):91. https://doi.org/10.1007/s00425-020-03372-8. PMID: 32236850

Ledoux L (1965) Uptake of DNA by living cells. Prog Nucleic Acid Res Mol Biol 4:231–267

Lee M-P, Yeun L-H, Abdullah R (2006) Expression of *Bacillus thuringiensis* insecticidal protein gene in transgenic oil palm. Electron J Biotechnol 9:117–126

Liu K (1997) Soybeans: chemistry, technology, and utilization. Springer, Singapore

Ma JK, Drake PM, Christou P (2003) The production of recombinant pharmaceutical proteins in plants. Nat Rev Genet 4(10):794–805

Maheshwari P, Kovalchuk I (2014) Genetic engineering of oilseed crops. Biocatal Agric Biotechnol 3:31–37

Maroti G et al (2011) Natural roles of antimicrobial peptides in microbes, plants and animals. Res Microbiol 162(4):363–374

Masani MYA, Noll GA, Parveez GKA et al (2014) Efficient transformation of oil palm protoplasts by PEG-mediated transfection and DNA microinjection. PLoS One 9:1–11

Murphy DJ (1998) Impact of genomics on improving the quality of agricultural products. In: Dixon GK, Copping LG, Livingstone D (eds) Genomics: commercial opportunities from a scientific revolution. Society of Chemical Industry, Oxford, pp 199–210

Narwez I et al (2018) Usage of the heterologous expression of the antimicrobial gene *afp* from *Aspergillus giganteus* for increasing fungal resistance in olive. Front Plant Sci 9:680

Parveez GKA (2003) Novel products from transgenic oil palm. AgBiotechNet 113:1–8

Passricha N et al (2016) Assessing zygosity in progeny of transgenic plants: current methods and perspectives. J Biol Methods 3(3):e46

Perl A, Galili S, Shaul O, Ben-Tzvi I, Galili G (1993) Bacterial dihydrodipicolinate synthase and desensitized aspartate kinase: two novel selectable markers for plant transformation. Biotechnology 11:715–718

Rakosy-Tican E, Aurori CM, Dijkstra C, Thieme R, Aurori A, Davey MR (2007) The usefulness of the *gfp* reporter gene for monitoring *Agrobacterium* mediated transformation of potato dihaploid and tetraploid genotypes. Plant Cell Rep 26(5):661–671

Rana SS, Rana MC (2016) Principles and practices of weed management. Department of Agronomy, College of Agriculture, CSK Himachal Pradesh Krishi Vishvavidyalaya, Palampur, 138 p. https://doi.org/10.13140/RG.2.2.33785.47207

Rapp W (2002) Development of soybeans with improved amino acid composition. In: 93rd AOCS Annual Meeting and Expo, May 5–8, 2002. Montreal. American Oil Chemists' Society Press, Champaign, pp 79–86

Roesler K, Shintani D, Savage L, Boddupalli S, Ohlrogge J (1997) Targeting of the Arabidopsis homomeric acetyl-coenzyme A carboxylase to plastids of rapeseeds. Plant Physiol 113:75–81. [PMC free article]

Saini G, Rana S (2018) Transgenic herbicide resistance crops. Lecture 13 Transgenic herbicide resistance crops (Agron 606, Advances in Weed Management). https://doi.org/10.13140/RG.2.2.14194.61127

Sanford JC, Skubik KA, Reisch BI (1985) Attempted pollen-mediated plant transformation employing genomic donor DNA. Theor Appl Genet 69:571–574

Sato S, Newell C, Kolacz K et al (1993) Stable transformation via particle bombardment in two different soybean regeneration systems. Plant Cell Rep 12:408–413

Shimada TL, Shimada T, Hara-Nishimura I (2010) A rapid and nondestructive screenable marker, FAST, for identifying transformed seeds of *Arabidopsis thaliana*. Plant J 61(3):519–528

Shimada T, Ogawa Y, Shimada T, Hara-Nishimura I (2011) A non-destructive screenable marker, OsFAST, for identifying transgenic rice seeds. Plant Signal Behav 6(10):1454–1456

Shintani D, DellaPenna D (1998) Elevating the vitamin E content of plants through metabolic engineering. Science 282:2098–2100

Tan H, Yang X, Zhang F et al (2011) Enhanced seed oil production in canola by conditional expression of *Brassica napus* LEAFY COTYLEDON1 and LEC1-LIKE in developing seeds. Plant Physiol 156:1577–1588

Tepfer D (1984) Transformation of several species of higher plants by *Agrobacterium rhizogenes*: sexual transmission of the transformed genotype and phenotype. Cell 37:959–967

Villanueva-Mejia D, Alvarez JC (2017) Genetic improvement of oilseed crops using modern biotechnology. In: Advances in seed biology. IntechOpen, pp 295–317. https://doi.org/10.5772/intechopen.70743

Wang H, Russa ML, QiL S (2016) CRISPR/Cas9in genome editing and beyond. Annu Rev Biochem 85:227–264

Weber G, Monajembashi S, Wolfrum J, Greulich KO (1989) A laser microbeam as a tool to introduce genes into cells and organelles of higher plants. Ber Bunsenges Phys Chem 93:252–254

Xu QF, Tian F, Chen X, Li LC, Lin ZS, Mo Y et al (2005) Molecular test and aphid resistance identification of a new transgenic wheat line with the GNA gene. J Triticeae Crops 25(3):7–10

Yang H, Singsit C, Wang A, Gonsalves D, Ozias-Akins P (1998) Transgenic peanut plants containing a nucleocapsid protein gene of tomato spotted wilt virus show divergent levels of gene expression. Plant Cell Rep 17:693–699

Yang H, Ozias-Akins P, Culbreath A, Gorbet D, Weeks J, Mandal B (2004) Field evaluation of tomato spotted wilt virus resistance in transgenic Peanut (*Arachis hypogaea*). Plant Dis 88:259–264. Google Scholar

Younis A, Siddique MI, Kim CK, Lim KB (2014) RNA interference (RNAi) induced gene silencing: a promising approach of hi-tech plant breeding. Int J Biol Sci 10(10):1150–1158

Zhang H-X et al (2001) Engineering salt-tolerant *Brassica* plants: characterization of yield and seed oil quality in transgenic plants with increased vacuolar sodium accumulation. PNAS 98(22):12832–12836

Zhou J, Li D, Wang G, Wang F, Kunjal M, Joldersma D, Liu Z (2020) Application and future perspective of CRISPR/Cas9 genome editing in fruit crops. J Integr Plant Biol 62(3):269–286. https://doi.org/10.1111/jipb.12793. Epub 2019 Apr 19. PMID: 30791200; PMCID: PMC6703982

Chapter 17
CRISPR Applications in Crops

Noha Alqahantani, Bayan Alotaibi, Raghdah Alshumrani, Muruj Bamhrez, Turgay Unver, and Huseyin Tombuloglu

Contents

Abbreviations

AMT	*Agrobacterium*-mediated T-DNA delivery
CRISPR	Clustered regularly interspaced short palindromic repeats
crRNA	CRISPR ribonucleic acid
dCas9	Dead CRISPR associated protein 9
gRNA	Guide ribonucleic acid
HDR	Homology-directed repair
NHEJ	Nonhomologous end joining
PAMs	Protospacer-adjacent motifs

N. Alqahantani · B. Alotaibi · R. Alshumrani · M. Bamhrez · H. Tombuloglu (✉)
Department of Genetics Research, Institute for Research and Medical Consultations (IRMC), Imam Abdulrahman Bin Faisal University, Dammam, Saudi Arabia
e-mail: htoglu@iau.edu.sa

T. Unver
Ficus Biotechnology, Ostim Teknopark, Ankara, Turkey

H. Tombuloglu et al. (eds.), *Oil Crop Genomics*,
https://doi.org/10.1007/978-3-030-70420-9_17

RNP	Ribonucleoprotein
tracrRNA	Trans-activating crRNA

17.1 Introduction

The term CRISPR stands for "clustered regularly interspaced short palindromic repeat," and was first discovered in *Streptococcus pyogenes*. It enables bacterial immunity against viral infections (Sapranauskas et al. 2011). CRISPR works along with Cas9 protein to defend against diseases such as viruses and invasive phages and conjugate plasmids by introducing site-specific double-strand breaks in target DNA (Jiang et al. 2017; Khanzadi and Khan 2020). For many decades, scientists were wondering about the repetitive sequence in the prokaryotic DNA, which is globally referred to now as CRISPR. CRISPR identifies an adaptive immune system expressed by microorganisms like archaea and bacteria (Khanzadi and Khan 2020). These microorganisms have a cellular memory captured from past invaders inside their genome. Therefore, as the virus infiltrates the host genome again, the host can recognize and fight back the invader. In such a manner, the acquired sequences allow the prokaryotes to identify pathogens like plasmid or viral attackers as non-self-cells, causing the degradation of the invading series. These processes are recognized as an adaptive immune system for prokaryotes (Thurtle-Schmidt and Lo 2018). This discovery leads the researcher to develop CRISPR\Cas9 technology as a genome-editing tool to modify genes in animal zygotes and human cells, even plant cells, so it is used to regulate, modify, or label genomic loci in a wide range of cells and species.

Moreover, this technology paved the way for basic research in biotechnology and clinical applications to lead outstanding discoveries and innovations (Liang et al. 2015; Doudna and Charpentier 2014). There have been several critical findings that let the CRISPR systems exist as genome-editing technology. The first finding was the observation of high similarity in some specific regions called protospacer-adjacent motifs (PAMs). This sequence is crucial to the CRISPR system (Deveau et al. 2008; Adli 2018). The other findings divulged that a CRISPR system was transportable from one bacterium to multiple bacterial strains (Adli 2018). So, this ability allows using it as a potent RNA-guided DNA targeting platform for genome editing (Jiang et al. 2017).

17.2 Mechanism of CRISPR/Cas9

In the beginning, researchers used the CRISPR system to edit the eukaryotes genomes. Then, they have identified protein Cas9, which has a function in CRISPR as well as in CRISPR RNA (crRNA) and the trans-activating CRISPR RNA (tracrRNA) complex. Purified Cas9 has been used to cut or edit DNA as a template

(Thurtle-Schmidt and Lo 2018; Jinek et al. 2012). This technology relies on the RNA-guided (gRNA) platform, noncoding RNA sequences combined with the Cas9 enzyme to help it by binding to the complementary targeted DNA sequences. Once it is activated, the gRNA as a leader determines the specific site to cut or edit the target genetic site. The capability of cutting and pasting cas9 and gRNA led to a considerable enthusiasm among biologists, as it enables adequate and accurate genome editing in different model organisms.

Furthermore, CRISPR/Cas9 genome-editing technology lets inventing revolutionary tools to help to improve treatment for genetic disorders and to allow easy applications in genome editing. CRISPR/Cas9 is not restricted in genetic fields; it also has many benefits in biomedical studies (Wang et al. 2016a, b, c). CRISPR/Cas9 depended on the DNA repair process and double-strand break (DSB). In the endogenous CRISPR/Cas9 system, mature CRISPR (crRNA) works along with trans activating CRISPR RNA (tracrRNA) to create a tracrRNA::crRNA complex, which guides Cas9 to the targeted site. TracrRNA is moderately complementary to crRNA and participates in crRNA evolution. Then CRISPR/Cas9-mediated gene-specific cleavage at the target site includes a DNA sequence protospacer fitting or matching the crRNA and a small protospacer-adjacent motif (PAM) (Zhang et al. 2014).

Cas9 stimulates two DNA repairing pathways: (i) homology-directed repair (HDR) and the nonhomologous end joining (NHEJ). NHEJ is the most complicated pathway but also the most common one (Fig. 17.1). The broken ends in NHEJ get

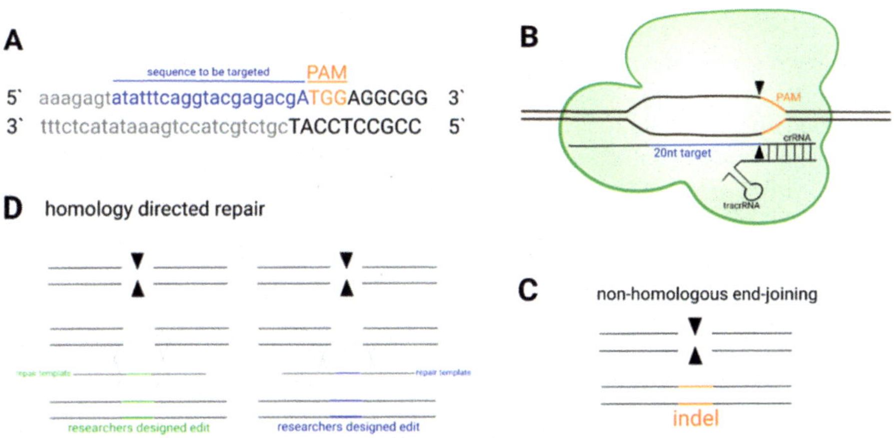

Fig. 17.1 Double-strand breaks repairing pathways. The primary goal depends on Cas9 nuclease cleavage of the target strand. After double-strand break, two different pathways activate in order to repair the broken ends in DNA strands by HDR and NHEJ. (**a**) The sequence of the target genome locus in rapport to PAM (5'-NGG-3′). (**b**) Assembly of gRNA and Cas9 protein causing cleavage in the target gene associated with PAM as an indicator to the target. (**c**) NHEJ results in spontaneous deletions, insertion, and indels. (**d**) HDR implements to repair prototype containing the desired edit in which the HDR cell repair system corrects the triggered double-strand breakage. (Adapted from Thurtle-Schmidt and Lo (2018))

back together through canonical NHEJ (c-NHEJ), a ligate, or basically "glue" that rejoins the fractured ends. Additionally, this rejoining can occur through the alternative end-joining pathway (alt-NHEJ), if one strand of the DNA on each side of the breakage is truncated to repair the lesion (Thurtle-Schmidt and Lo 2018; Wyman and Kanaar 2006; Butt et al. 2020). NHEJ produces minor insertions, deletions, and baseline substitutions and thus is suitable for targeted mutagenesis and targeting functional knockout genes. The HDR needs a repair prototype, from which information is copied into the broken interface (Butt et al. 2020).

17.3 Genome Editing in Plants

The world's population, along with the consumption of food resources, is constantly growing. The population pressure motivates biotechnologists to find alternative solutions. Therefore, there is a severe need to increase agricultural productivity and sustainability. The best way to do so is to find new creative innovations in crop breeding technology to increase crop yields (Chen et al. 2019). CRISPR-based plant-editing strategy depends on targeting the DNA double-strand breaks (DSBs) by using an engineered nuclease, activating cellular DNA repair mechanisms. It mainly uses CRISPR\Cas9, uniquely RNA-guided engineered nucleases, to edit the plant (Bortesi and Fischer 2015). Some of the most important reasons to apply CRISPR/Cas9 technology are that it is easy, cheap, and extremely versatile.

CRISPR/Cas9 is a game-changing technology prepared to revolutionize scientific research and plant production (Belhaj et al. 2015). Recently, researchers have been interested in directed evolution instead of depending on the natural genetic variation of crops by making mutations into the targeted gene. Also, CRISPR-Cas-mediated directed evolution (CDE) to improve resistance to biotic and abiotic stress has been proposed (Butt et al. 2020).

CRISPR/Cas9-based genome editing opens a new era due to its applications in plants. This phenome has contributed to increasing nutrition value, crop productivity, disease resistance, and biotic and abiotic adaptation factors (Zhu et al. 2016). The success rate of CRISPR/Cas9 editing engineering is high due to RNA-guided Cas9 rather than Cas9 protein. Guide RNA is the main factor that dramatically overcomes the limitations in TALENs and zinc finger tools. Some of CRISPR applications are (i) Chromosomal deletion, (ii) activating or repression of specific genes, (iii) mutation at multiple locations, and (iv) genome editing (Baltes et al. 2015). These applications can be performed in monocot and dicot species.

17.4 CRISPR Construct Delivery Methods for Plant Cells

The quality of the gene editing in the genome is significantly affected by the genetic transformation method used. The delivery of CRISPR/Cas9 component to plant cells can be either direct or indirect. Direct delivery methods are particle bombardment and polyethylene glycol (PEG)–based delivery of Cas9 with guide RNA vectors, or RNPs in the protoplast. The indirect delivery method is associated with *Agrobacterium*-mediated T-DNA transfer (Singh and Dhar 2020). In general, *Agrobacterium* or physical methods, including the protoplast-based polyethylene glycol and biolistic callus methods, delivers Cas9 and gRNAs into plant cells (Yin et al. 2015). Transient transformation methods that are most used include *Agrobacterium*-mediated transformation, plant tissue bombardment with DNA-coated gold particles, and the transfection of the protoplast (Pitzschke and Persak 2012).

The CRISPR/Cas9 is transferred to the target cells by particle bombardment or *Agrobacterium*-mediated T-DNA delivery using conventional DNA transformation techniques, choosing a gene as a marker. DNA is incorporated into the plant genome and then expressed to edit the target gene (Fig. 17.2a). An additional delivery tool is achieved by transgene-free editing, which is accomplished by subjecting CRISPR reagents to transient gene expression (Fig. 17.2b). Although subjecting the DNA of CRISPR/Cas9 to transient expression effectively decreases transgenic integration, it cannot eliminate it completely. However, the damaged fragments of the DNA can be incorporated into the genome of the plants. Nevertheless, by particle bombardment, some scientists delivered Cas9 and sgRNAs in vitro into the immature embryos of the wheat, then produced DNA-free edited wheat (Fig. 17.2c) (Zhang et al. 2016). An effective DNA-free genome-editing method was developed by utilizing Cas9/sgRNA RNP in the plant to avoid the drawbacks of Cas9/sgRNA-mediated plasmid and mRNA expression (Fig. 17.2c) (Chen et al. 2019).

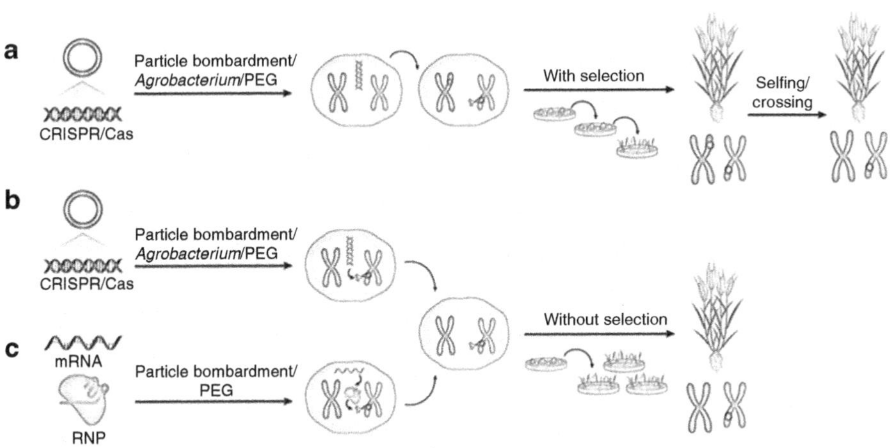

Fig. 17.2 Delivery strategies of the CRISPR/Cas9 system to plants. (**a**) Particle bombardment, *Agrobacterium*, and PEG-mediated delivery methods on selecting the edited plant based on antibiotic or herbicide resistance. Plants free of transgenes can be acquired via self-seeding as well as crossing through genetic segregations. (**b** and **c**) Transient delivery systems. CRISPR/Cas9 reagents involve DNA, RNA, and RNPs. Those reagents will degrade after transient expression. Whereas the plants edited could be renewal without using strain selection

17.4.1 *Agrobacterium-Mediated T-DNA Delivery*

The general approach for the generation of transgenic plants is *Agrobacterium*-mediated T-DNA transformation, which is the most utilized method for delivering DNA cassettes of CRISPR-Cas9 expression to the plant cells. For instance, genome editing by the CRISPR-Cas9 is done exclusively via the floral dip method in the model plant *Arabidopsis*. However, tobacco genome editing starts with the somatic cell transformation via infiltration of *Agrobacterium* into tissues accompanied by T0 plant regeneration utilizing standard tissue cultivation methods. The transformation of matured embryonic rice by the *Agrobacterium, which is* extremely useful, makes this tool a common and successful platform for CRISPR-Cas9 application. Another success with maize embryo transformation is also achieved by *Agrobacterium* (Lowder et al. 2016).

Agrobacterium transformation occurs via activation of plant cell division by a specific gene encoded by DNA fragments called transfer DNA (T-DNA), which resides in the tumor-inducing plasmid (pTi) of *Agrobacterium*. Another segment of the plasmid, namely virulence (*vir*), creates T-complex and facilitates membrane transportation. Exudate from injured plants induces the expression of genes in the *vir* region. Phenolics, such as acetosyringone, are the most active inducers present in wound exudate, but sugars and acidic pH intensify the reaction. T-complex consists of a copy of the single strand from the T-DNA (T-strand) with one of the molecules of VirD2 protein binding covalently to 5′ end. The T-complex transporter mediates the T-complex movement to the cells of the plant from the bacterium, and it is composed of 12 membrane-associated proteins that are specific to virulence. In the plant cell, the T-complex introduced into the nucleus, where the T-strand is present, efficiently incorporated into the plant chromosomes (Zupan et al. 2000).

Applications of this technique in crops include giving resistance to the powdery mildew in wheat (Wang et al. 2014), generating herbicide-resistant flax (Sauer et al. 2016) and maize (Svitashev et al. 2015), increasing tolerance in rice for bispyribac sodium (Endo et al. 2016; Li et al. 2016a, b), generating chlorsulfuron-resistant soybean (Li et al. 2015), developing bacterial blight-resistant grass (Jiang et al. 2013), and increasing blast-resistance of rice (Wang et al. 2016a, b, c).

Despite its common use in plant transformation, *Agrobacterium*-mediated CRISPR applications are minimal due to limiting factors for many other plant species. We expect the utmost attention in improving the successful methods of *Agrobacterium*-based transformation in the rebellious plants. One essential benefit of genome editing using CRISPR/Cas9 systems is that transient expression can often achieve desirable results. CRISPR/Cas9 transgenes do not need to be integrated immediately to the genomes of plants. That characteristic allows different potential options to deliver CRISPR/Cas9 reagent (Lowder et al. 2016). Besides, it is easy to execute leaf infiltration (typically tobacco) with *Agrobacterium*. However, several species of the plants are hostile to this *Agrobacterium* form, which can affect the behavior of many proteins of the plant. That part must be well considered when interpreting the information gathered from stress signaling components in *Agrobacterium*-mediated transformations (Pitzschke and Persak 2012).

17.4.2 Protoplast Transfection

The transfection of the protoplasts involves the addition of extracellular nucleic acid in polyethylene glycol (PEG) or protoplasts mediated by electroporation. PEG-mediated method of transfection is commonly used for the introduction of DNA or RNA into protoplasts (Ren et al. 2020). Expression plasmids could transmit the cassettes of CRISPR/Cas9 to the cells of the plant. It is usually accomplished via PEG protoplast transformation and the delivery of coated particles into the plant cell. Transgenes could be expressed effectively in CRISPR/Cas9, whether they are embedded in chromosomes or expressed incidentally inside the nucleus. Both methods, designed for multiple reasons, were typically versatile in producing the required CRISPR/Cas9 reagents. The protoplast method is primarily utilized for the rapid CRISPR/Cas9 action test within the cells of the plant, so these tests were routinely performed in *Arabidopsis*, tobacco, rice, etc. They can generally be applied almost on every plant (Ren et al. 2020).

On the contrary, biolistic delivery, which depends on gene guns, has been utilized mainly for transforming the tissues of the plant to regenerate plants edited afterward. Gene gun is the gene delivery method based on the formation of charged DNA with gold or tungsten particles aiming to penetrate the plant cell wall. This method has been successfully applied in plenty of plant crops, including rice, maize, wheat, and soybean. It could be difficult and time-consuming to rebuild certain plants using protoplasts and normal somatic cells. Developing plasmid delivery strategies focused on regeneration would be an essential goal for opening up CRISPR/Cas9 transmission to those intransigent species of plant (Lowder et al. 2016).

17.4.3 Particle Bombardment

Particle bombardment is an alternative method of gene transfer in cases where the other methods are inefficient. This is because it can facilitate the delivery of DNA into intact plant cells without biological constraints or host constraints (Angulo-Bejarano et al. 2019). Genome editing of CRISPR/Cas9 ribonucleoprotein complexes (RNPs) by biolistic delivery (also known as particle bombardment) to plant cells eliminates off-target mutations and prevents the incorporation of foreign DNA fragments, thereby providing an effective and accurate tool for precision crop breeding (Liang et al. 2019). The use of biolistic delivery of CRISPR/Cas9 RNPs into embryo cells of maize has been very successful, and plants have been regenerated in nonselective culture conditions. These results indicate that these DNA-free editing methods with CRISPR/Cas9 or RNPs should be applicable to different crop species and provide reliable genome editing for crop improvement (Liang et al. 2018). Particle bombardment is more advantageous than other transformation methods because both organized tissues and single cells can be utilized as transformative goals (Altpeter et al. 2005). The major feature of particle bombardment against *Agrobacterium*-mediated T-DNA delivery is that there are no biological compatibility problems found with that kind of biologic vector. Particle bombardment has been shown to be useful in the plant kingdom for conifer, dicots, and monocot transformation.

17.5 Agricultural Applications of CRISPR/Cas9

Crops are an essential part of our life. It provides the food we eat and fuel we consume as well as other products which are essential for us to live in this world. Population growth brings more demand for crops. For the long term, there are requirements for economic growth to manage crop sustainability and, most importantly, enhance gain productivity. But today's agricultural system cultivators face a lot of stress, such as heat, global warming, soil pollution, pesticides, and insecticides, etc. The CRISPR/Cas9 is the leading biotechnology for advanced breeding, that is why scientists and researchers have developed it. The CRISPR/Cas9 is a very precise and targeted technology that we can use to enhance the efficiency of the products which are put on the market. It has helped us improve the quality of agriculture and make it agronomically sustainable.

Depending on the bacterial adaptive immune system, CRISPR/Cas9 system integrates the endonuclease enzyme that produces double-strand breaks from RNA. By doing so, naturally occurring mutations can be reversed (Corte et al. 2019). For several agricultural applications, CRISPR/Cas9 was used to identify valuable alleles that produce beneficial phenotypes and to apply genome engineering to improve plant properties, such as yield improvements, disease resistance, drought resistance, and pest resistance (Sedeek et al. 2019).

Due to the power and efficiency outcome it provides, CRISPR/Cas9 technology has achieved what the science community had been looking forward to (Govindan and Ramalingam 2016). CRISPR/Cas9 mechanism works as an artificial modification in the genome regardless of whether it is DNA, RNA, or on the protein level. This allows the application to be successfully performed in many species, including crop plants, for which changing and modifying the genome is challenging (van Esse et al. 2020; Feng et al. 2013; Shabir 2020).

The genome-editing tools vary in method and delivery. In addition to CRISPR/Cas9, there are other tools for genome modification that assist our understanding of the engineering applications such as meganucleases, transcription activator-like effector nucleases (TALENs), oligonucleotide-direct mutagenesis (site-directed mutagenesis), and zinc-finger nucleases (ZFNs) (van Esse et al. 2020). These technologies serve for customizing and manipulating the genome target code by adding, deleting, or replacing the cell's DNA sequence. They also enable identifying the map of disease agents in organisms, the function of specific genes, and regulatory elements (Li et al. 2020). After an illustration of what genetic engineering has changed, now it is time to highlight the applications of CRISPR/Cas9 in genome editing in plants. The applications of CRISPR/Cas9 systems in some agricultural plants are summarized in Table 17.1.

17.5.1 CRISPR/Cas9 on Yield Improvement

Yield is a quantitative attribute which is regulated by several genes. The quantitative trait loci (QTLs), which are controlling vital traits in plants, has been demonstrated in the crop improvement program (Jaganathan et al. 2018). Recently, CRISPR-based-QTL editing approach was used to determine grain size (*GS3*) and grain number QTLs (*Gn1a*) in rice in order to identify the rare yield mutations (Shen et al. 2018). In addition, there are four regulator genes (*Gn1a*, *DEP1*, *GS3*, and *IPA1*) that identify a negative regulator yield in rice. Knocking out these genes (*gn1a*, *dep1*, and *gs3*) by CRISPR/Cas9 enhanced yield parameters (improved grain number, dense, erect panicles, and larger grain size) in the T_2 generation (Li et al. 2016a, b). Another CRISPR/Cas9 application was achieved in tomato for parthenocarpy. Knocking out the *SlAGAMOUS-LIKE 6* (*SlAGL6*) gene by CRISPR/Cas9 resulted in parthenocarpy trait in tomato. By exposing it to heat stress, the mutant plant interestingly produced the fruits (Klap et al. 2017). Another study was conducted by Zheng et al. (2020) using CRISPR/Cas9 system in order to improve plant architecture and increase yield in rapeseed (*Brassica napus* L.). Knock-out of double *BnaMAX1* homologs by CRISPR/Cas9-targeted mutagenesis resulted in semi-dwarf phenotypes in rapeseed. Also, increased branching with more siliques enhanced the yield per plant relative to wild type (Zheng et al. 2020).

17.5.2 CRISPR/Cas9 to Improve Disease Resistance

Crops yields are endangered by biotic stresses such as bacteria, viruses, and fungi. These microorganisms impair the plants by carrying infections with the possibility of transmitting to humans and animals as well. CRISPR/Cas9 technology has been used in several genetic modifications to improve plant disease resistance. In citrus, bacterial resistance is acquired by altering the promoter sequence of a gene *CsLOB1* (Jia et al. 2017; Peng et al. 2017). An improved resistance to powdery mildew disease, by targeting the gene *SlMlo1*, is achieved thanks to CRISPR/Cas9 system. They used tomelo-free tomatoes (Nekrasov et al. 2017). In addition, cotton leaf curl virus (*CLCuKoV*) and tomato yellow leaf curl virus (*TYLCV*) inducing cotton leaf curl disease (CLC) can be interfered by CRISPR/Cas9 technology (Borrelli et al. 2018; Wang et al. 2016a, b, c). Five forms of genome editing in grape (*Vitis vinifera*)

Table 17.1 Applications of CRISPR/Cas9 systems in some agricultural plants

Plant type	Targeted Gene	Results	Stress Type	Reference
Camelina sativa	All three *FAD2* homeologs	Decrease polyunsaturated fatty acid concentrations with an improvement in the oleic acid rate	Abiotic stress: Increase polyunsaturated fatty acids	Jiang et al. (2017) and Morineau et al. (2017)
Citrus	*CsLOB1*	Bacterial resistance and increase disease resistance	Biotic stress: bacteria	Jia et al. (2017) and Peng et al. (2017)
Cotton	*MYB25* *FAD2*	Efficient and spesific transformation. Increased oleic acid	-	Chen et al. (2021); Li et al. (2017)
Grape	*WRKY52*	Higher resistance to Botrytis cinerea	biotic stress: *Botrytis cinerea*	Wang (2018)
Maize	replacement of *ARGOS8* with *GOS2* promoter	Increases growth of yield under drought stress	Abiotic stress: drought	Shi et al. (2017)
Maize	*ZmIPK1A*, *ZmIPK* and *ZmMRP4*	PSY1 and mutant carotenoid biosynthesis (psy1) which results in white kernels and albino seedlings	Abiotic stress: Phytic acid production	Liang et al. (2014) and Zhu et al. (2016)
Rice	*OsRAV2*	Salt resistance	Abiotic stress: high Salt concentration	Duan et al. (2016)
Rice	*SBEI, SBEIIb*	Increase amylose concentration	Improve nutritional value	Sun et al. (2017)
Rice	*Gn1a, DEP1, GS3, IPA1*	Strength, improve grain quantity, thick erect panicles, and bigger grain size	Abiotic stress: decrease the yield	Li et al. (2016a, b)
Rice	*OsCYP71A1*	Increase the resistance to planthoppers and stem borers	Biotic stress: Pests	Lu et al. (2018)
Rice	*OsERF922*	Enhanced resistance to mold disease	Biotic stress: fungus	Wang et al. (2016a, b, c)
Tomato	*SlAGL6*	Producing parthenocarpy fruits under heat stress tolerance	Abiotic stress: heat	Klap et al. (2017)
Tomato	*SlMlo1*	Tomelo transgene-free tomato	Biotic stress: fungus	Nekrasov et al. (2017)
Tomato	*TYLCV*	Increase virus resistance	Biotic stress: virus	Ali et al. (2015)
Tomato	*SlMAPK3*	Increase the drought resistance	Abiotic stress: drought	Wang et al. (2017a, b)
Tomato	changing the *T317A* into the *ALC* gene	Increase the drought resistance	Abiotic stress: drought	Yu et al. (2017)

have been identified (Malnoy et al. 2016). The purified CRISPR/Cas9 ribonucleoproteins (RNPs) mechanism based on the delivery of the particles in grape protoplasts is efficient against the powdery mildew vulnerability gene *MLO-7*. The *VvWRKY52* targeted mutagenesis, a transcription factor gene, has elucidated its role in the responses to biotic stress. Additionally, the grape knock-out of *VvWRKY52* improved fungal infection resistance to disease (*Botrytis cinerea*) (Wang et al. 2018).

17.5.3 CRISPR/Cas9 to Increase Drought Tolerance

The CRISPR/Cas9 by gene modifier or suppression increases plant resistance to abiotic stress. The studies have shown that figuring out the regulatory process of *Slmapk3* gene in tomato plants can increase the drought resistance of tomato. The mutated tomato also exhibits high yield with increased drought tolerance (Wang et al. 2017a, b). The shelf-life of tomatoes increases evenly by converting *T317A* into the *ALC* gene (Yu et al. 2017). Genome editing also increases drought resistance in maize, but the *ARGOS8* encodes a negative ethylene response controller and shows low levels. However, researchers replaced the promoter *ARGOS8* with the promoter *GOS2*, which contributed to increased yield under drought stress (Chen et al. 2019). In maize, the modification of *SDN-3* to the specific genomic *Argos8* (also defined as *Zar8*) conferred constitutive transcription of the endogenous gene and increased maize yield in drought conditions (Gao 2018). Moreover, the study finds a gene that is responsible for increasing phytohormone (ABA) drought tolerance in association with other 14 genes that encode receptors binding to the activation (APA) signal, and 13 *PYR1* genes that change the APA passage to minimize yield loss and increase plant growth during drought stress (Scheben et al. 2017). CRISPR/Cas9 system is used to generate the *slnpr1* mutants and to determine the possible regulatory mechanism mediated by *SlNPR1* through analyzing stomatal closure, membrane damage, the activity of antioxidant enzymes, and drought-related gene expression (Li et al. 2019). They compared the drought tolerance of *SlNPR1* mutants (L16, L21, and L62) and wild plants (WT) at physiological and molecular levels. These findings provide details on the fundamental regulating process for the *SlNPR1* regulation of drought in tomato plants.

17.5.4 CRISPR/Cas9 to Improve Resistance Against Pests

A pest is any type of insects, animals, plants, and pathogenic agents or even humans that damage plants or products of plants. CRISPR/Cas9 technology in genetic engineering has been scientifically developed to improve plant resistance to pests. Besides, it is discovered that the distribution of *OsCYP71A1* blocked serotonin biosynthesis that increases salicylic acid, which is important to increase resistance in rice against planthoppers and stem borers (Chen et al. 2019). It is shown that the

CRISPR/Cas9 technology worked well on the *abdominal-A* gene for the abdominal segments of *Plutella xylostella* L., which is a highly insecticide-resistant agricultural pest (Huang et al. 2016). Soybean (*Glycine max*) is among the essential crops of seed oil with a high protein quality. Homologous genetic factor substitution (Avr4/6) with a marker gene (NPT II) induced by the CRISPR/Cas9 system suggested the contribution of the virulence gene for the recognition of the pathogen by plants containing the *R*, *Rps4*, and *Rps6* soy gene loci. Moreover, in the subsequent T2 generation, the CRISPR knockout of the soybean-flowering time gene *GmFT2* was efficiently heritable, with a homozygous *GmFT2* mutant showing late flowering in both long and short-day situations (Cai et al. 2018).

17.6 Conclusion

The CRISPR/Cas9 technology has a major impact on plant breeding by both working on plant characteristics, such as grain size and quality, and generating high mutations that are suitable for the creation of gene alleles. The scientists and researchers highly anticipated generating novel variations in several plant types. Also, the gene editing mediated by CRISPR/Cas9 has significant success in the plants. Besides, the CRISPR/Cas9 technology can increase yield by improving quality through working on germination, development, root growth, and resistance to biotic and abiotic stress.

References

Adli M (2018) The CRISPR tool kit for genome editing and beyond. Nat Commun 9(1):1–1

Ali Z, Abulfaraj A, Idris A, Ali S, Tashkandi M, Mahfouz MM (2015) CRISPR/Cas9-mediated viral interference in plants. Genome Biol 16:238

Ali Z, Ali S, Tashkandi M, Zaidi SSEA, Mahfouz MM (2016) CRISPR/Cas9-mediated immunity to geminiviruses: differential interference and evasion. Sci Rep 6(1):1–13

Altpeter F, Baisakh N, Beachy R, Bock R, Capell T, Christou P, Fauquet C (2005) Particle bombardment and the genetic enhancement of crops: myths and realities. Mol Breed 15(3):305–327

Angulo-Bejarano PI, Sharma A, Paredes-López O (2019) Factors affecting genetic transformation by particle bombardment of the prickly pear cactus (*O. ficus*-indica). 3 Biotech 9(3):98

Baltes NJ, Hsieh TF, Lowder LG, Paul III JW, Qi Y, Tang X, Zhang Y (2015) A CRISPR/Cas9 toolbox for multiplexed plant genome editing and transcriptional regulation. Doctoral dissertation, NC Docks

Belhaj K, Chaparro-Garcia A, Kamoun S, Patron NJ, Nekrasov V (2015) Editing plant genomes with CRISPR/Cas9. Curr Opin Biotechnol 32:76–84

Borrelli VMG, Brambilla V, Rogowsky P, Marocco A, Lanubile A (2018) The enhancement of plant disease resistance using CRISPR/Cas9 technology. Front Plant Sci 9:1245

Bortesi L, Fischer R (2015) The CRISPR/Cas9 system for plant genome editing and beyond. Biotechnol Adv 33(1):41–52

Butt H, Zaidi SSEA, Hassan N, Mahfouz M (2020) CRISPR-based directed evolution for crop improvement. Trends Biotechnol 38(3):236–240

Cai L, Zhang L, Fu Q, Xu ZF (2018) Identification and expression analysis of cytokinin metabolic genes IPTs, CYP735A and CKXs in the biofuel plant Jatropha curcas. PeerJ 6:e4812

Chen K, Wang Y, Zhang R, Zhang H, Gao C (2019) CRISPR/Cas genome editing and precision plant breeding in agriculture. Annu Rev Plant Biol 70:667–697

Chen Y, Fu M, Li H, Wang L, Liu R, Liu Z, et al (2021). High oleic acid content, nontransgenic allotetraploid cotton (Gossypium hirsutum L.) generated by knockout of GhFAD2 genes with CRISPR/Cas9 system. Plant Biotech J 19:424-426

Corte LED, Mahmoud LM, Moraes TS, Mou Z, Grosser JW, Dutt M (2019) Development of improved fruit, vegetable, and ornamental crops using the CRISPR/cas9 genome editing technique. Plan Theory 8(12):1–22

Deveau H, Barrangou R, Garneau JE, Labonté J, Fremaux C, Boyaval P, Moineau S (2008) Phage response to CRISPR-encoded resistance in *Streptococcus thermophilus*. J Bacteriol 190(4):1390–1400

Doudna JA, Charpentier E (2014) The new frontier of genome engineering with CRISPR-Cas9. Science 346(6213)

Duan Y-B, Li J, Qin R-Y, Xu R-F, Li H, Yang Y-C (2016) Identification of a regulatory element responsible for salt induction of rice OSRAV2 through ex situ and in situ promoter analysis. Plant Mol Biol 90:49–62

Endo M, Mikami M, Toki S (2016) Biallelic gene targeting in rice. Plant Physiol 170(2):667–677

Feng Z, Zhang B, Ding W, Liu X, Yang DL, Wei P, Zhu JK (2013) Efficient genome editing in plants using a CRISPR/Cas system. Cell Res 23(10):1229–1232

Gao C (2018) The future of CRISPR technologies in agriculture. Nat Rev Mol Cell Biol 19(5):275–276

Govindan G, Ramalingam S (2016) Programmable site-specific nucleases for targeted genome engineering in higher eukaryotes. J Cell Physiol 231(11):2380–2392

Huang Y, Chen Y, Zeng B, Wang Y, James AA, Gurr GM, Yang G, Lin X, Huang Y, You M (2016) CRISPR/Cas9 mediated knock-out of the abdominal-A homeotic gene in the global pest, diamondback moth (*Plutella xylostella*). Insect Biochem Mol Biol 75:98–106

Jaganathan D, Ramasamy K, Sellamuthu G, Jayabalan S, Venkataraman G (2018) CRISPR for crop improvement: an update review. Front Plant Sci 9(7):1–17

Jia H, Zhang Y, Orbović VXJ, White FF, Jones JB, Wang N (2017) Genome editing of the disease susceptibility gene Cs LOB 1 in citrus confers resistance to citrus canker. Plant Biotechnol J 15(7):817–823

Jiang W, Zhou H, Bi H, Fromm M, Yang B, Weeks DP (2013) Demonstration of CRISPR/Cas9/sgRNA-mediated targeted gene modification in Arabidopsis, tobacco, sorghum and rice. Nucleic Acids Res 41(20):1–12

Jiang WZ, Henry IM, Lynagh PG, Comai L, Cahoon EB, Weeks DP (2017) Significant enhancement of fatty acid composition in seeds of the allohexaploid, *Camelina sativa*, using CRISPR/Cas9 gene editing. Plant Biotechnol J 15:648–657

Jinek M, Chylinski K, Fonfara I, Hauer M, Doudna JA, Charpentier E (2012) A programmable dual-RNA–guided DNA endonuclease in adaptive bacterial immunity. Science 337(6096):816–821

Khanzadi MN, Khan AA (2020) CRISPR/Cas9: nature's gift to prokaryotes and an auspicious tool in genome editing. J Basic Microbiol 60:91–102

Klap C, Yeshayahou E, Bolger AM, Arazi T, Gupta SK, Shabtai S, Usadel B, Salts Y, Barg R (2017) Tomato facultative parthenocarpy results from SlAGAMOUS-LIKE 6 loss of function. Plant Biotechnol J 15:634–647

Li Z, Liu ZB, Xing A, Moon BP, Koellhoffer JP, Huang L, Ward RT, Clifton E, Falco SC, Cigan AM (2015) Cas9-guide RNA directed genome editing in soybean. Plant Physiol 169(2):960–970

Li J, Meng X, Zong Y, Chen K, Zhang H, Liu J, Li J, Gao C (2016a) Gene replacements and insertions in rice by intron targeting using CRISPR–Cas9. Nat Plants 2(10):1–6

Li M, Li X, Zhou Z, Wu P, Fang M, Pan X (2016b) Reassessment of the four yield-related genes Gn1a, DEP1, GS3, and IPA1 in rice using a CRISPR/Cas9 system. Front Plant Sci 7:377

Li C, Unver T, Zhang B (2017) A high-efficiency CRISPR/Cas9 system for targeted mutagenesis in Cotton (Gossypium hirsutum L.). Sci Rep 7:43902

Li R, Liu C, Zhao R, Wang L, Chen L, Yu W, Zhang S, Sheng J, Shen L (2019) CRISPR/Cas9-mediated SlNPR1 mutagenesis reduces tomato plant drought tolerance. BMC Plant Biol 19(1):1–13

Li H, Yang Y, Hong W, Huang M, Wu M, Zhao X (2020) Applications of genome editing technology in the targeted therapy of human diseases: mechanisms, advances and prospects. In: Signal transduction and targeted therapy (vol 5, issue 1). Nature Publishing Group, London, pp 1–23

Liang Z, Zhang K, Chen K, Gao C (2014) Targeted mutagenesis in Zea mays using TALENs and the CRISPR/Cas system. J Genet Genom 41:63–68

Liang P, Xu Y, Zhang X, Ding C, Huang R, Zhang Z et al (2015) CRISPR/Cas9-mediated gene editing in human tripronuclear zygotes. Prot Cell 6(5):363–372

Liang Z, Chen K, Zhang Y, Liu J, Yin K, Qiu JL, Gao C (2018) Genome editing of bread wheat using biolistic delivery of CRISPR/Cas9 in vitro transcripts or ribonucleoproteins. Nat Protoc 13(3):413–430

Liang Z, Chen K, Gao C (2019) Biolistic delivery of CRISPR/Cas9 with ribonucleoprotein complex in wheat. In: Qi Y (ed) Plant genome editing with CRISPR systems, Methods in Molecular Biology, vol 1917. Humana Press, New York

Lowder L, Malzahn A, Qi Y (2016) Rapid evolution of manifold CRISPR systems for plant genome editing. Front Plant Sci 7:1683

Lu HP, Luo T, Fu HW, Wang L, Tan YY et al (2018) Resistance of rice to insect pests mediated by suppression of serotonin biosynthesis. Nat Plants 4:338–344

Malnoy M, Viola R, Jung MH, Koo OJ, Kim S, Kim JS et al (2016) DNA- free genetically edited grapevine and apple protoplast using CRISPR/Cas9 ribonucleoproteins. Front Plant Sci 7:1904

Morineau C, Yannick B, Frédérique T, Lionel G, Zsolt K, Fabien N et al (2017) Selective gene dosage by CRISPR-Cas9 genome editing in hexaploid Camelina sativa. Plant Biotechnol J 15:729–739

Nekrasov V, Wang C, Win J, Lanz C, Weigel D, Kamoun S (2017) Rapid generation of a transgene-free powdery mildew resistant tomato by genome deletion. Sci Rep 7(1):1–6

Peng A, Shanchun C, Tiangang L, Lanzhen X, Yongrui H, Liu W et al (2017) Engineering canker-resistant plants through CRISPR/Cas9-targeted editing of the susceptibility gene CSLOB1 promoter in citrus. Plant Biotechnol J 5:1509–1519

Pitzschke A, Persak H (2012) Poinsettia protoplasts-a simple, robust and efficient system for transient gene expression studies. Plant Methods 8(1):14

Ren R, Gao J, Lu C, Wei Y, Jin J, Wong SM, Yang F (2020) Highly efficient protoplast isolation and transient expression system for functional characterization of flowering related genes in Cymbidium orchids. Int J Mol Sci 21(7):2264

Sapranauskas R, Gasiunas G, Fremaux C, Barrangou R, Horvath P, Siksnys V (2011) The Streptococcus thermophilus CRISPR/Cas system provides immunity in *Escherichia coli.* Nucleic Acids Res 39(21):9275–9282

Sauer NJ, Narvaez-Vasquez J, Mozoruk J, Miller RB, Warburg ZJ, Woodward MJ, Gocal GF (2016) Oligonucleotide-mediated genome editing provides precision and function to engineered nucleases and antibiotics in plants. Plant Physiol 170(4):1917–1928

Scheben A, Wolter F, Batley J, Puchta H, Edwards D (2017) Towards CRISPR/CAS crops – bringing together genomics and genome editing. New Phytol 216(3):682–698

Sedeek KEM, Mahas A, Mahfouz M (2019) Plant genome engineering for targeted improvement of crop traits. Front Plant Sci 10:1–16

Shabir PA (2020) CRISPR/Cas9-mediated genome editing in medicinal and aromatic plants: developments and applications. In: Medicinal and aromatic plants. Academic, pp 209–221

Shen L, Wang C, Fu Y, Wang J, Liu Q, Zhang X et al (2018) QTL editing confers opposing yield performance in different rice varieties. J Integr Plant Biol 60:89–93

Shi J, Gao H, Wang H, Lafitte HR, Archibald RL et al (2017) ARGOS8 variants generated by CRISPR-Cas9 improve maize grain yield under field drought stress conditions. Plant Biotechnol J 15:207–216

Singh V, Dhar PK (2020) Genome engineering via CRISPR-Cas9 system. Academic, Cambridge

Sun Y, Jiao G, Liu Z, Zhang X, Li J, Guo X et al (2017) Generation of high-amylose rice through CRISPR/Cas9-mediated targeted mutagenesis of starch branching enzymes. Front Plant Sci 8:298

Svitashev S, Young JK, Schwartz C, Gao H, Falco SC, Cigan AM (2015) Targeted mutagenesis, precise gene editing, and site-specific gene insertion in maize using Cas9 and guide RNA. Plant Physiol 169(2):931–945

Thurtle-Schmidt DM, Lo TW (2018) Molecular biology at the cutting edge: a review on CRISPR/CAS9 gene editing for undergraduates. Biochem Mol Biol Educ 46(2):195–205

van Esse HP, Reuber TL, van der Does D (2020) Genetic modification to improve disease resistance in crops. New Phytol 225(1):70–86

Wang Y, Cheng X, Shan Q, Zhang Y, Liu J, Gao C, Qiu JL (2014) Simultaneous editing of three homoeoalleles in hexaploid bread wheat confers heritable resistance to powdery mildew. Nat Biotechnol 32(9):947–951

Wang H, La Russa M, Qi LS (2016a) CRISPR/Cas9 in genome editing and beyond. Annu Rev Biochem 85:227–264

Wang F, Wang C, Liu P, Lei C, Ha W, Gao Y et al (2016b) Enhanced rice blast resistance by CRISPR/Cas9-targeted mutagenesis of the ERF transcription factor gene OsERF922. PLoS One 11(4):e0154027

Wang Y, Liu X, Ren C, Zhong GY, Yang L, Li S, Liang Z (2016c) Identification of genomic sites for CRISPR/Cas9-based genome editing in the *Vitis vinifera* genome. BMC Plant Biol 16(1):1–7

Wang L, Chen L, Li R, Zhao R, Yang M, Sheng J, Shen L (2017a) Reduced drought tolerance by CRISPR/Cas9-mediated SlMAPK3 mutagenesis in tomato plants. J Agric Food Chem 65(39):8674–8682

Wang Y, Meng Z, Liang C, Meng Z, Wang Y, Sun G, Lin Y (2017b) Increased lateral root formation by CRISPR/Cas9-mediated editing of arginase genes in cotton. Sci China Life Sci 60(5):524

Wang X, Tu M, Wang D, Liu J, Li Y, Li Z, Wang X (2018) CRISPR/Cas9-mediated efficient targeted mutagenesis in grape in the first generation. Plant Biotechnol J 16(4):844–855

Wyman C, Kanaar R (2006) DNA double-strand break repair: all's well that ends well. Annu Rev Genet 40:363–383

Yin K, Han T, Liu G, Chen T, Wang Y, Yu AYL, Liu Y (2015) A geminivirus-based guide RNA delivery system for CRISPR/Cas9 mediated plant genome editing. Sci Rep 5:14926

Yu QH, Wang B, Li N, Tang Y, Yang S, Yang T, Asmutola P (2017) CRISPR/Cas9-induced targeted mutagenesis and gene replacement to generate long-shelf-life tomato lines. Sci Rep 7(1):1–9

Zhang F, Wen Y, Guo X (2014) CRISPR/Cas9 for genome editing: progress, implications, and challenges. Hum Mol Genet 23(R1):R40–R46

Zhang Y, Liang Z, Zong Y, Wang Y, Liu J, Chen K, Gao C (2016) Efficient and transgene-free genome editing in wheat through transient expression of CRISPR/Cas9 DNA or RNA. Nat Commun 7(1):1–8

Zheng M, Zhang L, Tang M, Liu J, Liu H, Yang H, Hua W (2020) Knockout of two Bna MAX 1 homologs by CRISPR/Cas9-targeted mutagenesis improves plant architecture and increases yield in rapeseed (Brassica napus L.). Plant Biotechnol J 18(3):644–654

Zhu J, Song N, Sun S, Yang W, Zhao H, Song W, Lai J (2016) Efficiency and inheritance of targeted mutagenesis in maize using CRISPR-Cas9. J Genet Genom 43(1):25–36

Zupan J, Muth TR, Draper O, Zambryski P (2000) The transfer of DNA from *Agrobacterium tumefaciens* into plants: a feast of fundamental insights. Plant J 23(1):11–28

Chapter 18
Applications of CRISPR/Cas9 in Oil Crops to Improve Oil Composition

Samira Smajlovic, Azra Frkatovic, Hussein Sabit, Huseyin Tombuloglu, and Turgay Unver

Contents

Samira Smajlovic and Azra Frkatovic contributed equally with all other contributors.

S. Smajlovic
Faculty of Science, Department of Biology, University of Zagreb, Zagreb, Croatia

A. Frkatovic
Genos Glycoscience Research Laboratory, Zagreb, Croatia

H. Sabit · H. Tombuloglu (✉)
Department of Genetics Research, Institute for Research and Medical Consultations (IRMC), Imam Abdulrahman Bin Faisal University, Dammam, Saudi Arabia
e-mail: htoglu@iau.edu.sa

T. Unver
Ficus Biotechnology, Ostim Teknopark, Ankara, Turkey

H. Tombuloglu et al. (eds.), *Oil Crop Genomics*,
https://doi.org/10.1007/978-3-030-70420-9_18

18.1 Introduction

Comprehensive information on the subject is given in **Chap. 17**.

18.2 CRISPRed Oil Crops

18.2.1 Soybean

The Δ12-fatty acid desaturases-2 (FAD2) is a hydroponic transmembrane protein located at the endoplasmic reticulum. It is mainly functional in the conversion from oleic acid to linoleic acid (Jin et al. 2001; Wang et al. 2015). Oleic acid is the most crucial and fruitful molecule in the enhancement of the nutritional value and shelf life of soybean oil. In addition, it plays a role in reducing the risk of type II diabetes (Palomer et al. 2017; al Amin et al. 2019). For the conversion of oleic acid to linoleic acid, the plant tissues express *FAD2* gene, but the highest expression is observed in seeds (Wang et al. 2015). A mutation was created in the soybean (*Glycine max* L.) *FAD2–2* gene using the CRISPR-Cas9 nuclease system. After construct preparation (pCas9-AtU6-sgRNA), the vector was transferred to soybean cotyledon via *Agrobacterium*-mediated transformation. The transgenic plants have low *FAD2–2* expression in the seeds. Near-infrared spectroscopy (NIR) indicated considerable enhancement in the content of oleic acid up to 65.5%. In opposite, the linoleic acid content was recorded as the lowest which was 16% (al Amin et al. 2019). The knockout of two *FAD2* genes namely *GmFAD2-1A* and *GmFAD2–2A*, by CRISPR led to a significant increase in the oleic acid composition of soybean oil. It was 17.1% in wild type, however determined as 73.5% after CRISPR application. In opposite, the linoleic acid content was decreased from 62.91% to 12.23% in the T_2 generation. The oil compositions are similar in the T_3 generation; the oleic acid was 72.02% and the linoleic acid was 17.27% (Wu et al. 2020).

CRISPR/Cas9 system was also successfully used to introduce mutation in three lipoxygenase genes (*GmLox1*, *GmLox2*, and *GmLox3*) in soybean (Wang et al. 2020). Lipoxygenase, an enzyme that catalyzes the oxygenation of polyunsaturated fatty acids such as arachidonic acid and linoleic acid, is responsible from the beany flavor of soybean which restricts human consumption. Lipoxygenase-free new mutant lines (triple mutants *gmlox1gmlox2gmlox3*) showed reduced lipoxygenase activity in T_1 and T_2 generations, however no significant differences ($p > 0.05$) were observed in the seed oil or protein content of *Gmlox* mutants in comparison to the wild-type control (Wang et al. 2020). The applications of CRISPR/Cas9 system to improve oil production is summarized in Table 18.1.

Table 18.1 Applications of CRISPR/Cas9 system to improve oil production

Plant	Targeted gene	Result	Reference
Soybean	*FAD2–2*	Increase in oleic acid content	al Amin et al. (2019)
Rapeseed	*FAD2*	Increase in oleic acid content	Okuzaki et al. (2018)
	LPAT	Enlarged oil bodies, decrease in oil content, and increase in accumulation of starch in mature seeds	Zhang et al. (2019)
	ALC	Increase shatter resistance of seeds	Braatz et al. (2018)
Oil palm	*EgEMLP*	Resistance towards fungus *G. boninense*	Budiani et al. (2020)
Linseed (*flax*)	*EPSPS*	Recovery of entire plants from altered protoplasts without utilizing selection; introduction of the resistance to the herbicide glyphosate in flax; normal Mendelian segregation of EPSPS edits in descendants	Sauer et al. (2016)
Camelina	*FAE1*	Reduction in the content of very-long-chain fatty acids	Ozseyhan et al. (2018)
	CsFAD2	Increase in oleic acid level from 16% to >50% and total monounsaturated fatty acid level (18:1, (20:1, 22:1) increased from 32% to >70%	Jiang et al. (2017)
	CsFAD2	Raise in the oleic acid content, increased from 10% to 62% of total fatty acids	Morineau et al. (2017)
Coconut	*PTI5/ NBS-LRR-type RGAs*	Potential genes to be targeted by CRISPR/ Cas9 system for coconut root wilt disease (RWD) resistance	Haque et al. 2018
Mustard	*BcFLA1*	Shorter hair roots compared to control roots; significant reduction of the transcript compared to control; phenotypic evaluation of gene lesions by CRISPR/ Cas9 system which don't require germline transmission made possible	Kirchner et al. (2017)
Opium poppy	*4′OMT2*	Applicability of the CRISPR/Cas9 system shown for medicinal aromatic plants for the first time to manipulate metabolic pathways	Alagoz et al. (2016)
Jatropha	*JcCYP735A*	Concentrations of tZ and tZ-riboside reduced fundamentally in the JcCYP735A mutants, which indicated seriously hindered development	Cai et al. (2018)

18.2.2 Rapeseed

Rapeseed (Brassica napus) is considered a valuable source of edible oil and a raw material source for biodiesel production. *FAD2* gene manipulation in rapeseed led to a significant enhancement of oleic acid accumulation (Okuzaki et al. 2018). Another example of gene manipulation in rapeseed is lysophosphatidic acid acyltransferase (LPAT), which is a key component of the Kennedy pathway catalyzes fatty acid chains into 3-phosphoglycerate (Chapman and Ohlrogge 2012). CRISPR/Cas9 editing of the *BnLPAT2* and *BnLPAT5* genes resulted in enlarged oil bodies, decrease in oil content, and increase in seed starch content, thereby confirming role of *BnLPAT2* and *BnLPAT5* in biosynthesis of oil (Zhang et al. 2019). CRISPR/Cas9 approach was also applied to increase shatter resistance of rapeseed during mechanical harvest and was done by transformation with construct targeting *ALCATRAZ* (*ALC*) gene. ALC has a function in development of valve margin and greatly contributed to seed shattering and proof of successful knockout of this gene would increase shatter resistance (Braatz et al. 2018).

18.2.3 Cotton

Cotton (*Gossypium hirsutum*) is an important plant grown primarily to benefit from its fiber and seeds. The seeds are recognized as a valuable source of edible oil and protein. The effectiveness of the CRISPR/Cas9 system in targeting a gene in the genome of cotton has been proven in many studies, such as the one using CRISPR/Cas9 to knockout *green fluorescent protein* (*GFP*) gene in transgenic cotton plants (Janga et al. 2017). Other studies also concerned with introducing mutations in specific genes to test the efficacy and specificity of CRISPR/Cas9-induced modifications, in *GhCLA1* and *GhVP* genes (Chen et al. 2017), and *GhMYB25-like A* and *GhMYB25-like D* genes (Li and Zhang 2019) and have obtained transgenic plants with no off-target mutations, thereby indicating that CRISPR/Cas9 system is a specific and efficient tool for targeted mutagenesis in cotton genome.

In addition, CRISPR/Cas9 approach was applied in editing of agronomically important genes in cotton, that is, *MYB25-like* (Li et al. 2017), *ARG* (a gene-encoding arginase) (Wang et al. 2017). However, there is still lack of efficient gene-editing system that can straightforwardly be applied and adopted by laboratories with basic facilities and edit other genes of potential breeding significance.

18.2.4 Melon

Melon (*Cucumis melo*) has specific biological properties and represents a great economical value due to its fruit. CRISPR/Cas9 has not been performed to enhance melon fruit properties (i.e., oil content), but an efficient protocol was developed to modify the melon genome via CRISPR/Cas9 system. The applied experimental procedure generated multiallelic mutations in phytoene desaturase gene of melon (*CmPDS*) and albino phenotype could easily be detected in a few weeks (Hooghvorst et al. 2019).

18.2.5 Oil Palm

Oil palm (*Elaeis guineensis*) species belong to the Arecaceae family and are perennial tropical trees representing an efficient vegetable oil resource. Even though CRISPR/Cas9-mediated genome targeting in oil palm is a great challenge due to its huge and complex genome, CRISPR/Cas9 system was deployed in performing "*Early methioninelabeled polypeptide 1*" (e.g., *EMLP* gene) knockout in oil palm with aim of conferring a putative defense-related trait toward pathogenic fungus *G. boninense* to produce resistant oil palm plants (Budiani et al. 2020).

18.2.6 Linseed (Flax)

Sauer et al. (2016) reported an oligonucleotide-directed mutagenesis method that is based on simultaneous delivery of single-stranded oligonucleotides (ssODN) and a site-specific DNA double-strand breaker to generate genome modifications. By doing this, an herbicide-resistant flax (*Linum usitatissimum*) was generated by targeting the *5′-ENOLPYRUVYLSHIKIMATE-3-PHOSPHATE SYNTHASE* (*EPSPS*) gene. EPSPS alters happened at adequate frequency that they could recover entire plants from altered protoplasts without utilizing selection. These plants were in this way resolved to be resistant to the herbicide glyphosate in greenhouse splash tests. Descendants (C1) of these plants showed the normal Mendelian segregation of EPSPS alters. Their discoveries show the huge capability of utilizing a genome-altering platform for exact, reliable attribute advancement in commercially related agronomic plant such as flax.

18.2.7 Coconut

Coconut palms (*Cocos nucifera* L.) are significant constituent in the worldwide agrarian engineering, supporting customers and businesses globally. Continuous exploration in breeding strategies to generate resistant coconut is as of now the significant hope to diminish the effects of the "Lethal yellowing disease (LYD)" of coconut palms. However, the endeavors are being outpaced by latest reports across areas that recount of expanded occurrence of this disease. Thus, Ekhorutomwen et al. (2016) reported the requirement to utilize incorporated control measures or energize preliminaries on the utilization of novel ecological methods, such as the CRISPR/Cas9 system or entomopathogens in controlling the issues of LYD among other control measures. If used, CRISPR/Cas9 might deal with the lethal phytoplasma by knocking-out its DNA from either host plants or insect vectors. Moreover, in their review, Haque et al. (2018) reported transcriptional activator *PTI5-like* (root wilt disease – RWD – resistance) and *NBS-LRR type RGAs* (coconut root/wilt disease) genes as potential genes to be targeted by CRISPR/Cas9 system for enhanced quality in addition to improved tolerance to exterior stressors.

18.2.8 Mustard

Kirchner et al. (2017) showed the effective utilization of CRISPR/Cas9 in hairy roots of Ethiopian mustard (*Brassica carinata*). The results on the root hair phenotype revealed shorter hair roots compared to control roots, significant reduction of the transcript compared to control, etc. They chose this plant because of assortments of this species were revealed to vary in their answer to phosphate and nitrogen deficit concerning the formation of root hair. They effectively induced site-directed deletions and insertions in mustard by utilizing a combination of the CRISPR/Cas9 and hairy root transformation methods. The modification in *BcFLA1* gene – a consistently expressed gene depending on tissue type that has role in root hair architecture – gave them the certainty to legitimately endeavor a phenotypic evaluation of the gotten roots. Their research makes the way easier for exploratory techniques including the phenotype evaluation of gene modification created by CRISPR/Cas9 system which does not require germline transmission.

18.2.9 Opium Poppy

In their study, Alagoz et al. (2016) knock-out 4′-O-methyltransferase-2 (*4′OMT2*) gene in opium poppy (*Papaver somniferum* L.) by using type II CRISPR/SpCas9 method. The gene is functional in the biosynthesis of benzylisoquinoline alkaloids (BIAs). Regardless of its relevance as an oil crop, opium poppy as well biosynthesize numerous benzylisoquinoline alkaloids (BIAs) commonly utilized in medicine, such as anticancer drug and morphine production. The study for the first time revealed the applicability of CRISPR/Cas9 approach in a medicinal aromatic plant, to manipulate metabolic pathways.

18.2.10 Jatropha

Jatropha (*Jatropha curcas* L.) seed oil is counted as a possible source of bioenergy which might substitute fossil fuels. Still, Jatropha seed yield is short and has hitherto be enriched. Lately, it has been reported that cytokinins can meaningfully improve seed yield. Among other things, in their study Cai et al. (2018) analyzed the function of *JcCYP735A*, a cytokinin metabolic gene, by utilizing CRISPR/Cas9 system to produce Jatropha transformants with *JcCYP735A* knocked out. The results showed that the abundance of *trans*-zeatin (tZ) and tZ-riboside reduced fundamentally in the *Jccyp735a* mutants, which indicated seriously hindered development. These discoveries will be useful for future investigations of cytokinin metabolism-related genes, and understanding the function of cytokinins in Jatropha development and growth.

18.2.11 Camelina

Camelina (*Camelina sativa* (L.) Crtz.), a member of the Brassicaceae family, is also known as false flax or gold-of-pleasure. It is accepted as a potential biofuel crop due to its favorable attributes such as a short period of growth cycle (80–100 days), increased resistance to pests, and high resistance to water deficiency and cold temperatures (Pilgeram et al. 2007; Yuan and Li 2020). Camelina seed has an oil content of ~35% of its dry mass. The seed oil composes of rich polyunsaturated fatty acids (PUFA, ~50%), especially α-linolenic acid (18:3, ω-3; 33–35%). Also, its seeds accumulate high protein content (30%). For this reason, Camelina can be used as healthy food as well as for feeding purposes (Ozseyhan et al. 2018; Yuan and Li 2020). In addition, camelina seeds hold a considerable quantity of very-long-chain fatty acids (VLCFAs), such as eicosenoic (20:1) and erucic (22:1) acids. For health, the amount of VLCFAs in Camelina seeds should be reduced. For this purpose, three *Fatty Acid Elongase* (*CsFAE1-A, CsFAE1-B, CsFAE1-C*) genes were simultaneously knocked out via CRISPR/Cas9 technology (Ozseyhan et al. 2018). The knock-out *fae1* mutations reduced the content of VLCFAs in camelina seed without a side effect on agronomic traits.

In a different study, the knockout of three *CsFAD2* homologs by CRISPR/Cas9 led to increase oleic acid level from 16% to >50%; and MUFAs (18:1, 20:1, 22:1) from 32% to >70%. In opposite, the editing dramatically decreased the linoleic and linolenic acid contents (Jiang et al. 2017). In addition, Morineau et al. (2017) knocked out three, two, and one isologues *CsFAD2* by CRISPR/Cas9. The oleic acid content was raised from 10% to 62% of total fatty acids.

18.3 Future Perspective

Other oilseed crops are of interest for genome-editing due to their oil profile, yield, or production of specialty fatty acids. These include castor beans, olive, sesame, argan, and sunflower but due to years needed for some these plants to mature or complexity of their genomes, the CRISPR/Cas9 genome-editing methods are slow to implement. Furthermore, deficiency of fast and effective transformation protocols as well as lack of genome sequence information of other relevant non-model plants such as shea nut or jojoba make use of CRISPR/Cas9 system limited due to inability to identify the presence of any potential off-target effects which are one of the major drawbacks of CRISPR/Cas9 technology.

References

Al Amin N, Ahmad N, Wu N, Pu X, Ma T, Du Y, Bo X, Wang N, Sharif R, Wang P (2019) CRISPR-Cas9 mediated targeted disruption of FAD2–2 microsomal omega-6 desaturase in soybean (*Glycine max. L*). BMC Biotechnol 19:9

Alagoz Y, Gurkok T, Zhang B, Unver T (2016) Manipulating the biosynthesis of bioactive compound alkaloids for next-generation metabolic engineering in opium poppy using CRISPR-Cas9 genome editing technology. Sci Rep 6:1–9

Braatz J, Harloff HJ, Mascher M, Stein N, Himmelbach A, Jung C (2018) CRISPR-Cas9 targeted mutagenesis leads to simultaneous modification of different homoeologous gene copies in polyploid oilseed rape (*Brassica napus*). Plant Physiol 174:935–942

Budiani A, Nugroho IB, Sari DA, Palupi I, Putranto RA (2020) CRISPR/Cas9-mediated knockout of an oil palm defense-related gene to the pathogenic fungus *Ganoderma boninense*. Indones J Biotechnol 24:101–105

Cai L, Zhang L, Fu Q, Xu ZF (2018) Identification and expression analysis of cytokinin metabolic genes IPTs, CYP735A and CKXs in the biofuel plant *Jatropha curcas*. PeerJ 6:e4182

Chapman KD, Ohlrogge JB (2012) Compartmentation of triacylglycerol accumulation in plants. J Biol Chem 287:2288–2284

Chen X, Lu X, Shu N, Wang S, Wang J, Wang D, Guo L, Ye W (2017) Targeted mutagenesis in cotton (*Gossypium hirsutum L.*) using the CRISPR/Cas9 system. Sci Rep 7:44304

Ekhorutomwen OE, Eziashi EI, Adeh SA, Omoregie KO, Okere CI (2016) Insight on the possibility to control lethal yellowing disease (LYD) in coconut palms using CRISPR/Cas9 system and bio-control (entomopathogens) among other measures. Int J Agric Earth Sci 2:24–34

Haque E, Taniguchi H, Hassan MM, Bhowmik P, Karim MR, Śmiech M, Zhao K, Rahman M, Islam T (2018) Application of CRISPR/Cas9 genome editing technology for the improvement of crops cultivated in tropical climates: recent progress, prospects, and challenges. Front Plant Sci 9:617

Hooghvorst I, López-Cristoffanini C, Nogués S (2019) Efficient knockout of phytoene desaturase gene using CRISPR/Cas9 in melon. Sci Rep 9:17077

Janga MR, Campbell LAM, Rathore KS (2017) CRISPR/Cas9-mediated targeted mutagenesis in upland cotton (*Gossypium hirsutum L.*). Plant Mol Biol 94:349–360

Jiang WZ, Henry IM, Lynagh PG, Comai L, Cahoon EB, Weeks DP (2017) Significant enhancement of fatty acid composition in seeds of the allohexaploid, *Camelina sativa*, using CRISPR/Cas9 gene editing. Plant Biotechnol J 15:648–657

Jin UH, Lee JW, Chung YS, Lee JH, Yi YB, Kim YK, Hyung NI, Pyee JH, Chung CH (2001) Characterization and temporal expression of a ω-6 fatty acid desaturase cDNA from sesame (*Sesamum indicum l.)* seeds. Plant Sci 161:935–941

Kirchner TW, Niehaus M, Debener T, Schenk MK, Herde M (2017) Efficient generation of mutations mediated by CRISPR/Cas9 in the hairy root transformation system of *Brassica carinata*. PLoS One 12(9):e0185429

Li C, Zhang B (2019) Genome editing in cotton using CRISPR/Cas9 system. In: Methods in Molecular Biology. Humana Press Inc, Clifton, pp 95–104

Li C, Unver T, Zhang B (2017) A high-efficiency CRISPR/Cas9 system for targeted mutagenesis in Cotton (*Gossypium hirs*utum L.). Sci Rep 7:43902

Morineau C, Bellec Y, Tellier F, Gissot L, Kelemen Z, Nogué F, Faure J-D (2017) Selective gene dosage by CRISPR-Cas9 genome editing in hexaploid *Camelina sativa*. Plant Biotechnol J 15:729–739

Okuzaki A, Ogawa T, Koizuka C, Kaneko K, Inaba M, Imamura J, Koizuka N (2018) CRISPR/Cas9-mediated genome editing of the fatty acid desaturase 2 gene in Brassica napus. Plant Physiol Biochem 131:63–69

Ozseyhan ME, Kang J, Mu X, Lu C (2018) Mutagenesis of the *FAE1* genes significantly changes fatty acid composition in seeds of Camelina sativa. Plant Physiol Biochem 123:1–7

Palomer X, Pizarro-Delgado J, Barroso E, Vázquez-Carrera M (2017) Palmitic and oleic acid: the Yin and Yang of fatty acids in type 2 diabetes mellitus. Trends Endrocrinol Metab 29:135–200

Pilgeram AL, Smith DC, Boss D, Dale N, Wichman S, Lamb P, Lu C, et al (2007) Camelina sativa: a Montana omega-3 and fuel crop. In Issues in New Crops and New Uses (Janik J, Whipkey A eds) pp 129–131. Alexandra, VA: ASHS Press

Sauer NJ, Narváez-Vásquez J, Mozoruk J, Miller RB, Warburg ZJ, Woodward MJ, Mihiret YA, Lincoln TA, Segami RE, Sanders SL, Walker KA (2016) Oligonucleotide-mediated genome editing provides precision and function to engineered nucleases and antibiotics in plants. Plant Physiol 170:1917–1928

Wang Y, Zhang X, Zhao Y, Prakash CS, He G, Yin D (2015) Insights into the novel members of the FAD2 gene family involved in high-oleate fluxes in peanut. Genome 58:375–383

Wang Y, Meng Z, Liang C, Meng Z, Wang Y, Sun G, Zhu T, Cai Y, Guo S, Zhang R, Lin Y (2017) Increased lateral root formation by CRISPR/Cas9-mediated editing of arginase genes in cotton. Sci China Life Sci 60:524

Wang J, Kuang H, Zhang Z, Yang Y, Yan L, Zhang M, Song S, Guan Y (2020) Generation of seed lipoxygenase-free soybean using CRISPR-Cas9. Crop J 8:432–439

Wu N, Lu Q, Wang P, Zhang Q, Zhang J, Qu J, Wang N (2020) Construction and analysis of GmFAD2-1A and GmFAD2-2A soybean fatty acid desaturase mutants based on CRISPR/Cas9 technology. Int J Mol Sci 21:1104

Yuan L, Li R (2020) Metabolic engineering a model oilseed Camelina sativa for the sustainable production of high-value designed oils. Front Plant Sci 11:11

Zhang K, Nie L, Cheng Q, Yin Y, Chen K, Qi F, Zou D, Liu H, Zhao W, Wang B, Li M (2019) Effective editing for lysophosphatidic acid acyltransferase 2/5 in allotetraploid rapeseed (Brassica napus L.) using CRISPR-Cas9 system. Biotechnol Biofuels 12:1–18

Chapter 19
Economics of Oil Plants: Demand, Supply, and International Trade

Ghulam Mustafa and Asim Iqbal

Contents

19.1 Introduction

Plants are vital to the environment and maintain the atmosphere. The crop plants particularly provide food for consumption to human. These include vegetables, pulses, cereals, spics, condiment, aromatic and medicinal plants, etc. Crop plants along with insects, birds, animals, and other living creatures are vital for sustaining human life. Oil plants are an important group of crop plants whose oil is not only use for human consumption but it is also an engine to global economy. Oil plants has become necessary part of people dietary and lifestyle. Oil extraction from the plants are not only use for cooking but they are also use in transportation as biofuel and other commercial purposes. It is the necessary part of the people food expenditure. Ali et al. (2008) found that household food expenditures are significantly affected by edible oil; for example, edible oil deficit have negative and significant

G. Mustafa (✉) · A. Iqbal
Department of Economics and Business Administration, Division of Arts and Social Sciences, University of Education, Lahore, Pakistan
e-mail: ghulam.mustafa@ue.edu.pk

H. Tombuloglu et al. (eds.), *Oil Crop Genomics*,
https://doi.org/10.1007/978-3-030-70420-9_19

long-run association with food expenditure and if 1% edible oil prices increase there would decrease of food expenditure by 0.14%.

Lennerts (1983) identified 40 different oil plants whose oil seeds can be used for human consumption. Oil crops are grown around the globe under different agroclimatic zones and are crucial for many economies of the world through production, consumption and trade. These crops can be grouped in following ways;

1. Perennial crops: These include perennial tree crops such as oil palms and coconuts.
2. Annual or biennial crops: These include rapeseed, groundnut, sunflower, and soybean.
3. Cotton and corn crops: Fiber and food from cotton and corn are not only distinct benefit of the cotton and corn but their embryo is a by-product whose oil can be used in varieties consumption and commercial purposes.
4. Minor oil crops: Some of the oilseeds have little share in world trade but they have key role in local markets through provision of raw materials for special products like linseed, castor, safflower, and tung nut. Many of the oilseeds are used for both human and animal consumptions.

Oilseeds are important source of dietary proteins for animals in unprocessed and processed forms. On the one hand oilseeds directly can be used for animal feed in unprocessed form, while on the other hand these can be used as cakes or meals in processed form.

19.2 Green Revolution and Oil Crops

Byerlee et al. (2017) reported that the oil crops (green) revolution consisted of two crops (palm oil and soy), such as Green Revolution which comprised on two cereal crops – wheat and rice. Oil crops showed a remarkable expansion in term of growth, trade, and production and surpassed other agricultural and food crops. However, other agricultural and food crops primarily grown for home consumption while oil crops revolution characterized by increase production and trade of animal feed, high protein meal for food, vegetable oils, and a variety of industrial uses, such as biodiesel, cosmetics, and lubricants. In some ways or other, oil crops or its by-products are being used by whole world consumer. For instance, over 60% of soy and 75% of the palm oil production are exported and made the soy complex more valuable commodity than wheat in global agricultural market while oil palm stood at third place.

An additional comparison of Green Revolution between oil and cereal crops is the production characteristics. For instance, main driving force of the growth in cereal grain production is the yield gains and this was the reason cropped area for wheat, rice and maize grew at the rate for 0.10%, 0.48%, and 1.31%, respectively, during 1991 to 2012. On the other hand, main production characteristics of oil crops is the area expansion and that is why it annually grew at the rate of 4.65% for oil palm and 3.25% for soy during the same period. Oil palm has the growing period of 25 years and this is the main reason of relatively high rate of growth for this perennial crop that also restricts introduction of high yield and new varieties for oil palm to increase yield further (Naylor 2016). Whereas most of the growth in soy production has occurred through its importance regarding trade and foreign earnings. During last 25 years, area expansion of soy has occurred mainly in northern Argentina and Brazil, and Malaysia and Indonesia for oil palm. Although some small countries (Colombia, Bolivia, Thailand, and Paraguay) have also emerged as important producers of oil crops.

Most important thing regarding green revolution of oil crop is that it occurred mainly in tropical areas where lands are relatively cheaper and heavy investments are required. Thus, multinational companies, large landlords, and agribusiness established this industry with their heavy investments. That is why share landholdings and hence production and trade of oil crops is greater for large business tycoons as compared to small producers. Albeit tiny growers also prevail in oil farming industry, equatorial oil farming structure in blunt compose production scheme of Green Revolution. That enlist relatively tens of millions of small growers far and near the Asia and Latin America.

19.3 Demand of Oil Crops

Research on demand for agricultural commodities have long been studies to understand the household consumption behavior (Huang and Lin 2000; Kruse 2010; Kumar et al. 2011; Huang et al. 2015). Demand is associated with some factors that affect the demand. Like any demand-related studies oil crops do have some factors. These factors are population, urbanization, income, price, and health and nutrition values.

19.3.1 Population and Urbanization

Food demand and population growth has long been studied during the Malthus times. In twenty-first century, population is increasing with rapid rate and about half of the population lives in cities. Asia is about one-third of the total global population (4.4 billion or 60% of the global population). It is predicted that population of sub-Saharan Africa would be double by end 2050 as it has highest growth and fertility rate than that of rest of the world (PRB 2015). It means both Asia and Africa have had rapid urbanization rate. For instance, as per United Nation reports, urbanization in small- to medium-sized cities has accelerated from 0.5 million to 5 million residents. The mega cities are also on rise and it is estimated that 40 mega cities worldwide will have population exceeding ten million each. Several of these cities will be near seaboard, providing the incentives for strengthen global swap (UN 2014). One such impressive shift in comprehensive food economy of one last quartile is due to demand-driven elements establishing trade flow and pricing strategies. The globalization and free trade demographic cultures will set the pattern of worldwide food systems.

The increasing trend of urbanization means that food producers have become largely different from food consumers. Minimizing trade difficulties and upgradation in transportation and communication means to have expanded global supply chain of agricultural and food products all around the world. The outcome has been extended homogeneity in the production of food supply globally. For instance, Khoury et al. (2014) find out that 90% of calories, proteins, and fats has used worldwide is participated by 50 crop products in recent times. Their study suggests that major oil crops like soy and oil palm are among the world's most globally used commodities.

Since 1970 the average global supply of fat has increased by 20 g capita^{-1} day^{-1} particularly due increase in the demand of vegetable oil. Thus, dietary fat supply surpassed dietary protein supply in last 5 decades. The proportion of energy contributed by dietary fats has increased in almost all regions particularly in industrialized regions where it has escalated to 30%. On the whole, edible oils consumption has expanded at three times the rate of cereals since 1990. On one the hand, it provided essential nutrition of recommended 15–20% calories from fat to extremely poor countries such as those in South Asia and sub-Sahara Africa, and on the other hand, it created the obesity issue where fat intake exceeded 35% of calorie such as in Western Europe and North America (FAO 1994). Asia at present rationalizes more than a half of nutritive oils requirement including highly populated Asian countries like China, India, and Indonesia. Besides, the sub-Saharan African countries' demand for nutritive oil is low at present but soon due demographic changes and high population growth rate the demand will increase by significant numbers.

Mostly increase in consumption of biofuels and animal feed accompanies with higher growth in the oil crops particularly in emerging economies (Rueda and Lambin 2014). For instance, increase in the feed demand of soy has expanded its growth which further processed into high protein meal. Currently China alone accounts 60% share of imports in the world trade of soy to meet its edible oil demand and remaining 40% it produces from its own livestock sector (Byerlee et al. 2017), while Brazil is the world largest producer of soy. Palm oil crop has high oil content as compared to meal content so its demand is higher in biodiesel extraction; however, some countries like the US, Argentina, and Brazil are also using residual soy oil from feed processing for biodiesel. Despite biofuels are constituting merely 13% of the global cooking oil consumption, but they account for nearly half the increase in vegetable oil consumption from 2003 to 2012. If the countries with high GDP per capita and rich in the production of edible oil ensure the biodiesel consumption that will lead to more economic growth in these nations. High growth without incremental cost.

19.3.2 Income

Income is the major determinant of demand. Income not only shifts the demand curve for particular product (crops oil in this study) to right, but it also shifts the consumption of the other products. It increases the utility and hence the welfare of any society. Rising per capita income of developing countries in last 6 decades has escalated the demand for crops oil. Guo et al. (2000) examined the relationship between energy from fat and incomes of the countries through food balance data. They found that income required to gain energy from fat in 1990 was about half as compared to same energy requirement in 1962. For instance, in 1990, a poor country could get energy requirement of 20% from fat with per capita GNP (gross national product) of 750 USD; however, a diet deriving 20% of energy from fat was associated with a GNP of $1475 in 1963. Such alteration was predominantly outcome of growing consumption of edible oil by significantly low-income countries and fewer rise in middle- and high-income nations. By 1990, the low-income countries experienced edible fats as a source of energy relatively higher than that of animal fats nutrition. Changes in edible vegetable oil supply, in prices, and in consumption equally affected rich and poor countries, although the net impact was relatively much greater in low-income countries. An equally large and important shift in the proportion of energy from added sugars in the diets of low-income countries was also a feature of the nutrition transition (Drewnowski and Popkin 1997). Thus, it has a lot of demand in developing world where there is population rapidly growing. Income growth is the most important determinant of demand of oil crops. For instance, demand elasticity with respect to income is found in the range of 0.4–0.5 on average for vegetable oil and this elasticity is even higher for lower-income consumer with coefficient value of 0.406 (40.6%) (Green et al. 2013). The main reason of high elasticity is the continuously declining trend of vegetable oil prices for the last 3–4 decades. This gave rise to the consumption of vegetable oils, especially as the low-earning bunches have comparably flatter demand curve which explains the highly price elastic demand curve.

Investigation of customers' routine geared toward understanding the connection between education or income they have and the amount they spend on various commodities and their spending patterns reveals many things. Research work in China has shown that as income of individuals increase, the pattern of purchasing also shifts toward different amounts of commodities as consumers prefer to buy more and most importantly in diverse pattern of purchasing behavior. Extra income of rich and poor in China affects rich and poor differently. Overall analyses showed that fat intake increased by the poor more as compared to rich counterpart due to increase in income (Popkin 2001). This might be due to health consciousness (obesity) associated with higher income leads to better awareness and hence reduction in fat intake up to threshold level. However, what that threshold level is, still needs to be addressed in future research. Consequently, it can be seen from above discussion that rising incomes in the developing world have also led to an increase in the availability and consumption of energy-dense high-fat diets. However, in rich countries it may decreased to a threshold level but this may vary from country to country and family to family. Ahmed et al. (2019) also showed that consumption of fat has reduced due to increase in per capita income in UAE.

Investigation through data in different countries, family units, and time intervals implement further intuitions about income–oil relationships. Byerlee et al. (2017) reported that per capita GDP vis-à-vis per capita vegetable oil consumption for 50 nations whose populations exceed 20 million habitats. They showed that income–consumption association levels off with rise of income. The fit of the curve is significant but far from perfect, demonstrates that for a given amount of income and living style, gastronomic plays a vital role in dietary oil consumption. The income elasticity coefficient measured across the full range of 50 countries was found 0.44. Seven countries with real GDP per capita explaining about half the variance – a correlation of 0.7. In some sense, this constant elasticity formulation can be thought of as the edible oil income elasticity for the world.

Moreover, Byerlee et al. (2017) has shown that the elasticities of log of GDP per capita against income of 90 countries by combining with purchasing power parity (PPP) per capita GDP data for 2013. Their findings indicated that with incomes of US$4000 per capita, a 10% increase in per capita income would result in about a 5% increase in oil consumption, whereas at US$40,000 per capita, the response would be only about 2%. Some variation appears across countries, but the underlying regularity is impressive.

19.3.3 Prices

Prices has definite link with demand of oil crops. Conventional law of demand states that, ceteris paribus, when price of any product decreases, its demand increases if the good is normal. For example, Kumar et al. (2011) found that peanut and liquid butter oil prices has significantly negative impact on demand of peanut and liquid butter, respectively. Usually there are little vegetable oil price variations and hence markets are working well with little trade barriers, however, because of improvements in supply chain and reduction of transportation costs, prices have declined from past 7 decades. For instance, Fry (2011) found that the real price of palm oil decreased to one-third from 1950 to 2000, that is, 2% reduction annually since 1950. Similarly, Overseas Development Institute (2015) reported that the cooking oil prices reduced about 50% from 1980 to 2006 in China.

Frequently used concept of demand analysis of any commodity is cross prices elasticities (i.e., change of palm oil quantity demanded with respect to soy oil price). In this regard, there is high degree of correlation among the four major oils, that is, palm, rapeseed, soybean, and sunflower. For instance, soybean typically trades slightly higher than it major competitor, palm oil. Brümmer et al. (2015) are of the opinion that sticky nature of these four oils prices are due to the movements in exchange rates. High levels of correlations among four oils are due to the fact that they have widespread substitution possibilities in consumption. Olive oil is the exception and does not give clear link of being substituted with other oils in the sense that it trades on very high prices and only account for 2.5% of total edible oil consumption Byerlee et al. (2017).

By reversing the tendency noticed during the 2016/2017 (September/October) term, the first term of 2017/2018 had witnessed increased oilseeds and oil meal prices. On the other hand, the prices of vegetable oils were easing as demonstrated by FAO's cost indicators following the oilseed complex (Fig. 19.1). As the 2017/2018 period opened, anticipation of significant reduction in Argentina's soybean production strained the world perspective for meals and oilseeds. This relapse in one of the principle suppliers of soybean products in world trading markets, which concurred with scarce availability for both different sources and protein meals, caused an ascending movement in global oil meals and oilseeds prices by April 2018. FAO's price index about oilseeds and oil meals had gone up by 22-month and 40-month highs, respectively.

The world also witnessed the downfall in the prices of vegetable oil during the last term of 2017, in response to expectations of higher aggregate output in 2017/2018. Vital stats of market compel prices were (i) ongoing improvement in Southeast Asia's palm oil gains, which mixed with low demand in global market, predicted broad stock scales in Indonesia and Malaysia; and (ii) high production of soybean in America and other soybean-rich countries, which led toward significant availability of soya oil. Opposite to this scenery, FAO's price index of vegetable oil shown downward slope nearing 2.5 years depressed in June 2018.

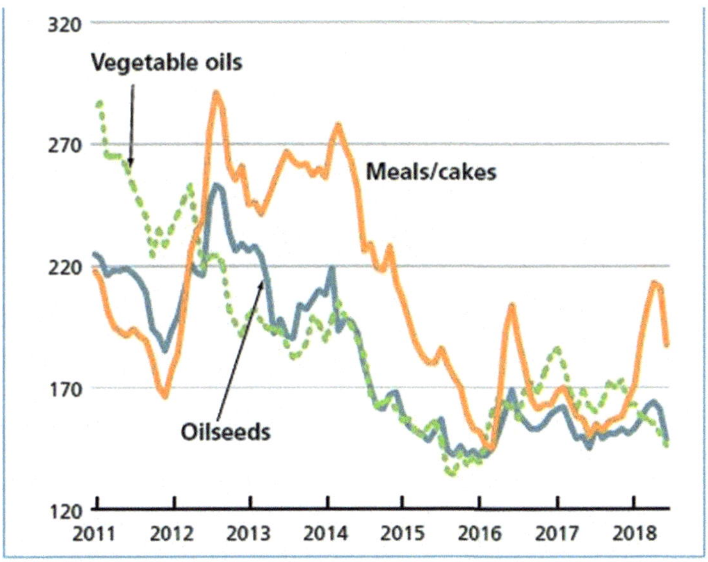

Fig. 19.1 FAO monthly international price indices for vegetable oils, meals/cakes, and oilseeds (2002–2004 = 100). (Source: FAO 2018)

The trade rivalry between two giants of global trade China and the United States of America in first quarter of 2018 led to an instability in the global market. As the USA is the world biggest supplier of soybeans and China is biggest purchaser, so mere the announcement that China will impose trade barriers in form of tariff on soybean trade from the USA initiated applying forcefully downward trend in global soybean and meal prices. In June 2018, China actually imposed tariff on soybean trading and that particular initiative led to spillover effect on overall world oil crops production.

19.3.4 Health and Nutrition

Health and nutrition are the major determinants of shaping demand of oil crops. They influence both the total consumption of oils and also the specific oil being consumed. Consumption of oils can be seen from two angles with respect to health and nutrition. On the one hand, too little vegetable oils create health and nutrition problem through undernourishment, while on the other hand, too much edible oils generate the issue of another health issue – obesity. In both cases, demand for oils for dietary and calories requirement would increase in this century. For instance, Etilé and Oberlander (2019) estimated that the prevalence of obesity has increased three times since 1975. They reported that about 266 million men and 375 million women were obese in 2014, which corresponds respectively to 10.8% and 14.9% of the world's men and women as compared with 3.2% and 6.4% in 1975. The condition will be worse when one sees the projections of Depenbusch and Klasen (2019) where they found that calorie requirements would be increased by 61.05% between 2010 and 2100. Further, they estimated that increases in height and BMI could add another 18.73% regarding calories requirements. Therefore, it can easily be seen that the demand of oil plants would be increased for calories intake.

The contribution of the vegetable oil in total calories intake varies from region to region and even varies within the same region. For instance, Zhai et al. (2014) found that vegetable oil in total calories consumption was 6% in China. Similarly, they found that Brazil had 13% of total calories coming from vegetable oil followed by Nigeria (12%), Indonesia (9%), India (6%), and Ethiopia (3%). Moreover, the average absolute number of calories per day coming from oils in the United States (676) and the European Union (485) were more than double those found in most other countries (Blasbalg et al. 2011). Additionally, Blasbalg et al. also found that in the USA calories intake rose from 0.09 kg/capita to 11.64 kg/capita during 1909–1999.

Vegetable oil consumption has a great role to play regarding elimination of chronic undernourishment. FAOSTAT found that global calories has increased to 278 (26%) from 1991 to 2011. For instance, undernourished people are highest in South Asia, while it has the lowest average per capita calories intake. However, still in this region vegetable oil consumption increased to 32%. For the poorest people in the region, whose intake of oils and fats was low to begin with, increased vegetable oil consumption is likely to have been important in improving their nutritional status. This role of vegetable oils in improving diets is largely obscured by the current debate about high fat consumption. Obesity is sometimes a problem even in poor societies, where undernutrition persists. In middle- and upper-income countries, cardiovascular diseases are the primary cause of death, and since the 1950s, dietary fat has been examined closely as a major cause of heart disease (Daud et al. 2012).

The health effects of vegetable oil (particularly tropical oils) do not depend entirely on fat or cholesterol contents, but rather on what they replace in or add to the diets of different groups and subgroups (Micha and Mozaffarian 2010). There are different school of thoughts regarding the consumption of olive oil, sunflower oil, rapeseed oil, soybean oil, and palm oil and their effects with respect to cholesterol level. For instance, one group of authors are on the opinion that in low-income countries where palm oil consumption is at peak, the main reason of cardiovascular diseases are due bad health facilities and not due to palm oil consumption (Basu et al. 2013; Chen et al. 2011). Other school of thoughts are in favour of consumption of sunflower oil and suggest that sunflower oil has a lesser effect on cholesterol (Fattore et al. 2014; Fattore and Fanelli 2013). Most of the authors are agreed that vegetable oils are not substantially different in their impacts if consumed in moderation.

19.4 Profitability Analysis of Oil Crops

Profitability is another factor that increased the demand, supply, and trade of oil crops. Most farmers in producing their crops consider the availability of resources to determine production capacity. Pargar (2017) stated that resource utilization plays an important role in the production process. Therefore, using of appropriate resources produced optimal production which gains high profit. Companies (farmers) need to increase the utilization of resources by determining their level of productivity to earn more profits (Fakorede et al. 2014). In this regard, oil crops producers proved themselves to be more efficient and earning more profit as compared to other crops producers. From many decades, particularly after green revolution, returns from oil crops have substantially increased as compared to other crops. Byerlee et al. (2017) compared profitability analysis of cotton, soybeans, maize, and sorghum crops. They found that oil crops are more profitable as compared to other crops. For instance, they reported that soybeans provided a return of 27% over total cost and proved to be a more profitable crop as compared to other summer crops. they even estimated that sorghum and maize showed negative returns from 2007 to 2011. Only cotton was more profitable than soybean; however, cotton crop requires larger investment and longer growing season.

One of the indicators to measure the profitability is the resource–cost ratio (RCR) that measures the cost of domestic resources (labor, land, and capital) to generate savings in one unit of foreign exchange. In this regard, soybean seems to be only crop that is efficient economically as indicated by Gulati and Kelley (1999). They found that RCR value of soybean was 0.84, which showed an efficient production of soybean. Although the findings are outdated, but Byerlee et al. (2017) found that there are no major changes in the underlying technology that would alter them fundamentally. They conducted competitiveness of Indian soybean production through comparative cost of production and found that the cost of production and price f soybeans received by farmers in Madhya Pradesh (India) averaged about 5% less than that in Iowa (Byerlee et al. 2017; Iowa State University 2015). Growth in total factor productivity may also be negative in India as its costs were up to 20% higher than in Brazil (Kurup et al. 2015). It means higher costs further lessen the competitiveness of Indian soybean.

Despite soybeans, producing others oil crops is also very low cost and profitable venture. For instance, Konwar et al. (2019) find that rapeseed production is a profitable venture under intensification, particularly when it is sown by 25 cm × 25 cm geometry. By this way, it gives higher yields, returns, and profits. Among other oil crop production, groundnut is also found to be a profitable business in various countries. For instance, Iliyasu et al. (2008) found that groundnut production gives gross margins of about N43.34 per kg and the return on investment is about 40%. Similarly, Abubakari et al. (2019) find that groundnut production is profitable venture in Ghana. Sunflower is another important oil crop that is a profitable business. For instance, Das and Rout (2018) showed that sunflower has a bright future and net return per hectare is higher than other crops. Oil palm industries have been the mainstay for Malaysia and Indonesia from many decades and proved itself a most profitable venture not only for large state companies but for smallholders as well. For instance, Hidayat et al. (2016) finds that oil palm is one of the most profitable venture for farmers if we consider monetary returns in relation to investments as compared to crops like rubber, cassava, and rice (Subervie and Vagneron 2013; Brandi et al. 2015). A case study of profitability of palm oil smallholders is presented in Box 19.1, using new techniques and standards that can increase the returns of small producers.

Box 19.1: Smallholders' Profitability of Palm Oil

Hidayat et al. (2016) examines the advantages of palm oil certification with the help of Financial Cost Analysis (CBA) and evaluations of net present values. For awareness of financing in certification adoption can be used by the decision makers or certification contributors to bring in more smallholders and to develop certification much more beneficial for generally unsafe smallholders. In the self-support plot, certification is not beneficial for scheme smallholder and just beneficial for independent smallholders and when they start to collect premium price. In premium price eliminates, the independent smallholders may need improbable big premium fee for certification to still be beneficial in this case. After taking palm oil certification by the farmer it came to know that organization of the farmers near miller companies put up effectively as compared to before certification. So contributing to organization should be impressive form of government participation.

Expanding understanding of global buyers about continued hurdles linked to the production of agricultural commodities has assisted to the exposure of private sustainable certification standard, such as roundtable on sustainable palm oil (RSPO). The excellence can be spotted in governance model and it can substitute steering tools for governmental ordinance to get the better of the downside effects of agriculture production. The RSPO is one of the most valuable organizations for sustainable certification, which was formed in 2004 and started targeting large-scale production. However, 42% of the Indonesian palm oil manufacturers are smallholders who combined hold 4.42 million ha of oil palm plantation (statistik_indoneria, 2014). Moreover palm oil certification has likely positive effects on smallholders' living standards. However, Brandi et al. (2013) found that certification does not upgrade oil palm smallholders in Indonesia. Has two types, the scheme smallholders, who bind to a palm oil company gives the farmer with technical encouragement support, and the self-sufficient smallholders, who control freely and without help of palm oil companies. Successes to financial advantages, certification certainly gives to nonfinancial features. These features added to case of trading fresh fruit bunch (FFB), involvement of farmer organizations, provided knowledge and training, greater protection, and healthy surrounding traditions and verity. After collaborating with RSPO, the independent smallholders perceived greater access to miller companies, which made it easy to sell their FFB; however, few schemes are already contractually bound to a miller company for the moment they participated with Nucleus Estate Smallholder (NES) scheme.

Both self-sufficient and scheme smallholders claim that certification increases the tradeoff of knowledge and involvement of smallholder in farmer organizations. In certified schemes, smallholders take part more commonly in farmer organizations meeting than the noncertification scheme smallholders.

Likewise, after becoming certified, the participation of certified independent smallholders in farmer organization increases. By these regular meetings, members have the chance to get proper knowledge of things which are managed by the farmer organizations, which take part in visibility and liability, and about current developments in, or influence on, the palm oil sector. Moreover, certification is trusted to improve the safety and health of both independent and smallholder farmers. For instance, through certification famers could save US$26 to US$120 in case of accidents that may occur in the absence of certification. Additionally, they can recover around US$11.67–$158.13 for medical expenses. Certification also generates knowledge about the importance of environmental conservation (Brandi et al. 2013). Palm oil madrib (a large strengthened vein along the midline of a leaf) in the plantations organized by the certified farmers in a separate way, planting bamboo or other trees across the river, and not apply chemical substances across the river side to minimize erosion and pollution of waterways. Nearly all the certified scheme and self-sufficient smallholders apply soil and water conservation knowhow, which they assess as a positive effect of certification. Saving biodiversity is one of the major purposes of RSPO and was between other reasons scare by unlawful hunting practices. Therefore, adopting new techniques such as certification in oil crops give both financial and nonfinancial benefits.

19.5 Role of Oil Crops in Poverty Alleviation

Shifting diet habits from animal protein toward oil crops (particularly soy and oil palm), urbanization, fuel demand, and income growth in many developing countries have set the stage for the oil crop revolution. In the case of oil palm, majority of planters are big business tycoons and investors. About 85% of total production occurs in Malaysia and Indonesia and majority of oil palm states owned by these investors as substantial capital investments are required for replanting of 25-year production cycle without any production in initial 3–5 years. Palm oil thus cannot deemed to be a yield that is produced for the poor by the poor. Although this industry is capital intensive, and it has absorbed at least 5 million small-scale producers around the globe, producing about 40% of the global oil palm output (Byerlee et al. 2017). For instance, Indonesia has involved smallholders in this industry from many decades with the sole purpose to reduce poverty. The main aims of Indonesian oil palm sector are to reduce poverty among the smallholders, to uplift impoverished areas, the provision of labor to large industries, the resettlement of poor farmers from densely populated islands, to take and develop the lands of indigenous groups without objection, and rural development. Similarly, Thailand has also developed its oil palm industry through smallholders. Although oil palm sectors played a significant role in poverty reduction, employment generation, tax revenues, rural development, and foreign exchange in both Malaysia and Indonesia but there is still a debate regarding its positive and negative role. The oil palm places positive roles on poor masses, particularly when low-income farmers capture value from the industry, and in return they contribute more in form of higher productions and yield. For instance, Blackman and Guerrero (2012) and McCarthy et al. (2012) were on the opinion that oil palm plantations contribute positively to the economic situation of smallholders by reducing unemployment and poverty, particularly in rural areas. However, it puts negative impacts when farmers' employment conditions are poor, and the land rights, environment resources, and social conditions of smallholder are violated (McCarthy et al. 2012). Farmers really hurt when they lack the capital and when crops are destroyed through surrounding ecosystems such as bad quality air from smoke.

In West and Central Africa, where the oil palm crop originated, small farmers often cultivate oil palm in a semi-wild state with other crops. Therefore, these crops act as a staple food for local growers with limited processing. Mega or multinational companies are hunting West and Central Africa for oil crops because of increased consumption and escalated population growth rate. However, poverty can only be reduced if development of oil crops in Africa is inclusive of smallholder, by reducing the deforestation rates, by mitigating ecosystem damages, protecting the land rights of farmers, and the macroeconomic spillovers from employment and trade in the sector.

Soy production, particularly large scale, is one of the world's agricultural activities that is least labor intensive. Although, palm oil is somehow more labor intensive as compared to soy production but it is still less labor intensive when we see cocoa and rubber production. Mostly oil palm is grown on tropical areas where population is sparsely located and thus it make this crop less labor intensive. However, major issue is to retain the labor for longer period of time for oil palm production. Labor-saving innovations are used widely for some operations, but harvesting is largely manual, so the minimum labor requirement is about one worker for every 10–12 ha (Byerlee et al. 2017). Therefore the gains to the poor households will occur through rural development and economic enlargement, forex earning, public sector expenditure by the government on agriculture sector, other social sector development expenditure such as on health education etc. as a substitute development model, small-scale industry of soy ranching at present moment raising in India due to efficient allocation of resources in less than 5 ha farms produce maximum soy output, amounting to about two-thirds production. Africa produces less than 1% output of soy for the overall world soy market (Naylor 2016). In Africa, the farmers who produce in bulk, rule the market, that is, the massive producers, who hold the norms of small-scale industry of soy, despite the fact that sub-Saharan African countries has more potential from what they producing at present.

Development is this particular region is restricted due to mismanagement, acid soil, improper supply chain management, as well as low demand of their products both for oil palm and soy, and from the situation it is very clear that the potential of high growth is available in this region from the production of these products, but the barriers to development is small numbers of farmers, uncertain food security, unidentified land rights, land hoardings, rainforests (because many of multinational firms are not in favor of planting soy and oil palm in these circumstances). Nonetheless most well-known foreign companies (e.g., Unilever, Cargill, Wilmar) are presently arguing "zero deforestation" to be conducted worldwide. Without the involvement of small farmers, the phenomena of zero deforestation will not work as many small households continue to increase their production levels in tropical forests and eject high carbon soil with flaming.

19.6 Oil Crops in Global Trade

The agriculture sector globally assumes a pivotal role in both developing and developed economies. It is also a fundamental source of employment, income, and food for the associated population. This sector is by and large connected with the production of vital food crops. At present, it incorporates beekeeping, dairy, forestry, fruit cultivation, mushroom, oil crops, poultry, and so forth.

Among the various agriculture commodities, oil crops are used to produce multiple oils for human utilization. There is a wide range of oilseeds whose oil can be consumed, but only a few have got importance in global trade. Oil crops are grown in the different corners of the world, depending on the agroclimatic conditions and are an essential part of agribusiness and international trade of many economies (Sharma et al. 2012).

The oilseed demand is rising due to the persistent increase in population in the world. This demand pressure has also led farmers to adopt new technologies, increase crop area, and utilization of high-yielding seed varieties. Moreover, there is a substantial difference among different continents and countries in terms of comparative advantage due to varied agroclimatic conditions and available production technologies. Such situations have escalated the overall global trade of oil crops.

In international trade, China is a dominant global player in terms of total aggregate trade of oilseeds during 2014–2018 and has a share of 31% (225.8 billion US dollars) (see Fig. 19.2 and Table 19.1). The USA and Brazil hold second and third positions in the list and contribute 28.6% (142.8 billion US dollars) and 25.1% (125.4 billion US dollars) respectively in the total trade. Moreover, aggregate trade statistics of oilseed during 2014–2018 reveal that China is a net importer, and the US is a net exporter (see Figs. 19.2 & 19.3 and Tables 19.2 & 19.3).

The major oil crops in the global trade include coconut, corn, cotton, groundnut, oil palm, olive, perennial tree crops, rapeseed, soybean, and sunflower. The soybean crop has a significant share among other oil crops in the international trade followed by rapeseed, sunflower and mustard, and groundnut. The cultivation origin of the soybean crop was the location nearby lakes and rivers in the central plains of China. Around 3000 years prior, it spread all through Asia, and the US started its production in the early twentieth century (Sanches et al. 2004). In 2017, the Asian continent exported approximately 80% of soyabean to the other regions in which China was the major contributor (63%) while North and South America were major importers of this crop (see Fig. 19.2 and Tables 19.1 & 19.2).

The cultivation origin of rapeseed was China before 1000BP and around 1500BP in India (Li 1980; Parkash 1980). Its cultivation spread throughout Europe in the middle ages. The production of rapeseed has been increasing due to the processing method, the use of high-yielding seeds, and technological improvements in the

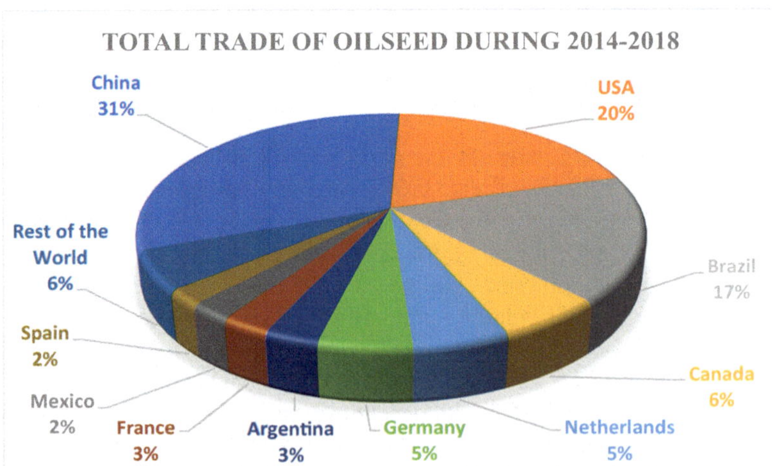

Fig. 19.2 Total trade of oilseed during 2014–2018

Table 19.1 Global leaders in total trade (export + import) of oilseed

Country	2014	2015	2016	2017	2018	Five years total	Five years average
China	49.0	42.6	41.0	47.2	46.1	45.2	225.8
USA	32.6	26.3	30.1	28.8	25.0	28.6	142.8
Brazil	24.0	21.5	19.9	26.3	33.8	25.1	125.4
Canada	8.5	7.6	8.0	8.8	8.7	8.3	41.6
Netherlands	8.1	6.9	7.3	7.8	7.6	7.5	37.6
Germany	7.6	6.6	6.9	7.5	7.9	7.3	36.5
Argentina	4.3	4.8	4.2	3.9	4.4	4.3	21.7
France	3.9	3.5	3.9	3.9	4.0	3.8	19.1
Mexico	3.7	3.1	3.1	3.4	3.7	3.4	17.0
Spain	3.5	3.0	2.9	3.0	3.2	3.1	15.6
Rest of the World	40.5	34.3	35.8	38.9	38.4	187.9	39.2

Source: UN COMTRADE and Authors' Compilation

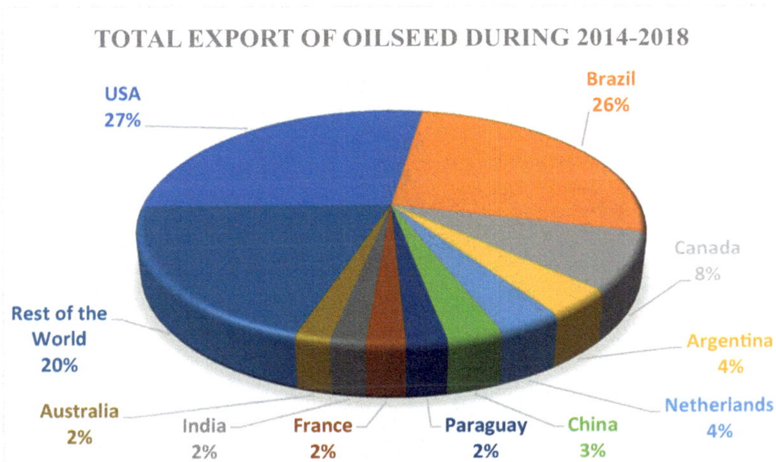

Fig. 19.3 Total export of oilseed during 2014–2018

agriculture sector. In 2017, Europe and Asia exported almost 91% of the total global exports of rapeseed to the other continents. Germany was the dominant individual country whose share was 23% (2.58 billion US dollars). The major importers were North America (46% – 4.68 billion US dollars) and Europe (46% – 5.19 billion US dollars). From North America, Canada was a major importer of rapeseed whose import was 45% (5.1 billion US dollars) of the total import (Fig. 19.4 and Tables 19.2 & 19.3).

According to Heiser (1955), the use of sunflower was observed in American Indians. Its cultivation started in New Mexico and Arizona around 3000 B.C. (Semelczi-Kovacs 1975). In 2017, the US was the main importing and exporting region of sunflower crop. Europe had an import share of about 68%, and export shares about 70% in international trade. Among the European countries,

Table 19.2 Top world exporters of oilseed

Country	2014	2015	2016	2017	2018	Five years total	Five years average
USA	28.9	23.6	27.7	26.4	22.4	129.0	25.8
Brazil	23.5	21.2	19.6	26.0	33.5	123.8	24.8
Canada	7.6	6.8	7.2	7.9	7.6	37.1	7.4
Argentina	4.2	4.7	3.8	3.1	1.8	17.7	3.5
Netherlands	3.4	2.8	3.2	3.5	3.6	16.6	3.3
China	3.1	2.9	2.7	2.6	2.7	14.1	2.8
Paraguay	2.4	1.7	1.9	2.2	2.3	10.5	2.1
France	2.0	1.8	1.9	2.0	2.2	10.0	2.0
India	2.2	1.7	1.7	1.8	1.6	8.9	1.8
Australia	2.0	1.8	1.5	2.0	1.6	8.9	1.8
Rest of the World	20.3	17.9	17.2	19.9	19.3	93.3	20.5

Source: UN COMTRADE and Authors' Compilation

Table 19.3 Top world importers of oil seed

Country	2014	2015	2016	2017	2018	Five years total	Five years average
China	45.9	39.7	38.3	44.5	43.4	211.8	42.4
Germany	6.1	5.3	5.5	5.9	6.2	29.1	5.8
Japan	5.3	4.9	4.5	4.7	4.7	24.1	4.8
Netherlands	4.6	4.1	4.1	4.3	4.0	21.0	4.2
Mexico	3.5	3.0	3.0	3.2	3.5	16.1	3.2
USA	3.7	2.7	2.4	2.5	2.5	13.8	2.8
Spain	2.8	2.3	2.3	2.4	2.4	12.3	2.5
Belgium	2.3	1.8	2.0	1.9	2.0	10.1	2.0
Turkey	2.3	1.9	1.8	1.9	1.9	9.8	2.0
France	1.9	1.7	2.0	1.8	1.8	9.2	1.8
Rest of the World	29.2	26.0	27.3	30.3	31.6	144.4	30.4

Source: UN COMTRADE and Authors' Compilation

Romania was a leading importer whose share of this crop in the total import was 17% (622 million US dollars). However, the leading exporter of sunflower seed was Turkey, which exported 7.9% (293 million U.S. dollars) of the total export of this crop (Fig. 19.5 and Table 19.3).

19.7 Conclusion

Oilseeds are an important group of crop plants whose oil can be used for food, animal feed, and chemicals. Oil plants include herbaceous plants (flax), trees (palm), and even fungi (Fusarium). Oil crops have been used from centuries and these have been the mainstay of many economies. These plants, on one hand, helped to reduce the food insecurity and on the other hand provide livelihood to hundreds of million

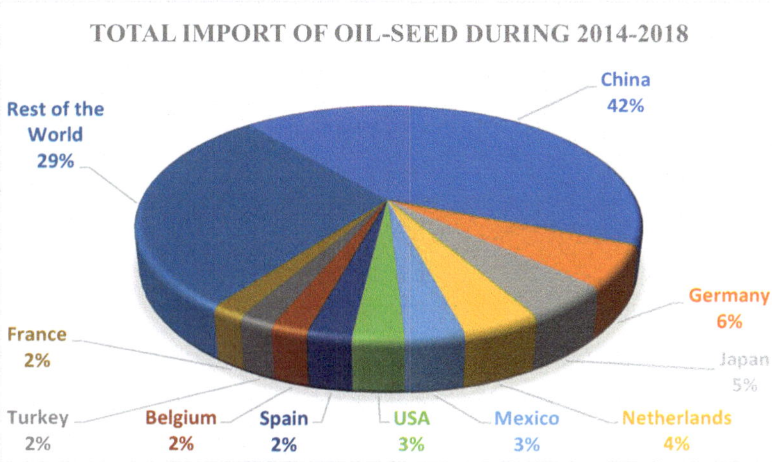

Fig. 19.4 Total import of oilseed during 2014–2018

households across the globe. Many economies, such as Malaysia, Indonesia, and some African countries are highly dependent on oil crops. Although multinational companies own major portion of oil crops, particularly palm oil and soy are dominated by these mega companies and states, however oil plants still help to reduce the poverty among the oil crops smallholders.

The oil crops (green) revolution consisted of two crops (palm oil and soy), like the Green Revolution which comprised two cereal crops – wheat and rice. This set the stage of oil crops development. Now oil crops are not only used for food and diets, but also for commercial purposes such as energy and biofuels. That is why demand for oil crop production increased after the 1970s. There are other important factors that shaped the demand of oil crops such as population, urbanization, income (GDP) of countries, price of oil crops and their substituted crops prices, and health and nutrition.

Among the various agriculture commodities, oil crops are used to produce multiple oils for human utilization. There is a wide range of oilseeds whose oil can be consumed, but only a few have got importance in global trade.

Although oil crops have become the necessary part of household food consumption across the globe and become a part of the societies' lifestyles, however the world is still witnessing shortages of oil crops. Productivities of oil crops are not matching the ever-increasing populations demand for food. Yields of oil crops are declining where rights of small farmers are being violated, where climate changes are affecting the crops and growers, and where there is shortage of labor for oil crops. Among the other issues, problem of obesity is the major one. Studies have found that excessive consumption of crops can cause cardiovascular diseases. However, other studies have found that developing countries' level of oil crops consumption increase with the increase of per capita income while in the developed

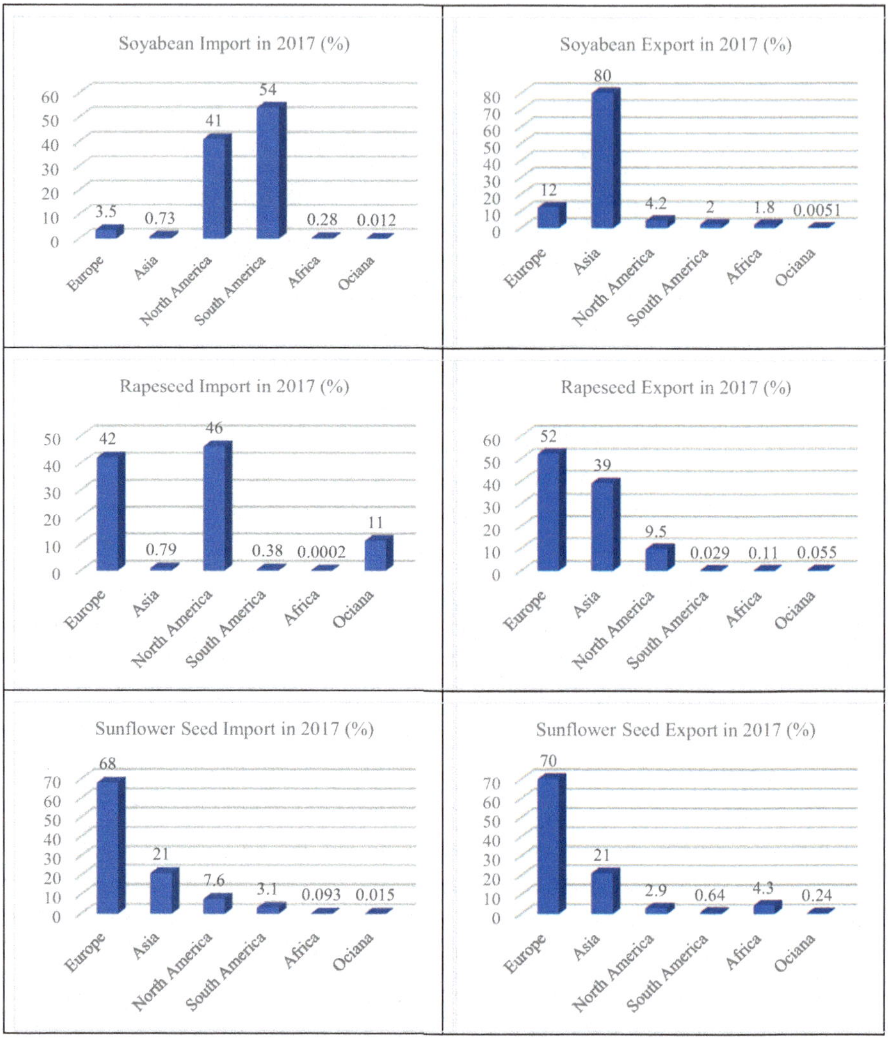

Fig. 19.5 Region-wise trade of selected oil crops

countries consumption level of crop oil decrease after a threshold level (u-shaped curve). Therefore, we recommend that future research is required to check the threshold level of income where the consumption of oil crops start declining.

References

Abubakari SB, Abdulai MHN, Anang BT (2019) Economic analysis of Groundnut production in Tolon District, Ghana. IJIRAD 3(1):192–200

Ahmed AE, Elbushra AA, Salih OA (2019) Food consumption patterns and trends in the Gulf Cooperation Council. Pak J Nutr 18:623–636. https://doi.org/10.3923/pjn.2019.623.636

Ali M, Arifullah S, Memon M.H, & Salam A (2008) Edible Oil Deficit and Its Impact on Food Expenditure in Pakistan. The Pakistan Development Review, 531-546. Retrieved from https://www.pide.org.pk/pdf/PDR/2008/Volume4/531–546.pdf on 18/03/2021

Basu S, Babiarz KS, Ebrahim S, Vellakkal S, Stuckler D, Goldhaber-Fiebert JD (2013) Palm oil taxes and cardiovascular disease mortality in India: economic-epidemiologic model. BMJ 347:f6048. https://doi.org/10.1136/bmj.f6048

Blackman A, Guerrero S (2012) What drives voluntary eco-certification in Mexico? J Comp Econ 40(2):256–268. https://doi.org/10.1016/j.jce.2011.08.001

Blasbalg TL, Hibbeln JR, Ramsden CE, Majchrzak SF, Rawlings RR (2011) Changes in consumption of omega-3 and omega-6 fatty acids in the United States during the 20th century. Am J Clin Nutr 93(5):950–962. https://doi.org/10.3945/ajcn.110.006643

Brandi C, Cabani T, Hosang C, Schirmbeck S, Westermann L, & Wiese H (2013) Sustainability certification in the Indonesian palm oil sector: benefits and challenges for smallholders (No. 74). Studies. Retrieved from https://www.econstor.eu/handle/10419/199199 on 17/03/2021

Brandi C, Cabani T, Hosang C, Schirmbeck S, Westermann L, Wiese H (2015) Sustainability standards for palm oil: challenges for smallholder certification under the RSPO. J Environ Dev 24(3):292–314. https://doi.org/10.1177/1070496515593775

Brümmer B, Korn O, Schlüßler K, Jamali Jaghdani T (2015) Volatility in oilseeds and vegetable oils markets: drivers and spillovers. J Agric Econ 67(3):685–705. https://doi.org/10.1111/1477-9552.12141

Byerlee D, Falcon WP, Naylor R (2017) The tropical oil crop revolution: food, feed, fuel, and forests. Oxford University Press, New York

Chen BK, Seligman B, Farquhar JW, Goldhaber-Fiebert JD (2011) Multi-country analysis of palm oil consumption and cardiovascular disease mortality for countries at different stages of economic development: 1980–1997. Glob Health 7(1):45. https://doi.org/10.1186/1744-8603-7-45

Das LK, Rout, RK (2018) Economic analysis of Sunflower Enterprise in Western Odisha. Int. J. Pure App. Biosci. 6(4): 498–505 http://doi.org/10.18782/2320-7051.6583 retrieved from http://www.ijpab.com/form/2018%20Volume%206,%20issue%204/IJPAB-2018-6-4-498-505.pdf on 17/03/2021

Daud ZAM, Kaur D, Khosla P (2012) Health and nutritional properties of Palm oil and its components. In: *Palm oil*. AOCS Press, pp 545–560. https://doi.org/10.1016/B978-0-9818936-9-3.50021-6

Depenbusch L, Klasen S (2019) The effect of bigger human bodies on the future global calorie requirements. PLoS One 14(12). https://doi.org/10.1371/journal.pone.0223188

Drewnowski A, Popkin BM (1997) The nutrition transition: new trends in the global diet. Nutr Rev 55(2):31–43

Etilé F, Oberlander L (2019) The Economics of diet and obesity: understanding the global trends. In Oxford research encyclopedia of economics and finance. https://doi.org/10.1093/acrefore/9780190625979.013.19

Fakorede DO, Babatunde AI, Ovat A (2014) Productivity increase by optimum utilization of machines and manpower energy. Int J Eng Res Dev 10(5):11–24

FAO (2018) Food outlook: biannual report on global food markets. Retrieved from http://www.fao.org/fileadmin/templates/est/COMM_MARKETS_MONITORING/Oilcrops/Documents/Food_outlook_oilseeds/FO_Oilcrops.pdf. Accessed on 09/03/2020)

Fattore E, Fanelli R (2013) Palm oil and palmitic acid: a review on cardiovascular effects and carcinogenicity. Int J Food Sci Nutr 64(5):648–659. https://doi.org/10.3109/09637486.2013.768213

Fattore E, Bosetti C, Brighenti F, Agostoni C, Fattore G (2014) Palm oil and blood lipid–related markers of cardiovascular disease: a systematic review and meta-analysis of dietary intervention trials. Am J Clin Nutr 99(6):1331–1350. https://doi.org/10.3945/ajcn.113.081190

Food and Agriculture Organization of the United Nations (1994) Expert's recommendations on fats and oils in human nutrition. Agriculture and Consumer Protection Department. http://www.fao.org/docrep/ t4660t/ t4660t02.htm (accessed 31 January 2018).

Fry J (2011) A global challenge: markets and oil palm expansion. http://static.zsl.org/files/session-1-2-james-fry-a-global-challenge-markets-oil-palm-expansion-1463.pdf

Green R, Cornelsen L, Dangour AD, Turner R, Shankar B, Mazzocchi M, Smith RD (2013) The effect of rising food prices on food consumption: systematic review with meta-regression. BMJ 346:f3703. https://doi.org/10.1136/bmj.f3703
Gulati A, Kelley T (1999) Trade liberalization and Indian agriculture: cropping pattern changes and efficiency gains in semi- arid tropics. Oxford University Press, New York
Guo X, Mroz TA, Popkin BM, Zhai F (2000) Structural change in the impact of income on food consumption in China, 1989–1993. Econ Dev Cult Chang 48(4):737–760
Heiser CB (1955) Origin and development of the cultivated sunflower. Am Biol Teach 17:161–167
Hidayat NK, Offermans A, Glasbergen P (2016) On the profitability of sustainability certification: An analysis among Indonesian palm oil smallholders. J Econ Sustain Dev 7(18):45–62
Huang KS, Lin BH (2000) Estimation of food demand and nutrient elasticities from household survey data (No. 1488-2016-123635)
Huang J, Yang J, Deng X, Wang J, Rozelle S (2015) Urbanization, food production and food security in China (No. 331-2016-14070)
Iliyasu A, Wulet I, Yusuf K (2008) Profitability analysis of groundnuts processing in Maiduguri Metropolitan Council of Borno State, Nigeria. Niger J Basic Appl Sci 16(2):253–256
Iowa State University (2015) Ag decision maker. File A1–21. Iowa State University of Science and Technology. http://www.extension.iastate.edu/agdm/crops/pdf/a1-21.pdf. Accessed 29 Feb 2020
Khoury CK, Bjorkman AD, Dempewolf H, Ramirez-Villegas J, Guarino L, Jarvis A et al (2014) Increasing homogeneity in global food supplies and the implications for food security. Proc Natl Acad Sci 111(11):4001–4006
Konwar K, Bharadwaj ASK, Boruah P (2019) Productivity and profitability of rapeseed (Brassica rapa var. dichotoma) under system of rapeseed intensification. J Oilseed Brassica 10(2):106–111. Retrieved from http://srmr.org.in/ojs/index.php/job/article/view/348/225
Kruse J (2010) Estimating demand for agricultural commodities to 2050. Global Harvest Initiative, Washington, DC
Kumar P, Kumar A, Shinoj P, Raju SS (2011) Estimation of demand elasticity for food commodities in India §. Agric Econ Res Rev 24(1):1–14
Kurup S, Jha G, Singh A (2015) Technical and efficiency changes in oilseed sector in India: Implications for policy (No. 1008-2016-80244). https://ageconsearch.umn.edu/record/212017/. Accessed on 29 Feb 2020
Lennerts L (1983) Oelschrote, oelkuchen, pflanzliche Oele und Fette, Herkunft, Gewinning, Verwendung, Bonn 1983. Alfred Strothe, Hannover
Li CS (1980) Classification and evaluation of mustard crops (Brassica juncea) in China. Cruciferace Newslett 5:33–36
McCarthy JF, Gillespie P, Zen Z (2012) Swimming upstream: local Indonesian production networks in "Globalized" Palm oil production. World Dev 40(3):555–569. https://doi.org/10.1016/j.worlddev.2011.07.012
Micha R, Mozaffarian D (2010) Saturated fat and cardiometabolic risk factors, coronary heart disease, stroke, and diabetes: a fresh look at the evidence. Lipids 45(10):893–905. https://doi.org/10.1007/s11745-010-3393-4
Naylor RL (2016) Oil crops, aquaculture, and the rising role of demand: a fresh perspective on food security. Glob Food Sec 11:17–25. https://doi.org/10.1016/j.gfs.2016.05.001
Overseas Development Institute (2015) The rising cost of a healthy diet: changing relative prices of foods in high- income and emerging economies. Overseas Development Institute, London
Pargar F (2017) Resource optimization techniques in scheduling: applications to production and maintenance systems. University of Oulu, Oulu
Popkin BM (2001) Nutrition in transition: the changing global nutrition challenge. Asia Pac J Clin Nutr 10:S13–S18. https://doi.org/10.1046/j.1440-6047.2001.0100s1S13.x
Prakash S (1980) Cruciferous oilseeds in India. In: Tsunoda S, Hinata K, Gomez-Campo C (eds) Brassica Crops and Wild Allies. Japan Scientific Press, Tokyo, pp 151–163

PRB (Population Reference Bureau) (2015) Population data sheets. http://www.prb.org/Publications/Datasheets/2015/2015-world-population-data-sheet.aspx. Accessed 7.02.20

Rueda X, Lambin EF (2014) Global agriculture and land use changes in the twenty-first century. In: The evolving sphere of food security, vol 319. Oxford University Press, Oxford

Sanches AC, Michellon E, Roessing AC (2004) Os limites de expansão da soja. Informe GEPEC 9(1)

Semelczi-Kovacs A (1975) Acclimatization and dissemination of the sunflower in Europe. (In German). Acta Ethnogr Acad Sci Hung 24:47–88

Sharma M, Gupta SK, Mondal AK (2012) Production and trade of major world oil crops. In: Technological innovations in major world oil crops, vol 1. Springer, New York, pp 1–15

Subervie J, Vagneron I (2013) A drop of water in the Indian Ocean? The impact of GlobalGap certification on lychee farmers in Madagascar. World Dev 50:57–73. https://doi.org/10.1016/j.worlddev.2013.05.002

UN COMTRADE – United Nations Commodity Trade Statistics Database (2015) https://comtrade.un.org/. Accessed 5 Mar 2020

United Nations (UN) (2014) World urbanization prospects: the 2014 revision. UN Department of Economic and Social Affairs, Population Division, New York

Zhai FY, Du SF, Wang ZH, Zhang JG, Du WW, Popkin BM (2014) Dynamics of the Chinese diet and the role of urbanicity, 1991–2011. Obes Rev 15:16–26. https://doi.org/10.1111/obr.12124

Chapter 20
Production and Trade of Oil Crops, and Their Contribution to the World Economy

Dilek Tokel and Bedriye Nazli Erkencioglu

Contents

20.1 Introduction

Since the birth of civilization, plants have taken an important place in human life as fundamental nutrients. Among these, crop plants, especially cereals, oilseeds, pulses, vegetables, aromatic and medicinal plants, etc., are the main sources of nutrition of many different types of living organisms in the nature. Oilseed crops belonging to various families are important as they are used both as oil source and raw material in various industries (Maheshwari and Kovalchuk 2016). Currently, about 40 different oilseed crops are consumed, but only a few have commercial significance. The most preferred oilseed crops are soybean, sunflower, groundnut,

D. Tokel (✉)
Department of Economics, Faculty of Economics, Marmara University, Istanbul, Turkey
e-mail: dozyigit@marmara.edu.tr

B. N. Erkencioglu
Department of Medical Pathology, Faculty of Medicine, Istinye University, Istanbul, Turkey

H. Tombuloglu et al. (eds.), *Oil Crop Genomics*,
https://doi.org/10.1007/978-3-030-70420-9_20

sesame, and castor since they are renewable energy sources and associated with power generation (Efe et al. 2018; Qi et al. 2020). Sunflower, soybean, and canola crops provide reduction in irrigation as well as in input costs. The oil obtained from the seeds is divided into three major groups. Among them, annual oilseeds include soybean, rapeseed, sunflower, and groundnut. Perennial tree crops create a group that includes oil palm and coconut. Finally, the last group includes cotton and corn. Oilseeds other than these are also used as animal nutrients in addition to human food ingredient, although they have a small share in trade (Sharma et al. 2012; Waseem et al. 2017).

20.2 Latest Trends in Oilseed Production

In the last decade, developments in the fields of biotechnology and genetic engineering as well as conventional breeding provide high-quality seed production (Rai et al. 2019). Therefore, there is a significant increase in annual oilseed production. In addition, oilseeds occupy a large area in world agriculture and are grown in an area of approximately 180 million hectares. The annual average oilseed production has been increasing since 1997, especially in the last 2 years. Considering the data of the last 5 years, it is seen that total production of oilseed reached 606.15 million metric tons (mmt) in June 2020, while 573.75 mmt in 2016/2017. In global distribution of oilseeds, soybean, rapeseed, peanut, cottonseed, and sunflower seed are prominent (Table 20.1). Soybean has the highest contribution in the world economy since the production reached 362.85 mmt in the first 6 months of 2020, accounting for approximately 60% of total oilseed crop production, followed by rapeseed (12%), sunflower seed (9%), peanut (8%), cottonseed (7%), palm (3%), and copra (1%), respectively. However, when looking at vegetable oil production, it is seen that palm oil has the largest contribution in the economy with a share of 36%, by leaving soybean behind (Table 20.2). Additionally, as of June 2020, palm

Table 20.1 Annual oilseed production in the world in mmt

Oilseed Crop	2016/17	2017/18	2018/19	2019/20	Jun 2020/21
Soybean	348,30	341,74	360,26	335,35	362,85
Rapeseed	69,49	75,02	72,61	68,20	70,79
Sunflower seed	48,23	47,85	50,54	54,98	56,78
Peanut	45,26	47,09	47,09	46,05	46,06
Cottonseed	39,09	45,24	43,28	44,96	43,64
Palm kernel	17,85	19,39	20,17	19,85	20,28
Copra	5,52	5,94	5,98	5,86	5,77
Total	**573,75**	**582,26**	**599,92**	**575,23**	**606,15**

Source: USDA (2020a)

Table 20.2 Annual vegetable oils production in the world in mmt

Oilseed Crop	2016/17	2017/18	2018/19	2019/20	Jun 2020/21
Palm	65,34	70,58	74,02	72,27	74,60
Soybean	53,82	55,09	55,64	56,60	58,70
Rapeseed	27,55	28,06	27,68	27,31	27,36
Sunflower seed	18,19	18,53	19,34	21,25	21,48
Palm	7,83	8,53	8,87	8,73	8,91
Peanut	5,72	5,92	5,87	6,09	5,94
Cottonseed	4,38	5,09	4,97	5,12	5,03
Coconut	3,41	3,67	3,74	3,62	3,58
Olive	2,61	3,27	3,25	3,12	3,03
Total	**188,85**	**198,72**	**203,38**	**204,09**	**208,63**

Source: USDA (2020a)

production is followed by soybean (28%), rapeseed (13%), sunflower (10%), palm kernel (4%), peanut (3%), cotton (2%), coconut (2%), and olive (1%) production, respectively (Table 20.2). Overall, a rising trend in oilseed and vegetable oils has been observed since 2016 (USDA 2020a).

20.3 Cultivation and Use of Oilseed Crops

The most important product obtained from oilseed plants is oil, which can be used as both food and feedstock. Oilseed crops are the main sources of biodiesel manufacture, which are alternative to fuel in the oil industry (Valladares-Diestra et al. 2020). They can also be used as an alternative to vegetable oils, and thanks to biotechnology, crops can be transformed to provide the desired oil composition feature (Waseem et al. 2017). In general, oilseed production takes place in temperate regions, primarily in America and Europe. The United States, which had the most oilseed cultivation by 2019, left its place to Brazil in 2019 and in the first 6 months of 2020. As of June 2020, 22% of total oilseed crop cultivation is performed in Brazil. Brazil is followed by the US (20%), China (10.2%), Argentina (9.7%), and India (6%), respectively (Table 20.3) (USDA 2020a).

On the other hand, when looking at vegetable oil production, Indonesia (24%) ranks first in June 2020. Indonesia is followed by China (13%), Malaysia (10%), European Union (9%), the US (6%), Argentina (5%), and Brazil (4.76%), respectively (Table 20.4).

Table 20.3 Major oilseed-producing countries in mmt

Country	2016/17	2017/18	2018/19	2019/20	Jun 2020/21
Others	176,91	187,34	188,78	184,17	189,36
US	126,94	131,48	130,72	107,00	123,20
Brazil	117,59	125,81	123,93	129,15	135,68
Argentina	60,16	42,61	61,00	55,08	59,31
China	55,09	59,60	59,95	62,63	61,90
India	37,05	35,43	35,55	37,20	36,71
Total	**573,75**	**582,26**	**599,92**	**575,23**	**606,15**

Source: USDA (2020a)

Table 20.4 Major vegetable oil–producing countries in mmt

Country	2016/17	2017/18	2018/19	2019/20	Jun 2020/21
Others	51,77	54,33	56,44	58,36	59,17
Indonesia	41,10	45,08	47,22	48,34	49,43
China	26,76	27,77	26,43	26,73	27,75
Malaysia	21,13	22,02	23,24	20,72	21,60
EU	18,07	19,04	18,92	18,14	18,14
US	11,43	12,11	12,20	12,49	12,62
Argentina	9,87	8,78	9,48	9,47	9,97
Brazil	8,73	9,60	9,47	9,85	9,95
Total	**188,85**	**198,72**	**203,38**	**204,09**	**208,63**

Source: USDA (2020a)

20.4 Oilseed Crops

20.4.1 Soybeans

Soybean (*Glycine max* L.), a vegetable crop belonging to the Fabaceae family, grows up to 1.5 meters in length, with a seed coat of different colors, depending on the variety that seed does not germinate if it is cracked. The rapid increase in the use of soybean oil and meal demand caused increase in soybean production in recent years (Miranda et al. 2019). Soybeans are also used as protein supplement, cooking oil, flour, infant formula, and in pharmaceutical industry (Sánchez-Duarte et al. 2019). The oleic acid found in soybean is one of the reasons why the plant has industrial importance (Waseem et al. 2017).

Soybean oilseed US farm prices increased from $309 dollars in 2018/19 period to $316 dollars per metric ton as of May 2020. Although soybean oil price has declined since the 2016/17 season, the average price increased by $40 dollars per metric ton compared to the previous year and reached $557 dollars in the sixth month of 2020. In the first 6 months of 2020, Brazil ranked first by producing 131,000 metric tons (tmt) soybean and this is 36.1% of the total production of the

Table 20.5 Soybean production in tmt

Country	2016/17	2017/18	2018/19	2019/20	Jun 2020/21
Brazil	114,600	122,000	119,000	124,000	131,000
US	116,931	120,065	120,515	96,676	112,264
Argentina	55,000	37,800	55,300	50,000	53,500
China	13,596	15,283	15,967	18,100	17,500
India	10,992	8350	10,930	9300	10,500
Paraguay	9163	10,478	8850	9900	10,250
Canada	6597	7717	7267	6000	6150
Others	21,419	20,051	22,428	21,371	21,864
Total	**348,298**	**341,744**	**360,257**	**335,347**	**362,848**

Source: USDA (2020a)

Table 20.6 Soybean oil: world supply in tmt

Country	2016/17	2017/18	2018/19	2019/20	Jun 2020/21
China	15,770	16,128	15,232	15,680	16,755
US	10,035	10,783	10,976	11,154	11,276
Brazil	7755	8485	8150	8500	8640
Argentina	8395	7236	7190	7950	8385
EU	2736	2841	2964	3002	2983
India	1620	1386	1730	1495	1692
Mexico	820	937	1100	1110	1145
Others	6692	7292	7582	7704	7825
Total	**53,823**	**55,088**	**55,644**	**56,595**	**58,701**

Source: USDA (2020a)

world. Production in Brazil is projected to increase even more in the next decade, as cultivated areas increase with crop intensification. Furthermore, tariffs introduced on the US by China will provide an advantage for the Brazilian soybean in this race. Currently, Brazil has the major contribution to the oilseed economy and it is followed by the US, Argentina, China, India, Paraguay, and Canada (Table 20.5). However, China makes the biggest economic contribution to soybean oil production (Table 20.6). In the first 6 months of 2020, China, which produces 16.755 tmt, is followed by the US, Brazil, Argentina, EU, India, and Mexico. Also, according to the latest data, soybean oil production has increased in all countries except EU (USDA 2020a).

In the last 3-year period, countries exporting the highest amount of soybean oilseeds are South America, Brazil, North America, the US, and Argentina. The top countries in soybean oilseed import are East Asian countries, followed by China, EU, and Southeast Asian countries. In soybean oil trade, as of June 2020, while Argentina (51%), Brazil (9%), the US (8%), EU (7%), and Paraguay (%6) are among the countries that export the most; India (29%), China (11%), Algeria

(7%), and Bangladesh (7%) are the leading countries that import the most (USDA 2020a).

20.4.2 Rapeseed

Brassicaceae family member rapeseed is an important source of protein derived from the plant. It is the most produced oilseed crop after soybean. It is a well-known bioenergy crop and its oil contains crucial fatty acids such as palmitic, erucic, oleic, and eicosenoic acid (Konuskan et al. 2019). The most commonly cultivated species that also used in vegetable oil production are *Brassica rapa* L., *Brassica juncea* (L.) Czern., *Brassica napus* L., and *Brassica carinata* A. Braun. However, seeds produced mainly from *B. napus* and *B. rapa* are preferred in the world trade of rapeseed in Canada and Australia. Other areas where rapeseed can be used are the production of wood fuel pellets of rapeseed cake and sawdust mixture (Azargohar et al. 2019).

Rapeseed prices increased from $420 dollars per metric ton in the previous year to an average of $428 dollars during the first 5 months of 2019/20 period. The total harvested rapeseed area in the period of 2020/21 is 35.57 mmt and the total production is 70.79 mmt. However, when compared to the 2017/18 period, a decreasing trend is observed in the production and cultivated area. It is observed that rapeseed oil production has also decreased after 2017/18 period. As shown in Table 20.7, the countries have the highest contribution to the world economy in rapeseed oil production are EU countries, China, Canada, India, and Japan, respectively (Table 20.7) (USDA 2017, 2020a).

Despite the decrease in its production, it is seen that rapeseed oil import and export have increased in the last 5-year period. As of June 2020, the top countries in rapeseed oil export are Canada (64%), EU (4%), China (0.1%), India (0.05%), and Japan (0.01%). China (32%) has the biggest share in rapeseed oil imports, followed by EU (4%), India (2%), and Japan (1%) (USDA 2020a).

Table 20.7 Rapeseed oil: world supply in tmt

Country	2016/17	2017/18	2018/19	2019/20	2020/21
China	6669	6745	6425	5928	5967
India	2200	2100	2622	2660	2584
Canada	4020	4020	4048	4350	4350
Japan	1061	1075	1045	990	990
EU	10,199	10,450	9823	9395	9363
Other	3838	4060	3717	3988	4106
Total	**27,987**	**28,450**	**27,680**	**27,311**	**27,360**

Source: USDA (2017, 2020a)

20.4.3 Cotton

Cotton crop (*Gossypium* spp.), which is an important source of textile and food industries, is also important with its seeds besides its fiber (Tausif et al. 2018). Although genus *Gossypium* consists of 51 species, there are four species of cotton cultivated; *Gossypium herbaceum* L., *Gossypium arboretum* L., *Gossypium hirsutum* L., and *Gossypium barbadense* L. Among these, *G. hirsutum* L. is native to Mexico and Central America (Iqbal et al. 2001; Dogan et al. 2012). Its fibers are used in textile products, cordage, and it is now grown in more than 90 countries in the world (Ozyigit 2012; Rauf et al. 2019). Cotton seeds are used in industrial cellulose production, as well as for culinary purposes (Dogan et al. 2012). Content of cottonseed is rich in proteins (high in alanine), palmitic, and oleic acids (56%, 23%, and 17%, respectively). However, this feature cannot be used due to the toxic defense mechanism called gossypol, which exists against various insect pests in cotton crops (Krist 2020).

Following the sharp decline in cottonseed oil US farm prices in 2017, prices have increased in the recent years. In the first 6 months of 2020, the price of cottonseed oil per metric ton has reached $843 dollars. As of June 2020, the countries providing the major contribution to the world economy in cotton production are India, China, US, Brazil, Pakistan, Turkey, and Uzbekistan (Table 20.8). As shown in Table 20.9, as of 2020, countries with the largest share in production of cottonseed oil are China (27%) and India (27%), followed by the US (5%), Turkey (4%), and the EU (1%). Although cottonseed oilseed production increased in 2017/2018, production has been around 44 mmt in recent years. Also, a gradual increase has been observed in cottonseed oilseed import and export since 2017/18 period (USDA 2020b).

20.4.4 Palms

Palm (*Areca catechu* L.) is one of the most important sources of vegetable oil, usually grown in tropical areas. Approximately 36% of the overall vegetable oil production in the world is obtained from palms (Sharma et al. 2012; Tani et al. 2020).

Table 20.8 Cotton: World supply in tmt

Country	2016/17	2017/18	2018/19	2019/20	Jun 2020/21
India	5879	6314	5617	6641	6205
China	4953	5.987	6042	5933	5770
US	3738	4555	3999	4336	4246
Brazil	1528	2007	2830	2874	2613
Pakistan	1676	1785	1655	1350	1372
Turkey	697	871	816	751	718
Uzbekistan	811	840	713	762	708
Total	**3943**	**4628**	**4161**	**4127**	**4221**

Source: USDA (2020b)

Table 20.9 Cottonseed oil: World supply in tmt

Country	2016/17	2017/18	2018/19	2019/20	Jun 2020/21
China	1115	1251	1374	1356	1339
India	1160	1320	1106	1390	1390
Turkey	187	250	7	202	191
US	246	343	110	222	236
EU	38	44	13	55	57
Others	1656	1808	1779	1890	1819
Total	**4402**	**5016**	**5943**	**5115**	**5032**

Source: USDA (2017, 2020a)

Table 20.10 Palm oil: World supply in tmt

Country	2016/17	2017/18	2018/19	2019/20	Jun 2020/21
Indonesia	36,000	39,500	41,500	42,500	43,500
Malaysia	18,858	19,683	20,800	18,500	19,300
Thailand	2500	2780	3000	2800	3100
Colombia	1146	1627	1632	1529	1670
Nigeria	990	1025	1015	1015	1015
Others	5845	5960	6077	5927	6013
Total	**65,339**	**70,575**	**74,024**	**72,271**	**74,598**

Source: USDA (2020a)

The total palm kernel production in 2018 was 18,111 mmt in worldwide and palm oil prices reached an average of $628 dollars per metric ton in the 2019/2020 period, despite the lowest level of the last decade in December 2018 ($489 dollars per metric ton). The highest production was made in Indonesia (10,330 mmt) and Malaysia (4800 mmt). As of 2020, the countries with the highest contribution to the world economy in palm oil production are Indonesia (58%), Malaysia (26%), Thailand (4%), Colombia (2%), and Nigeria (1%), respectively. When the data of the last 5 years is analyzed, it is seen that Indonesia is also the first in palm oil export in worldwide, followed by Malaysia, Guatemala, and Colombia.

There is an overall increase in total palm oil production, import, and export in the world, excluding the 2019/20 period. As of June 2020, the countries with the largest amount of palm oil imports are India, China, EU, Pakistan, and Bangladesh (USDA 2020a) (Table 20.10).

20.4.5 Sunflower

Sunflower (*Helianthus annuus* L.) originates from Central and North America and is the third most grown oilseed crop in the world after soybean and rapeseed (Ozyigit et al. 2006; Wales et al. 2019). It is mostly grown under rain-fed conditions and water requirement is much higher compared to other oil crops. Sunflower seed is

Table 20.11 Sunflower seed oil production in tmt

Country	2016/17	2017/18	2018/19	2019/20	2020/21
Argentina	1353	1465	1425	1350	1413
Russia	4192	4130	4875	5782	5804
Turkey	761	822	1022	1055	1065
Ukraine	6351	5590	6364	7055	7160
EU	3338	3465	3670	3676	3676
Others	2223	2238	1979	2334	2361
Total	**18,218**	**17,710**	**19,335**	**21,252**	**21,479**

Source: USDA (2017, 2020a)

used for food, coffee, fuel, livestock, medicinal purposes, pulmonary afflictions, and manufacturing cosmetics (Ozyigit et al. 2002). The total area of sunflower seed harvested in the world was 26,668,101 ha in 2018 and 51,954 mmt was produced in total (FAO 2020a, b).

In the last 5 years, there has been an increase in the harvesting area of sunflower seed and also in the sunflower seed oil production. Ukraine has been in the first place in the production of sunflower seed oil for the last 5 years, and has approximately 33% of total production. After Ukraine, Russia, EU, Argentina, and Turkey are the highest sunflower seed oil–producing countries, respectively (Table 20.11) (USDA 2017 2020a).

In recent years, there has been a more significant increase in sunflower seed prices compared to other oilseeds. Sunflower seed US farm prices per metric ton have increased from $389 dollars on an average in 2018/19 period to $421 dollars in the first 6 months of 2020. Moreover, sunflower seed oil prices reached the highest level in the last 10 years in the 2019/2020 period, reaching an average of $1527 dollars per metric ton. Overall, sunflower seed oil imports and exports have been increasing since 2012, excluding the 2017/18 period. The countries that have the biggest share in sunflower seed oil export in 2020/21 period are Ukraine, Russia, and Argentina, while the EU countries made the major import in the same period (USDA 2020a).

20.4.6 Peanut

Peanut, also known as groundnut (*Arachis hypogaea* L.) originates from South America (Chen et al. 2019). The crop is adapted to well-drained, loose soils and both tropical and temperate regions. Its seeds are important sources of oil and protein. It can be used as cooking oil as well as peanut butter, cosmetics, plastics, dyes, and textile materials. Although it ranks sixth in vegetable oil production in the world, it is fourth in oilseed production. While the area with shell groundnuts harvested in the world was 28,515,387 ha in 2018, total groundnuts production is 45,950 mmt. China ranked first in 2018 in groundnut production with 17,332 mmt. China was followed by India, Nigeria, and Sudan (FAO 2020b).

Table 20.12 Peanut oil production in tmt

Country	2016/17	2017/18	2018/19	2019/20	2020/21
China	2864	2960	2928	3040	2928
India	1240	1190	1090	1139	1106
Turkey	7	7	7	7	7
US	129	142	99	115	110
EU	12	12	13	13	13
Others	11,611	1631	1734	1771	1779
Total	**5863**	**5942**	**5871**	**6085**	**5943**

Source: USDA (2017, 2020a)

Although peanut oilseed prices reached $508 dollars per metric ton in 2017/18, it fell to $445 dollars in the first 6 months of 2019/20. When peanut oil prices are analyzed, it is seen that the prices are fluctuated in the range of $1350–1450 dollars as of 2018/19 period. Average prices per metric ton have declined since the 2016/17 season and reached $1334 in the 2019/20 period. As of 2020, China, India, Turkey, and the US have the major share in peanut oil production (Table 20.12). Also, peanut oil production has not increased in any of these countries in recent years. In 2020/21 period, most of the peanut oil import belonged to China, and it is 65% of total import. However, the country with the biggest contribution to export is India (11%) (USDA 2020a).

20.4.7 Coconut

Coconut (*Cocos nucifera* L.) mainly grows in coastal areas, under high temperature and humid conditions. It is believed to be originated from Asia. Copra, which is the dried meat or kernel of coconut palm, is the fruit of the plant (Stein et al. 2015). One of the most important features of coconut oil is that it has a high melting point as well as a pleasant smell. Moreover, it has a certain oxidation resistance and it is rich in short-chain fatty acids. It is easy to digest, can be used as an oil source in foods such as infant milk and ice cream, and also it can be converted into hard butter in order to be used in confectionery (Mahisanunt et al. 2019). The most important share in the world economy in coconut supply belongs to Indonesia and the Philippines. While total area of coconut harvested was 12,381,051 ha in 2018, 61,865 mmt of coconut was produced in total. Approximately one-third of total production belongs to Indonesia (FAO 2020b).

A slight change was seen in overall copra oilseed production from 2016 to 2020/21 period. As of June 2020, the production decreased by 0.9 mmt compared to the previous year and became 5.77 mmt. The average price of copra oilseed per metric ton increased by $98 dollars compared to the previous year and reached $581 dollars in 2019/20 period. Moreover, coconut oil US farm price reached the highest level in 2016/2017, reaching an average of $1621 per metric ton. However, in the following years, there was a sharp decline in prices and decreased to $643 dollars

Table 20.13 Olive oil production in tmt

Country	2016/17	2017/18	2018/19	2019/20	2020/21
China	5	5	4	5	5
Turkey	180	200	183	235	190
US	15	16	16	16	16
EU	1750	1800	2400	2000	2050
Others	537	682	647	859	771
Total	**2487**	**2703**	**3250**	**3115**	**3032**

Source: USDA (2020)

per metric ton in June 2019. Prices increased again in 2019/20 period, and reached an average of $869 dollars as of May 2020. On the other hand, considering the production of coconut oil, it is seen that, despite the slight increase in the 2018/19 period, it has been on average around 3.50 mmt in the last 5 years. Although coconut oil import has increased in recent years, its export has decreased since 2018/19 period (USDA 2020a).

20.4.8 Olive

Olive tree (*Olea europaea* L.) is a perennial crop that originates from the Eastern Mediterranean region (Bourgeon et al. 2018). This region has a mild winter and a hot, dry summer (Villa et al. 2020). Olive harvested area in the world in 2018 was 10,513,320 ha and total olive production was 21,066 mmt. Spain, which is one of the most olive-producing countries in the same year, ranks first with 9819 mmt of production (FAO 2020b). Spain is followed by Italy, Morocco, Turkey, and Greece. On the other hand, when looking at olive oil production, it is seen that the EU is the first in the last 5-year period and followed by Turkey and the US (Table 20.13) (USDA 2017, 2020a).

There is a downward trend in both olive oil production and olive oil export and import in the world in 2020/21 period. As of 2020, the EU, the major olive oil exporter, while Turkey ranks second, and the US third. In olive oil import, the US ranks first, followed by EU and China (USDA 2020a).

20.5 Conclusion and Future Perspective

In this chapter, annual oil crop and oilseed production, export and import information, and their contributions to the world economy are presented, especially in the last 5 years. With the new biotechnological methods, it is aimed to improve oil crop varieties and increase yields by transferring the desired features to crops. In recent years, the demand for oil crop has been increasing since it can be used as vegetable oils, livestock feeds, as well as in pharmaceuticals, biofuels, and various industries. Also, oilseed crop cultivated areas have increased in order to meet this demand. The contribution of oil crop production to the world economy is also gaining more and more importance. Oil crop export and import, which is carried out in significant amounts annually, also contribute to the national economies by increasing the trade between the countries.

References

Azargohar R, Nanda S, Kang K, Bond T, Karunakaran C, Dalai AK, Kozinski JA (2019) Effects of bio-additives on the physicochemical properties and mechanical behavior of canola hull fuel pellets. Renew Energy 132:296–307. https://doi.org/10.1016/j.renene.2018.08.003

Bourgeon O, Pagnoux C, Mauné S, Vargas EG, Ivorra S, Bonhomme V et al (2018) Olive tree varieties cultivated for the great Baetican oil trade between the 1st and the 4th centuries AD: morphometric analysis of olive stones from Las Delicias (Ecija, Province of Seville, Spain). Veg Hist Archaeobotany 27(3):463–476. https://doi.org/10.1007/s00334-017-0648-5

Chen Y, Wang Z, Ren X, Huang L, Guo J, Zhao J et al (2019) Identification of major QTL for seed number per pod on chromosome A05 of tetraploid peanut (*Arachis hypogaea* L.). The Crop Journal 7(2):238–248. https://doi.org/10.1016/j.cj.2018.09.002

Dogan I, Ozyigit II, Demir G (2012) Mineral element distribution of cotton (*Gossypium hirsutum* L.) seedlings under different salinity levels. Pak J Bot 44(SI):15–20

Efe S, Ceviz MA, Temur H (2018) Comparative engine characteristics of biodiesels from hazelnut, corn, soybean, canola and sunflower oils on DI diesel engine. Renew Energy 119:142–151. https://doi.org/10.1016/j.renene.2017.12.011

FAO (2020a) Food and agriculture organization of the United Nations. Available from: http://www.fao.org/land-water/databases-and-software/crop-information/sunflower/en/. Last accessed on 7 Jul 2020

FAO (2020b) Food and agriculture organization of the United Nations. Available from: http://www.fao.org/faostat/en/#data/QC/visualize. Last accessed on 7 Jul 2020

Iqbal MJ, Reddy OUK, El-Zik KM, Pepper AE (2001) A genetic bottleneck in the evolution under domestication of upland cotton *Gossypium hirsutum* L. examined using DNA fingerprinting. Theor Appl Genet 103(4):547–554. https://doi.org/10.1007/PL00002908

Konuskan DB, Arslan M, Oksuz A (2019) Physicochemical properties of cold pressed sunflower, peanut, rapeseed, mustard and olive oils grown in the Eastern Mediterranean region. Saudi Journal of Biological Sciences 26(2):340–344. https://doi.org/10.1016/j.sjbs.2018.04.005

Krist S (2020) Cottonseed oil. In: Vegetable fats and oils. Springer, Cham, pp 281–287. https://doi.org/10.1007/978-3-030-30314-3_43

Maheshwari P, Kovalchuk I (2016) Genetic transformation of crops for oil production. In: McKeon TA, Hayes DG, Hildebrand DF, Weselake RJ (eds) Industrial oil crops. AOCS Press, Urbana, pp 379–412. https://doi.org/10.1016/B978-1-893997-98-1

Mahisanunt B, Hondoh H, Ueno S (2019) Effects of tripalmitin and tristearin on crystallization and melting behavior of coconut oil. J Am Oil Chem Soc 96(4):391–404. https://doi.org/10.1002/aocs.12202

Miranda C, Culp C, Škrabišová M, Joshi T, Belzile F, Grant DM, Bilyeu K (2019) Molecular tools for detecting Pdh1 can improve soybean breeding efficiency by reducing yield losses due to pod shatter. Mol Breed 39(2):27. https://doi.org/10.1007/s11032-019-0935-1

Ozyigit II (2012) Influence of levothyroxine sodium on growth and uptake of some mineral elements in cotton (*Gossypium hirsutum* L.). Pakistan Journal of Botany 44(SI):101–104

Ozyigit II, Bajrovic K, Gozukirmizi N, Semiz BD (2002) Direct plant regeneration from hypocotyl and cotyledon explants of five different sunflower genotypes (*Helianthus annuus* L.)

from Turkey. Biotechnology and Biotechnological Equipment 16(1):8–11. https://doi.org/10.1080/13102818.2002.10819148

Ozyigit II, Gozukirmizi N, Semiz BD (2006) Callus induction and plant regeneration from mature embryos of sunflower. Russ J Plant Physiol 53(4):556–559. https://doi.org/10.1134/S1021443706040194

Qi W, Lu H, Zhang Y, Cheng J, Huang B, Lu X et al (2020) Oil crop genetic modification for producing added value lipids. Crit Rev Biotechnol:1–10. https://doi.org/10.1080/07388551.2020.1785384

Rai SK, Bawa V, Dar ZA, Sofi NR, Mahdi SS, Qureshi AMI (2019) Use of modern molecular biology and biotechnology tools to improve the quality value of oilseed brassicas. In: Qureshi A, Dar Z, Wani S (eds) Quality breeding in field crops. Springer, Cham, pp 255–266. https://doi.org/10.1007/978-3-030-04609-5_13

Rauf S, Shehzad M, Al-Khayri JM, Imran HM, Noorka IR (2019) Cotton (Gossypium hirsutum L.) breeding strategies. In: Al-Khayri J, Jain S, Johnson D (eds) Advances in plant breeding strategies: industrial and food crops. Springer, Cham, pp 29–59. https://doi.org/10.1007/978-3-030-23265-8_2

Sánchez-Duarte JI, Kalscheur KF, Casper DP, García AD (2019) Performance of dairy cows fed diets formulated at 2 starch concentrations with either canola meal or soybean meal as the protein supplement. J Dairy Sci 102(9):7970–7979. https://doi.org/10.3168/jds.2018-15760

Sharma M, Gupta SK, Mondal AK (2012) Production and trade of major world oil crops. In: Gupta S (ed) Technological innovations in major world oil crops, vol 1. Springer, New York, pp 1–15. https://doi.org/10.1007/978-1-4614-0356-2_1

Stein HH, Casas GA, Abelilla JJ, Liu Y, Sulabo RC (2015) Nutritional value of high fiber co-products from the copra, palm kernel, and rice industries in diets fed to pigs. Journal of Animal Science and Biotechnology 6(1):56. https://doi.org/10.1186/s40104-015-0056-6

Tani N, Hamid ZAA, Joseph N, Sulaiman O, Hashim R, Arai T et al (2020) Small temperature variations are a key regulator of reproductive growth and assimilate storage in oil palm (*Elaeis guineensis*). Sci Rep 10(1):1–11. https://doi.org/10.1038/s41598-019-57170-8

Tausif M, Jabbar A, Naeem MS, Basit A, Ahmad F, Cassidy T (2018) Cotton in the new millennium: advances, economics, perceptions and problems. Text Prog 50(1):1–66. https://doi.org/10.1080/00405167.2018.1528095

USDA (2017) United States department of agriculture, oilseeds: world markets and trade, Published on Dec, 2017. Available from: https://downloads.usda.library.cornell.edu/usda-esmis/files/tx31qh68h/4t64gn49r/hd76s053w/oilseed-trade-12-12-2017.pdf. Last accessed on 7 Jul 2020

USDA (2020a) United States department of agriculture, oilseeds: world markets and trade, Published in June 2020, Available from: https://downloads.usda.library.cornell.edu/usda-esmis/files/tx31qh68h/1257bd51b/fj236p26g/oilseeds.pdf. Last accessed on 18 June 2020

USDA (2020b) United States department of agriculture, cotton: world markets and trade, Published on Jul, 2020. Available from: https://apps.fas.usda.gov/psdonline/circulars/cotton.pdf. Last accessed on 11 Jul 2020

Valladares-Diestra K, de Souza Vandenberghe LP, Soccol CR (2020) Oilseed enzymatic pretreatment for efficient oil recovery in biodiesel production industry: a review. Bioenergy Res. https://doi.org/10.1007/s12155-020-10132-9

Villa M, Santos SA, Sousa JP, Ferreira A, da Silva PM, Patanita I et al (2020) Landscape composition and configuration affect the abundance of the olive moth (Prays oleae, Bernard) in olive groves. Agric Ecosyst Environ 294:106854. https://doi.org/10.1016/j.agee.2020.106854

Wales N, Akman M, Watson RH, Sánchez Barreiro F, Smith BD, Gremillion KJ et al (2019) Ancient DNA reveals the timing and persistence of organellar genetic bottlenecks over 3,000 years of sunflower domestication and improvement. Evol Appl 12(1):38–53. https://doi.org/10.1111/eva.12594

Waseem S, Imadi SR, Gul A, Ahmad P (2017) Chapter 1: Oilseed crops: present scenario and future prospects. In: Ahmad P (ed) Oilseed crops: yield and adaptations under environmental stress. Wiley, Chichester. https://doi.org/10.1002/9781119048800

Index

H. Tombuloglu et al. (eds.), *Oil Crop Genomics*,
https://doi.org/10.1007/978-3-030-70420-9

H

I

J

Printed by Printforce, the Netherlands